Green Networking and Communications

ICT for Sustainability

Green Networking and Communications

ICT for Sustainability

Shafiullah Khan

Jaime Lloret Mauri

CRC Press
Taylor & Francis Group
Boca Raton   London   New York

CRC Press is an imprint of the
Taylor & Francis Group, an **informa** business

CRC Press
Taylor & Francis Group
6000 Broken Sound Parkway NW, Suite 300
Boca Raton, FL 33487-2742

© 2014 by Taylor & Francis Group, LLC
CRC Press is an imprint of Taylor & Francis Group, an Informa business

No claim to original U.S. Government works

Printed on acid-free paper
Version Date: 20130508

International Standard Book Number-13: 978-1-4665-6874-7 (Hardback)

Library of Congress Cataloging-in-Publication Data

Green networking and communications : ICT for sustainability / edited by Shafiullah Khan and Jaime Lloret Mauri.
    pages cm
Includes bibliographical references and index.
ISBN 978-1-4665-6874-7
    1. Telecommunication--Energy conservation. 2. Computer networks--Environmental aspects. 3. Sustainable engineering. 4. Green technology. I. Khan, Shafiullah.

TK5102.86.G755 2014
621.39028'6--dc23   2013016362

Visit the Taylor & Francis Web site at
http://www.taylorandfrancis.com

and the CRC Press Web site at
http://www.crcpress.com

Contents

SECTION I Energy Efficiency and Management in Wireless Networks

SECTION II Cellular Networks

SECTION III   Smart Grid

Contributors

Taufik Abrão
Department of Electrical Engineering
State University of Londrina
Londrina, Brazil

Adnan Abu-Dayya
Qatar Mobility Innovations Center
Qatar Science and Technology Park
Doha, Qatar

Mohamed-Slim Alouini
Computer, Electrical and Mathematical
    Sciences and Engineering Division
King Abdullah University of Science and
    Technology
Thuwal, Saudi Arabia

Sebastián Andrade-Morelli
Polytechnic University of Valencia
Valencia, Spain

Artur Miguel Arsénio
IST Taguspark Campus
Porto Salvo, Portugal

Rachad Atat
Radio Communication Systems
Royal Institute of Technology
Kista, Sweden

Bernard Cousin
Institute for Research in Computer Science and
    Random Systems
University of Rennes—Beaulieu Campus
Rennes, France

Luca de Nardis
DIET Department
Sapienza University of Rome
Rome, Italy

Raúl Duque
Telefónica I+D
Madrid, Spain

Amin Ebrahimzadeh
Department of Electrical Engineering
Sahand University of Technology
Tabriz, Iran

Zabih Ghassemlooy
Faculty of Engineering and Environment
Northumbria University
Newcastle upon Tyne, United Kingdom

Hakim Ghazzai
Computer, Electrical and Mathematical
    Sciences and Engineering Division
King Abdullah University of Science and
    Technology
Thuwal, Saudi Arabia

Emilio Granell Romero
Polytechnic University of Valencia
Valencia, Spain

Cédric Gueguen
Institute for Research in Computer Science and
    Random Systems
University of Rennes—Beaulieu Campus
Rennes, France

Yoram Haddad
Department of Computer Science
Jerusalem College of Technology
Jerusalem, Israel

Oliver Holland
Institute of Telecommunications
King's College London
London, United Kingdom

Felipe Jiménez
Telefónica I+D
Madrid, Spain

Akil Jrad
Center Azm for Research in Biotechnology and
    Its Applications
The Lebanese University
Tripoli, Lebanon

Mohammad-Ali Khalighi
Fresnel Institute
University Campus Saint Jerome
Marseille, France

Adrian Kliks
Poznan University of Technology
Poznan, Poland

Paul Okuthe Kogeda
Faculty of ICT
Tshwane University of Technology
Pretoria, South Africa

M. Bala Krishna
University School of Information and
    Communication Technology
GGS Indraprastha University
New Delhi, India

Peter M. Krummrich
Technical University of Dortmund
Dortmund, Germany

It Ee Lee
Faculty of Engineering and Environment
Northumbria University
Newcastle upon Tyne, United Kingdom

Carlos León
Department of Electronic Technology
University of Seville
Seville, Spain

Jaime Lloret
Polytechnic University of Valencia
Valencia, Spain

Víctor López
Telefónica I+D
Madrid, Spain

Jorge López Vizcaíno
Huawei Technologies Duesseldorf GmbH
Munich, Germany

Dimitrios Makrakis
School of Electrical Engineering and
    Computer Science
University of Ottawa
Ottawa, Ontario, Canada

Konstantinos P. Mantzoukas
School of Electrical and Computer Engineering
National Technical University of Athens
Athens, Greece

Gil Mardon
DIWEL
Cintré, France

Antonio Martín
Department of Electronic Technology
University of Seville
Seville, Spain

Moshe Timothy Masonta
Council for Scientific and Industrial Research
Meraka Institute
Pretoria, South Africa

Natalie Matta
Charles Delaunay Institute
Troyes University of Technology
Troyes, France

Arturas Medeisis
Telecommunications Engineering Department
Vilnius Gediminas Technical University
Vilnius, Lithuania

Fisseha Mekuria
Council for Scientific and Industrial Research
Meraka Institute
Pretoria, South Africa

Leïla Merghem-Boulahia
Charles Delaunay Institute
Troyes University of Technology
Troyes, France

Hussein T. Mouftah
School of Electrical Engineering and
    Computer Science
University of Ottawa
Ottawa, Ontario, Canada

Wai Pang Ng
Faculty of Engineering and Environment
Northumbria University
Newcastle upon Tyne, United Kingdom

Thomas Otieno Olwal
Council for Scientific and Industrial Research
Meraka Institute
Pretoria, South Africa

Athanasios D. Panagopoulos
School of Electrical and Computer Engineering
National Technical University of Athens
Athens, Greece

Khalid A. Qaraqe
Department of Electrical and
Computer Engineering
Texas A&M University at Qatar
Doha, Qatar

Akbar Ghaffarpour Rahbar
Department of Electrical Engineering
Sahand University of Technology
Tabriz, Iran

Rana Rahim-Amoud
Center Azm for Research in Biotechnology and
Its Applications
The Lebanese University
Tripoli, Lebanon

Stavros E. Sagkriotis
School of Electrical and Computer Engineering
National Technical University of Athens
Athens, Greece

Sandra Sendra Compte
Polytechnic University of Valencia
Valencia, Spain

Erchin Serpedin
Electrical and Computer Engineering Department
Texas A&M University
College Station, Texas

Muhammad Zeeshan Shakir
Department of Electrical and
Computer Engineering
Texas A&M University at Qatar
Doha, Qatar

Tiago Silva
Instituto Superior Técnico
Technical University of Lisbon
Lisbon, Portugal

Álvaro Ricieri Castro e Souza
Department of Electrical Engineering
State University of Londrina
Londrina, Brazil

Hina Tabassum
Computer, Electrical and Mathematical
Sciences and Engineering Division
King Abdullah University of Science and
Technology
Thuwal, Saudi Arabia

Idelfonso Tafur Monroy
Department of Photonics Engineering
Technical University of Denmark
Lyngby, Denmark

Binod Vaidya
School of Electrical Engineering and
Computer Science
University of Ottawa
Ottawa, Ontario, Canada

Barend J. Van Wyk
Faculty of Engineering and the
Built Environment
Tshwane University of Technology
Pretoria, South Africa

Jean Paul Vuichard
France Télécom
Cesson Sévigne, France

Elias Yaacoub
Qatar Mobility Innovations Center
Qatar Science and Technology Park
Doha, Qatar

Han Yan
France Télécom
Cesson Sévigne, France

Yabin Ye
Huawei Technologies Duesseldorf GmbH
Munich, Germany

Section I

Energy Efficiency and Management in Wireless Networks

1 Peer-to-Peer Content Sharing Techniques for Energy Efficiency in Wireless Networks with Fast Channel Variations

Rachad Atat, Elias Yaacoub, Mohamed-Slim Alouini, and Adnan Abu-Dayya

CONTENTS

1.1   INTRODUCTION

According to the International Telecommunication Union, information and communication technology (ICT) was emitting 0.83 $GtCO_2e$ (gigatons of carbon dioxide equivalent), contributing to around 2%–2.5% of global greenhouse gas (GHG) emissions in 2007 [1]. With the continuous growth of ICT, especially in developing countries, the GHG emissions are expected to grow at double the rate over the next 10 years [1]. The Global e-Sustainability Initiative research is estimating a 72% increase in ICT energy usage from 2007 to 2020 with around 1.43 $GtCO_2e$ emissions in 2020 [1]. In addition, the telecommunications industry is witnessing an explosive increase in data traffic especially with the introduction of wireless modems and smart phones and with the presence of more than one billion wireless subscribers today. The data traffic volume is increasing by a factor of 10 every 5 years, leading to an increase of 16%–20% in energy consumption every 5 years [2]. For instance, in India, the mobile telecom industry is considered the fastest-growing sector with 584.3 million subscribers in 2010–2011 with an annual growth rate of 49.15%. It is estimated that the energy consumption of the Indian Mobile Telecom Industry was 163 PJ (petajoules) with 52.66 million tons emissions of carbon dioxide (CO_2) in 2010–2011 [3]. A user who travels a distance of 25 km using public transport such as car or train can result in 1.22 kg of CO_2 emissions, compared to 0.11 kg of CO_2 emissions for 1 hour of video conferencing with two laptops [4]. A talk of 2 minutes per day on the phone can produce 47 kg CO_2e (equivalent) per year, with a total of 125 million tons of CO_2e produced by mobile phones in 1 year [5].

While some electromagnetic radiations occur naturally in the universe, such as radiation from the sun, others are created by humans, such as from power lines, electrical devices, and telecommunication antennas [6]. As the wireless technology is moving toward 4G, mobile terminals (MTs) (laptops, smart phones, personal digital assistants, etc.) are required to support the increase in demand for multimedia services. Since 4G necessitates high data rates with high quality of service (QoS), power requirements of mobile devices are increasing. Furthermore, the MTs are powered by rechargeable batteries, which make energy a critical resource as battery is considered to be the top concern from a consumer point of view, the fact that refrains the end users from using advanced multimedia applications more frequently on their mobile devices [1]. With the support of multimedia services, high power is expected to be drained from the MT batteries since these MTs should generally be active for a long period to download large data sizes. On the other hand, battery capacity is limited and the progress of creating large battery capacity is very slow and increasing by only 80% within the last 10 years, compared to the processing performance, which is increasing at a double rate every 18 months [1]. Moreover, with existing battery technologies, no breakthrough is expected in the near future, unless new innovations and technologies are invented [1]. This huge gap between the limitations of battery life and the high energy consumption requirements of MTs constitute a barrier for future wireless communication systems. It is evident that nowadays, there is a need to limit the energy consumption of MTs in order to limit the CO_2 emissions in mobile communication industries and to prolong the operational times of the MTs. Extensive research is foreseen to shift toward a green paradigm in wireless networks by finding power-efficient solutions for the multistandard MTs that can act on several wireless standards (universal mobile telecommunications system [UMTS], long-term evolution [LTE], WiFi, Bluetooth, etc.). The green solutions will aim not only at reducing the pollution and waste emissions, especially the undesirable CO_2 footprint, but also in achieving higher capacity and throughput with low latency. Also, by minimizing the energy consumption, less interference in the network is expected with longer MT battery lifetime [1]. For instance, Nokia was able to reduce the CO_2 emissions in their facilities by 17% in 2011 as compared to 2006 [7].

Intermediate relays can reduce communication distance and thus the emitted power, which results in reduced electromagnetic emissions [8]. Relays for the purpose of minimizing the total power consumption of the network nodes, maximizing network lifetime, extending the coverage, and expanding the capacity in wireless systems have been investigated in many research works such as those in Refs. [9–14]. This led to the investigation of heterogeneous network architectures where the MTs would be active on two wireless interfaces: one to communicate with a wireless base station (BS) on the long range (LR) such as WiMAX, LTE, or UMTS, and one to communicate with other MTs on the short range (SR) using technologies such as Bluetooth, wireless local area network (WLAN), or ultra wideband. One of the beneficial side effects of this peer-to-peer (P2P) collaboration is a reduced electromagnetic emission, which will contribute to the field of green communications.

In this chapter, a cooperative content distribution (CCD) framework is considered, where the data should reach all MTs, including the relay. In fact, one of the MTs is selected as relay in order to achieve energy minimization in a given cooperative cluster of MTs. An MT is selected as relay when it leads to the optimal energy-minimizing solution during content distribution. We analyze the optimal performance in the case of unicasting by the BS to the relay in Section 1.3.5 and the case where the BS performs multicasting to all MTs on the LR in Section 1.3.6. When LR multicasting is adopted, the BS multicasts the content to all MTs using a threshold transmission rate. MTs having achievable LR rates larger than the threshold can receive the content correctly on the SR. Afterward, they send the content to the MTs that failed to receive on the LR, using SR multicasting. Furthermore, implementation methods taking into account practical constraints are presented.

The contribution of this chapter will create a strong impact on new wireless network architectures that aim at reducing the total energy consumption of MTs and thus reducing the CO_2 emissions by creating an environmentally friendly network. The chapter is fully aligned with several European projects under seventh framework (FP7) such as C2POWER (Cognitive Radio and Cooperative Strategies for Power Saving), GREENET, EARTH, ECONET, and TREND. The theme of this chapter is also aligned with several projects under COST actions such as IC1004 (Cooperative Radio Communications for Green Smart Environments) and IC0804 (Energy Efficiency in Large-Scale Distributed Systems) [15]. The rest of the contributions of the chapter is organized as follows. Research works on cooperative networks and mobile relays for the purpose of energy minimization in wireless networks are presented in Section 1.2. The proposed energy-efficient cooperative models are presented in Section 1.3, particularly in Sections 1.3.5 and 1.3.6. In Section 1.3.7, practical implementation aspects of the proposed methods are discussed. Next, in Section 1.3.8, simulation results are presented to show the efficacy of the proposed cooperative models. Section 1.4 applies the proposed cooperative models in high-speed train systems. Future research directions are discussed in Section 1.5. Finally, conclusions are drawn in Section 1.6.

1.2 BACKGROUND/RELATED WORK

In this section, a deep insight into background work related to cooperative networks and mobile relays for the purpose of energy savings in wireless networks is presented, along with a summary highlighting the differences between the existing work and the proposed cooperative models.

1.2.1 An Overview on Cooperative Networks for Energy Minimization

New protocols and network architectures are suggested in Refs. [16–22] that aim at increasing throughput and capacity of the network as well as the performance, with benefits including reduction of energy consumption. A CCD protocol that has benefits in terms of throughput and energy consumption in wireless ad hoc networks is proposed in Ref. [16]. The protocol consists of finding the maximum number of close neighbors to limit interference and long routes. CCD reduces the number of unnecessary retransmissions that are usually due to interference and collisions, and results show that peers consume roughly the same amount of energy.

Since 3G multicast is limited by the worst channel rate of the multicast group, it is suggested in Ref. [17] that 3G BS delivers packets to proxy mobile devices that have better channel quality, and the proxy then forwards the packets to receivers through IEEE 802.11.

P2P video distribution schemes have been proposed in Refs. [18–20] for the purpose of energy minimization. Few peers pull video descriptions from BS in Ref. [18], and then they share the streams with nearby neighbors via a free broadcast channel such as WiFi or Bluetooth. The streaming cost and bandwidth cost would decrease while scalability increases since a single broadcast may cover a large number of users without the need for mobile forwarding. By taking turns in being video pullers, mobiles can share streaming cost more fairly.

In Ref. [19], a group of mobile users interested in the same video such as downloading a football game pull one of the video parts from the server that holds the content and then share it with other MTs via their SR wireless interface such as Bluetooth. This scenario proved to be more energy efficient as compared to the one where each of the MTs has to pull the whole video from the server. By exchanging the video parts or descriptions over SR links, the video parts would require less reception power, making these links more energy efficient. The video of interest would be encoded into multiple descriptions and the MT would pull part of these descriptions and share them with others.

In Ref. [20], the MT would be using two wireless interfaces: WLAN to communicate on the LR with the server and Bluetooth to communicate with other MTs on SR links. Measurement showed that the reception over WLAN is 33 times that required by transmission or reception over Bluetooth. Results showed that with three peers, power consumption over the WLAN interface is three times less than the case of one peer. The power reduction gain reached 16% with three cooperating devices.

In Ref. [21], the concept of P2P data exchange on uplink channels in a UMTS network is presented. Users interested in downloading multimedia files will use the cooperative P2P, so that multimedia traffic is shifted away from the downlink, making its capacity available for other services. Results showed overall downlink throughput gain up to 85% by using the P2P technique. These results demonstrated the ability to interconnect wireless devices on unoccupied uplink carrier frequencies for content distribution. A new message notification protocol is presented in Ref. [22] to enable MTs to cooperatively share a single notification channel. By doing so, telecommunication charges will be reduced and battery lifetime will be increased. Results showed that the suggested protocol is able to achieve significant energy savings for MTs.

A new network architecture referred to as cellular controlled P2P network is suggested in Ref. [23] to implement efficient error recovery in multicast scenarios for wireless networks. In this new architecture, MTs communicate with each others to perform cooperative retransmissions using their SR wireless interfaces. This new scheme is compared to non-cooperative error-recovery schemes such as automatic repeat request and layered forward error correction, and results showed that the suggested scheme is more energy efficient.

Recently, network coding has been introduced in order to correct packet losses during SR communications [24,25]. In Ref. [24], SR broadcasting is used to save LR bandwidth, instead of having the BS unicast the data to each MT on the LR. A retransmission of a lost packet would usually benefit one MT. With network coding, linear combinations of video packets are used. Hence, a single retransmission can correct the losses of several MTs. In Ref. [24], 50% overhead is used for network coding. In this chapter, energy is considered, and the proposed methods are assumed aware of the channel state information (CSI), conversely to Ref. [24]. The obtained solutions represent a benchmark to which the performance of practical implementations can be compared. Therefore, comparing the optimal solution of this chapter to a scenario where network coding is used to correct losses in a practical scenario with imperfect CSI, in addition to optimizing overhead retransmissions, is an interesting topic for future research. In Ref. [25], cooperation between two groups of MTs is considered, with each group assumed to be interested in a different content. Inside each cooperative group or cluster, the optimal set of mobiles that can distribute the content with minimal energy is assumed to be determined by exhaustive search. In this chapter, if CSI information is available, the optimal solution at a given fading realization is to send the content to a unique mobile.

1.2.2 MOBILE RELAYS FOR ENERGY-EFFICIENT NETWORKS

With the limitation of battery life of mobile devices, many studies in the literature focus on minimizing the energy consumption in order to extend their lifetime and operational time. In wireless sensor networks (WSNs), sensor nodes are powered by batteries and have limited processing and memory capabilities. Sensor nodes can be deployed in battlefields, dense jungles, and so on, for environment monitoring, which means they must collect large amount of data to be transmitted to a BS. A moderate-size WSN can gather up to 1 Gb/year from a biological habitat [26]. Many studies aim at overcoming the main challenge that faces WSN, which is reducing the total energy consumption through the deployment of mobile relays. Mobile relays can assist static sensor nodes in forwarding data to destination, which will improve the lifetime of the network of up to four times over static network [26,27]. In Ref. [9], the energy balanced adaptive clustering hierarchy scheme is proposed to be used in WSN, where sensor nodes send messages to an elected cluster head (CH), which compresses them and sends them to the BS via a relay node. The relay node is chosen such that the distance from the relay to the BS is less than that from the CH to the BS. The second criterion in choosing the relay node is by selecting the one that has the highest residual energy. A hybrid network architecture for efficient utilization of battery-powered handheld devices is suggested in Ref. [28], where an ad hoc link exists between mobile station (MS) and relay station (RS) and is referred to as high data rate link, while the direct cellular link between MS (or RS) and BS is referred to as low data rate link. Power consumption is investigated on four different uplink and downlink traffic paths, assuming that MS and RS are battery powered, while the BS has no power constraints.

Many studies focused on maximizing the device lifetime in relay networks (e.g., see Refs. [10–14]). An algorithm to select the best relays with minimum power consumption that would form cooperative links to establish a route from source to destination is proposed in Ref. [10]. For the cooperative link, total power consumption, which is taken to be the summation of the transmit power of the source node and the transmit power of the relay node, is minimized subject to a target bit error rate (BER). Then, the relay node that minimizes the total power consumption is selected and one hop cooperative route from source to destination is established. An optimal power allocation in case of multiple relays is suggested in Ref. [11]. The formulated optimization problem is based on maximizing the network lifetime at each transmission stage where channels are slowly varying over time by applying the optimal power allocation strategy. This scheme is termed perceived lifetime algorithm. The total power per transmission is minimized by giving more weight to the relay with smaller residual energy. In Ref. [12], power-aware reputation system (PARS) is proposed to make nodes announce the true residual energy. PARS detects nodes that misreport a low energy and then isolates them and notifies all neighboring nodes via a broadcast message about the misreporting node so that it can be isolated. This scheme proved to have better system performance. In Ref. [13], the average energy consumed in the network to transmit message from source to destination via intermediate cooperative relays is minimized subject to a target outage probability. An algorithm equivalent to shortest path computation is proposed with link cost being the energy consumed for transmitting a message from one node to another. In Ref. [14], suboptimal algorithms are proposed to increase device lifetime by exploiting the cooperative diversity and taking both location and energy advantages under BER constraint. Results showed that the device lifetime in a cooperative network increases twice longer than that in a non-cooperative network.

Access services such as Bluetooth, WLAN, and ad hoc networks are stimulating the growth of wireless traffic. As traffic is becoming more bursty and unevenly distributed leading to congestion in the network, a new network architecture iCAR is suggested in Ref. [29]. The proposed network scales with the number of mobile hosts and dynamically balances the load by placing ad hoc relaying systems (ARSs) at strategic positions to divert traffic from a congested cell to a non-congested cell. iCAR improves the mobile hosts' battery life and transmission rate. Each ARS has two interfaces: one to communicate with a base transceiver station, and one to communicate with a mobile host or another ARS.

1.2.3 Summary of Chapter Contributions

Table 1.1 summarizes the differences between the existing literature and the proposed cooperative models in this chapter.

TABLE 1.1

Differences between the Existing Literature and the Proposed Cooperative Models

	Existing Literature	The Proposed Models
Lifetime maximization considering residual energy	[9–14,26,27]	The MTs have comparable stored energy and energy minimization is done on the basis of the power drained from their batteries
New protocols and architecture leading to reduction in energy for MTs	[16,17]	Objective is the minimization of energy of MTs
Non-optimal energy minimization in CCD	[18–22]	Optimal energy minimization in CCD

1.3   ENERGY-EFFICIENT SR COOPERATION OVER WIRELESS NETWORKS

In this section, the energy consumption of the two CCD system models is derived and presented along with the optimal solutions. Hence, the contributions of this chapter can be summarized as follows:

- Formulating the energy minimization problem and presenting the optimal solution in closed form.
- Comparing the proposed methods to the recent literature and showing that they lead to significant energy savings.
- Proposing a suboptimal method for the LR unicasting scenario that approximates the optimal solution with significantly reduced overhead.
- Discussing other practical implementation aspects of the proposed methods. Although they have some practical limitations, they constitute a benchmark to which practical suboptimal solutions can be compared.
- Applying the proposed cooperative models in high-speed train systems, which lead to huge savings in the energy consumption from the batteries of the MTs inside the train.

The system model is depicted in Figure 1.1. The model consists of a certain number of cooperating MTs in the range of a BS, interested in the same content. The BS is connected via a wired local area network to the server that holds the content. Without cooperation, the BS sends the content to the MTs via either unicasting or multicasting on the LR. In the proposed approach, neighboring MTs are grouped into cooperating clusters. In each cluster, an MT is selected as relay to distribute the content to the cluster members via cooperative SR multicasting. We study the optimal relay selection in a cluster of size K MTs. The relay is selected in order to optimize a certain performance metric. In this chapter, the metric considered is energy minimization during the content distribution. When the BS uses unicasting, it sends the data to the relay, which in turn multicasts it to the other MTs. This leads to saving wireless resources since the BS is sending the content once instead of sending it K times. This approach will be referred to as Cooperative System Model 1. When the BS uses multicasting, it sends the data at a certain rate. Among the MTs that are able to receive the data correctly, a relay is selected, and this relay multicasts the data to the other MTs. This approach will be referred to as Cooperative System Model 2.

1.3.1   CHANNEL MODEL

The channels on the LR and SR links are assumed to be orthogonal and are modeled by pathloss, shadowing, and fading. Thus, the received power P_r can be linked to the transmitted power P_t by a pathloss model as in [30]:

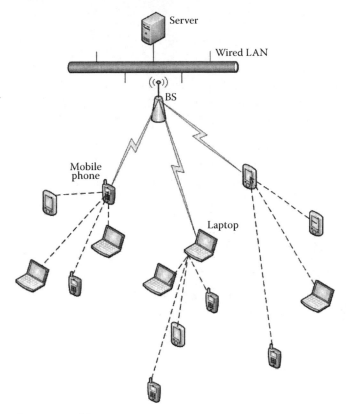

FIGURE 1.1 General system model.

$$\frac{P_r}{P_t}(dB) = \underbrace{10 \log_{10}\kappa - 10\nu \log_{10} d}_{\text{distance-based pathloss}} + \underbrace{h_{dB} + f_{dB}(a)}_{\text{random variables}}, \tag{1.1}$$

where κ is a unitless constant that depends on the antenna characteristics and the average channel attenuation, ν, is the path loss exponent, d is the distance where the received power is calculated, h is a Gaussian random variable representing shadowing or slow fading having a zero mean and a variance $\sigma^2_{h_{dB}}$, and f is a random variable representing Rayleigh fading with a Rayleigh parameter a.

We consider a block fading model where the fast fading remains constant for a fixed time T_{dec}, which is the channel decorrelation time. Then, the channel conditions change and remain constant for another T_{dec}, and so on.

1.3.2 DATA RATES

Denoting by P_t the transmit power the sender is transmitting with and by σ^2 the thermal noise power, then the received signal-to-noise ratio (SNR) γ can be calculated as $\gamma = P_r/\sigma^2$. Given the target BER P_e and the SNR, the bit rates on the LR and SR links can be calculated according to the following equation:

$$R = W\log_2(1 + \beta\gamma), \tag{1.2}$$

In Equation 1.2, W is the passband bandwidth of the channel and β is called the SNR gap. It indicates the difference between the SNR needed to achieve a certain data transmission rate for a practical M-QAM system and the theoretical limit (Shannon capacity) [30,31]. It is given by $\beta = -1.5/\ln(5P_e)$.

1.3.3 PARAMETERS AND VARIABLES

The parameters that affect the energy consumption in the studied scenarios are defined as follows:

- K: the number of requesting MTs.
- S_T: the size in bits of the content of common interest to be distributed to all MTs.
- $R_{L,k}$: transmission rate on the LR when unicasting is used from the BS to MT k.
- $R_{L,Th}$: threshold rate used for transmission on the LR when the BS communicates with the MTs via multicasting.
- $R_{S,kj}$: transmission rate on the SR links from MT k to MT j.
- $P_{L,Rx,k}$: power consumed by MT k during reception on the LR link.
- $P_{S,Rx,k}$: power consumed by MT k during reception on the SR links.
- $P_{S,Tx,k}$: power consumed by MT k while transmitting on the SR links.
- $S_R(n)$: remaining data, in bits, at the nth channel realization.
- n_T: number of channel variations until the whole content of size S_T is distributed. In other words, we have $S_R(n_T) > 0$ and $S_R(n_T + 1) = 0$.

It should be noted that $P_{L,Rx}$, $P_{S,Rx}$, and $P_{S,Tx,kj}$ correspond to the power consumed by the MT, that is, drained from its battery, during reception, and during transmission, respectively. They are not to be confused with P_r and P_t, the respective receive and transmit powers over the air, measured at the antenna. It should be noted that $P_{S,Tx,kj}$ can be expressed as

$$P_{S,Tx,kj} = P_{S,Tx,0} + P_{t,kj}, \tag{1.3}$$

where $P_{S,Tx,0}$ corresponds to the power consumed by the circuitry of the MTs during transmission on the SR links and $P_{t,kj}$ corresponds to the power transmitted over the air on the SR links from MT k to MT j.

1.3.4 NON-COOPERATIVE SCENARIO

In the case of no cooperation between MTs with LR unicasting, the BS unicasts the content on the LR links to each MT separately. In fast fading scenarios, fading might change before the content distribution process is complete, that is, where T_{dec} is not large enough to distribute the whole content, and hence the time needed to distribute the content is larger than T_{dec}. The number of data bits available at each fading realization n is given by

$$S_R(n) = S_T - \sum_{y=1}^{n-1} R_{L,k}(y)T_{dec}. \tag{1.4}$$

Hence, the data bits sent to MT k during T_{dec} will either be $R_{L,k}(n)T_{dec}$ if there are sufficient bits or $S_R(n)$ if the remaining data bits are less than the amount that can be transmitted during the nth channel realization; that is, $S_R(n) < R_{L,k}(n)T_{dec}$. The energy consumption of MT k at the nth channel realization when no cooperation takes place is

$$E_{\text{No-coop},k,n} = \min(R_{L,k}(n)T_{dec}, S_{R,k}(n)) \frac{P_{Rx}}{R_{L,k}(n)}, \tag{1.5}$$

where $S_{R,k}$ is used instead of S_R to denote that a different number of bits can be transmitted on the LR for each MT k since when no cooperation is present, LR transmission is not dedicated to a single MT $k^*(n)$ and thus the number of remaining bits would vary between MTs.

In case of no cooperation between MTs with LR multicasting, the BS multicasts the data on the LR using the rate achievable by the user having the worst channel conditions. Hence, the energy consumption of MT k at the nth channel realization when no cooperation takes place is given by

$$E_{\text{No-coop},k,n} = \min(R_{\text{L},k}(n)T_{\text{dec}}, S_{R,k}(n)) \frac{P_{\text{L,Rx},k}}{\min_{j} R_{\text{L},j}}. \tag{1.6}$$

The total energy consumption in the case of no cooperation at the nth channel realization is expressed as

$$E_{\text{No-coop},n} = \sum_{k=1}^{K} E_{\text{No-coop},k,n}. \tag{1.7}$$

1.3.5 COOPERATIVE SYSTEM MODEL 1

In this cooperative model, the BS sends the data to a single relay for each cooperating cluster. The BS selects the relay in order to achieve reduced energy consumption. For each T_{dec}, the BS would select the optimal relay k^* that would minimize the energy. The fading fluctuations will generally lead to variations at each iteration of the MTs selected as relays for transmission on the SR and hence will lead to fairness in energy consumption among different MTs. The data bits sent to k^* during T_{dec} will either be $R_{\text{L},k^*(n)}(n)T_{\text{dec}}$ if there are sufficient bits or $S_R(n)$ if the remaining data bits are less than the amount that can be transmitted during the nth channel realization; that is, $S_R(n) < R_{\text{L},k^*(n)}(n)T_{\text{dec}}$. To minimize the total energy during content distribution, the energy is minimized at each fading realization. The energy consumed when selecting k as relay in the nth channel realization is given by

$$E_{\text{coop},k}(n) = \min(R_{\text{L},k(n)}(n)T_{\text{dec}}, S_R(n))$$

$$\cdot \left(\frac{P_{\text{L,Rx},k}}{R_{\text{L},k}(n)} + \frac{P_{\text{S,Tx},k}}{\min_{i \neq k} R_{\text{S},ki}(n)} + \sum_{j=1, j \neq k}^{K} \frac{P_{\text{S,Rx},j}}{\min_{i \neq k} R_{\text{S},ki}(n)} \right). \tag{1.8}$$

Hence, in order to minimize energy consumption inside the cooperative cluster, the BS should select MT k^* as relay, with k^* given by

$$k^*(n) = \arg \min_{k} \left(\frac{P_{\text{L,Rx},k}}{R_{\text{L},k}(n)} + \frac{P_{\text{S,Tx},k} + \sum_{j=1, j \neq k}^{K} P_{\text{S,Rx},j}}{\min_{i \neq k} R_{\text{S},ki}(n)} \right). \tag{1.9}$$

When the MTs have similar SR wireless interfaces, then $P_{\text{S,Rx},j} = P_{\text{S,Rx}} \; \forall j$, and Equation 1.9 becomes

$$k^*(n) = \arg \min_{k} \left(\frac{P_{\text{L,Rx},k}}{R_{\text{L},k}(n)} + \frac{P_{\text{S,Tx},k} + (K-1)P_{\text{S,Rx}}}{\min_{i \neq k} R_{\text{S},ki}(n)} \right). \tag{1.10}$$

1.3.6 COOPERATIVE SYSTEM MODEL 2

In the case of Cooperative System Model 1, the BS needs information about the MTs that form each cluster in order to select the best relay for that cluster. Thus, the implementation of Cooperative

System Model 1 requires some knowledge to be available at the BS about the SR distribution and the achievable SR rates of the various MTs. Multicasting allows the BS to deal with this problem by transmitting at the LR to all MTs without the need to feedback any SR information to the BS. However, non-cooperative LR multicasting is limited by the performance of the worst-case user, as discussed previously. To allow SR collaboration in the presence of multicasting, and to achieve gains compared to the non-cooperative scenario, Cooperative System Model 2 is based on threshold-based multicasting: the BS multicasts on the LR on the basis of a threshold rate $R_{L,Th}$ achievable by a threshold SNR γ_{Th}. Thus, an MT receives successfully from the BS if it has an SNR $\gamma_k \geq \gamma_{Th}$. These MTs will send the data via SR collaboration to their neighbor MTs having $\gamma_k < \gamma_{Th}$, which did not receive successfully on the LR. It should be noted that with Cooperative System Model 2, all MTs will consume the energy of reception on the LR, $E_{L,Th}$, since they will all keep their LR wireless interface active trying to receive the content from the BS. In the case of SR multicasting, energy minimization is reached when the MT that received successfully on LR and that can multicast with the least energy consumption on the SR is selected to perform SR multicasting. Denoting this MT at the nth channel realization by $\beta(n)$, it can be determined as follows:

$$\beta(n) = \arg \min_{k \in U^*(n)} \left[\frac{(P_{S,Tx,kj}(n) + P_{S,Rx})}{\min_{j \notin U^*(n)} R_{S,kj}(n)} \right], \tag{1.11}$$

where $U^*(n)$ is the set of MTs that received successfully on the LR at the nth channel realization, and may be given as

$$U^*(n) = \{k \mid \gamma_k(n) \geq \gamma_{Th}(n)\}. \tag{1.12}$$

In the case of adaptive rate control, the MT transmit power is constant; that is, $P_{t,kj} = P_t$ and $P_{S,Tx,kj} = P_{S,Tx}$. Consequently, the rate $R_{S,kj}$ on the link between MTs k and j is the rate achievable with the transmit power P_t. It is varied adaptively depending on the channel conditions between MTs k and j. High data rates result in low energy per bit consumption, thus leading to a gain in total energy consumption. For example, the WLAN technologies use rate control [32].

Considering the rate adaptive example, Equation 1.11 becomes

$$\begin{aligned} \beta(n) &= \arg \min_{k \in U^*(n)} \left[\frac{(P_{S,Tx} + P_{S,Rx})}{\min_{j \notin U^*(n)} R_{S,kj}(n)} \right] \\ &= \arg \min_{k \in U^*(n)} \left[\frac{1}{\min_{j \notin U^*(n)} R_{S,kj}(n)} \right]. \\ &= \arg \max_{k \in U^*(n)} \left\{ \min_{j \notin U^*(n)} R_{S,kj}(n) \right\} \end{aligned} \tag{1.13}$$

The result of Equation 1.13 indicates that the MT that successfully received on the LR and that has the maximum multicasting rate on the SR will be selected to send the data on the SR to the other MTs. In the example of Figure 1.2, MT 1 can transmit at the rate of $\min\{R_{S,13} = 1 \text{ Mbps}, R_{S,14} = 2 \text{ Mbps}\} = 1$ Mbps and MT 2 can transmit at the rate of $\min\{R_{S,23} = 2 \text{ Mbps}, R_{S,24} = 3 \text{ Mbps}\} = 2$ Mbps. Thus, MT 2 would multicast to others on the SR since it can transmit at a higher rate than MT 1.

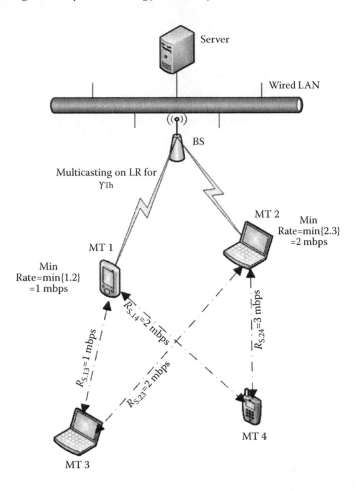

FIGURE 1.2 Example of Cooperative System Model 2.

Defining the indicator variable $\Gamma_k(n)$ as

$$\Gamma_k(n) = \begin{cases} 1 & k = \beta(n) \\ 0 & \text{otherwise} \end{cases}. \tag{1.14}$$

The indicator variable $\delta_k(n)$ is defined as follows:

$$\delta_k(n) = \begin{cases} 1 & \gamma_k(n) \geq \gamma_{Th}(n) \\ 0 & \gamma_k(n) < \gamma_{Th}(n) \end{cases}, \tag{1.15}$$

where $\delta_k(n) = 1$ indicates that MT k received successfully on the LR at the nth channel realization.

The energy consumed by each MT k when the optimal solution of Equation 1.13 is implemented at the nth channel realization is given by

$$E_{k,n} = \frac{\min(R_{L,Th}(n)T_{dec}, S_R(n))}{R_{L,Th}(n)} P_{L,Rx}$$
$$+ \frac{\min(R_{L,Th}(n)T_{dec}, S_R(n))}{\min_{j \in U^*(n)} R_{S,kj}(n)} P_{S,Tx}\Gamma_k(n) \qquad (1.16)$$
$$+ \frac{\min(R_{L,Th}(n)T_{dec}, S_R(n))}{\min_{j \in U^*(n)} R_{S,\beta(n)j}(n)} P_{S,Rx}(1 - \delta_k(n))$$

The first term in Equation 1.16 corresponds to the energy consumed on the LR, regardless if MT k receives successfully or not, since its LR wireless interface will be active in order to try to receive the data from the BS. The second term corresponds to the energy consumed by MT k during transmission on the SR. The multiplication by $\Gamma_k(n)$ indicates that k will not transmit on the SR unless it is the MT $\beta(n)$ that leads to optimal energy consumption. The last term corresponds to the energy consumed by k during reception on the SR. The multiplication by $(1 - \delta_k(n))$ indicates that there is no need for MT k to receive on the SR if it successfully received the data on the LR.

1.3.7 PRACTICAL IMPLEMENTATION ASPECTS

In this section, some limitations of the proposed energy-minimizing solutions are discussed and possible practical solutions to overcome these limitations are outlined.

1.3.7.1 LR Approximation of the Optimal Cooperative System Model 1

The optimal solution assumes that the BS is aware of the CSI, and hence of the achievable rates $R_{S,kj}$ on the SR links in addition to the CSI and rates $R_{L,k}$ on the LR link. In practice, the CSI on the LR links is known by using feedback from the MTs. This is common in state-of-the-art mobile communications technologies. However, assuming the BS is aware of all the CSI on the SR links between all MTs poses some practical limitations, owing to the considerable feedback needed from the MTs.

In this section, this problem is overcome by selecting the MT that has the best CSI on the LR to distribute the data on the SR. Then, the energy consumption is compared to the optimal case. In other words, the BS sends the data to the MT satisfying

$$k^*_{(LRapp)} = \arg\min_k \left(\frac{1}{R_{L,k}} \right) = \arg\max_k R_{L,k}. \qquad (1.17)$$

Then, MT $k^*_{(LRapp)}$, which is referred to as CH, distributes the data on the SR independently from the BS using an appropriate SR technology. As can be seen from Equation 1.17, the BS needs only the CSI on the LR links and thus the overhead associated with the feedback of SR CSI to the BS is eliminated. When the CH is selected, it broadcasts pilot signals to the MTs. These pilot signals are used by the MTs to determine their CSI on the link with the CH. This CSI is fed back to the CH, similarly to the CSI feedback on the LR to the BS. Using this information, the CH determines the achievable rates of all MTs and transmits to the MTs with the achievable rates on appropriate channels via SR multicasting. Thus, instead of having each MT exchange CSI information on the SR with $(K - 1)$ other MTs, it only exchanges information with one MT, the CH.

1.3.7.2 Practical Implementation of the Cooperative System Model 2

To implement the optimal solution in the case of Cooperative System Model 2, all MTs need to exchange CSI information to determine the MT that will be multicasting the content on the SR links at each fading realization. From Equation 1.13, it can be concluded that MTs $\notin U^*$ need to receive control signals from MTs $\in U^*$ and then measure their corresponding CSI and send it as feedback to the MTs $\in U^*$. Each of these latter MTs can determine, on the basis of CSI measurements received, the minimum rate that it can transmit with on the SR. Then, these MTs $\in U^*$ need to communicate between each other to determine the MT with the highest minimum rate as specified by $\beta(n)$. Thus, a high overhead is associated with this scenario since all MTs $\in U^*$ need to communicate between each other to determine the MT that will multicast. A possible solution is that instead of having each MT $\in U^*$ communicate with all other MTs $\in U^*$, each of these MTs could report the SR multicast rate it can achieve to the BS, which can then select $\beta(n)$ to multicast on SR links. In that way, the overhead could be reduced as compared to the case where all MTs $\in U^*$ should communicate together.

1.3.8 Results and Analysis

This section presents the simulation results for the methods discussed in the previous sections. The simulation parameters are presented in Table 1.2. Channel parameters are obtained from Ref. [33], whereas energy consumption parameters are taken as in Ref. [34], where measurements are made for 3G communications on the LR, and 802.11 b on the SR.

The rate adaptive approach is considered. MTs are assumed to be uniformly distributed in a rectangular area of size 20 meters × 20 meters whose origin is at a distance $d_{LR} = 400$ meters from the BS. Unless otherwise specified, we set the SR bandwidth to $W_{SR} = 1$ MHz and the LR bandwidth to $W_{LR} = 5$ MHz.

The energy results of the proposed methods are presented and compared to the non-cooperative scenarios. In addition, they are compared to the widely used approach that divides the content into equal parts to be sent on the LR, one to each MT within a cooperative cluster (e.g., as in Refs. [18–20,35,36]). This method will be referred to in the sequel as the "equal parts" method.

The normalized energy results of the Cooperative System Model 1 are shown in Figure 1.3, whereas those of the Cooperative System Model 2 case are shown in Figure 1.4. These results are plotted on two different figures since for the Cooperative System Model 1, the normalization is with respect to the non-cooperative LR unicasting whereas the normalization for the Cooperative System Model 2 is done with respect to non-cooperative LR multicasting.

Figures 1.3 and 1.4 show that the optimal solutions achieve significant savings compared to the non-cooperative scenarios, since the value of η is significantly smaller than 1, where $\eta = E_{coop}/E_{No\text{-}coop}$. As the number of MTs increases, the normalized energy decreases. The optimal

---

TABLE 1.2
Simulation Parameters

Parameter	Value
Pathloss constant (κ)	−128.1 dB
Pathloss exponent (ν)	3.76
Shadowing standard deviation $\left(\sigma_{h_{dB}}\right)$	8 dB
Channel decorrelation time (T_{dec})	10 ms
Power consumption during transmission on the SR ($P_{S,Tx}$)	1.425 J/s
Power consumption during reception on the SR ($P_{S,Rx}$)	0.925 J/s
Power consumption during reception on the LR ($P_{L,Rx}$)	1.8 J/s
BS transmit power on the LR	5 W
Total transmit power of each MT on SR	0.125 W
BER (P_e)	10^{-6}
Content size (S_T)	1 Mbit
Receiver noise figure	7 dB

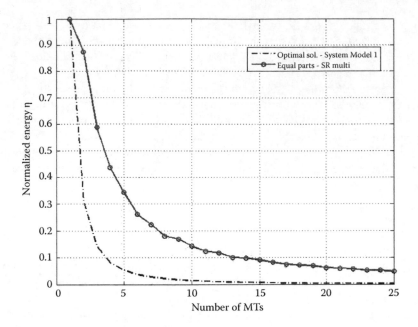

FIGURE 1.3 Normalized energy consumption versus the number of MTs for Cooperative System Model 1.

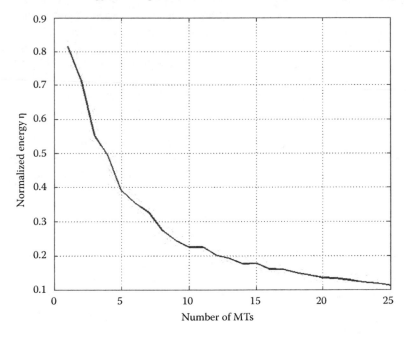

FIGURE 1.4 Normalized energy consumption versus the number of MTs for Cooperative System Model 2.

plots of Cooperative System Model 1 converge to a stable value when the number of cooperating MTs exceeds 10, whereas convergence is slower in the case of Cooperative System Model 2. Furthermore, the optimal solution outperforms the approach of Refs. [19,20,35,36] as shown in Figure 1.3.

Figure 1.5 shows a comparison of the energy consumption between the optimal solution and the LR approximation. A slight increase in the energy consumption for the LR approximation method

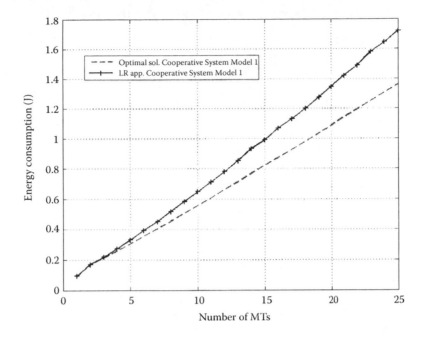

FIGURE 1.5 Energy consumption versus the number of MTs: comparison of the optimal solution and the LR approximation.

can be seen. However, this increase is negligible compared to the non-cooperative scenario. In fact, plotting the non-cooperative case in Figure 1.5, or plotting the normalized energy consumption, makes the plots of the LR approximation and of the optimal solution appear to overlap.

It should be noted that, although the performance is close to optimal, $k^*_{(LRapp)}$ is different than k^* most of the time. In fact, Figure 1.6 shows that the probability of having $k^*_{(LRapp)} = k^*$ decreases as the number of MTs increases.

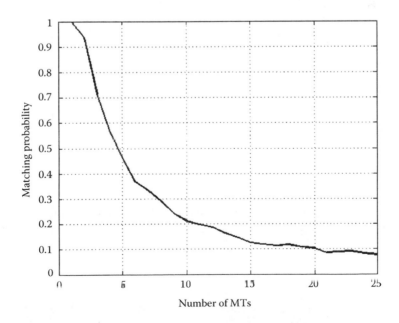

FIGURE 1.6 Probability of selecting the same MT by both the optimal solution and LR approximation.

1.4 PRACTICAL APPLICATION IN HIGH-SPEED TRAIN SYSTEMS

In this section, performance enhancements of cellular networks for passengers in high-speed railway systems in terms of savings in the energy consumption from the batteries of the MTs are investigated. In high-speed train systems, passengers may want to have Internet access to be able to browse a website, read/send emails, download multimedia services, and so on [37], initiating the investigation of new network architectures that can provide the passengers with high-speed mobile data services. While the global system for mobile communications-railway (GSM-R) is used to exchange train control information (location, schedule, speed, etc.) with a maximum data rate of 200 kbps, LTE network constitutes the next generation wireless communication system for high-speed railways. LTE meets increasing bandwidth demand with high spectrum efficiency and latency. LTE can achieve a higher data rate of up to 300 Mbps with 20 MHz of channel bandwidth [38]. There exist various technologies for train-to-land connections, namely, the satellite, which is suitable for tracks without obstructions; WiFi, which is suitable for train journeys with multiple stops; 2G/EDGE, which is suitable for low-bandwidth applications; and 3G, which is suitable in urban areas. However, LTE is best suitable for high-speed broadband data [39]. An extension to LTE is LTE-R (LTE-railway), which is based on the standard of LTE and SAE (system architecture evolution) and will constitute the next-generation wireless communication system for high-speed railways since it was shown to provide good performance with advanced channel estimation and disperse deployed antennas on train [40].

However, many problems and challenges arise despite the wireless broadband technology in use. When the MT communicates directly with the BS, it will experience a severe degradation in the QoS since the wireless signal has to travel through the train and penetrate through the metalized windows, which dramatically reduces and weakens the wireless link quality [38]. In addition, as the train moves at a high speed, the high bandwidth that is required to support multimedia applications (videos, pictures, audio, etc.) decreases, and users may not be able to establish a direct link with the outside world unless the passenger carries equipment that has direct access to a satellite node [37,41].

In this section, the cooperative system models are applied in train systems, where an antenna relay is installed on top of each car of the train (on the ceiling), and the closest wireless BS in the vicinity of the train sends the content of interest to the relays by means of unicasting or multicasting using LTE technology. Then, each relay multicasts the content to the MTs that are located inside the train car using WLAN 802.11a, which is perceived as the hottest candidate for high-speed trains to provide in-train coverage as it can achieve a peak data rate of 54 Mbps [38]. This network approach aims at minimizing the total energy consumption of the MTs to get the whole content when compared to the case when the BS transmits the content directly to the MTs in the train. Thus, the BS does not need to communicate with the hundreds of passengers in the train, which reduces radio resource management control significantly. Moreover, it will help in avoiding the radio signal propagation losses and low QoS and maintaining a stable high-speed wireless link between the relay and the MTs inside the train car.

The system model is depicted in Figure 1.7. Cellular coverage inside the train is ensured by LTE BSs deployed parallel to the train path. These BSs could be colocated with GSM-R BSs. Typical deployments consist of having a separation distance d_{BS} between BSs on the order of 7–15 km along the railroad. However, distinction should be made between GSM-R used for railway control, which is out of the scope of this chapter, and LTE, which is used to ensure cellular connectivity to passengers inside the train. The BS should communicate a content of common interest (files, live video, pictures, etc.) to the relays that sit on top of each car, which transmit the data to the MTs using IEEE 802.11a on the SR.

The relay performs heterogeneous communications using two antennas (or two sets of antennas in case MIMO communication is used, which could be an interesting future extension of this work): one antenna located outside the train car, used for communication with the BS on the LR cellular links, and another antenna inside the train car, used to communicate with the MTs inside the train

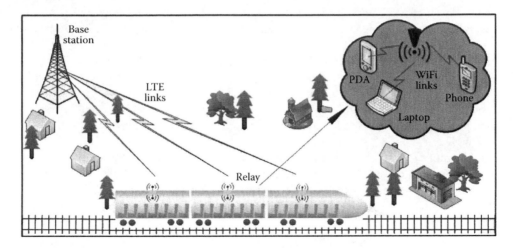

FIGURE 1.7   Train system model.

using WLAN. In a given car, relay k serves a number M_k of MTs belonging to the train passengers. LTE communication is considered between the BS and the relays. The SR links between the relays and the MTs inside the car use the WLAN 802.11a protocol, which uses the orthogonal frequency division multiplex technique including 8 different data rates and 64 carriers. It supports a bandwidth of 16.6 MHz, with a carrier spacing of 312.5 kHz and a data rate ranging between 6 and 54 Mbps [42,43]. In this section, the downlink direction is studied. Unicasting and multicasting on the LR LTE links are investigated. On the SR, it is considered that multicasting is performed by the relays inside each train car. This corresponds to a scenario where the train passengers are interested in a similar content. For example, this could be the case of entertainment services provided by the railroad operator in cooperation with the mobile operator: live news websites, broadcast of a live sports match, and so on.

In the channel model, spatial shadowing correlation is taken into account since shadow fading values depend on the fixed location of obstacles [44]. Spatial correlation can be described as a measure of how fast the local mean power evolves as the vehicle moves along a certain route [45]. The correlated shadowing model of Refs. [45,46] is applied, where the shadowing correlation is expressed as

$$\Lambda_\xi(\Delta d_T, \Delta d_R) = \exp\left(-\frac{\Delta d_R + \Delta d_T}{d_{cor}}\ln 2\right),\tag{1.18}$$

where Δd_R and Δd_T represent the movements of receiver and transmitter, respectively, and d_{cor} is the decorrelation distance taken to be 20 meters for urban environment and 5 meters for the indoor environment. As the train moves, the model of Equation 1.18 can be applied to determine the correlation between shadowing values at the different train positions, in addition to the correlation between shadowing values of the same moving train on the link with the BS (by setting $\Delta d_T = 0$ since the BS is fixed).

Given the transmit power $P_{s,d}^{(x)}$ that source s is using in order to transmit to destination d over subcarrier x, the channel gain $H_{s,d}^{(x)}$ of the channel between s and d over subcarrier x, the antenna gain of the source G_s, the antenna gain of the destination G_d, and the thermal noise power σ^2, the received SNR $\gamma_{s,d}^{(x)}$ on the link between s and d over subcarrier x can be calculated as

$$\gamma_{s,d}^{(x)} = \frac{P_{s,d}^{(x)}G_s G_d H_{s,d}^{(x)}}{\sigma^2}.\tag{1.19}$$

In the presence of relays, the source is either the BS transmitting to the relays (destination), or it could be the relay transmitting to the MTs inside the train car. In the traditional case without relays, the source is the BS and the destinations are the MTs.

In this section, discrete sets of modulation and coding schemes (MCSs), as is the case in practical standards, are considered. Thus, the discrete rates used in LTE and 802.11a are adopted. The 14 MCSs used in LTE can be found in Ref. [47] and the 8 MCSs used in 802.11a in addition to their corresponding data rates can be found in Refs. [42,43]. Assuming L possible discrete rates such that $r_1 < r_2 < \cdots < r_L$, then rate r_l is used between s and d over subcarrier x if the SNR is above a certain threshold η_l; that is, $\eta_l \leq \gamma_{s,d}^{(x)} < \eta_{l+1}$.

Slight variations on the energy consumption formulations of both cooperative system models that assume that relays are powered by the electricity of the train are defined as follows. For Cooperative System Model 1, Equation 1.8 becomes

$$E_{\text{coop},k,n} = P_{\text{S,Rx}} \sum_{j=1}^{M_k} \frac{\min(R_{\text{L},k}(n)T_{\text{dec}}, S_R(n))}{R_{\text{S},kj}(n)}. \tag{1.20}$$

For Cooperative System Model 2, Equation 1.16 becomes

$$E_{\text{coop},k,n} = P_{\text{S,Rx}} \sum_{j=1}^{M_k} \frac{\min(R_{\text{L,Th}}(n)T_{\text{dec}}, S_R(n))}{R_{\text{S},kj}(n)}. \tag{1.21}$$

1.4.1 LTE Resource Allocation

In this section, we present the resource allocation techniques adopted on the LTE LR communications for each of the investigated scenarios. We denote by $r_{d,y}$ the achievable rate of a destination d over RB y. The destination could be a relay in the proposed approach or an MT in the traditional approach.

LTE Resource Allocation with LR Unicasting: In the case of LR unicasting, we consider that one RB is allocated to each destination according to Algorithm 1.

Algorithm 1

LTE resource allocation with unicasting

   1: **while** All destinations have not been assigned an RB **do**
   2: Find the pair (Destination d^*, RB y^*) such that:

$$(d^*(n), y^*(n)) = \left\{ \arg\max_{d,y} r_{d,y}(n) \right\} \tag{1.22}$$

   3: Mark RB $y^*(n)$ as occupied and
   4: Mark destination $d^*(n)$ as served
   5: Set: $R_{\text{L},d^*}(n) = r_{d^*,y^*}(n)$
   6: Repeat Equation 1.22 for the remaining RBs and destinations
   7: until all destinations are served or all RBs are allocated
   8: **end while**

This approach allocates RBs to destinations in a way to maximize LR performance by selecting the best RB for each destination.

LTE Resource Allocation with Threshold-Based LR Multicasting: We define $U_y^*(n) = \{d \mid r_{d,y}(n) \geq R_{L,Th}(n)\}$, that is, the set of destinations that can receive the data successfully at the rate $R_{L,Th}(n)$ when RB y is used for multicasting by the BS. The notation $|\cdot|$ is used to denote set cardinality. Resource allocation is performed according to Algorithm 2.

Algorithm 2

LTE resource allocation with threshold-based LR multicasting

1: **for** $y = 1 \rightarrow N_{RB}$ **do**
2: $N_{served,y} \leftarrow \left| U_y^*(n) \right|$
3: **end for**
4: $y^*(n) \leftarrow \arg\max_y N_{served}, y$
5: **for all** d **do**
6: **if** $d \in U_{y^*}^*(n)$ **then**
7: Set $R_{L,Th,d}(n) = R_{L,Th}(n)$
8: **else**
9: Set $R_{L,Th,d}(n) = 0$
10: **end if**
11: **end for**

This algorithm finds, for each RB, the number of destinations $N_{served,y}$ that can successfully receive the data when RB y is used for threshold-based multicasting at a rate $R_{L,Th}(n)$. The algorithm finds the RB y^* such that when the data are multicast on that RB, the maximum number of destinations can successfully receive this data on the LR. In the proposed approach, destinations are relays. In the traditional approach, destinations are MTs.

LTE Resource Allocation with LR Multicasting: The transmission on a given RB is limited by the rate achieved by the destination having the worst channel conditions on that RB. Thus, the BS performs multicasting on the RB having the highest minimum rate, that is, according to the following:

$$y^*(n) = \left\{ \arg\max_y \left(\min_d r_{d,y}(n) \right) \right\}. \tag{1.23}$$

1.4.2 Results and Analysis

In the simulations, we assume that the train moves along a track with a speed of 250 km/hour. The train consists of 10 cars, each of 20 meters in length and 5 meters in width. A relay is placed in the middle of the ceiling of each train car, in which 20 MTs are uniformly distributed inside. Furthermore, we consider a file of size $S_T = 1$ Mbit to be transmitted to all requesting MTs and the time where fading is considered constant is taken to be $T_{dec} = 10$ ms. In addition, we consider an 802.11a bandwidth of $W_{SR} = 16.6$ MHz on the SR and an LTE bandwidth on the LR of $W_{LR} = 5$ MHz. With $W_{LR} = 5$ MHz, the bandwidth is subdivided into $N_{RBs} = 25$ RBs of 12 subcarriers each [48,49]. We consider BS transmit power to be subdivided equally among all RBs. Channel parameters are obtained from Ref. [33]: $\kappa = -128.1$ dB, $\upsilon = 3.76$, and $\sigma_\xi = 8$ dB. It is assumed that the BS is located 2 km away from the train and has a transmit power of 5 W (watts), equally divided among the LTE RBs. Moreover, we consider that the BS antenna gain is 15 dBi and that of the relay is 6 dBi. Furthermore, relays are powered by the train power and thus their energy consumption is not considered. In the case of LR unicasting without relays, one RB is dedicated to each MT. However, the number of MTs that can be simultaneously served cannot exceed $N_{RBs} = 25$. When the number of MTs increases beyond this limit, performance degrades significantly, as many MTs do not have an RB allocated to them. Therefore, in Figure 1.8, we compare the results to a "fictitious" scenario where we assume that an RB is available for each MT (200 RBs in total). This could correspond in practice to an LTE-advanced (LTE-A) deployment with carrier aggregation, where two 20 MHz bandwidth slots can be aggregated to lead to a total of 40 MHz, subdivided into 200 RBs. However, even in this extreme scenario, the case of 5 MHz bandwidth and 25 RBs performs better in the presence of relays. In Figure 1.8, the scenario of unicasting without relays in the case of 25 RBs assumes that the LR interface is put to sleep when an MT is not allocated an RB on the LR. If this is not the case, the plot for unicasting without relays in Figure 1.8 would correspond only to the case of LTE-A with 200 RBs, and the energy consumed in the case of 25 RBs would be orders of magnitude higher when the number of MTs increases, since most of the MTs would be spending energy most of the time without actually receiving any data. The case of LR multicasting leads to higher energy consumption since the transmission occurs at the lowest LR rate, whereas with unicasting, each MT receives at the rate it can achieve, which leads to lower overall energy consumption. The superiority of the proposed approach is obvious, where the cooperative system models lead to the same results, since the interest is in the energy consumed by the MTs (not relays). Hence, the energy is drained from the MTs' batteries only during SR multicasting regardless of the LR transmission method. Consequently, only one plot for the scenarios with relays is presented in Figure 1.8. From Figure 1.8, it can be noted that cooperative schemes achieve significant energy savings, around 3.45 J (joules) with 200 MTs, whereas the energy for multicasting without relays reaches 740 J.

The results of varying the distance between the BS and the train were investigated but are not shown because of space limitations. These results show that increasing the distance leads to

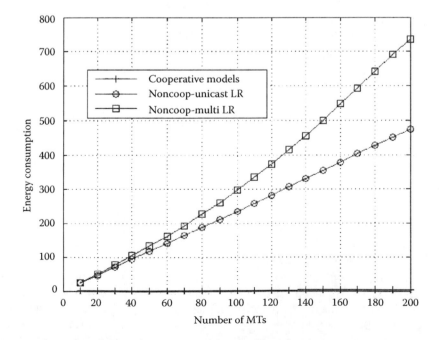

FIGURE 1.8 Energy consumption in joules versus the number of MTs.

increased energy consumption for the cases without relays, since a larger distance to the BS leads to lower achievable rates and higher energy consumption. However, the energy consumption for the scenarios with relays remains the same as in Figure 1.8, regardless of the LR distance, since MTs receive the content via SR multicasting, and the distances between the MTs and relays inside each train car remain the same regardless of the LR distance to the BS.

1.5 FUTURE RESEARCH DIRECTIONS

This section describes future research directions and possibilities that can be pursued as a continuation of the work presented in this chapter.

1.5.1 GROUPING OF MTs INTO COOPERATING CLUSTERS

The derivations presented in this chapter assume the MTs form a single cooperating cluster. In practice, this might not be the case. However, a clustering approach can be implemented, and then the proposed methods can be implemented in each cooperating cluster independently.

In the case of LR unicasting, an initial clustering phase at the start of the content sharing approach is needed. In this phase, each MT detects its neighbors, that is, the MTs that are close enough and that it can successfully communicate with at rates higher than its LR rate. It then sends a list of those to the BS, which can in this case group the MTs into cooperating clusters and then communicate with the CH k^*, selected either according to the optimal approach of Section 1.3.5 or according to the suboptimal LR approximation approach of Section 1.3.7.1. Clustering variations would typically occur at a relatively slow scale (even if fading varies at a fast scale), and hence the initial cluster formation phase does not have to be repeated frequently.

For Cooperative System Model 2, the MTs do not need to feedback any information to the BS about their neighbors and hence clustering can be performed in a distributed way. MTs that are close enough to be able to successfully communicate with each other form a cooperating cluster. The MTs within each cluster that receive the content from LR multicasting can then distribute the content on the SR according to the approach described in Section 1.3.6 or the practical approach of Section 1.3.7.2.

1.5.2 IMPACT OF MT RESIDUAL ENERGY

In this chapter, energy consumption was considered, and it was implicitly assumed that all MTs have enough energy in their batteries to be able to participate in the content distribution process or receive the content directly on the LR links without any cooperation. In practice, different MTs can have different levels of available energy in their batteries. The investigated problem can be easily extended to take into account the stored energy in each MT's battery. This can be done by considering an energy consumption cost $B(k)$ for each MT k. In this case, Equation 1.8 becomes

$$
E_{\text{B,coop}}(n) = \min(R_{\text{L},k(n)}(n)T_{\text{dec}}, S_R(n))
$$
$$
\cdot \left(\frac{B(k)P_{\text{L,Rx},k}}{R_{\text{L},k}(n)} + \frac{B(k)P_{\text{S,Tx},k}}{\min\limits_{i \neq k} R_{\text{S},ki}(n)} + \sum_{j=1, j \neq k}^{K} \frac{B(j)P_{\text{S,Rx},j}}{\min\limits_{i \neq k} R_{\text{S},ki}(n)} \right), \tag{1.24}
$$

where $E_{\text{B,coop}}$ is the weighted energy in the cooperative scenario, with the weights being the consumption costs related to the energy stored in the battery of each MT.

Following the same approach as in Section 1.3.5, the optimal solution becomes

$$k^*(n) = \arg\min_k \left(\frac{B(k)P_{L,Rx,k}}{R_{L,k}(n)} + \frac{B(k)P_{S,Tx,k} + \sum\limits_{j=1,j\neq k}^{K} B(j)P_{S,Rx,j}}{\min\limits_{i\neq k} R_{S,ki}(n)} \right). \qquad (1.25)$$

Different methods can be used to set the costs. The main guidelines are that the cost should be low when a lot of energy is available in the battery of MT k, and it should increase fast as the battery becomes more and more depleted. A possible cost function can be expressed as

$$B(k) = \exp\left[\left(\frac{B_F(k)}{B_L(k)} - 1 \right) \right], \qquad (1.26)$$

where $B_F(k)$ is the energy stored in the battery of node k when the battery is full and $B_L(k)$ is the current energy level available in the battery of node k when the content distribution process starts. Thus, when a node has its battery full, then $B_L(k) = B_F(k)$ and $B(k) = 1$. When all nodes are in this situation, then Equation 1.24 becomes equivalent to Equation 1.8. On the other hand, when $B_L(k) \rightarrow 0$, then $B(k) \rightarrow \infty$. Hence, the purpose of Equation 1.25 is to perform energy-efficient content distribution while avoiding to drain the batteries of the users in critical energy conditions.

1.5.3 MAC LAYER ISSUES

The actual implementation of the optimal solutions of Sections 1.3.5 and 1.3.6 necessitates a large overhead to schedule SR communications between MTs. Hence, when implemented in practice, the MAC energy consumption might eliminate the energy savings on the physical layer taken into account.

However, in Refs. [20,24], actual measurements were performed and SR cooperation was shown to be beneficial, although the impact of the MAC layer of Bluetooth and WLAN is part of the measured results. Hence, conversely to the optimal solution, the LR approximation of the LR unicasting scenario lends itself to efficient practical implementation. In fact, it does not require additional overhead on the LR beyond the usual CSI feedback implemented in most cellular systems. IEEE 802.11 in ad hoc mode or Bluetooth unicasting can then be implemented on the SR with their standard MAC protocols and the results of Refs. [20,24] confirm that this would not eliminate all the gains achieved on the physical layer.

1.5.4 FUTURE WORK FOR THE RAILWAY APPLICATION

This section describes the future research directions that can enhance the contributions of Section 1.4. In this chapter, the same propagation model was used for three types of communication links in Section 1.4: BS-Relay links, Relay-MT links, and BS-MT links. The first two types correspond to the scenario with relays, whereas the last type corresponds to the traditional approach. More accurate results can be obtained if each of these link types is described by a dedicated propagation model, since the propagation characteristics differ between each of them. In practice, the BS-Relay and Relay-MT propagation models would be expected to be line of sight or near line of sight, which leads to better results for the proposed approach than those presented in this chapter. For the BS-MT link in the absence of relays, penetration loss inside the train should be added to the absence of line of sight, which would make the performance worse compared to the results presented in this chapter. Thus, overall, the assumptions of Section 1.4 are in favor of the traditional scenario and lead to pessimistic results concerning the proposed approach. Using more realistic assumptions, or deriving more accurate propagation models, would only increase the level of enhancements obtained with the proposed method.

Another important problem in high-speed cellular communications is the Doppler shift that affects the handover process. The Doppler effect was not considered in this chapter. However, efficient techniques to overcome this problem and perform LTE handover in high-speed railway systems have been recently investigated [50]. In addition, there is no significant Doppler effect in the Relay-MT links: their relative movement is too small since both are inside the train car. Furthermore, the Doppler in the BS-Relay links is easier to mitigate in the railway communication scenario than in other high-speed scenarios, since the speed and track of the train are known and predictable, in addition to the locations of the BSs along the rail track. Thus, advanced techniques can be applied both at the BSs and relays to reduce the impact of the Doppler shift. The effect of the Doppler shift will be more significant on the BS-MT links in the absence of relays.

1.6 CONCLUSIONS

With the paradigm shift toward 4G, power requirements of mobile devices are increasing as they are required to support the increase in demand for multimedia applications, which necessitate higher data rates with high QoS. The increase in energy consumption is accompanied by an increase of pollution and waste emissions, especially the undesirable CO_2 footprint in mobile communication industries. Furthermore, mobile devices that are powered by batteries for their power supply are confronted with limitations of battery capacity. In an attempt to overcome these problems, cooperative power-saving strategies between neighboring nodes using low-power SR communications have been suggested in this chapter. Results showed that significant energy savings are achieved compared to the non-cooperative scenarios and other previous related work consisting of dividing the content into equal parts among users. Moreover, the practical implementation aspects of the proposed methods were analyzed. Insights on the impact of MT residual energy and energy consumption in the MAC layer were discussed. In addition, the suggested cooperative models were applied in high-speed train systems using LTE on LR and 802.11a on SR links. LTE resource allocation on the LR BS-Relay links was considered, in the case of both unicasting and multicasting. Relays were shown to lead to huge savings in the energy consumption of MTs.

ACKNOWLEDGMENTS

The authors would like to thank King Abdullah University of Science and Technology and the Qatar National Research Fund for their support that made this research work possible.

AUTHOR BIOGRAPHIES

Rachad Atat received his BE degree in computer engineering with distinction from the Lebanese American University, Byblos, Lebanon, in 2010. He obtained his MSc in electrical engineering from King Abdullah University of Science and Technology (KAUST), Saudi Arabia, in 2012 after being awarded the KAUST Discovery Scholarship Award. He is currently pursuing his PhD at the Royal Institute of Technology, Sweden. His research interests are in the general area of wireless communications with emphasis on energy-efficient networks, LTE network, cooperative communications, relays, heterogeneous networks, and cognitive radios. He is a member of the Order of Engineers in Beirut and a member of the IEEE Communications Society.

Elias Yaacoub received his BE degree in Electrical Engineering from the Lebanese University in 2002, his ME degree in computer and communications engineering from the American University of Beirut (AUB) in 2005, and his PhD degree in electrical and computer engineering from the AUB in 2010. He worked as a research assistant in the AUB from 2004 to 2005 and in the Munich

University of Technology in spring 2005. From 2005 to 2007, he worked as a telecommunications engineer with Dar Al-Handasah, Shair and Partners. Since November 2010, he has been a research scientist at the QUWIC (QU Wireless Innovations Center), rebranded to QMIC (Qatar Mobility Innovations Center) in July 2012. His research interests include wireless communications, antenna theory, sensor networks, energy efficiency in wireless networks, video streaming over wireless networks, and bioinformatics.

Mohamed-Slim Alouini was born in Tunis, Tunisia. He received the Diplome d'Ingenieur from the Ecole Nationale Supérieure des Telecommunications (TELECOM Paris Tech) and the Diplome d'Etudes Approfondies (D.E.A.) in electronics from the Université Pierre et Marie Curie in Paris, both in 1993. He received his MSEE degree from the Georgia Institute of Technology in the United States in 1995 and a doctorate in electrical engineering from the California Institute of Technology in 1998. He also received the Habilitation degree from the Université Pierre et Marie Curie in 2003. Dr. Alouini started his academic career at the Department of Electrical and Computer Engineering of the University of Minnesota in the United States in 1998. In 2005, he joined the Electrical and Computer Engineering program at Texas A&M University at Qatar. He has been a professor of electrical engineering at King Abdullah University of Science and Technology since 2009. His research interests include statistical characterization and modeling of fading channels, performance analysis of diversity combining techniques, MIMO (multiple input–multiple output) and multi-hop/cooperative communications systems, capacity and outage analysis of multiuser wireless systems subject to interference or jamming, cognitive radio systems, and design and performance evaluation of multi-resolution, hierarchical, and adaptive modulation schemes. Dr. Alouini has published several papers on the above subjects, and he is coauthor of the textbook *Digital Communication over Fading Channels* published by Wiley Interscience. He is a fellow of the Institute of Electrical and Electronics Engineers (IEEE), a member of the Thomson ISI Web of Knowledge list of Highly Cited Researchers, and a co-recipient of best paper awards in eight IEEE conferences (including ICC, GLOBECOM, VTC, and PIMRC).

Adnan Abu-Dayya received his PhD in electrical engineering from Queens University, Canada, in 1992. He then worked as a manager at Nortel Networks in Canada in the advanced technology group and as a senior consultant at the Communications Research Center in Ottawa, Canada. Before moving to Qatar in March 2007, he worked for 10 years at AT&T Wireless in Seattle, USA, where he served in a number of senior management positions covering product innovations and emerging technologies, systems engineering, and product realization. He was also responsible for developing and licensing the extensive patent portfolio of AT&T Wireless. From April 2007 to December 2008, he was the chairman of the Electrical Engineering Department at Qatar University (QU). He was appointed as the executive director of the QUWIC (QU Wireless Innovations Center) in December 2008 (QUWIC was rebranded to QMIC [Qatar Mobility Innovations Center] in July 2012). Dr. Abu-Dayya has more than 20 years of international experience in the areas of wireless/telecomm R&D, innovations, business development, and services delivery. He has many issued patents and more than 50 publications in the field of wireless communications.

REFERENCES

1. A. Radwan, M. Albano, and J. Rodriguez. Green communications: The C2POWER approach. *ICaST: ICST's Global Community Magazine*, November 3, 2010. Available at: http://icast.icst.org/2010/11/green-communications-c2power-approach.
2. B. Badic, T.O'Farrell, P. Loskot, and J. He. Energy efficient radio access architectures for green radio: Large versus small cell size deployment. In *IEEE 70th Vehicular Technology Conference (VTC 2009 Fall)*, Anchorage, Alaska, September 2009.
3. S.S. Krishnan, N. Balasubramanian, and A. Murali Ramakrishnan. Energy consumption and CO_2 emissions by the Indian Mobile Telecom industry. In *The 4th Annual International Conference on Next Generation Infrastructures*, Virginia Beach, Virginia, November 2011.
4. Swisscom. Carbon footprint fact sheet-conferencing services. http://www.swisscom.ch/dam/swisscom/en/ghq/verantwortung/documents/Klimabilanz_Conferencing_Services_en.pdf, last accessed: January 3, 2013.

5. Strategic Environmental Communication (StrategiKas). CO_2 and cell phones. August 2011. http://estrategiks.blogspot.com/2011/08/co2-and-mobile-phones.html#!/2011/08/co2-and-mobile-phones.html, last accessed: January 3, 2013.

6. C. Oliveira, D. Sebastiao, G. Carpinteiro, L.M. Correia, C.A. Fernandes, A. Serralha, and N. Marques. The moniT project: Electromagnetic radiation exposure assessment in mobile communications. *IEEE Antennas and Propagation Magazine*, 49(1):44–53, February 2007.

7. A. Shetty. Nokia achieved 17% reduction in CO_2 emissions in 2011. *Tech 2*, May 2012. Article link: http://tech2.in.com/news/general/nokia-achieved-17-reduction-in-co2-emissions-in-2011/311182, last accessed: January 3, 2013.

8. M. Kubisch, S. Mengesha, D. Hollos, H. Karl, and A. Wolisz. Applying ad-hoc relaying to improve capacity, energy efficiency, and immission in infrastructure-based WLANs. TKN Technical Report (TKN-02-012), July 2002.

9. S. Huwang, J. Xing, D. Zhang, X. Luo, and J. Zhang. An energy-balanced relaying communication protocol based on power and distance cooperation. In *The 2008 International Conference on Embedded Software and Systems ICESS*, Chengdu, Sichuan, China, July 2008.

10. Z. Sheng, Z. Ding, and K. Leung. Distributed and power efficient routing in wireless cooperative networks. In *IEEE International Conference on Communications (ICC)*, Dresden, Germany, June 2009.

11. M. Hajiaghayi, M. Dong, and B. Liang. Energy-aware power allocation for lifetime maximization in single-source relay cooperation. In *25th Bienniel Symposium on Communications*, Ontario, Canada, May 2010.

12. D. Lu, H. Wu, Q. Zhang, and W. Zhu. PARS: Stimulating cooperation for power-aware routing in ad-hoc networks. In *IEEE International Conference on Communications ICC'05*, pages 3187–3191, Seoul, Korea, May 2005.

13. R. Madan, N. Mehta, A. Molisch, and J. Zhang. Energy-efficient decentralized cooperative routing in wireless networks. *IEEE Transactions on Automatic Control*, 54(3):512–527, March 2009.

14. T. Himsoon, P. Siriwongpairat, Z. Han, and K. Liu. Lifetime maximization via cooperative nodes and relay deployment in wireless networks. *IEEE Journal on Selected Areas in Communications*, 25(2):306–317, February 2007.

15. Cost Europe. Cost actions. 2012. http://www.cost.eu/domains_actions/ict/Actions, last accessed: January 3, 2013.

16. P. Michiardi and G. Urvoy-Keller. Performance analysis of cooperative content distribution in wireless ad hoc networks. In *Fourth Annual Conference on Wireless on Demand Network Systems and Services*, Obergurgl, Tyrol, Austria, January 2007.

17. R. Bhatia, L. Li, H. Luo, and R. Ramjee. ICAM: Integrated cellular and ad hoc multicast. *IEEE Transactions on Mobile Computing*, 5(8):1004–1015, August 2006.

18. M. Leung and G. Chan. Broadcast-based peer-to-peer collaborative video streaming among mobiles. *IEEE Transactions on Broadcasting*, 53(1):350–361, March 2007.

19. E. Yaacoub, L. Al-Kanj, Z. Dawy, S. Sharafeddine, and A. Abu-Dayya. Energy efficient fair content distribution over LTE networks using bluetooth piconets. In *26th Wireless World Research Forum (WWRF)*, Doha, Qatar, April 2011.

20. M. Ramadan, L. Zein, and Z. Dawy. Implementation and evaluation of cooperative video streaming for mobile devices. In *IEEE International Symposium on Personal, Indoor, and Mobile Radio Communications (PIMRC)*, Cannes, French Riviera, France, September 2008.

21. L. Popova, T. Herpel, and W. Koch. Efficiency and dependability of direct mobile-to-mobile data transfer for UMTS downlink in multi-service networks. In *Proceedings of IEEE Wireless Communication and Networking Conference (WCNC'07)*, Hong Kong, March 2007.

22. D. Zhu and M. Mutka. Cooperation among peers in an ad hoc network to support an energy efficient IM service. *Journal of Pervasive and Mobile Computing*, 4(3):335–359, June 2008.

23. Q. Zhang, F.H.P. Fitzek, and V.B. Iversen. Design and performance evaluation of cooperative retransmission scheme for reliable multicast services in cellular controlled P2P networks. In *Proceedings of IEEE International Symposium on Personal, Indoor and Mobile Radio Communications*, September 2007.

24. P. Vingelmann, F.H.P. Fitzek, M.V. Pedersen, J. Heide, and H. Charaf. Synchronized multimedia streaming on the iPhone platform with network coding. *IEEE Communications Magazine*, 49(6):126–132, June 2011.

25. L. Militano, F.H.P. Fitzek, A. Iera, and A. Molinaro. Group interactions in wireless cooperative networks. In *IEEE Vehicular Technology Conference (VTC) – Spring 2011*, Budapest, Hungary, May 2011.

26. F. El-Moukaddem, E. Torng, G. Xing, and S. Kulkarni. Mobile relay configuration in data-intensive wireless sensor networks. In *IEEE 16th International Conference on Mobile Adhoc and Sensor Systems*, pages 80–89, Macau, China, October 2009.

27. W. Wang, V. Srinivasan, and K.-C. Chua. Extending the lifetime of wireless sensor networks through mobile relays. *IEEE/ACM Transactions on Networking*, 16(5):1108–1120, October 2008.
28. Z.H. Abbas and F.Y. Li. Power consumption analysis for mobile stations in hybrid relay-assisted wireless networks. In *5th IEEE International Symposium on Wireless Pervasive Computing (ISWPC)*, pages 16–21, Modena, Italy, May 2010.
29. H. Wu, C. Qiao, S. De, and O. Tonguz. Integrated cellular and ad hoc relaying systems: iCAR. *IEEE Journal on Selected Areas in Communications*, 19(10):2105–2115, October 2001.
30. A. Goldsmith. *Wireless Communications*. Cambridge University Press, Cambridge, UK, 2005.
31. X. Qiu and K. Chawla. On the performance of adaptive modulation in cellular systems. *IEEE Transactions on Communications*, 47(6):884–895, June 1999.
32. I. Haratcherev, R. Lagendijk, K. Langendoen, and H. Sips. Hybrid rate control for IEEE 802.11. In *Proceedings of International Symposium on Mobility Management and Wireless Access*, Philadelphia, PA, October 2004.
33. 3rd Generation Partnership Project (3GPP). 3GPP TR 25.814 3GPP TSG RAN physical layer aspects for evolved UTRA, v7.1.0, 2006.
34. K. Mahmud, M. Inoue, H. Murakami, M. Hasegawa, and H. Morikawa. Energy consumption measurement of wireless interfaces in multi-service user terminals for heterogeneous wireless networks. *IEICE Transactions on Communications*, E88-B(3):1097–1110, March 2005.
35. F. Albiero, M. Katz, and F. Fitzek. Energy-efficient cooperative techniques for multimedia services over future wireless networks. *IEEE International Conference on Communications (ICC'08)*, Beijing, China, May 2008.
36. F. Albiero, M. Katz, and F. Fitzek. Overall performance assessment of energy-aware cooperative techniques exploiting multiple description and scalable video coding schemes. In *Communication Networks and Services Research Conference (CNSR'08)*, Halifax, Nova Scotia, Canada, May 2008.
37. K.-D. Lin and J.-F. Chang. Communications and entertainment onboard a high-speed public transport system. *IEEE Wireless Communications*, 9(1):84–89, February 2002.
38. Y. Zhou, Z. Pan, J. Hu, J. Shi, and X. Mo. Broadband wireless communications on high speed trains. In *2011 20th Annual Wireless and Optical Communications Conference (WOCC)*, New Jersey, USA, April 2011.
39. J. Garstenauer. GSM-R evolution towards LTE. In *2010 Institution of Railway Signal Engineers (IRSE) International Convention*, New Delhi, India, October 2010.
40. K. Guan, Z. Zhong, and B. Ai. Assessment of LTE-R using high speed railway channel model. In *2011 Third International Conference on Communications and Mobile Computing (CMC 2011)*, Qingdao, China, April 2011.
41. F. Greve, B. Lannoo, L. Peters, T. Leeuwen, F. Quickenborne, D. Colle, F. Turck, I. Moerman, M. Pickavet, B. Dhoedt, and P. Demeester. FAMOUS: A network architecture for delivering multimedia services to FAst MOving USers. In *Wireless Personal Communications*, New Jersey, USA, April 2011.
42. Rohde and Schwarz. WLAN 802.11p measurements for vehicle to vehicle (V2V) DSRC. Rohde and Schwarz Application Note, September 2009.
43. IEEE 802.11. Wireless LAN medium access control (MAC) and physical (PHY) layer specifications, amendment 6: wireless access in vehicular environments, 2010.
44. I. Forkel, M. Schinnenburg, and M. Ang. Generation of two-dimensional correlated shadowing for mobile radio network simulation. In *Proceedings of the 7th International Symposium on Wireless Personal Multimedia Communications (WPMC 2004)*, Padova, Italy, September 2004.
45. Z. Wang, E. Tameh, and A. Nix. Joint shadowing process in urban peer-to-peer radio channels. *IEEE Transactions on Vehicular Technology*, 57(1):52–64, January 2008.
46. K. Yamamoto, A. Kusuda, and S. Yoshida. Impact of shadowing correlation on coverage of multihop cellular systems. In *IEEE International Conference on Communications (ICC 2006)*, pages 4538–4542, Istanbul, Turkey, June 2006.
47. 3rd Generation Partnership Project (3GPP). 3GPP TR 36.942 3GPP TSG RAN evolved universal terrestrial radio access (E-UTRA) radio frequency (RF) system scenarios, version 8.1.0, Release 8, 2008.
48. 3rd Generation Partnership Project (3GPP). 3GPP TS 36.211 3GPP TSG RAN evolved universal terrestrial radio access (E-UTRA) physical channels and modulation, version 8.3.0, Release 8, 2008.
49. 3rd Generation Partnership Project (3GPP). 3GPP TS 36.213 3GPP TSG RAN evolved universal terrestrial radio access (E-UTRA) physical layer procedures, version 8.3.0, Release 8, 2008.
50. W. Luo, R. Zhang, and X. Fang. A CoMP soft handover scheme for LTE systems in high speed railway. *Eurasip Journal on Wireless Communications and Networking*, 2012:196, June 2012.

2 FIREMAN: Foraging-Inspired Radio-Communication Energy Management for Green Multi-Radio Networks

Thomas Otieno Olwal, Barend J. Van Wyk,
Paul Okuthe Kogeda, and Fisseha Mekuria

CONTENTS

2.1 INTRODUCTION

In the past decade, the remarkably rapid evolution of wireless networks into the regime of the next-generation heterogeneous broadband and mobile networks has triggered the emergence of multi-radio wireless infrastructures. Infrastructures of these types have been expected to integrate the future internet of people, technologies, content, and clouds into a common digital information society [1]. As a result, the move will eventually witness a harmonious coexistence of many wireless technologies in the same constrained radio resource environment in order to provide ubiquitous and seamless broadband services. To achieve this goal, the multi-radio networking technologies have to be designed in such a way as to ensure that they are self-organized, self-configured, reliable, and robust with a capacity to sustain high traffic volumes and long "online" time [2].

Such complex functional and structural features stemming from the multi-radio networks will, however, essentially cause unnecessary energy consumption in future networks [1]. Thus, it follows that the need to reduce the energy consumption in ICT industries becomes relevant in order to mitigate the adverse impacts of energy consumption on the economy, environment, and ICT markets. To address this challenge, many studies have proposed several green strategies for wireless networking technologies and protocols [3]. For example, green strategies have been recently exploited to design energy-efficient residential gateways [4]. The gateways employ appropriate home networking interfaces and service logic to allow homeowners to perform personalized, pervasive programming of the energy consumption of home devices such as electrical, communication, and audiovisual equipment. The green networking research has also been considered, in order to address issues of autonomous link rate adaptation, interface proxying, and energy awareness infrastructures and applications [3].

In a bid to contribute to the autonomy of energy-efficient architectures capable of supporting green heterogeneous wireless infrastructures and applications, this chapter proposes a novel energy management solution known as the Foraging-Inspired Radio-Communication Energy Management (FIREMAN) method. The FIREMAN method integrates the optimal transmission energy allocation with the energy-saving efforts in multi-radio networks, so as to ensure a substantial energy consumption reduction in a random wireless ecosystem [5]. The main concept has been coined from the field of behavioral ecology, or foraging theory, in which a solitary forager in an ecosystem makes optimal decisions that maximize its rate of energy gain, thereby improving its survival probability and lifetime in a random environment [6]. Using this bio-inspired methodology, a solitary forager represents a foraging-inspired radio energy (FIRE) resource manager while the so-called nutrients or prey mimic the radio communication energy resources that the radio interfaces need in order to exchange packets in a wireless link. The FIREMAN method involves the development of a prey model algorithm whereby the radio communication energy resources (energy link costs) are encountered randomly by the radio interfaces since the wireless links are stochastic in nature. In this manner, the algorithm maximizes an energy-aware throughput (EAT) or communication profitability experienced in every link. The profitability is described by a set of feasible foraging behaviors consisting of optimizing resource preference rates and allocation times that are capable of improving the energy consumption [7].

In order to minimize the multi-radio function complexities, the FIREMAN method is coordinated by an autonomous foraging radio resource allocation (AFRRA) protocol module built from an energy-aware multi-radio unification protocol [8]. This module virtualizes functions of multiple medium access control (MAC) and PHY layers so that the application layers can only visualize the homogenous single-radio networks rather than the complex heterogeneous wireless platforms. The performance of the developed FIRE manager has been extensively validated through computer simulations. This chapter has also provided future plans for prototype development suited for future networks. To our best knowledge, this work can be viewed as an early contribution toward the application of the foraging theory of nutrients optimization to the field of green wireless networking research.

The remainder of this chapter is organized as follows: Related work in the field of radio energy management is discussed in Section 2.2. An AFRRA protocol will be presented in Section 2.3. In Section 2.4, the FIREMAN problem is formulated and a corresponding FIREMAN algorithm is developed in Section 2.5. Section 2.6 provides the throughput and energy efficiency performance evaluation. Future research directions and conclusions regarding the FIREMAN method are presented in Sections 2.7 and 2.8, respectively.

2.2   RELATED WORK

The development of the FIREMAN method has been prompted by a number of experimental results stemming from measurements of the energy consumption behaviors in real WiFi networks [9,10]. In Gomez et al. [9], the actual impact between the traffic and power consumption for a typical wireless local area network (WLAN) (the IEEE 802.11b/g) access point (AP) was measured experimentally. The experimental results showed a significant impact of different traffic sizes on the power consumption pattern of the wireless devices, both at the interface level, with respect to the power expenditure for transmission and reception, and at the device level, with respect to the energy spent for processing the application traffic. In Carvalho et al. [10], an investigative study was conducted on the energy consumption of IEEE 802.11 cards when nodes were in contention for channel access under saturation conditions. In such scenarios, the study found that the radio's transmit mode had marginal impact on the overall energy consumption, while other modes (receive and idle) were responsible for most of the energy consumption. It was also noted that the energy link cost to transmit useful data increased almost linearly with the network size. Transmitting large payloads was more energy efficient under saturated conditions than small payloads.

The exploitation of the multi-channel MAC layer to provide an energy- and spectrum-efficient throughput had stimulated a flurry of research activities in power-saving MAC protocols and algorithms [8,11–14]. Manweiler and Choudhury [15] proposed the SleepWell energy-saving mechanism that achieved energy efficiency by evading the network contention in WiFi networks. Different APs adjusted their activity cycles to minimally overlap with others and consequently to regulate the sleeping window of their clients in such a way that different APs were active or inactive during non-overlapping time windows. The SleepWell was implemented on a test-bed platform of eight laptops and Android phones. An evaluative study over a wide variety of scenarios and traffic patterns (YouTube, Pandora, FTP, Internet radio, and mixed) showed a significant energy gain with a practically negligible loss of performance. The SleepWell enforced energy efficiency through scheduling the activity cycles of APs during non-overlapping time windows to evade network contention, while the proposed FIREMAN method achieves energy efficiency by enforcing the non-foraging interface cards (NFICs) to go to the "doze mode," while the low-power foraging interface card (FIC) indicates traffic belonging to the target receivers.

Anastasi et al. [11] presented an analytical model of a power-saving mode (PSM) aimed at reducing the energy consumption caused by the networking activities in IEEE 802.11 standard technologies. According to the IEEE 802.11 PSM algorithm, a mobile device is left in the active mode only for the time necessary to exchange data; it is turned to the sleep mode as soon as it becomes idle. In connecting to the infrastructure 802.11 WLAN or the WiFi hotspot, the PSM algorithm was achieved by exploiting the role of the AP, whereby each client station inside the hotspot informed the AP whether it utilized the PSM algorithm or not. As the AP relayed every frame from or to any client station, it buffered frames addressed to the client stations operating in the PSM while they were sleeping. Once during every beacon interval, usually 100 ms, the AP broadcasts a beacon frame containing the traffic indication map (TIM). The TIM indicated identifications of PSM stations whose application frames were buffered at the AP. The PSM stations were then synchronized with the AP and woken up to receive beacons. If these PSM stations were indicated in the TIM, they could download the application frames. Even though the PSM reduced the sensing or contention time, the TIM window was made static and only the energy consumption in the transmit mode was taken into account. Thus, Moshe et al. [12] extended this energy-saving scheme to another method known as the LESS-MAC where the TIM window was made dynamic with respect to different payload sizes. Through simulations, the LESS-MAC was shown to save energy in the idle mode with minimal additional functionalities as in Ref. [11]. Moreover, the dynamic TIM window contributed a greater energy saving in the transmit and receive modes. The FIREMAN algorithm autonomously adapts the energy link costs in a random wireless environment such that the per-link EAT is maximized.

Recently, the IEEE 802.11 PSM method has been extended by Kumruzzaman [13] to perform a TDMA-based energy-efficient cognitive radio MAC (ECR-MAC) protocol. In this protocol, ad hoc nodes were allowed to dynamically negotiate multi-channels such that multiple radio communications could take place in the same region simultaneously, each in a different channel. In this way, the licensed primary users could coexist with non-licensed users in an interference-free and ad hoc-based multi-channel cognitive radio environment. To achieve the goal of reducing the idle time, the protocol divided time into fixed-time intervals using beacons and had a small window at the start of each interval to indicate the application traffic and negotiate channels. This protocol is complementary to the FIREMAN method except that the idle time and the energy link costs are minimized by the FIREMAN in order to realize better energy efficiency.

To express this concisely, these studies have focused on single radio-based power-saving mechanisms and not on the multi-radio wireless networks. The FIREMAN method, on the other hand, seeks to address the energy-efficient issues in multi-radio wireless networks wherein a very large percentage of the energy consumption arises. In Wang et al.'s research study [5], an opportunistic spectrum access and adaptive power management under the setting of multi-radio nodes and multi-channel WLANs was proposed. This power-saving multi-channel MAC (PSM-MMAC) protocol aimed at reducing the collision probability and the waiting time in the "awake" state of a node under the distributed coordination function (DCF) mode [11,16]. The protocol allowed for the estimation of the number of active links; the selection of appropriate channels, radios, and power states (i.e., awake or doze state), given the link estimates, queue lengths, and channel conditions; and the optimization of the medium access probability in the p-persistent carrier sense multiple access (CSMA) used in the data exchange. The simulation and analytical results showed an improved throughput, delay, and energy-efficient performance. However, several drawbacks associated with the PSM-MMAC were found: the default radio interface consumed a substantial amount of energy when estimating the number of active links and communicating the default channel to the rest of a dense network; there was no guarantee that the default radio interface was operating in low-power modes during the ad hoc traffic indication message (ATIM) window; the protocol considered the energy saving in the transmit mode only and not in the receive mode; and the PSM-MMAC did not provide a transmission energy control strategy; instead, all awake radios exchanged application frames using high transmission power levels. In contrast, the FIREMAN algorithm utilizes the AFRRA protocol to perform the channel negotiation and traffic indication with the neighboring nodes during the traffic and radio resource allocation window when the link is in both transmit and receive modes. The default radio interface (FIC) is enforced to operate in a low-power mode to exchange control packets only while other radios use power-controlled levels to exchange the application traffic.

The implementation of the FIREMAN method is closely related to the one proposed by Lymberopoulos et al. [17] in which the authors presented an energy-efficient multi-radio platform. In this case, an examination of the efficient interfaces between the multiple heterogeneous radios and one or more processors on a single sensor node for energy efficiency was performed. The authors focused on the effect of the application-level parameters such as packet payload sizes and the packet transmission period on the energy consumption of CSMA protocols having multiple transmission attempts compatible with 802.15.4 and 802.11 MAC layer specifications. However, unlike the FIREMAN approach, the proposed platform did not suggest a unified layer to conceal the complex functions of the multiple radio interfaces and MACs from the upper layers.

To conceal this complexity, studies [1,8,18,19] proposed an autonomous transmission energy adaptation for multi-radio multi-channel wireless mesh networks (WMNs). The transmission energy was dynamically adapted asynchronously or synchronously by each radio interface. The interfaces were coordinated by a power selection multi-radio multi-channel unification protocol (PMMUP) [8]. The transmission energy adaptations were based on the locally residing energy in a node, the amount of local queue load, the quality of the links, and the interference conditions in the wireless medium [20,21]. The authors divided the WMN into a set of orthogonal unified channel graphs (UCGs) whereby each radio interface of each node was tuned to a unique graph. The PMMUP first set initial unification variables such as energy reserves and channel states from other UCGs; radio interfaces then predicted channel states of the wireless medium; the PMMUP updated the unification variables, and finally, radio interfaces computed optimal transmission energy levels on the basis of the predicted states [19]. However, effects of queue and link instability on the link-level energy consumption were not discussed by the PMMUP method.

In response to this gap, Olwal et al. [14,22] modeled the inter-channel and co-channel interference, energy consumption at the queues, and the network connectivity problems as a joint queue-perturbation and weakly coupled (SPWC) systems. A Markov chain model was developed to describe the steady-state probability distribution behavior of the queue energy and buffer state variations in multi-radio nodes. The impact of such queue perturbations on the transmission probability using some transmission energy values was analyzed. The simulation results indicated that the proposed power control method converged at the steady state. Although the SPWC system was energy aware, it was computationally complex and did not address the dynamic channel negotiations jointly with the energy efficiency. Consequently, the proposed FIREMAN method has been developed. It simplifies the architectural and functional designs by exploiting the benefits of the AFRRA protocol, of searching for and optimizing the locally available energy resources in a random wireless environment.

2.3   AFRRA PROTOCOL

The AFRRA protocol is an extension of the virtual MAC developed in Refs. [1,18]. The AFRRA is a software module that controls functions of the FICs and the NFICs in a highly random wireless ecosystem. The AFRRA protocol dynamically adapts the radio communication energy, channel negotiations, and the energy-saving mechanisms to achieve a foraging energy-efficient network. The protocol assumes that every radio interface card in the "awake" state consumes a significant amount of the energy resource and that when in the doze state it consumes a very low energy resource. In the awake state, an interface may be in one of the three different modes: the transmit, receive, and idle or sense modes [5].

2.3.1   Bio-Inspired MAC Firmware Architecture

Consider a bio-inspired MAC firmware architecture shown in Figure 2.1, having an autonomous foraging resource allocation message (AFRRAM) window during which the channel negotiation or

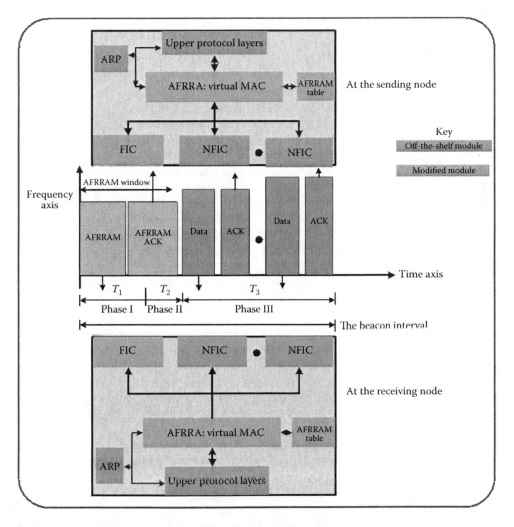

FIGURE 2.1   The proposed bio-inspired MAC firmware architecture.

contention (in Phase I) and the energy link cost estimation (in Phase II) are performed and information are stored in the AFRRAM table. The first phase aims to form a link layer connection of nodes. When a node has data packets destined for another node or AFRRA, it may transmit an AFRRAM or, as it is conventionally known, an ATIM via the awake default FIC to the intended receiver. The default FIC listens on the foraging frequency channel (FFC) during the AFRRAM window (meant for exchanging control messages only). Upon receiving an AFRRAM, the intended receiver FIC will reply with an AFRRAM-ACK message before data download commences. The transmission or retransmission of the AFRRAM follows the normal DCF access procedure [16].

In the second phase, the optimal link resources consisting of the energy costs and the frequency channels using the FIREMAN algorithm are determined. After the end of the current AFRRAM window, any node, neither having sent an AFRRAM nor having received an AFRRAM via its FIC containing its own address, the initial frequency channel and power settings in the awake state during the AFRRAM window will enter the doze state. Any node that has sent an AFRRAM or received an AFRRAM containing its own address during the AFRRAM window will remain in the awake state until the end of the next AFRRAM window.

The third phase covers the exchange of the application data packets. These nodes in the awake state transmit/receive the application data packets and acknowledgements using the awakened NFICs. At the end of the beacon interval, all NFICs switch to the doze mode until the next packet arrivals.

2.3.2 The FIREMAN Protocol

Figure 2.2 illustrates the bio-inspired radio resource (channel and energy link costs) allocation scheme in a random environment. Initially, at least one FIC, say a0 of node A, is awake at the default to estimate the energy link costs and allocate the frequency channels to the link. At the sending AFRRA, when this AFRRA has packets destined for another AFRRA in the network, it wakes up a default FIC and tunes the rest of the interfaces to the doze mode. The awakened FIC selects an initial random frequency channel using a low-power mode, c0, as a default FFC from a pool of channels and an initial lowest possible power setting, p0, from a pool of the off-the-shelf power levels. The sending FIC advertises, to all its neighbors listening on all channels via their FICs: the AFRRAM consisting of the selected FFC, the selected power setting, the residual energy in the awake state, and the MAC address of the intended receiving AFRRA. If received correctly, the receiving FICs reply with AFRRAM-ACK indicating to the sending FIC that they have received the sent AFRRAM containing their own address, the power setting, and the sender's residual energy. If this does not occur, either the link is busy or the selected power setting cannot reach the intended receiver. If the link is busy, the sending FIC switches to the doze mode and re-advertises after a random back-off period. If the power setting is too low, then the AFRRAM is re-sent with a power setting level that is incrementally higher than the previous level. It should be noted that during the re-advertisements, different commodity power settings are selected, while the FFC and the address of the intended AFRRA are kept the same. Incrementing the transmission mode power settings increases the chances of reaching the intended receiver on subsequent re-transmissions. If a particular FFC has been grabbed, no other neighboring link that hears that particular FFC should grab the reserved FFC for the transmission of its AFRRAM until the next AFRRAM window, thereby avoiding collisions among neighboring transmitters. In Phase II of the proposed MAC firmware architecture and on the basis of the successfully exchanged AFRRAM, the first task of each sending–receiving FIC pair or link is to find a set of power settings and, thus, the estimate of the energy link costs. The second task that they have is to find a clumped patch of the frequency channel available for occupation in the network. The optimal resource type is free to be chosen, from a set of discrete power levels and the frequency channel cluster, from the unused spectrum in the available ISM bands.

At the intended receiving AFRRA, the FIC listens omni-directionally to receive FFCs, power settings, and the residual energy from all possible sending AFRRAs in the awake state. In the

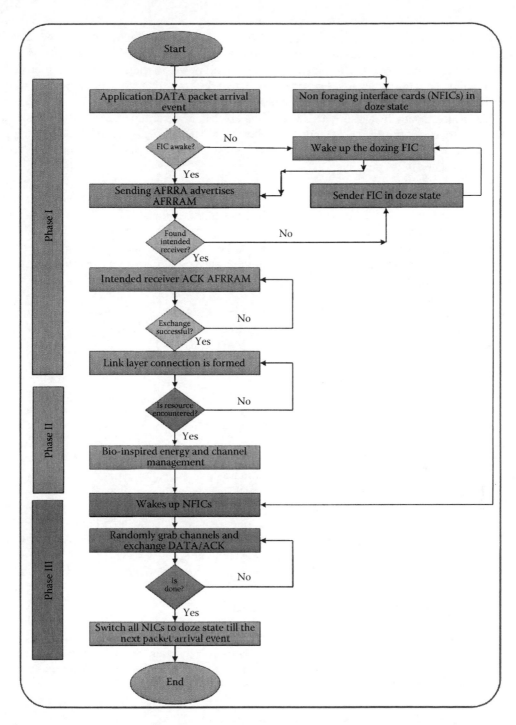

FIGURE 2.2 The proposed bio-inspired radio resource allocation scheme.

meantime, the NFICs at the intended receiver node are compelled to enter the doze mode by the AFRRA in order to save the energy. If the FIC can hear or detect or sense the frequency channels from the sending AFRRAs, this means that the said FIC is listening on that FFC and that the FFC is temporarily reserved for a period of the AFRRAM window. No other listening, neighboring links are able to grab it. The FIC at the intended receiver grabs a frequency channel it detects if, and only if, that frequency channel contains the least amount of energy link costs and its own MAC address; otherwise, it rejects any other frequency channels heard from the neighborhood. If the AFRRAM is successfully received, then the FIC of the intended receiver replies with the AFRRAM-ACK message to the sending AFRRA. It should be noted that all power settings (prey types) sent to the intended neighbors but not received (detected) imply that they cannot guarantee the neighborhood connectivity. All frequency channels (patch types) heard by the neighbors that do not contain their own MAC addresses imply that such neighbors cannot pick them up as they are already occupied as FFCs. Thus, they are ignored. The intended neighboring AFRRA using AFRRAM-ACK messages replies to every sending AFRRA through the corresponding FICs and FFCs only; meanwhile, all NFICs are switched to the doze state.

At the link layer, the intended receiving AFRRA and the sending AFRRA pair estimates the energy link costs (prey-type executions) for every "hello packet" successfully sent and received at different radiated power levels by the FICs. On the basis of the estimates of each energy link cost, the AFRRA link computes the EAT and the foraging energy efficiency (FEE) that varies randomly from one energy link cost to the other, depending on the channel conditions. The optimal energy link cost (computed from the electronic energy of the receiver and transmitter interfaces, and the link-radiated power) that yields the highest profitability measured in terms of the EAT compared to the FEE functions is selected by the AFRRA.

After the end of the current AFRRAM window, the AFRRA wakes the NFICs and switches them randomly to the corresponding non-overlapping channels to exchange the application data with the optimally computed energy link costs.

2.4 PROBLEM FORMULATION

The software AFRRA module represents the biological forager while the radio communication energy and channels it encounters are the tasks or resources it must optimally choose and determine for what period to allocate the links. In order to adopt the bio-inspired resource allocation algorithms, Stephens and Krebs [23] described two popular models. The first model was known as the prey model, which inspired the argument that the radio communication energy (i.e., prey) comes randomly in lumps or as individual quantities that have fixed resource allocation times (i.e., the processing-time constraints are such that $\tau_i^- = \tau_i^+ > 0$ for each resource type i). Thus, allocating optimal radio communication energy to a certain link follows the prey model algorithm. The AFRRA (i.e., forager) has only to decide whether to allocate the radio communication energy optimally or to ignore the allocation process and proceed to the next lump [7]. The second model was referred to as the patch model (i.e., a cluster of preys), which assumes that the AFRRA allocates the link to every encountered resource type (i.e., preference constraints $p_i^- = p_i^+ = 1$ for each resource type i), but each encountered resource is seen as a clumped patch of prey with decreasing marginal returns (e.g., owing to the depletion of prey within the patch or set of prey). Since the channels in WiFi networks are randomly encountered, allocating the link to an optimal channel follows the patch model algorithm whereby grabbing of one channel from the pool decreases the number of channels for other users within the same wireless environment. The AFRRA must decide on a length of time to process every patch on the basis of a constant preference probability [6].

2.4.1 OBJECTIVE FUNCTION

Suppose that the AFRRA can complete allocating $n \in \{1,2,...\}$ discrete types of distinct radio resources (i.e., energy link costs and channels) to the radio links. For a resource of type $i \in \{1,2,..., n\}$, AFRRA allocates each radio link $p_i \in [0,1]$ fraction of the encountered energy and channel type i and spends an average of $\tau_i \geq 0$ time allocating each selected type i. This implies that the complete radio resource allocation behavior of the sending link is described by vectors $\vec{p} \triangleq [p_1, p_2, ..., p_n]^T$

and $\vec{\tau} \triangleq [\tau_1, \tau_2, \ldots, \tau_n]^T$ for all the resources. Constraints on the feasible behaviors are defined by constants $p_i^-, p_i^+ \in [0, 1]$ and $\tau_i^-, \tau_i^+ \in \Re_{\geq 0}$ for each resource type $i \in \{1, 2, \ldots, n\}$ so that the feasible set of behaviors becomes a *convex separable polyhedron* [24]:

$$\Gamma \triangleq \left\{ (\vec{p}, \vec{\tau}) \in [0, 1]^n \times \Re_{\geq 0}^n : p_i^- \leq p_i \leq p_i^+, \tau_i^- \leq \tau_i \leq \tau_i^+, i \in \{1, 2, \ldots, n\} \right\}. \tag{2.1}$$

The optimal behavior of the AFRRA is to maximize the generic *advantage-to-disadvantage function* of each radio interface as is inspired by the foraging theory [24],

$$J(\vec{p}, \vec{\tau}) \triangleq \frac{A(\vec{p}, \vec{\tau})}{D(\vec{p}, \vec{\tau})} \triangleq \frac{a + \sum\limits_{i=1}^{n} p_i a_i(\tau_i)}{d + \sum\limits_{i=1}^{n} p_i d_i(\tau_i)}, \tag{2.2}$$

where $a \in \Re$ and $d \in \Re$ are constants and $a_i : \left[\tau_i^-, \tau_i^+ \right] \mapsto \Re$ and $d_j : \left[\tau_j^-, \tau_j^+ \right] \mapsto \Re$ are functions of time τ_i associated with type $i \in \{1, 2, \ldots, n\}$. The type can be either any radio communication energy or any frequency channel. Then, the *advantage-to-disadvantage function* becomes the ratio of the expected link throughput to the expected radio communication energy consumed by that link. The function signifies the profitability of the AFRRA's decisions on energy link costs and channel allocations and is succinctly written as

$$J(\vec{p}, \vec{\tau}) \triangleq \frac{\sum\limits_{i=1}^{n} \lambda_i p_i^k \left\{ \left[w_i^k \log_2 \left(1 + \mathrm{SNR}_i^k(\tau_i) \right) - O_i^{\mathrm{rate}}(\tau_i) \right] \right\} - C^{\mathrm{search}}}{\sum\limits_{i=1}^{n} \lambda_i p_i^k \left[P_i^k(\tau_i) - O_i^{\mathrm{Pow}}(\tau_i) \right]}, \tag{2.3}$$

whereby, for each resource type $i \in \{1, 2, \ldots, n\}$ and a corresponding resource allocation duration τ_i, several notations can be defined: λ_i is the rate of encounter with each resource type i and w_i^k is the channel bandwidth of resource type i associated with the radio interface k. The channel bandwidth is defined as $w_i^k = 2\left(f_0^i - f_\ell^k \right)$, where $f_\ell^k = f_0^k - w/2$ is the lower frequency and f_0^ℓ is the middle frequency between the lower and the upper frequency bounds. The received signal-to-noise ratio from a resource type i is denoted as SNR_i^k, O_i^{rate} is the message overhead, O_i^{Pow} is the energy overhead, and P_i^k is the radio communication energy. The cost of searching for a certain resource type is denoted as C^{search} and is assumed fixed. In an analogy with Equation 2.2, Equation 2.3 is simplified as follows:

$$a \triangleq -C^{\mathrm{search}}, a_i(\tau_i) \triangleq \lambda_i \left[2\left(f_0^i - f_\ell^k \right) \log_2 \left(1 + \mathrm{SNR}_i^k(\tau_i) \right) - O_i^{\mathrm{rate}}(\tau_i) \right], d \triangleq 0,$$

$$d_i(\tau_i) \triangleq \lambda_i \left[P_i^k(\tau_i) - O_i^{\mathrm{Pow}}(\tau_i) \right].$$

$$b_i \triangleq \sum\limits_{\substack{j=1 \\ j \neq i}}^{n} \lambda_j \left[2\left(f_0^i - f_\ell^k \right) \log_2 \left(1 + \mathrm{SNR}_j^k \right) - O_j^{\mathrm{rate}} \right], e_i \triangleq \sum\limits_{\substack{i=1 \\ j \neq i}}^{n} \lambda_i \left[P_i^k - O_i^{\mathrm{Pow}} \right].$$

2.4.2 Radio Resource Decision Variables

The radio resource decision variables are the radio communication energy and channels. In wireless commodity devices, the number of radio resource types, n, is free to be chosen as any reasonable number of discrete resource values available. The resource types and actual link variables have a distribution that resembles [7]

$$i = \text{ceil} \ (n \times \exp \ (-R^k)), \tag{2.4}$$

where $k = 1, 2, \ldots, k$ is the radio interface zone number, i is the resource type and $R^k \in [0, R_{max}]$ is the resource variable belonging to the kth radio interface zone while the "ceil" is the standard ceiling function for converting non-integers to integers. The nonlinear relationship defines a large number of types for resource values within the range, thus providing a better accuracy near the real values. Equivalently, the resource values are derived by noting that if $y = b^x$, then $x = \log_b y$ so that

$$R^k = -\log_e \ (i/n) \Rightarrow R^k = -\ln(i/n). \tag{2.5}$$

The resource allocate on times for each resource type using the function is denoted as

$$\tau_i = n + n \times \exp \ (-i). \tag{2.6}$$

The exponential characteristic of this function matches the distribution of processing times to the distribution of the resource types. Rates of encounter λ_i with different resource types are usually estimated in real time [7]. At any given instant, an estimate $\hat{\lambda}_i$ of the rate of encounter with type i is calculated as the number of times type i has been encountered by the AFRRA divided by the amount of time the AFRRA has spent searching for resources that it should allocate. Once the relationship between the radio resource value and type, the processing time function, τ_i, and the objective function are determined, the FIREMAN algorithm executes an optimal radio communication energy management behavior that maximizes the objective function.

2.5 THE FIREMAN ALGORITHM

From the simplified objective function in Equation 2.2, feasible solutions are obtained when the relevant assumptions are made. For example, the function d_i is constant and possibly zero; the constant $d \neq 0$; if $d_i \neq 0$, then it has the same sign as d; if d is positive, then a_i has a maximum, and if d is negative, then a_i has a minimum (i.e., function a_i/d has a maximum). The probability of allocating or processing each radio communication energy type is the decision variable for the AFRRA when applying the prey model to implement the FIREMAN algorithm. Thus, the AFRRA chooses p_i^k that maximizes J defined in Equation 2.3. Let J be rewritten as

$$J(\vec{p}, \vec{\tau}) = \frac{a + p_i^k \lambda_i a_i + b_i}{d + p_i^k \lambda_i d_i + e_i}, \tag{2.7}$$

where b_i is the summation of all terms in the numerator not involving the energy resource type i and e_i is a similar variable for the denominator. To obtain the value of p_i^k at which J is maximum, we differentiate J with respect to p_i^k,

$$\frac{\partial J}{\partial p_i^k} = \frac{\lambda_i a_i \left(d + p_i^k \lambda_i d_i + e_i \right) - \lambda_i d_i \left(a + p_i^k \lambda_i a_i + b_i \right)}{\left(d + p_i^k \lambda_i d_i + e_i \right)^2}$$

$$= \frac{\lambda_i a_i e_i - \lambda_i d_i b_i}{\left(d + p_i^k \lambda_i d_i + e_i \right)^2}. \tag{2.8}$$

By viewing Equation 2.8, it is noted that if the numerator is negative, then J is maximized by choosing the lowest possible p_i^k. Alternatively, if the numerator is positive, then J is maximized by choosing the highest possible p_i^k. However, we know that $0 \le p_i^k \le 1$. Thus, p_i^k that maximizes J is either $p_i^k = 1$ or $p_i^k = 0$ for each $i \in \{1, 2, \ldots, n\}$. The decision depends directly on the sign of $a_i e_i - d_i b_i$. This type of decision is referred to as the *zero–one rule*, which is summarized as

$$\begin{aligned} &\text{set } p_i = 0 \text{ if } a_i / d_i < b_i / e_i \\ &\text{set } p_i = 1 \text{ if } a_i / d_i > b_i / e_i \end{aligned}. \tag{2.9}$$

Here, a_i / d_i is the profitability that results from processing resource type i and b_i / e_i is the alternative profitability resulting from searching for and processing other resource types.

Using this rule, an AFRRA either processes energy of type $i \in \{1, 2, \ldots, n\}$ every time it encounters it or never processes it at all. The question is: which radio communication energy level should the AFRRA process and which level should it ignore? The answer for "which it should not" must account for the missed opportunity. That is, if it profits the AFRRA more when it allocates the energy of type i than that of searching for and allocating the energy of other types, then the AFRRA should process the energy of type i and ignore other types. Conversely, if more benefits are likely to derive through processing other energy types other than those of type i, then the AFRRA should ignore type i.

To process multiple types, the radio communication energy levels are first ranked or sorted according to their profitability such as that $a_1 / d_1 > a_2 / d_2 > \ldots > a_n / d_n$. If type j is included in the AFRRA's "resource allocation pool" (those types that the AFRRA will process, once encountered), then all types with profitability greater than that of type j will be included in this pool as well. After ranking the resource types by profitability, types are included in the pool iteratively, starting with the most profitable type (i.e., when $i = 1$) until the following condition is attained:

$$\frac{\sum_{i=1}^{j} \lambda_i a_i}{\sum_{i=1}^{j} \lambda_i d_i} > \frac{a_{j+1}}{d_{j+1}}. \tag{2.10}$$

The highest j that satisfies Equation 2.10 is the least profitable resource type that is included in the pool. That is, if resource types in the environment are ranked according to profitability with $i = 1$ being the most profitable, and if type $j + 1$ is the least profitable type such that the AFRRA will benefit more from searching for and processing types with profitability higher than the profitability of $j + 1$, then resource types 1 through j should be processed when encountered and all other resources should not. If Equation 2.10 does not hold for any j, then all resources should be processed when encountered. The most profitable type j is substituted into Equation 2.5 as the optimal radio communication energy.

2.6 PERFORMANCE EVALUATION

To evaluate the performance of the FIREMAN method, the above algorithm was validated using the MATLAB® simulation tool. The tool has a computational capability to simulate realistic physical channel characteristics and radio link energy costs of a distributed small number of nodes. To assess the impact of the FIREMAN on the network topology, 10 stationary wireless multi-radio nodes with a maximum transmission range of 500 m were uniformly placed in a 1000 m × 1000 m area. Each node had up to four radio interfaces with one interface acting as a default FIC for exchanging control messages and others operating as the NFICs for exchanging the application traffic. Following the proposed FIREMAN algorithm and AFRRA protocol, the interfaces were each tuned to non-overlapping UCGs of frequency spectrum available between 2.412 GHz and 2.484 GHz [16]. The orthogonal channel numbers 1, 6, 11, and 14 of channel widths of 20 MHz each in the IEEE 802.11 b/g were considered [16]. Depending on the phases within a beacon time interval, certain radio interfaces were set to either doze or active states. Application packets arrived at each MAC and PHY layers' queue following a Poisson process [14]. In each arrival, the sender node sent an AFRRAM to the intended receiver during the AFRRAM window. For each arriving packet, time was divided into identical beacon intervals of typically 1 s. At the start of each beacon interval, all nodes stayed awake via their FICs for the duration of an AFRRAM window. During the AFRRAM window, the FIC executed traffic indication, the FIREMAN algorithm, and the channel negotiation mechanism.

The radio communication energy in transmit and receive modes was evaluated from the FIREMAN algorithm. The radio communication energy in a link in the transmit mode was considered as the energy link cost. A link was said to be in the transmit state if the sending interface was transmitting packets (control or data) to a receiving interface connected to it on the same physical link. That is, the sum of the radio transmit, receive, and the device-pair electronics' energy constitutes the transmit energy consumption of a link. The energy per link in the receive mode was the sum of the receive and device-pair electronic energy. A link was said to be in the receive, idle, or doze state, respectively, if any two devices were receiving or idle listening or dozing with respect to the neighbor transmissions in the direction of the same virtual link. The performance evaluation concerning the energy consumption after executing the FIREMAN algorithm was performed for a duration sufficiently long for the output statistics to stabilize (i.e., 60 s). Each datum point in the plots was the result of averaging four data points from four simulation runs, whereby each run represented a different randomly generated network topology of the same number of the nodes. The rest of the system performance was generated from parameters specified in Table 2.1.

In Figure 2.3, the average energy types and link costs distributions at the four radio interface zones are shown with a 95% level of confidence. As the encountered link cost increases, the energy types drop from some high values and become constant thereafter. Conversely, the increase in types causes the link cost to show an inverse response with the link costs. This is because, given the available transmission energy settings of a commodity WiFi device, the exponential type distribution function provides an inverse relation with the energy settings, for example, at the third zone, 0 mJ (millijoules) (type 11) to 100 mJ (type 1) of the multi-radio IEEE 802.11b/g. Type 11 signifies the least energy cost consumed by the link, while Type 1 shows the highest energy cost consumed by the same link. The exponential type distribution function was chosen because of its ability to define a large number of energy types for energy link costs with small order of magnitudes (i.e., millijoules) as compatible with the most WLAN commercial devices. The exploitation of a large number of types gives the forager a set of alternative choices for making more accurate decisions in the foraging-inspired resource optimizations [7].

In Figure 2.4, the average performance of the FIREMAN method with a 95% confidence level, when applied to the radio communication energy allocation in multi-radio network, is depicted. Figure 2.4a illustrates the effect of the radio communication energy on the EAT performance. The EAT performance mimics the foraging profitability function, where the biological forager increases

TABLE 2.1
Performance Evaluation Parameters

Parameter	Definition and Description	Specification
Transmission rate	Basic interface rate for both the AFRRAM and DATA exchanges	2 Mbps
Payload length	Fixed, 456 of DATA, 16 of UDP, 40 of IP	512 bytes
Buffer length	Fixed	50 bytes
Beacon interval	Fixed, AFRRAM window $(T_1 + T_2)$, DATA (T_3) window	T_1, max = 1 ms
	T_1: The channel negotiation window	T_2, max = 330 ms, depending on
	T_2: The AFRRAM exchange window	the AFRRAM traffic in the
	T_3: The payload data exchange window	medium
	T_4: The doze mode window. Randomly chosen if the start of	T_3 = Variable, depending on the
	the next beacon or the traffic load delays by over 10 ms	data traffic in the medium
	(approximately 1% of the beacon interval)	$T_1 + T_2 + T_3 = 1000$ ms
Adjustment factor	Adjustable AFRRAM window	Varied from 1.2 to 1.5
Electronics (Tx and Rx)	Transmission and reception electronic energy consumption	50 μJ/bit
No. of active links	Active links per node	Varied from 2 to 4
	Active links per network of 10 nodes	Varied from 20 to 40
No. of channels	Non-interfering channels in the network (2412, 2437, 2462, 2484 MHz)	Varied from 1 to 4
No. of interfaces per node	Total number of interfaces per node is at most the sum of incoming and outgoing active links per node	Varied from 2 to 4
Traffic load or the UDP test traffic	Load injected to each link	Varied from 1 to 10 packets/s/link Constant bit rate (CBR)
MAC overhead	24 bytes of PLCP header (which is transmitted at 1 Mbps) + 20 byte MAC frame header	48 bytes
DIFS	Distributed inter-frame space	50 μs
SIFS	Short inter-frame space	10 μs
Back-off slot time	Time taken in low transmission energy state when a collision is detected	20 μs

its nutrient value (in kilocalories) by spending its time searching for certain prey or nutrient types that can provide high nutrient contents. As the radio communication energy increases, the EAT drops linearly, rapidly, owing to the increase in the energy cost of communicating packets in the network. The NFIC zones have higher profitability than the FIC zone as the energy cost increases, because the NFIC zones perform overhead free, data exchanges with the controlled radio communication energy, while the FIC zone exchanges overhead control messages. Specially, at 10 mJ, the $NFIC_1$ zone provides 70% more throughput profitability than the FIC zone, on the average.

Figure 2.4b portrays the effect of the radio communication energy on the FEE performance. The FEE performance mimics the foraging loss function, where the biological forager decreases or wastes kilocalories by spending its time searching for certain prey or nutrient types that can only provide low nutrient contents. As the radio communication energy increases, the FEE charged increases linearly, rapidly, owing to the increase in the cost of communicating packets in the network. The NFIC zones are more energy efficient than the FIC zone, because the NFIC zones not only use the controlled energy levels but also stay awake only on demand (when there are application packets destined to a certain receiver). Otherwise, all NFICs stay in the doze mode throughout the beacon interval. In contrast, the FIC zone stays awake to coordinate the exchange of control packets between the AFRRA pairs and only stays in the doze mode for short intervals when application

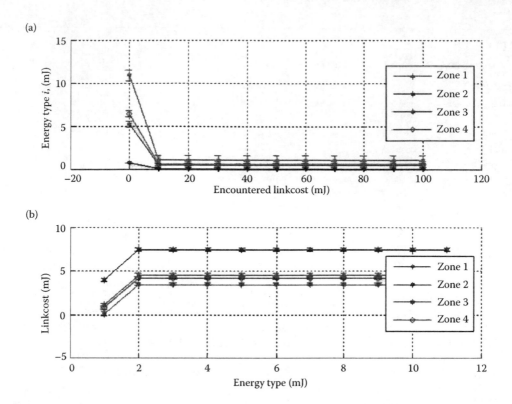

FIGURE 2.3   Average energy link cost types for multi-interface zones: (a) energy types and (b) energy link costs.

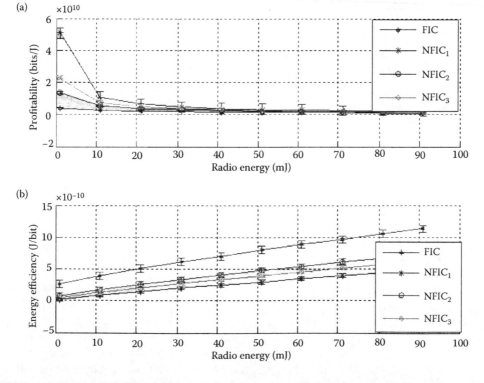

FIGURE 2.4   (a) Average EAT and (b) average foraging energy efficiency versus radio energy cost.

data are being exchanged. Specifically, at 60 mJ, the NFIC consumes 67% less energy than the FIC zone, on the average.

Figure 2.5 illustrates the impact of the traffic load offered to each link on the EAT performance and on the corresponding FEE performance. Figure 2.5a suggests that more traffic loading onto the link leads to a better EAT performance per link. The performance results agree with the theory that the offered load per link is directly proportional to the throughput in a lightly loaded network. The FIREMAN method was compared with the power-saving multi-channel medium access control (PSM-MMAC) protocol suggested in Ref. [5] and the singularly perturbed weakly coupled-based power selection multi-radio unification protocol (SPWC-PMMUP) proposed in Ref. [14]. It has been found that the FIREMAN method for a three-radio interface link outperforms the SPWC-PMMUP and PSM-MMAC methods tested under a similar number of radio interfaces, on the average. Specifically, at 10 packets, on the average, the FIREMAN method records 20% and 60% more EAT performance than those of the SPWC-PMMUP and PSM-MMAC methods, respectively. The findings are attributed to the reason that the FIREMAN algorithm is capable of making optimal decisions in a random wireless environment. It forces the FIC to exchange the control messages, while the NFICs exchange the application data packets on separate radio links and non-overlapping channels. In contrast, the PSM-MMAC protocol executes the RTS/CTS handshake at a full radiated energy when attempting to reduce the hidden terminal problems, at the expense of the increased message overheads. The SPWC-PMMUP method imposes some computational complexity when evaluating the queue perturbation and weak coupling coefficients. Increased computational time intervals leave less time available for the exchange of application data. Both the PSM-MMAC and SPWC-PMMUP methods assume static channel assignment irrespective of the channel qualities. Instead, the FIREMAN method has a quasi-static channel assignment whereby channels are

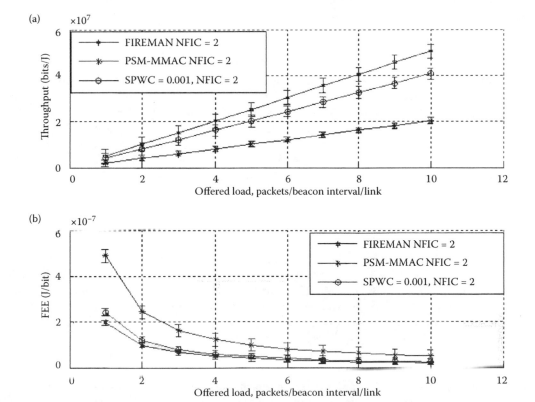

FIGURE 2.5 (a) EAT and (b) FEE versus offered traffic load per beacon interval per link.

assigned in every beacon interval but which dynamically changes with respect to the energy link costs (the link with the least energy cost is assigned a channel).

In Figure 2.5b, a corresponding average FEE performance is shown whereby the FIREMAN method indicates the best FEE compared to the other conventional methods. Specifically, at two packets, on the average, the FIREMAN method has 35% and 42% better FEE performance than the SPWC-PMMUP and PSM-MMAC methods, respectively. The reason is that the FIREMAN method forces the control messages' intervals to be as short as possible to allow longer intervals for the application data exchanges and to reduce the idle time of the FICs. All the NFICs are switched off until the energy is allocated and the channel is negotiated to save significant amounts of energy. Nodes stay awake only on demand; otherwise, they are switched off in the network. The FIREMAN method also ensures that soon after the data exchange and the current beacon interval have expired, all the radio interfaces are switched to doze mode until the next application packet arrives.

2.7   FUTURE RESEARCH DIRECTIONS

Future research involves the extension of the FIREMAN method to the implementation of joint dynamic energy and channel assignment in the wireless broadband networks. In such a design, the modified link layer firmware will be expected to execute the cross-layer energy management for lifetime maximization, while ensuring that the available frequency channels are dynamically assigned to the active communication links. These anticipated research studies will also involve the investigation of the energy and spectrum acquisition from a network environment by utilizing foraging search techniques. The central notion will be to have both resource acquisition and optimal management integrated into a single foraging life cycle, closely resembling the case of biological foragers who search for nutrients in a random environment. When the said biological foragers encounter nutrients, they decide whether to consume or ignore these on the basis of the perceived energy gain and lifetime maximization.

2.8   CONCLUSION

The chapter has proposed a FIREMAN protocol for green multi-radio networks. The protocol and algorithm designs of the FIREMAN, as motivated by a random wireless environment, have been presented here. The computer simulations have been used to validate the designs and have shown better throughput and energy efficiency performances than the conventional energy management methods. This result holds promise for the implementation of green heterogeneous wireless networks as discussed in Section 2.7. The future work of this study involves the software development of the prototype for real-life performance tests. The test findings will be used to scale the application of the FIREMAN to large heterogeneous wireless networks.

AUTHOR BIOGRAPHIES

Thomas Otieno Olwal obtained his PhD in computer science from the University of Paris-Est (Creteil, France) and a D Tech in electrical engineering from the University of Technology (South Africa) within a cotutelle agreement in 2010 and 2011, respectively. He is currently a full-time senior scientific researcher at the Department of Wireless Computing and Networking in the Council for Scientific and Industrial Research. He is also a part-time lecturer and external examiner of postgraduates at various South African universities. He is a member of the IEEE Communication and Computing Societies. His research and development interests include radio resource allocation using behavioral ecology methods (applied foraging theory) and free space optics communication R&D for developing regions. He has authored and coauthored four edited books concerning energy-efficient networks as well as more than 40 internationally recognized journal and conference papers.

Barend J. Van Wyk received a national higher diploma in tertiary education from Technikon Witwatersrand, a national higher diploma, a bachelor of technology, and a master of technology, all in electrical engineering, from Technikon Pretoria, a bachelor of commerce from the University of South Africa, a master of science from the University of Southern Mississippi, and a PhD in electrical and information engineering from the University of the Witwatersrand. After gaining extensive experience in the telecommunications and aerospace industries, he joined the Tshwane University of Technology as an associate professor in 2004. He is currently a professor associated with the French South African Technical Institute at TUT and the dean of the faculty of engineering and the built environment. He published more than 100 peer-reviewed conference and journal papers and his research interests include signal and image processing, pattern recognition, machine intelligence, and control.

Okuthe P. Kogeda was born in Kisumu, Kenya. He obtained a doctorate degree in computer science from the University of the Western Cape in Cape Town, South Africa, in 2009. He obtained his master of computer application from Dr. Babasaheb Ambedkar Marathwada University in Aurangabad, India, in 1999. His main areas of interest include biologically inspired modeling and simulation, Bayesian networks, cellular networks and OSS, mobile intelligent agents, wireless forensics, and data mining. He is currently a senior lecturer, chair of departmental research and innovation committee, and head of postgraduate section in the Computer Science Department at Tshwane University of Technology, South Africa. He was senior lecturer in the Computer Science Department at University of Fort Hare in Eastern Cape, South Africa, from 2009 to 2011. He was a lecturer in the Computer Science Department at the University of the Western Cape in Cape Town, South Africa, from 2004 to 2009. He was a lecturer at the University of Nairobi in Nairobi, Kenya, from 1999 to 2000. He has published over 20 internationally refereed conference and journal papers, was the author of a book chapter in *Cellular Networks*, and was the author of an edited book, *Modelling of Reliable Service Based Operations Support System*.

Fisseha Mekuria has PhD degrees in wireless communications and signal processing from Linköping University (LiTH.se) in Sweden. He has over 20 years of academic and industrial research experience in ICT and wireless communications. He was a senior researcher at Ericsson Mobile Platforms R&D Lab in Sweden for 10 years, during which time, he has developed over 12 US and European (EPO) patents in the areas of wireless computing platforms and communications networks. Dr. Mekuria played a crucial role in commercialization of research outputs as mentor in the IDEON Science Park in Sweden and has been the recipient of several research grants and IEEE nominations from the mobile and ICT industry for research and academic education in the mobile computing field. He is a visiting professor in wireless computing and ICT systems at several African universities and has been involved in research, teaching, development of curriculum, and organization of research groups and has been taking a leadership role in research and educational capacity building in Europe and several African countries. He played a leadership role in the building of mobile networks applications software development laboratories and mobile networking laboratories in resource-limited African countries. For this work, he has been nominated for the prestigious IEEE continuing education award. Several patents, IEEE/ACM publications, and book chapters in wireless communications has been the result of his work. He is the recipient of the Google Research Award 2009/2010 in Mobile Computing and a US-NSF research grant in collaboration with the University of California, Santa Barbara. His research interests are on future internetworking, next-generation wireless computing and green network technology systems, modeling and analysis of secure and cognitive wireless platforms and services, opportunistic TVWS broadband networks for health and education applications, and ICT for sustainable socioeconomic development. He is at present the principal researcher and research leader in wireless computing and networking at the Council for Scientific and Industrial Research, CSIR Meraka Advanced ICT Institute in Pretoria, South Africa, and holds an adjunct professor position at the University of Johannesburg, Faculty of Electrical and Electronics Engineering Science, South Africa.

REFERENCES

1. Olwal, T. O. Decentralised dynamic power control for wireless backbone mesh networks. PhD Thesis, University of Paris-EST Creteil (France) in consultation with the Tshwane University of Technology (South Africa), May 2010.
2. Comaniciu, C., Mandayam, N. B. and Poor, H. V. Radio resource management for green wireless networks. In *Proceedings of the IEEE Vehicular Technology Conference Fall*, pp. 1–5, Santa Cruz, California, USA, May 2009.
3. Bianzino, A. P., Chaudet, C., Rossi, D. and Rougier, J. L. A survey of green networking research. *IEEE Communications Surveys and Tutorials*, 14(1): 3–20, February 2012.
4. Tompros, S., Mouratidis, N., Hrasnica, H. and Gravras, A. A pervasive network architecture featuring intelligent energy management of households. In *Proceedings of the 1st ACM International Conference on Pervasive Technologies Related to Assistive Environments (PETRA'08)*, Athens, Greece, July 15–19.
5. Wang, J., Fang, Y. and Wu, D. A power saving multi-radio channel MAC protocol for wireless local area networks. In *Proceedings of the IEEE International Conference on Computer Communications (INFOCOM 2006)*, pp. 1–13, Barcelona, Spain, April 23–29.
6. Pavlic, T. P. and Passino, K. M. Generalizing foraging theory for analysis and design. *International Journal of Robotics Research*, 30(5): 505–523, April 2011.
7. Quijano, N. and Passino, K. M. Foraging theory for multi-zone temperature control. *IEEE Computational Intelligence Journals and Magazine*, 1(4): 18–27, November 2006.
8. Olwal, T. O., van Wyk, B. J., Djouani, K. Hamam, Y., Siarry, P. and Ntlatlapa, N. Autonomous transmission power adaptation for multi-radio multi-channel wireless mesh networks. In *Proceedings of the 8th International Conference on Ad Hoc, Mobile and Wireless Networks (ADHOC-NOW 2009)*, pp. 284–297, Springer-Verlag, Heidelberg.
9. Gomez, K., Riggio, R., Rasheed, T. and Granelli, F. Analysing the energy consumption behaviour of Wi-Fi networks. In *Proceedings of the Conference on Green Communications (GreenCom'2011)*, pp. 98–104, New York, USA, September 26–29, 2011.
10. Carvalho, M. M., Margi, C. B., Obraczka, K. and Garcia-Luna-Aceves, J. J. Modelling energy consumption in single-hop IEEE 802.11 ad hoc networks. In *Proceedings of the 13th International Conference on Computer Communications and Networks*, pp. 367–372, Chicago, IL, USA, October 11–13, 2004.
11. Anastasi, G., Conti, M., Gregori, E. and Passarella, A. Saving energy in Wi-Fi hotspots through 802.11 psm: Analytical model. *Journal of Linguistic Theory and Grammar Implementations*, 2(2): 24–26, June 2004.
12. Moshe, T. M., Olwal, T. O. and Ntlatlapa, N. An energy saving scheme for internet provision in rural Africa: LESS. In *Proceedings of the IST-Africa Conference*, Durban, South Africa, May 21–23, 2010.
13. Kumruzzaman, S. M. An energy-efficient multi-channel MAC protocol for cognitive radio ad hoc networks. *International Journal of Communication Networks and Information Security*, 2(2): 112–119, 2010.
14. Olwal, T. O., Djouani, K., Kogeda, O. P. and van Wyk, B. J. Joint queue-perturbed and weakly coupled power control for wireless backbone networks. *International Journal of Applied Mathematics and Computer Sciences*, 22(3): 749–764, 2012.
15. Manweiler, J. and Choudhury, R. R. Avoiding the rush hours: Wi-Fi energy management via traffic isolation, *IEEE Transactions on Mobile Computing*, 11(5): 739–752, May 2012.
16. Part 11: Wireless LAN medium access control (MAC) and physical layer (PHY) specifications, ANSI/IEEE St d802.11, 2012 edition. Available at http://ieeexplore.org/. Accessed on January 18, 2013.
17. Lymberopoulos, D., Priyantha, N. B., Goracko, M. and Zhao, F. Towards energy-efficient design of multi-radio platforms for wireless sensor networks. In *Proceedings of the ACM International Conference on Information Processing in Sensor Networks (IPSN'08)*, St. Louis, Missouri, USA, April 22–24.
18. Olwal, T. O., Djouani, K. and van Wyk, B. J. Optimal control of transmission power management in wireless backbone mesh networks. In *Wireless Mesh Networks*, Eds. Nobuo Funabiki, pp. 3–28, In Tech Open Access, January 2011, Vienna, Austria.
19. Olwal, T. O., van Wyk, B. J. and Djouani, K. A multi-radio multi-channel unification power control for wireless mesh network. *International Journal of Computer Sciences*, 5(1): 38–50, 2010.
20. Olwal, T. O., Djouani, K. and van Wyk, B. J. A multiple-state based power control for multi-radio multi-channel wireless mesh networks. *International Journal of Computer Sciences*, 4(1): 53–61, 2009.
21. Olwal, T.O., Van Wyk, B. J. and Djouani, K. Interference-aware power control for multi-radio multi-channel wireless mesh networks. In *Proceedings of the IEEE AFRICON Conference*, pp. 1–6, Nairobi, Kenya, September 23–25, 2009.

22. Olwal, T. O., van Wyk, B. J. and Djouani, K. Singularly-perturbed weakly-coupled based power control for multi-radio multi-channel wireless networks. *International Journal of Applied Mathematics and Computer Sciences*, 6(1): 4–14, 2010.

23. Stephens, D. W and Krebs, J. R. *Foraging Theory*, Princeton University Press, Princeton, NJ, 1986.

24. Pavlic, T. P. Optimal foraging theory revisited. Master's Thesis, The Ohio State University, Columbus, OH, 2007.

3 On the Design of Energy-Efficient Wireless Access Networks
A Cross-Layer Approach

*Konstantinos P. Mantzoukas, Stavros E. Sagkriotis,
and Athanasios D. Panagopoulos*

CONTENTS

3.1 INTRODUCTION TO CROSS-LAYER APPROACH FOR ENERGY-EFFICIENT WIRELESS ACCESS NETWORKS

Radio access technologies (RATs) of 3G emerging toward 4G offer high data rates for multimedia applications of mobile users, contrary to simple voice and SMS rate-limited applications of the past. The extra bandwidth exploitation, combined with the employment of high size modulation constellations and power link control to enhance spectral efficiency and mitigate interference effects, leads to extremely high levels of power consumption and consequently to high energy consumption at mobile users' devices [1]. In contrast, the improvement in battery technology is much slower, increasing by a modest 10% every 2 years, leading to an exponentially increasing gap between the demand for energy and the battery energy offered. The prospects are even worse in case of novel shrinking size mobile devices that possess batteries with limited energy resources. Therefore, battery life becomes a critical resource and certain aspects should be revisited in designing energy-efficient network protocols for the PHY layer up to the applications layer of network protocol stack.

Following the traditional layer-wise network analysis, protocols should be redesigned in each layer to accommodate energy efficiency, posing certain restrictions arising from efficient utilization of system bandwidth, till quality of services (QoS) constraints of users' applications. In Refs. [2–7], various aspects of energy efficiency have been investigated in different layers of protocol stack. Moreover, various cross-layer frameworks have been proposed that utilize the interaction among parameters that lie in different layers of network protocol stack (e.g., Refs. [8–14]). In Ref. [15], cross-layer analysis is extended to accommodate the interaction of key elements from medium access control (MAC) and PHY that affect mostly energy consumption.

The core functionality of the PHY layer for wireless access networks is link adaptation that includes instantaneous adaptation of transmission modes and power consumption to radio conditions of the receiver that vary in time of the order of some milliseconds as a result of fading, especially multipath fading. In the past, spectral efficiency was the dominant figure of merit concerning performance analysis of link adaptation procedures. Energy consumption was a feature if not totally neglected, certainly of secondary or even lower concern. In Refs. [16,17], link adaptation and especially adaptive modulation and coding (AMC) combined with the orthogonal frequency division multiple access (OFDMA) diversity technique is considered under energy efficiency maximization. The key element of redesigning AMC in order to increase energy efficiency could be summarized in the following: instantaneous transmission power should range within low values, leading to longer transmission intervals per bit of information, which in turn incurs less energy consumption. In other words, lower size constellations of modulation should be favored. Within this framework, the analysis is extended to tackle channel-selective fading [18] and to establish an efficient trade-off between energy efficiency and spectral efficiency [19].

The basic functionality of MAC is to share instantaneously radio resources among different users with diverse bandwidth requirements, following certain criteria for QoS such as throughput or delay and at the same time ensure high total performance. Apart from contention-based schemes that are associated mainly with wireless local area networks such as 802.11, the basic unit of MAC for HSDPA (high-speed downlink packet access), LTE (long-term evolution), and wireless metropolitan area networks such as WiMAX (802.16) is fast opportunistic channel-aware scheduling [20]. "Greedy" schedulers (such as MSNR [maximum signal-to-noise ratio], which bases its selection on the instantaneous SNR) not only maximize system throughput but also lead to starvation of users being distant from the base station (BS). A trade-off between high system throughput and fairness can be achieved by introducing appropriate balancing weights in the selection criterion [21]. One of the most well-known schedulers of this kind is proportional fair (PF) [22,23]. Moreover, energy-efficient schedulers such as the Lazy scheduler in Ref. [24] increase and equate users' transmission duration, decreasing energy consumption.

The remainder of this chapter is organized as follows. In Section 3.2, physical layer aspects of energy efficiency are presented in terms of link adaptation. AMC is combined with the diversity technique of OFDMA to provide an energy-efficient AMC rate function. The impact of multipath fading models such as Rayleigh and Nakagami-m to mean link energy efficiency is investigated. In Section 3.3, MAC layer aspects of energy efficiency are demonstrated in terms of PF scheduler. In Section 3.4 we present future research area directions in the field of cross-layer approach, and finally we conclude the book chapter.

3.2 PHYSICAL LAYER ASPECTS OF ENERGY EFFICIENCY

3.2.1 Single-User Performance

To capture the physical layer aspects of energy efficiency, we begin with link adaptation, the core functionality of the PHY layer. Link adaptation and especially AMC adjust various transmission resources to different intervals of instantaneous SNR at the receiver according to a target bit error rate BER_0. Apart from simple AMC that is based solely on modulation and channel coding, various multiple access techniques such as OFDMA combined with simple AMC give rise to diversity gains that can be further exploited in terms of energy efficiency. In this section, we investigate this diversity gain in terms of energy efficiency induced to the performance of a single-user uplink transmission that is served alone in the system. In the next section, we extend the analysis to multiple-users system transmissions, including the impact of schedulers.

3.2.1.1 Model Description for AMC

We consider a noise-limited cell and focus on uplink transmission. The time axis is divided into PHY frames of system-specific constant duration. We assume uncorrelated channels subject to quasi-static flat fading. The instantaneous SNR x_i at the receiver of BS induced by the transmission of user i is kept constant for the whole duration of a single frame, at a random value governed by the characteristics of the multipath fading. The value of average power $\bar{P}_{R_i}$ received at the BS owing to transmitter i depends on the user's distance d_i from the BS and is calculated through a well-accepted deterministic large-scale path loss attenuation model [25]; that is, $\bar{P}_{R_i} = \bar{P}_{T_i} A \cdot d_i^{-\gamma}$, where $\bar{P}_{T_i}$ is the average power transmitted by user i. Parameter A depends on the frequency of the carrier signal; that is, $A = 0.0015$ for 2 GHz and $\gamma = 3.8$ is the path attenuation exponent for NLOS attenuation in an urban area. Similarly, we define the average power channel gain $G_i = \bar{P}_{R_i}/\bar{P}_{T_i} = A \cdot d_i^{-\gamma}$. Users are associated with a number $L > 1$ of classes corresponding to distinct average power channel gains $\{G_k\}_{k=1}^L$ to accommodate the variety of radio conditions of different users on their uplink transmissions.

Apart from the difference in the average power channel gains, the same small-scale fading characteristics are assumed throughout all classes. Similar to the analysis in Ref. [21], we define the "normalized instantaneous power" values $y_i = P_{R_i}/\bar{P}_{R_i} = x_i/\bar{x}_i$ that are assumed independent and identically distributed (iid) with CDF $\hat{F}(y)$. Given $\hat{F}(\cdot)$, the CDF of the (un-normalized) power for a user within class $j = 1, \ldots, L$ can be written as $F_j(P_R) = \hat{F}\left(P_{R_i}/\bar{P}_{R_j}\right)$. The common CDF $\hat{F}(\cdot)$ allows the representation of heterogeneous radio conditions in concise terms [21] and is compatible with all fading models usually employed, including Rayleigh, Nakagami-m, and Ricean. For Rayleigh, $\hat{F}(y) = 1 - e^{-y}$, while for Nakagami-m, the CDF has the form $\hat{F}(y) = \Gamma(m)^{-1} \int_0^{my} t^{m-1} e^{-t}\, dt$ (depending on the common shape parameter m, but not on the average power channel gains $\{G_k\}_{k=1}^L$). Ricean fading can be handled similarly in view of their normalized nature.

The effect of AMC is captured by partitioning the range of instantaneous SNR values into K consecutive nonoverlapping intervals with successive boundary points $\{s\}_{k=1}^{K+1}$, where $s_1 = 0$ and $s_{K+1} = \infty$. Each such interval is associated with a different bitrate $\hat{r}_k$. The partition results from imposing the same target bit error rate (BER$_0$) threshold over all modes. Consequently, the rate function corresponding to AMC takes the piecewise-constant form

$$h(x) = \sum_{k=1}^K \hat{r}_k [u(x - s_k) - u(x - s_{k+1})], \tag{3.1}$$

where $u(\cdot)$ is the unit-step function. Note here that the AMC rate function should always be an increasing function of instantaneous SNR. The mean link rate of a class-i user is

$$\bar{r}_i = E[h(x_i)] = E[h(x_i y_i)] = \int_0^\infty h(\bar{x}_i y)\, d\hat{F}(y). \tag{3.2}$$

This is the highest possible throughput that the user can attain, when he is served alone in the system, subject to the radio conditions governing his class. Note that the average SNR $\bar{x}_i$ at the receiver of BS depends on the average transmitted power $\bar{P}_{T_i} = \bar{P}_{R_i}/G_i$.

3.2.1.2 Energy-Efficient Link Adaptation

We define the performance measure of energy efficiency $U_i(r_i)$ for an uplink transmission of a class-i user as the ratio of the bits transmitted by the class-i user to the total energy consumed at the transmitter of the class-i user within a frame [16]

$$U_i(r_i) \triangleq r_i / \left(P_{T_i}(r_i) + P_C \right) = r_i / \left(P_{R_i}(r_i) / G_i + P_C \right), \tag{3.3}$$

where P_C is constant power associated with the circuit of the transmitter and is independent of rate. Work in Ref. [17] provides a rigorous treatment of power associated with circuit energy consumption. In general, $P_{T_i}(r_i)$ and, consequently, $P_{R_i}(r_i)$ are increasing functions of instantaneous rate r_i. The higher the value that $U_i(r_i)$ possesses, the higher the energy efficiency of the uplink transmission of user i becomes. The instantaneous rate $\hat{r}_k$ for each of the K intervals of AMC SNR partition depends on the multiple access technique applied. For the OFDMA technique,

$$\hat{r}_k = \frac{c_k \cdot cr_k \cdot l_{up} \cdot b_k \cdot W}{l_{up} + l_{dn}}, \tag{3.4}$$

where c_k is the number of parallel subchannels transmitted within the time duration of a frame associated with k transmission mode, l_{up} is the number of OFDMA symbols for uplink transmission, and l_{dn} is the number of OFDMA symbols for downlink transmission, considering TDD, which is the usual case in practice. Each subchannel has a bandwidth equal to W. The parameter cr_k is the code rate that is associated with the respective channel coding scheme used and b_k is the number of bits per symbol of the particular constellation size of modulation used. As in Ref. [26], we assume a Nyquist pulse-shaping filter with $T_s = c_k/W$, where T_s is the time duration of a symbol. We want to find values of $\hat{r}_k$ that maximize $U_i(\hat{r}_k)$ for each transmission mode. Thus, for a given transmission mode k, we want to find the optimal instantaneous rate $\hat{r}_k^*$ that maximizes $U(\hat{r}_k)$

$$\hat{r}_k^* = \arg\max_{\hat{r}_k} U(\hat{r}_k) = \arg\max_{\hat{r}_k} \hat{r}_k / \left(P_{R_i}(\hat{r}_k) / G_i + P_C \right). \tag{3.5}$$

If $P_{R_i}(r_i)$ is strictly concave and monotonically increasing, it can be easily proved, taking the derivative of $U_i(r_i)$ and setting it equal to 0, that the solution to Equation 3.5 is

$$\hat{r}_k^* = \frac{P_{R_i}\left(\hat{r}_k^*\right) / G_i + P_C}{P'_{R_i}\left(\hat{r}_k^*\right) / G_i}. \tag{3.6}$$

For contemporary systems that use LDPC or Turbo channel codes, we can approximate $P_{R_i}(\hat{r}_k) = P_{R_i}(r_i) = \left(2^{r_i/(aW)} - 1 \right) N_0 W$. The approximation is obtained by multiplying a constant $a(0 < a < 1)$ to the Shannon-type formula for capacity over the AWGN channel with channel bandwidth W, namely, $r_i = aW \log_2 \left(1 + P_{R_i}(r_i) / N_0 W \right)$. We can find a similar approach in Refs. [24,27]. The approximation is highly accurate for low till moderate SNR values. Equation 3.6 is a fixed-point equation. It can be easily seen that the right part of Equation 3.6 is always a monotonically decreasing function of r_i, by taking the derivative of it.

To exploit the diversity of the OFDMA technique, we find the optimal values c_k^* that are associated with the optimal value $\hat{r}_i^*$ derived by Equation 3.6 for different b_k. The optimal c_k^* is from Equation 3.4,

$$c_k^* = \frac{\hat{r}_k^*}{b^* \cdot cr_k \cdot l \cdot W} \frac{l_{up} + l_{dn}}{l_{up}}. \tag{3.7}$$

```
For k:=2:1:K
begin
   ⎧ while r̂*ₖ > r̂*ₖ₋₁
   ⎪ begin
   ⎨    ⎧ cₖ = cₖ + 1
   ⎪    ⎩ compute new r̂ₖ of near-optimal h**(x)
   ⎪ end
   ⎩
end
```

FIGURE 3.1   Algorithm for near-optimal AMC rate function $h**(x)$.

With Equations 3.4, 3.6, and 3.7 in mind, we can create AMC rate functions $h*(x)$ that are energy efficient for the OFDMA multiple access technique. In general, optimal values of $\hat{r}_k^*$ derived from Equation 3.6 do not follow an increasing order. To ensure that the deduced $h*(x)$ is an increasing function of instantaneous SNR, we slightly modify the number of subcarriers c_k for the transmission modes that attribute lower rates with higher SNR, applying the standard algorithmic procedure in Figure 3.1 that provides near-optimal energy-efficient AMC rate function.

If the while case is fulfilled, the new $\hat{r}_k$ is computed simply with the help of Equation 3.4.

3.2.1.3 Performance Analysis of Energy-Efficient Link Adaptation

To assess the trade-off between energy efficiency and spectral efficiency induced by near-optimal AMC rate function $h**(x)$, we obtain a spectral efficient AMC rate function $h(x)$ to compare both of them in terms of mean link rate (Equation 3.2) and energy efficiency (Equation 3.3). A similar approach is presented in Ref. [19]. In our case, a spectral efficient AMC rate function $h(x)$ is created by using c_1^*, associated with optimal $h*(x)$ as the same for all other transmission modes; that is, $\forall k$, $c_k = c_1^*$.

We choose $L = 2$ classes corresponding to power channel gains $G_1 = 1.33e - 14$ (equivalent distance from BS $d_1 = 810m$) and $G_2 = 2.5e - 14$ (equivalent distance from BS $d_2 = 685m$). From now on, we use indices 1 and 2 to refer to near-optimal energy-efficient functions of classes 1 and 2, that is, $h_1^{**}(x), h_2^{**}(x)$. Note here that P_{R_i} and consequently $P_{T_i} = P_{R_i}/G_i$ are different for $h_i^{**}(x)$ and $h_i(x)$ within the same class i. To compute the average $\bar{P}_{T_i}$ and average SNR $\bar{x}_i$, we need to take the average on $P_{R_i}(\hat{r}_k)$ values in terms of the standard values of s_k, where $P_{R_i}(\hat{r}_k) = s_k c_k N_0 W$. Moreover, to acquire the instantaneous values of P_{T_i} and instantaneous values of x_i, we use the Shannon-type formula approximation like this: $P_{T_i} = \left(2^{\hat{r}_k/(aW)} - 1\right) N_0 W / G_i$ and $x_i = \left(2^{\hat{r}_k/(aW)} - 1\right)$. Thus,

$$\bar{P}_{T_i} = \frac{1}{G_i} \sum_{k=1}^{K} \left(2^{\hat{r}_k/(aW)} - 1\right) N_0 W \left(\hat{F}\left(s_{k+1} c_{k+1} N_0 W / G_i\right) - \hat{F}\left(s_k c_k N_0 W / G_i\right)\right) \tag{3.8}$$

and

$$\bar{x}_i = \sum_{k=1}^{K} \left(2^{\hat{r}_k/(aW)} - 1\right) \left(\hat{F}\left(s_{k+1} c_k N_0 W / G_i\right) - \hat{F}\left(s_k c_k N_0 W / G_i\right)\right). \tag{3.9}$$

For $h_1^{**}(x)$, the average SNR becomes $\bar{x}_1 = 2.74$, and for $h_1(x)$, $\bar{x}_1 = 3.93$ for the Rayleigh Fading model assumed. For $h_2^{**}(x)$, the average SNR becomes $\bar{x}_2 = 6.92$, and for $h_1(x)$, $\bar{x}_2 = 12.49$ again for Rayleigh Fading.

TABLE 3.1
System Parameters

Carrier frequency	2 GHz
Subchannel bandwidth	200 kHz
Symbols number for uplink within a frame, l_{up}	50
Symbols number for downlink within a frame, l_{dn}	100
Time duration of symbol, T_s	0.001 s
Thermal noise power, N_0	−174 dBm/Hz
Circuit power, P_c	100 mW
Propagation model	Propagation model for NLOS in urban area
Fading	Flat fading

The parameters used in the performance analysis are presented in Table 3.1. We used code rates and modulation constellation sizes presented in Ref. [26] for LDPC channel codes. For the $\{s\}_{k=1}^{K+1}$ boundaries, we used the formula derived from the Shannon approximation for $P_{R_i}(\hat{r}_k)$ like this:

$$s_k = \left(2^{3\hat{r}_k/(ac_kW)} - 1\right), \text{ where } \frac{l_{up}+l_{dn}}{l_{up}} = 3.$$

In Table 3.2, we compare the $\{s\}_{k=1}^{K+1}$ boundary points derived this way to the respective boundary points presented in Ref. [26]. We see that the differences are small, except for the case of the highest instantaneous SNR values that are associated with the 16-QAM. However, because of unimportant variation in distance induced by even large deviation in high SNR values, our approximation is still good in the scope of this presentation, justified by clarity and tractability.

In Figure 3.2, both optimal $h_2^*(x)$ and near-optimal energy-efficient $h_2^{**}(x)$ AMC rate functions are illustrated. The differences between them are minor, except for the highest transmission mode. In Figure 3.3, $h_2^{**}(x)$ and spectral efficient $h_2(x)$ are illustrated. In Figures 3.4 and 3.5, energy efficiency

TABLE 3.2
Transmission Mode Parameters

Modulation order (M-QAM)	4	4	16	16
Code rate	1/2	3/4	1/2	3/4
Boundary points, S_k (dB)	2.28	5.34	7.96	12.67
Boundary points, S_k (dB) for LDPC	2.93	5.92	9.68	11.42

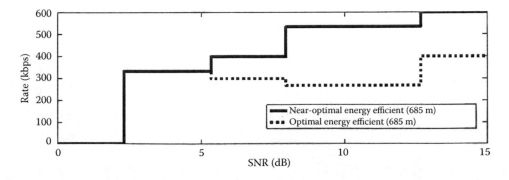

FIGURE 3.2 AMC rate functions of optimal (nonsolid line) and near optimal (solid line) for class 2.

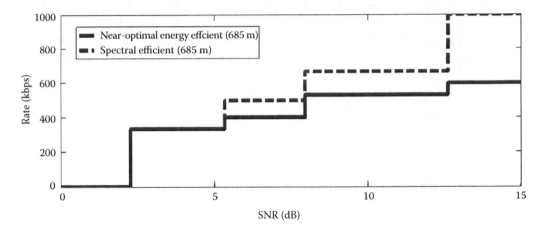

FIGURE 3.3 AMC rate functions of spectral efficient (nonsolid line) and near-optimal energy efficient (solid line) for class 2.

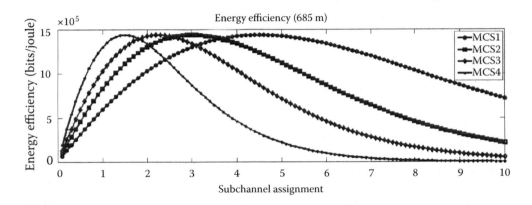

FIGURE 3.4 Energy efficiency (bits/joule) versus subchannel assignment for class 2.

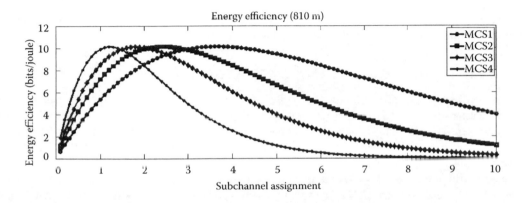

FIGURE 3.5 Energy efficiency (bits/joule) versus subchannel assignment for class 1.

TABLE 3.3
Fading Model Rayleigh ($m = 1$)

	Class 1	Class 2
Mean Link Rate (kbps)		
Near-optimal energy efficient	204.7	331.8
Spectral efficient	234.2	378.2
Mean Energy Efficiency (bits/joule)		
Near-optimal energy efficient	7.76E+5	1.03E+6
Spectral efficient	6.98E+5	7.60E+5

TABLE 3.4
Fading Model Nakagami ($m = 6$)

	Class 1	Class 2
Mean Link Rate (kbps)		
Near-optimal energy efficient	280.6	429.1
Spectral efficient	317.9	487.3
Mean Energy Efficiency (bits/joule)		
Near-optimal energy efficient	9.52E+5	1.21E+6
Spectral efficient	8.75E+5	1.04E+6

defined in Equation 3.3 is demonstrated as a function of subchannel assignment for different transmission modes within the two classes. The optimal assignments are equal to the ones obtained by the fixed-point equation in Equation 3.6. We point out that for both classes, transmission modes with lower s_k acquire more subchannels. Moreover, in class 2, the diversity induced by subchannel assignment is higher than that in class 1. Finally in Tables 3.3 and 3.4, we present the mean link rate and mean energy efficiency for the two classes for Rayleigh Fading and Nakagami-6. For class 1 and Rayleigh Fading, mean energy efficiency is increased by 11%, whereas mean link is decreased by 13%. For Nakagami-6, energy efficiency in increased by 9%, whereas mean link rate is decreased by 12%. For class 2 and Rayleigh Fading, mean energy efficiency is increased 36% and mean link rate is decreased by 12%. For Nakagami-6, mean energy efficiency is increased by 16% and mean link rate is decreased by 12%. It is clear that Rayleigh Fading provides higher mean energy efficiency for both classes especially for class 2, which is associated with lower distance to BS.

3.3   MAC LAYER ASPECTS OF ENERGY EFFICIENCY

In this section, we extend the analysis of single-user performance to a multiuser system performance with users facing heterogeneous radio conditions on their uplink transmissions. A fast scheduler selects a single user for service for each frame transmission. One of the most prominent schedulers that provide fairness in terms of long-term access rate opportunities is PF [21]. PF schedulers select users for transmission that have a maximum ratio of instantaneous rate to a mean rate attained within a time window. We compare the long-term energy consumption per packet for the near-optimal energy-efficient AMC rate function $h^{**}(x)$ and the spectral energy-efficient AMC rate function $h(x)$ for both classes of the previous section.

3.3.1 MODEL DESCRIPTION FOR SCHEDULING UNDER HETEROGENEOUS RADIO CONDITIONS

We consider N independent users. In each frame, a fast scheduler selects a single user for service. We use the vector $\vec{n} = (n_1,...,n_L)$ to denote the population of active users within each of the classes. When multiple active users are present, the achievable throughput depends on the distribution of users in classes and on the scheduling policy. Here, we consider schedulers maximizing some scaled version of the users' SNR, that is, scheduling according to $i = \arg \max_j \varphi_j x_j$, with equal φ_j for all users in the same class.

Consider the event that user i is selected for service. This is equivalent to having $y_j \leq y_i (\varphi_i \bar{x}_i)/(\varphi_j \bar{x}_j)$ for all $j \neq i$. Given the distribution of users to classes and the iid property of the normalized SNR values, the long-term throughput attributed to class-i user is [21]

$$\bar{r}_i(\vec{n}) = \int_0^\infty h(\bar{x}_i y) \prod_{\substack{1 \leq j \leq L \\ j \neq i}} \hat{F}\left(y \frac{\bar{x}_i \varphi_i}{\bar{x}_j \varphi_j} \right)^{n_j} \hat{F}(y)^{n_i-1} \, d\hat{F}(y). \tag{3.10}$$

The scheduling gain of a class-i user is defined as

$$g_i(\vec{n}) \triangleq \bar{r}_i(\vec{n})/\bar{r}_i . \tag{3.11}$$

The scheduling gain is the proportion of the highest possible throughput for a user in class i when this user is served under the considered scheduling policy as part of the population $\vec{n}$. Given Equation 3.11, the throughput of a single user in class-i and the total throughput can be expressed as $\bar{r}_i g_i(\vec{n})$ and $\sum_{l=1}^L n_l \bar{r}_l g_l(\vec{n})$, respectively.

The scheduling gain for any given class characterizes the fairness of the scheduler. If classes i and j feature $g_i(\vec{n}) < g_j(\vec{n})$, then, under a user population $\vec{n}$, the scheduler favors class j more than class i, because users of class j attain a greater fraction of their highest possible throughput. In this sense, the scheduler is perfectly fair when, for any $\vec{n}$, the scheduling gains of all classes take equal values. At the latter case, the throughput of the class-i user is proportional to his gain. The scheduling gain captures the combined effect of the scheduling policy (through φ_j), the heterogeneous population (through $\vec{n}$ and the average SNRs $\{\bar{x}_k\}_{k=1}^L$), the fading model (through $\hat{F}(\cdot)$), and rate function characteristics (through $h(\cdot)$). The impact of the AMC is accurately captured by using the precise rate function (Equation 3.1) within Equations 3.2 and 3.10.

The original PF proposal suggested weights inversely proportional to the throughput attained by the user within a given time window in the recent past. Here, we follow Refs. [28–30] and employ weights inversely proportional to the corresponding average SNRs; that is, $\varphi_i = 1/\bar{x}_i$ for all classes i. With this choice, Equation 3.10 simplifies to

$$\bar{r}_i(N) = \frac{1}{N} \sum_{k=1}^K \hat{r}_k \left(\hat{F}\left(s_{k+1}/\bar{x}_i\right)^N - \hat{F}\left(s_k/\bar{x}_i\right)^N \right) \quad \text{and} \quad \bar{r}_i = \bar{r}_i(1). \tag{3.12}$$

Moreover, in Ref. [21] a highly accurate approximation of Equation 3.11 has been derived in terms of scheduling gains for PF that simplifies to:

$$g_i(N) \approx \begin{cases} \dfrac{y_+/(1-p_+)}{N}, & \forall i, \text{ s.t. } \bar{x}_i \leq s_K/y_+, \\ 1/N, & \forall i, \text{ s.t. } \bar{x}_i \geq s_K/y_-. \end{cases} \tag{3.13}$$

where $p_+ \geq \int_{y_+}^{\infty} (1 - \hat{F}(y)) \mathrm{d}y$ (resp. $p_- \geq \hat{F}(y_-)$) are precision parameters. For relatively high values of $p_+ = 0.33$ and $p_- = 0.2$, the approximation yields $y_+ = 1.1$ and $y_- = 0.22$ for Rayleigh and $y_+ = 0.71$ and $y_- = 0.66$ for Nakagami-6 fading. Equation 3.11 facilitates the abstraction of the cross-layer interaction between MAC and PHY layers, as it captures immediately, in simple and concise terms, the effect of the fading model solely via parameters such as $y_+/(1 - p_+)$.

3.3.2 PF SCHEDULING PERFORMANCE ANALYSIS FOR NEAR-OPTIMAL ENERGY-EFFICIENT AMC RATE FUNCTION

The expression on scheduling gain (Equation 3.11) reveals the cross-layer interaction between PHY and MAC layers in terms of energy efficiency. The energy-efficient mean link rate $\bar{r}_i^{**}$ that adheres to the energy-efficient rate function $h^{**}(x)$, derived from the previous section, is associated with the energy-efficient class-i user throughput $\bar{r}_i^{*}(N)$, through a simple multiplication with constant scheduling gain $g_i(N)$.

In Figure 3.6, the PF class-i user throughput is illustrated as a function of the total number of users N, for both classes and both AMC rate functions, for the Rayleigh Fading model assumed. It is clear that the throughput of the spectral efficient AMC rate function is higher for both classes. Moreover, we consider the long-term energy consumption per average packet of 60 kB [31]. In Figure 3.7, the long-term energy consumption per packet is illustrated for both classes and both AMC rate functions. For the same number of users, energy consumption is higher in the case of

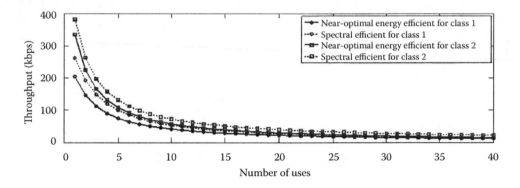

FIGURE 3.6   PF throughput for all classes and all AMC rate functions.

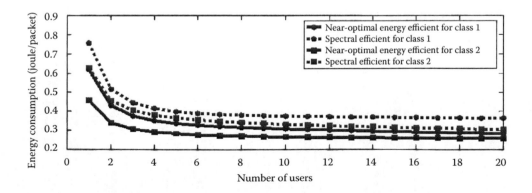

FIGURE 3.7   Long-term energy consumption as a function of N.

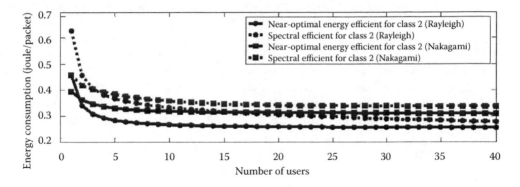

FIGURE 3.8   Long-term energy consumption as a function of N for Rayleigh and Nakagami-6 (class 2).

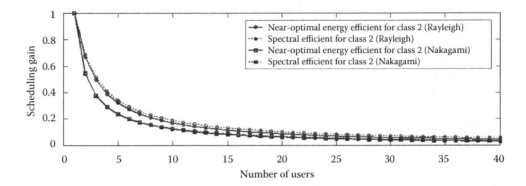

FIGURE 3.9   Scheduling gain as a function of users for Rayleigh and Nakagami-6 (class 2).

spectral efficient AMC rate functions for both classes. In Figure 3.8, long-term energy consumption is demonstrated for class-2 users for both Rayleigh and Nakagami-6. It is shown that Rayleigh outperforms Nakagami-6 in energy efficiency. The reason is that the scheduling gain in the latter case reduces to the round robin performance, that is, equal to $1/N$. However, the scheduling gain of Rayleigh is due to the approximation in Equation 3.13 equal to $1.6/N$. This phenomenon is highlighted in Figure 3.9.

3.4   FUTURE RESEARCH DIRECTIONS IN ENERGY-EFFICIENT WIRELESS ACCESS NETWORKS

This chapter has presented several aspects of cross-layer interaction between PHY and MAC layers for energy-efficient wireless access networks. It has been demonstrated that diversity techniques such as OFDMA can maximize energy efficiency in terms of link adaptation and especially AMC. The trade-off between increase in energy efficiency and loss of mean link rate is sufficient for real systems employment. It has been pointed out that multipath fading, especially Rayleigh fading that possesses PDF with high dynamic range, enhances energy efficiency contrary to the Nakagami-m fading model with a high number of parameter m, particularly $m = 6$.

Following the analytical approaches presented in the previous sections, several questions still exist, regarding the design and analysis of energy-efficient cross-layer between PHY and MAC layers. First of all, it is necessary to examine the impact of other opportunistic schedulers that may yield different scheduling gains for users with different radio conditions on their uplink, extending the analysis in Ref. [21]. The analysis could provide higher energy efficiency for over half of the cell users with no excessive performance degradation of throughput and mean delay, keeping maximum mean delay at a satisfying level.

Moreover, the impact of queuing dynamics as well as the design of a proper call admission control (CAC) in terms of energy efficiency should be investigated. Contrary to widely considered scenarios, the number of active users (i.e., users that have a packet in their buffer) is not constant with time. The number of active users is a random variable with a distribution dependent on load factor and the scheduling gain. The processor sharing model presented in Ref. [27] for channel oblivious scheduling of sharing equally a constant link, that is, round robin, could be properly modified to capture the dynamics of new schedulers such as PF, or some other scheduler more efficient in terms of energy consumption, as described in the previous paragraph. With these in mind, a proper CAC should be designed to meet the needs of throughput or delay of the various user classes without significant loss in spectral efficiency, applying robust engineering rules that depend on system specifications and not on traffic characteristics that vary generally with each different application.

Furthermore, traffic flows are in general divided into two types: elastic and streaming. Elastic traffic flows stand for applications such as WWW, FTP, and high data volume downlink traffic (e.g., video not in real time) wherein their throughput and mean delay can vary because of varying link capacity. There should be some minimum (or maximum) values for throughput (delay) to ensure QoS, but the more capacity exists in the wireless link, the more data of such flows can be delivered through it. On the contrary, streaming traffic flows such as real time video or VoIP have relatively constant rate dynamics that depend mostly on the application needs and not on the variation of the network link. Usually, streaming flows have higher mean duration than elastic users. This distinction among traffic flows produces several problems concerning an efficient CAC with robust engineering rules in terms of energy efficiency.

Apart from network dynamics in single cells, the future trend of 4G wireless networks, where different RATs are deployed covering the same area, providing different rate opportunities to multimode terminals, gives rise to vertical handover procedures of either elastic or streaming type of flows. Vertical handover at high traffic load, especially forced load balancing, can affect energy efficiency performance of partially or fully overlapped cells in many ways. For example, flows that are energy efficient in one RAT can increase severely energy consumption of mobile terminals if they are forced to be served by another RAT. This is obvious in hierarchical heterogeneous networks where mean link rates of the overlapping classes of different RATs have extremely different values.

Finally, another important and interesting issue is the investigation of more sophisticated schedulers that exploit frequency diversity in selective fading channels and can be employed with low computational cost and low signaling overhead. There are several works in prior ART that try to face this problem (e.g., Ref. [18]). However, all of them are based on favorable assumptions, for example, minimum OFDMA bandwidth allocation of the order of 10 kHz, producing a fine granularity of various selections, neglecting the impact of signal overhead, with respect to feedback channel data transmission that can have a tremendous and negative effect in total throughput.

The above examples consist only a subset of possible research directions in the field of energy-efficient cross-layer approach in wireless access networks.

3.5 CONCLUSIONS

The chapter has presented a basic overview of the key elements in the cross-layer design of energy-efficient communications in wireless access networks. Joint functionality potentials of MAC and PHY layers in order to enhance energy efficiency in wireless communication networks have been identified. The basic parameters of both layers that interact with each other, such as AMC and fading models from the PHY layers as well as the scheduling gain and throughput from the MAC layer, and that affect mostly the energy consumption were decoupled and investigated, and their joint impact on fundamental network functions was demonstrated with many examples and with an analytical performance model that facilitates abstraction in terms of the assumed fading model.

AUTHOR BIOGRAPHIES

Konstantinos P. Mantzoukas received his diploma degree in electrical and computer engineering from the National Technical University of Athens, Greece, in 2011. He specialized in telecommunications and computers networks. His diploma thesis focused on performance models of fourth-generation wireless systems. He is also an MBA candidate in techno-economic systems of the National Technical University of Athens with the participation of the University of Piraeus. He participated in the project "Benchmarking Measurements" of the Mobile Radiocommunications Laboratory at the National Technical University of Athens as telecommunications engineer.

Stavros E. Sagkriotis received his BS degree in informatics and telecommunications from the Department of Informatics and Telecommunications, University of Athens, Greece, in 2005. He received his MS degree in communication and network systems from the same department in 2008. In 2009, he received a four-year fellowship from the National Center for Scientific Research in "Demokritos" Institute of Informatics and Telecommunications to pursue his PhD in the field of performance analysis of 4G heterogeneous wireless access networks. His current scientific interests include performance analysis of wireless communication systems, queuing theory, applied probability, and resource allocation management. His masters thesis, as well as one of his recent publications, focuses on analytical performance analysis of fair schedulers in broadband radio access technologies such as mobile WiMAX.

Athanasios D. Panagopoulos was born in Athens, Greece, on January 26, 1975. He received his diploma degree in electrical and computer engineering (summa cum laude) and his Dr. Eng. degree from the National Technical University of Athens (NTUA), Athens, in July 1997 and in April 2002, respectively. From May 2002 to July 2003, he had served in the Technical Corps of Hellenic Army. From September 2003 to December 2008, he worked as assistant professor in the School of Pedagogical and Technological Education. From January 2005 to May 2008, he was the head of the Wireless and Satellite Division of Hellenic Authority for Communication Security and Privacy. Since May 2008, he has been a lecturer in the School of Electrical and Computer Engineering of NTUA. He has authored and coauthored more than 100 papers for international journals and transactions and more than 130 paper conference proceedings and 15 book chapters for international books. His research interests include radio communication systems design, wireless and satellite communications networks, and the propagation effects on multiple access systems and on communication protocols. He is a senior member of IEEE and a member of the Technical Chamber of Greece and also participates regularly in ITU-R Study Group 3 as Greek Delegate. Dr. Panagopoulos is the recipient of the URSI General Assembly Young Scientist Award in 2002 and 2005. He is vice chairman of the Greek IEEE Communication Chapter. He is the associate editor of *IEEE Transactions on Antennas and Propagation* and *IEEE Communication Letters*.

REFERENCES

1. K. Lahiri, A. Raghunathan, S. Dey, and D. Panigrahi, Battery-driven system design: A new frontier in low power design, In *Proceedings of the International Conference on VLSI Design*, Bangalore, India, pp. 261–267, Jan. 2002.
2. M. Pedram, Power optimization and management in embedded systems, In *Proceedings of the ASP-DAC 2001*, Yokohama, Japan, pp. 239–244, Feb. 2001.
3. L. Benini, A. Bogliolo, and G. De Micheli, A survey of design techniques for system-level dynamic power management, *IEEE Trans. VLSI Syst.*, vol. 8, no. 3, pp. 299–316, June 2000.
4. A. Raghunathan, N. Jha, and S. Dey, *High-Level Power Analysis and Optimization*. Norwell, MA: Kluwer Academic Publishers, 1998.
5. C. Schurgers, Energy-aware wireless communications, Ph.D. dissertation, University of California, Los Angeles, 2002.
6. IEEE, IEEE 802.16e-2004, Part 16: Air interface for fixed and mobile broadband wireless access systems—Amendment for physical and medium access control layers for combined fixed and mobile operation in licensed bands, Nov. 2004.

7. Y. Xiao, Energy saving mechanism in the IEEE 802.16e wireless man, *IEEE Commun. Lett.*, vol. 9, no. 7, pp. 595–597, July 2005.
8. G. Song and Y. Li, Cross-layer optimization for OFDM wireless networks—Part I: Theoretical framework, *IEEE Trans. Wireless Commun.*, vol. 4, no. 2, pp. 614–624, March 2005.
9. G. Song and Y. Li, Cross-layer optimization for OFDM wireless networks—Part II: Algorithm development, *IEEE Trans. Wireless Commun.*, vol. 4, no. 2, pp. 625–634, March 2005.
10. S. Shakkottai, T. S. Rappaport, and P. C. Karlsson, Cross-layer design for wireless networks, *IEEE Commun. Mag.*, vol. 41, no. 10, pp. 74–80, Oct. 2003.
11. M. van Der Schaar and N. Sai Shankar, Cross-layer wireless multimedia transmission: Challenges, principles, and new paradigms, *IEEE Wireless Commun.*, vol. 12, no. 4, pp. 50–58, Aug. 2005.
12. V. Srivastava and M. Motani, Cross-layer design: A survey and the road ahead, *IEEE Commun. Mag.*, vol. 43, no. 12, pp. 112–119, Dec. 2005.
13. X. Lin, N. B. Shroff, and R. Srikant, A tutorial on cross-layer optimization in wireless networks, *IEEE Journal on Selected Areas in Communications*, vol. 24, no. 8, pp. 112–119, Aug. 2006.
14. F. Foukalas, V. Gazis, and N. Alonistioti, Cross-layer design proposals for wireless mobile networks: A survey and taxonomy, *IEEE Commun. Surv. Tutor.*, vol. 10, no. 1, pp. 70–85.
15. G. W. Miao, N. Himayat, G. Y. Li, and A. Swami, Cross-layer optimization for energy-efficient wireless communications: a survey, *Wiley J. Wireless Commun. Mobile Comput.*, vol. 9, pp. 529–542, Apr. 2009.
16. G. Miao, N. Himayat, Y. Li, and D. Bormann, Energy efficient design in wireless OFDMA, In *Proceedings of the IEEE International Conference on Communications (ICC)*, May 2008.
17. S. Cui, A. J. Goldsmith, and A. J. Bahai, Energy-constrained modulation optimization, *IEEE Trans. Wireless Commun.*, vol. 4, no. 5, pp. 2349–2360, Sept. 2005.
18. G. Miao, N. Himayat, and Y. Li, Energy-efficient transmission in frequency-selective channels, in *Proceedings of the IEEE Globecom 2008*, pp. 1–5, Nov. 2008.
19. C. Xiong, G. Y. Li, S. Zhang, Y. Chen, and S. Xu, Energy- and spectral-efficiency tradeoff in downlink OFDMA networks, *IEEE Trans. Wireless Commun.*, vol. 10, no. 11, pp. 3874–3886, Nov. 2011.
20. T. Bonald, A score-based opportunistic scheduler for fading radio channels, In *Proceedings of the European Wireless Conference*, 2004.
21. S. Sagkriotis, K. Kontovasilis, and A. Panagopoulos, Proportional fair scheduling gains for AMC-aware systems under heterogeneous radio conditions, *IEEE Commun. Lett.*, vol. 16, no. 12, pp. 1984–1987, Dec. 2012.
22. A. Jalali, R. Padovani, and R. Pankaj, Data throughput of CDMA-HDR a high efficiency-high data rate personal communication wireless system, In *Proceedings of the IEEE VTC 2000*, pp. 1854–1858, 2000.
23. J. M. Holtzman, CDMA forward link waterfilling power control, In *Proceedings of the IEEE VTC 2000*, vol. 3, pp. 1663–1667, 2000.
24. B. Prabhakar, E. Uysal Biyikoglu, and A. El Gamal, Energy-efficient transmission over a wireless link via lazy packet scheduling, INFOCOM 2001. In *Proceedings of Twentieth Annual Joint Conference of the IEEE Computer and Communications Societies*, IEEE, vol. 1, pp. 386–394, 2001.
25. M. V. Clark, V. Erceg, and L. J. Greenstein, Reuse efficiency in urban microcellular networks, *IEEE Trans. Veh. Technol.*, vol. 46, pp. 279–288, May 1994.
26. F. Foukalas and E. Zervas, On cross-layer design of AMC based on rate compatible punctured turbo codes, *Int. J. Commun. Netw. Syst. Sci.*, vol. 3, pp. 256–265, March 2010.
27. T. Bonald, Flow-level performance analysis of some opportunistic scheduling algorithms, *Eur. Trans. Telecommun.*, vol. 16, no. 1, pp. 65–75, Jan. 2005.
28. J. G. Choi and S. Bahk, Cell-throughput analysis of the proportional fair scheduler in the single-cell environment, *IEEE Trans. Veh. Technol.*, vol. 56, no. 2, pp. 766–778, Mar. 2007.
29. D. Avidor, S. Mukherjee, J. Ling, and C. Papadias, On some properties of the proportional fair scheduling policy, in *Proceedings of the IEEE Symposium on Personal, Indoor and Mobile Radio Communications (PIMRC'04)*, vol. 2, (Barcelona, Spain), pp. 853–858, Sept. 2004.
30. L. Yang and M.-S. Alouini, Performance analysis of multiuser selection diversity, *IEEE Trans. Veh. Technol.*, vol. 55, no. 6, pp. 1848–1861, Nov. 2006.
31. S. Borst, User-level performance of channel-aware scheduling algorithms in wireless data networks, In *INFOCOM 2003. Twenty-Second Annual Joint Conference of the IEEE Computer and Communications. IEEE Societies*, vol. 1, pp. 321–331, Mar. 30–Apr. 3, 2003.

4 Green Home Network Based on an Overlay Energy Control Network

Han Yan, Cédric Gueguen, Bernard Cousin,
Jean Paul Vuichard, and Gil Mardon

CONTENTS

4.1 INTRODUCTION

In the last decade, there has been a proliferation of connected devices in the home environment. The number of connected devices has led to a sharp increase in energy consumption in the home [1]. A home network is a complex environment that contains several different types of devices: set-top box (STB), home gateway (HGW), PC, laptop, power line communication (PLC) plugs, and so on, with different kinds of connections: WiFi, Ethernet, and PLC.

Energy saving in this complex home network is crucial for the following reasons:

- First, each electricity generation system has a "carbon footprint" [2]. This means that all electricity generation systems generate carbon emissions, which are a major cause of global warming. Thus, for future generations, it is necessary to find an efficient energy-saving solution that will reduce carbon emissions.

- Second, the cost of electricity is not an insignificant part of the family's household budget. Especially for low-income families, it is always difficult to meet energy costs. According to a report [3], electricity costs may rise to more than one-fifth of a family's income. In order to improve quality of life, it is essential to reduce electricity bills by reducing energy consumption.

As seen, it is essential to provide an effective way of reducing power consumption for both environmental and economic reasons. The overlay energy control network (OECN) is proposed as a way of meeting the requirement to reduce energy consumption. This solution is based on a dedicated control network, in overlay with a typical home network. By sending overlay energy control messages, the OECN can turn off the devices or switch the devices to an ultra-low power consumption mode when they are not in operation.

Several studies have contributed to the topic of energy saving. In terms of the device system, dynamic power management is proposed [4–6]. Devices can be switched to a lower power mode when there is reduced demand for service. In addition, several algorithms have been proposed to minimize the energy consumption of device components, for instance, Ethernet links [7] and memory [8,9]. Since home network devices usually work together to offer multifunction, it is not sufficient to save energy at the level of each individual device. Consequently, the OECN provides a collaborative method by exchanging energy control messages to control the power states of home network devices.

On the network layer, Jeong et al. have proposed a power management algorithm. This reduces power consumption by reconfiguring the power control elements of each device [10]. Nevertheless, this study does not take into account the consumption associated with network connections, necessarily active in the proposed model. The authors also lack the consideration of the delay necessary for state changes. This delay can strongly influence the user-perceived experience. In this chapter, we propose a model based on an overlay network with very low energy consumption and evaluate the efficiency of our solution by analyzing not only the energy gain and cost gain but also the delay generated by the energy-saving solution. The evaluation is based on a home network that simulates real family life with variant day types.

With a home network, users want to be able to use the Internet at any time on any device when they need the network service. Moreover, home network devices work together to provide some collaborative services. For instance, a laptop user may want to watch a video saved on the STB hard drive, and these home network devices work together to offer the video playing service [11]. The OECN offers a collaborative way to immediately turn off/switch-to-sleep the devices when they are not interacting or in operation. With the help of the OECN, devices can be switched to an ultra-low-power state (soft-off state or sleeping state), instead of away state where the power consumption is much higher than the former two states.

We defined power states according to the Advanced Configuration and Power Interface standard [12]:

- G0 (Working State): The device is on and applications are executed.
- G0–1 (Away State or Idle State): This is a subset of the working state. The device is on but idle, and no applications are executed. We distinguish this state from the working state because this state consumes a lot of energy that is not required by the user.
- G1 (Sleeping State): The device is sleeping.
- G2 (Soft-off State): The device is turned off, but the power supply is still plugged in to the power source.
- G3 (Mechanical-off State): The power supply to the device has been completely removed.

Initially, the device needs a long waiting time, typically half an hour to 1 hour, to switch from G0–1 (away/idle state) to G1 (sleeping state). Moreover, users need to regulate the device manually to switch from G0–1 to G2 (soft-off state). We can gain energy if the device stays in G1 (sleeping state) or G2 (soft-off state) instead of G0–1 (away/idle state). In this chapter, we propose an always-on OECN that can switch devices from G0–1 to G1 much more quickly and from G0–1 to G2 automatically. This always-on overlay architecture consumes little energy since it is partly constructed on the ultra-low power consumption ZigBee modules. Each ZigBee module consumes about 18 to 120 milliwatt-hours per day [13], and it can turn on/turn off one device through the USB port. ZigBee is used to satisfy the need for a standard-based wireless network that has low power consumption, low data rates, and robust security. The other part of the overlay architecture is based on the Ethernet. Our system could wake the devices up by implementing the method of Wake-On-LAN, turn the devices off, and request the power state of the devices by the UPnP low power protocol [14]. UPnP low power protocol is defined to satisfy the demand of reporting and tracking the power states of the UPnP nodes. If possible, the UPnP low power protocol could also request the device to enter the sleeping state.

In this chapter, Section 4.2 sets out the architecture of our OECN. Section 4.3 describes the methodology used to evaluate our solutions and the simulations, and then presents the analysis of the results.

4.2 THE PROPOSED OECN

In a home network, there are many kinds of devices such as a HGW, STB, PLC plugs, PC, laptop, and so on. In order to reduce the overall energy consumption of the integral home network devices, a low power consumption control layer over this home network is proposed: this is an OECN. In this section, we present first the global architecture and the protocol stack of the OECN system; two solutions based on this system will be proposed.

4.2.1 GLOBAL ARCHITECTURE

The OECN is formed by at least one overlay energy control node connected to each home network device. The OECN power management coordinates the power states of all the home network devices. The overlay energy control nodes can exchange energy control messages. The devices can be turned on or turned off, or they can return to their power states when they receive the OECN messages.

We chose to implement the OECN power management in the HGW because we assumed that this device is always present and in active state in the home network to support VOIP (voice over Internet protocol) phone calls. At the home network level, all devices are interacting components that work together. By exchanging OECN messages, the devices centralize their information on the OECN management node. The OECN management collects the power information and controls the network devices (shown in Figure 4.1).

In the home network, all devices interact together to provide collaborative services. The network topology dependence, network traffic, and power state information of the devices are required by the OECN manager node. Meanwhile, the OECN manager node controls the integral home network based on this information.

When one device receives the power turn-on/turn-off message from ZigBee, it will be shut down by a command sent by ZigBee or woken up by WAKE-ON-USB. Since home network devices are generally equipped with a USB port, we assume that we can apply our solution to those USB-equipped devices.

The devices can be turned on or turned off, or they can return to their power states when they receive the OECN messages. These OECN messages can be sent in two ways. Therefore, the OECN can be implemented in two ways. This depends on the type of energy control node and the way overlay energy control messages are exchanged: if all OECN nodes are ZigBee nodes in the home network, this is a ZigBee Mandatory OECN Solution; if, however, one or more OECN nodes are devices that do not have ZigBee modules, this is a ZigBee Optional OECN Solution.

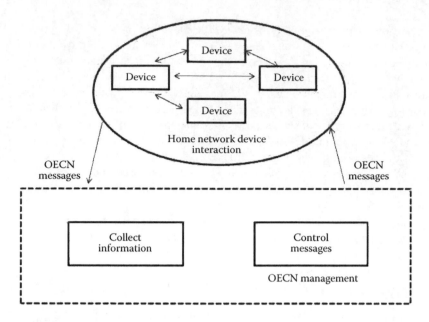

FIGURE 4.1   OECN management.

4.2.2   PROTOCOL STACK

In the OECN, the user device and the HGW (or any two user devices) can communicate through ZigBee network or any LAN (like Ethernet or WiFi). User devices that are capable of supporting a ZigBee module should have the function of powering the ZigBee module, exchanging the messages with the ZigBee USB/UART interface, and being powered on/off by the module. Between two user devices without ZigBee modules, the communication is realized by the UPnP low power messages over the Hypertext Transfer Protocol. Figure 4.2 shows the protocol stack of the proposed OECN.

4.2.3   ZIGBEE MANDATORY ENERGY-SAVING SOLUTION

When all the energy control nodes are ZigBee modules, this is called a ZigBee Mandatory energy-saving Solution (ZMS), as shown in Figure 4.3.

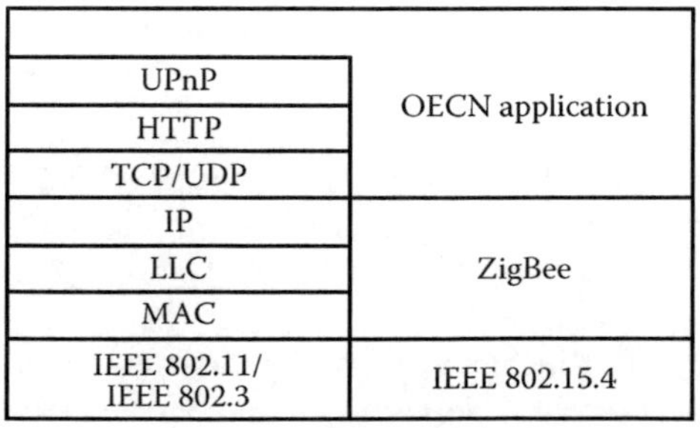

UPnP	OECN application
HTTP	
TCP/UDP	
IP	ZigBee
LLC	
MAC	
IEEE 802.11/ IEEE 802.3	IEEE 802.15.4

FIGURE 4.2   OECN protocol stack.

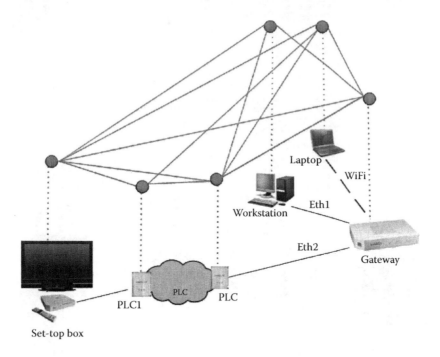

FIGURE 4.3 OECN architecture: ZMS.

There are several advantages to having ZigBee modules as control nodes. A device can be turned off and can also be started up by the ZigBee module connected to it, which is always on. Therefore, this device can go into an ultra-low power consumption state (soft-off state).

Although there are some significant advantages to using ZigBee modules as overlay energy control nodes, it is not possible to use this solution everywhere. As it might not be possible to connect a ZigBee module to the device or if the ZigBee transmission diameter is limited, we propose another alternative solution, namely, the ZigBee Optional energy-saving Solution (ZOS).

4.2.4 ZigBee Optional Energy-Saving Solution

Compared to the ZMS, the ZOS does not need each device to be fitted with a ZigBee module. When there is no ZigBee module on a device, the device itself becomes the energy control node and the energy control messages are sent via the data home network.

In Figure 4.4, the OECN is formed by ZigBee OECN nodes, one PLC plug and the STB.

The reasons why the devices (PLC plug and STB) become overlay energy control nodes are as follows:

- The STB is not equipped with a ZigBee module.
- The distance between the two PLC plugs is too great for the ZigBee transmission diameter.

On these non-ZigBee devices, the overlay energy control messages are sent through the data home network instead of a ZigBee network. Non-ZigBee devices can immediately go into a low-power state (sleeping state) when there is no executing application. However, we cannot switch the device to a soft-off state since the OECN cannot turn on a soft-off device without the help of ZigBee module. Although not all devices can be switched to an ultra-low power consumption state (soft-off

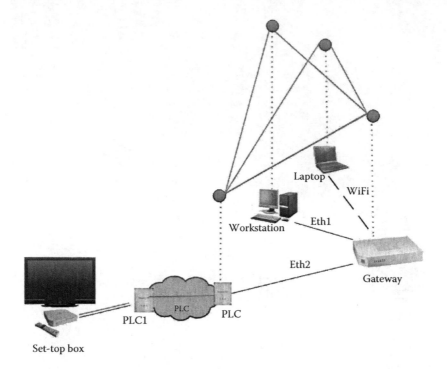

FIGURE 4.4   OECN architecture: ZOS.

state), we can still save energy using the ZOS by only leaving active those elements required for wake up (network interfaces, etc.).

4.3   SIMULATION AND ANALYSIS OF RESULTS

In order to demonstrate the efficiency of the two OECN solutions (ZMS and ZOS), we will compare the performance of these two proposed solutions with a traditional energy-saving solution called a self-controlled solution.

1. Self-controlled energy-saving solution: The device controls its own power state. This means that the device goes into the low power consumption state (sleeping state) by a user-defined condition (e.g., a 1-hour timer).
2. ZMS: The OECN manager controls all home network devices with a ZigBee module connected to each device. All overlay energy control messages will be transmitted by the ZigBee modules.
3. ZOS: The OECN manager controls devices in a hybrid way. The overlay energy control messages will be sent by the ZigBee overlay network or the data home network.

4.3.1   SIMULATION METHODOLOGY

In this section, we first build our device modeling, which is in the context of four different day types. Then, we are going to apply three solutions on one or several devices.

4.3.1.1   Device Utilization Modeling

We first generated the device modeling, and on top of that, we applied the energy-efficient solutions. Each device in the home network may be used at a random time. Thus, the device modeling is

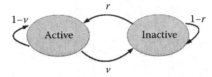

FIGURE 4.5 Device modeling from the user's perspective.

TABLE 4.1

Power Consumption

Device	Working (watt-hour)	Away (watt-hour)	Sleeping (watt-hour)	Soft-Off (watt-hour)
STB	21	19.2	13.5	2.5
PC	205	123.5	4.9	3.2
Laptop	79	54	5	2.5
PLC	6	3	2.6	0.15

---

expected to characterize its stochastic behavior. Here, we use a simple Markov process to describe each device. From the user's perspective, a home network device is either in operation (active) or not in operation (inactive). Figure 4.5 shows that the probability of one active device becoming inactive is v and the probability from inactive to active is r.

As the utilization rate of one device may vary during the day, the device will be represented by different r and v values over the 24-hour period. The device active utilization ratio is defined as

$$R_{\text{active}} = \frac{r}{r+v}. \tag{4.1}$$

This device active utilization ratio gives the normalized probability of being in active state. When the device is active, this corresponds to "working state." When the device is inactive, it can be in "away state," "sleeping state," or "soft-off state." The last two power states are low and ultra-low power consumption states. The different power states of an inactive device depend on the power-saving solution that we applied to this device. These different states have different power consumptions (watt-hour), as shown in Table 4.1.

Using the PC as an example, we can see that it consumes 205 watt-hours when it is in "working state." This is the average power consumption when a user uses the PC to download or play multimedia files. The PC consumes 123.5 watt-hours (away state) and 3.2 watt-hours (soft-off state). The difference in power consumption in these two states is significant. It proves that changing the power state when the device is inactive can effectively save energy.

4.3.1.2 Four Day Types

In order to make the device modeling realistic and adaptive to the family home network devices, the probabilities r and v are categorized into four different day types. We need to define a set of r and v values for one device in one day type. This is an example of four day types for a family of four. As a telecom operator, we have chosen these four day types according to our knowledge of our customers and how they use the devices. We also assumed that the home network devices are those shown in Figure 4.1.

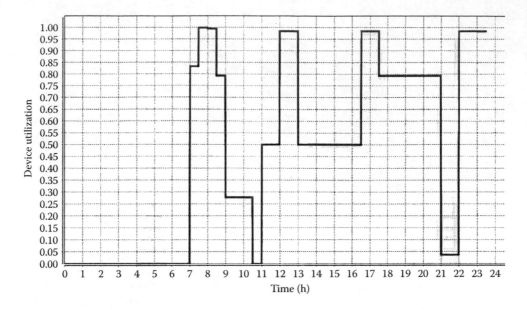

FIGURE 4.6   STB utilization ratio on day type 1.

1. Day type 1 (Working day): Parents go to work and children go to school. We will take the laptop as an example. The laptop has a high utilization ratio between 8 a.m. and 9 a.m. because one family member needs to check his or her e-mail. In the evening from 8 p.m. to 12 p.m., one family member wants to surf the Internet. Thus, the laptop also has a significant utilization ratio. Each home network device has its own utilization ratio at each time.

2. Day type 2 (Wednesday*): On this day type, the children stay at home and parents are at work. On this day type, the STB has a relatively high utilization ratio during the daytime. The laptop and PC are used at various times throughout the whole day.

3. Day type 3 (Weekend): All family members are at home. The laptop, PC, and STB are needed at different times over the weekend.

4. Day type 4 (Holiday day): The whole family is on holiday. Apart from the PC that is equipped with a home security camera, which is on, the other devices in the home network are turned off completely.

We use the STB on day type 1 as an example. As shown in Figure 4.6, all family members are at work or at school. The STB and television are turned on to watch the news when they get up. That is why we have high device utilization at 7 a.m. When the family members leave to go to school or work, the device utilization goes down. At 12 noon, we have peak device utilization, since the children return home to eat and watch television at the same time. In the evening, we also have high device utilization of the STB when every family member is at home. The device utilization of one device is relatively stable for each day type. Therefore, we can fix a set of device utilization values for each day type.

4.3.1.3   Application of the Three Solutions

As mentioned earlier, device utilization could be represented by a Markov process. Comparing this Markov process with the device power states, *active* corresponds to "working state," while *inactive* corresponds to "away state," "sleeping state," or "soft-off state." The selected state depends on which energy-efficient solution we apply.

---

* This study was conducted in France where pupils do not currently attend school on Wednesdays.

Each energy-efficient solution can therefore be defined by a finite-state machine. In the finite-state machine representation, we use the four power states that are cited in Section 4.1. From one power state to another, there is always a power and performance cost. A low-power state has low power consumption and a long transition time. Conversely, a high-power state has high power consumption and a short transition latency.

In our modeling, the abbreviations are defined in Table 4.2.

On the basis of the device utilization modeling, we can apply the three solutions to the device. Figure 4.7 shows the application of the self-controlled energy-saving solution. If a device is not being operated, the device will go into "away mode." Then, after the timer has timed out (for instance, T_a = 1 hour), the device will go into sleeping state. After staying in sleeping state for T_s, the device will be woken up by a user. Thus, we can define the total energy for the self-controlled solution as

$$E_{\text{self-controlled}} = T_w P_w + T_a P_a + T_s P_s. \tag{4.2}$$

Figure 4.8 shows that in the ZMS, if the device is not in operation, the device will be put into soft-off state immediately after utilization. In this case, the device does not need to go into "away state." The energy consumption for ZMS can be defined as

$$E_{\text{ZMS}} = T_w P_w + T_o P_o. \tag{4.3}$$

TABLE 4.2
Modeling Abbreviations

Abbreviations	Meaning
w, s, a, and o	Power state: working state, sleeping state, away state, and soft-off state.
$T_{sw}, T_{ws}, T_{aw}, T_{wa},$ $T_{as}, T_{wo},$ and T_{ow}	The transition time between two states. For instance, T_{sw} is the transition time from "sleeping state" to "working state."
$T_s, T_o, T_w,$ and T_a	The time spent in one state. For example, T_s is the time spent in sleeping state.
$P_s, P_o, P_w,$ and P_a	The power consumption in one state.
$P_{sw}, P_{ws}, P_{aw}, P_{wa},$ $P_{as}, P_{wo},$ and P_{ow}	The power consumption of each transition. For instance, P_{sw} is the transition power consumption from "sleeping state" to "working state."

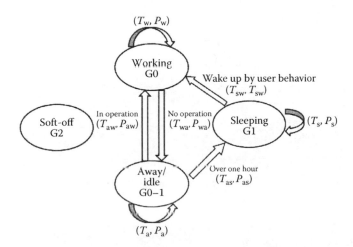

FIGURE 4.7 Self-controlled energy-saving solution.

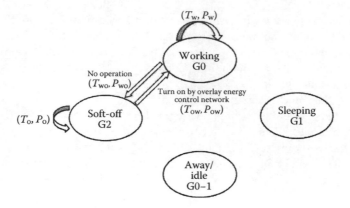

FIGURE 4.8 ZigBee Mandatory energy-saving solution.

Here, T_o in ZMS is equal to $T_a + T_s$ in the self-controlled solution. The ZOS is shown in Figure 4.9. If the device is not in operation, the device immediately goes into sleeping state. The energy consumption for ZOS is

$$E_{ZOS} = T_w P_w + T_s P_s. \tag{4.4}$$

Here, T_s in ZOS is equal to $T_a + T_s$ in the self-controlled solution. Since the transition power consumption is lower than the power consumption in each state, we ignore the power consumption on each transition.

In these three solutions, the energy consumed during the working state is the same. The greatest difference between the self-controlled solution and the OECN solutions is that the device does not need to remain in away state for the time defined by the timer. With the OECN solutions, we assume that the manager knows when a device will not be useful (could be switched off). For instance, the home network manager knows when the video broadcast is ending or when the Internet connection is closed. Thus, devices could go to low power consumption states immediately if they are not in operation. Since P_a is always bigger than P_s or P_o, energy is saved by putting devices into sleeping state and soft-off state instead of away state.

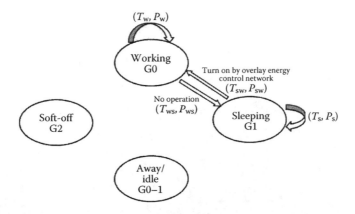

FIGURE 4.9 ZigBee Optional energy-saving solution.

4.3.2 Simulation on One Device and Analysis of Results

As presented above, we simulated the device utilization to evaluate these energy-saving solutions: self-controlled energy-saving solution, ZMS, and ZOS. To simulate the three solutions applied to one device, we used the following parameters:

- Power consumption of each device in different power states: P_s, P_o, P_w, and P_a, as presented in Table 4.1.
- Device utilization probabilities over 24 hours: A defined timer T_a for the self-controlled solution. In our simulations, we used a 1-hour timer.
- Number of simulation runs: 1000. We compared the results obtained by 10,000 runs and 1000 runs. The difference in results for laptop consumption on a day type 1 was lower than 1%. We can therefore assume that 1000 times is sufficient to obtain good accuracy.

4.3.2.1 Simulation on One Laptop on a "Weekend" Day Type

This is an example of one simulation based on one device in the home network. We first simulated laptop utilization on the weekend. Since all family members are at home on the weekend, they play video games and surf the Internet nearly all day.

The laptop utilization ratio is defined in Figure 4.10. At 9:30 a.m., there is a high probability that one family member at home is on the laptop until 11:30 a.m. After lunch, it is also quite probable that one family member turns on the laptop. We can see a utilization peak at 8:30 p.m., since the father checks his personal e-mail after dinner. We simulated the device usage 1000 times on the basis of the given probabilities. After each simulation of the device usage, we applied the three different energy-saving solutions to the device utilization. Figure 4.11 shows how the three power-saving solutions worked on this device for 1000 simulation runs. The three lines plot average power consumption.

The dark gray line is the power consumption of the device using the self-controlled energy solution. After operation, the device goes into away state for 1 hour and then goes into sleeping state.

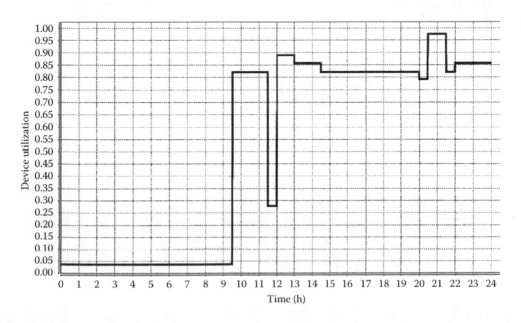

FIGURE 4.10 Device utilization ratio on the weekend.

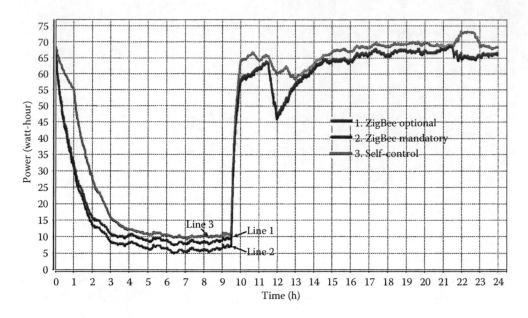

FIGURE 4.11   Device power consumption on the weekend.

The black line represents the power consumption of the device using the ZMS. The device goes into soft-off state immediately after operation.

The light gray line represents the power consumption of the device using the ZOS. This line sometimes overlaps with the ZMS line, because the power consumption in sleeping state and soft-off state is approximate. The device goes into sleeping state immediately after operation.

From Figure 4.12, we can clearly see that the power consumption in the sleeping state and soft-off state is less than that in the idle state, for instance, from 11:30 a.m. to 12:30 p.m. Thus, the energy gain comes from the times when device utilization changes from active to inactive.

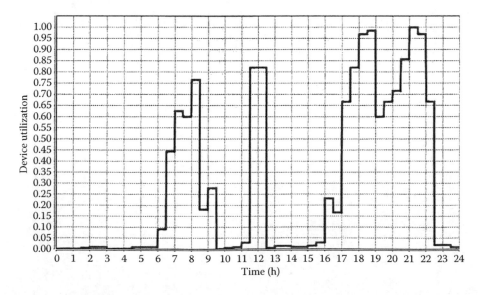

FIGURE 4.12   Device utilization on day type 3.

4.3.2.2 Simulation on One Laptop on a "Wednesday" Day Type

Energy is mostly gained when a device switches from active to inactive. In the first simulation example, there were fewer power state transitions. In this example, we simulated the same device on another day type where the power state transitions are more frequent.

On the Wednesday day type, the laptop is used at different times during the day. The stay time in active or inactive is more random and irregular than on the weekend day type. This is shown in Figure 4.13, where we have simulated laptop power consumption for a weekend day type.

The power consumption obviously decreases when the laptop utilization ratio decreases in Figure 4.12. From 9:30 a.m. to 10:30 a.m., there is an energy gain between the self-controlled solution and the OECN solutions. This is because the device is quickly turned off (ZMS) or switched to sleeping state (ZOS), instead of staying in away state (self-controlled solution). At 10:30 a.m., the self-controlled solution consumes almost the same energy as the ZOS since the device is in sleeping state in these two solutions. The ZMS can gain more energy than the other two solutions because the device is in soft-off state, which consumes less than the sleeping state. Our OECN solutions gain energy when there is a transition from active to inactive and while the device is in inactive state. There is a greater number of state transitions of power; more energy is gained.

4.3.3 SIMULATION ON ONE HOME NETWORK AND ANALYSIS OF RESULTS

On the basis of the home network shown in Figure 4.1, we will simulate the home network devices for 1 year (365 days).

4.3.3.1 Simulation Setup

In order to evaluate energy-efficient solutions in a home network environment, simulations are carried out for different day types. In 1 year, we will have 191 type 1 days, 96 type 2 days, 48 type 3 days, and 30 type 4 days. All these different days make up 1 year (365 days). This choice is representative on the basis of our own knowledge of our customers and how they use the devices.

The three energy-saving solutions are analyzed in three metrics:

- Annual energy consumption: Energy consumption is the energy used in the whole home network in 1 year. The power consumption of the devices is presented in Table 4.1. We take the energy consumption of the ZigBee modules into account for the OECN solutions.

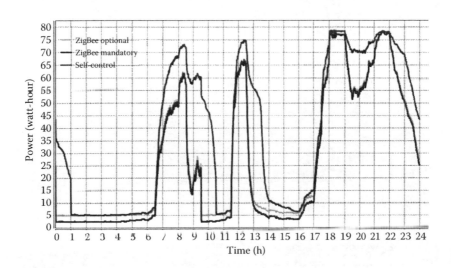

FIGURE 4.13 Device power consumption on day type 3.

- Daily delay: Daily delay is the cumulative waiting time every day. The waiting time is calculated from the moment that the device is requested to the moment that the device is in operation. It is the total duration of all transitions from inactive to active. For these energy-saving solutions, the delay for 1 day is calculated as

$$D_{\text{self-controlled}} = aT_{\text{aw}} + bT_{\text{sw}} \quad (4.5)$$

$$D_{\text{self-controlled}} = (a + b)T_{\text{ow}} \quad (4.6)$$

$$D_{\text{self-controlled}} = (a + b)T_{\text{sw}} \quad (4.7)$$

The "$a + b$" is the number of times that the device changes state from inactive to active in one day. a is the number of times that the device changes state and does not stay in the inactive state for more than 1 hour. b is the number of times that the device changes state and stays in the inactive state for more than 1 hour. The transition time is shown in Table 4.3. The home network delay is the sum of the delays for each home network device.

- Cost: The total monetary cost of the three solutions. We calculate the cost of electricity based on the European electricity tariff (for the year 2012) in Table 4.4. This tariff is cheaper during the night than during the daytime. For the self-controlled energy solution, we calculated the cost of electricity. For the ZMS, we calculated the cost of the electricity and the ZigBee modules. For the ZOS, we calculated the electricity cost of the electricity and the ZigBee module if the device has one. Otherwise, we just calculated the electricity cost of the device.

TABLE 4.3
Transition Delay

Device	Delay (s)		
	T_{aw}	T_{sw}	T_{aw}
HGW	0.01	1	40
STB	0.01	7	80
PLC	0.01	1	3
PC	0.01	4	30
Laptop	0.01	2	25

TABLE 4.4
Electricity and ZigBee Module Tariff

Cost	Tariff	
Electricity (€/kilowatt-hour)	Day rate: 0.1312	Night rate: 0.0895
ZigBee Module (€/unit)	6.05	

4.3.3.2 Analysis of Results

The results of the three energy-saving solutions simulated on one home network is analyzed in three dimensions: energy consumption, delay, and cost.

4.3.3.2.1 Energy Consumption

From Figure 4.14, we can see the energy consumption of one home network with different energy solutions applied. Compared with the self-controlled solution, we can gain 21.79% energy with the ZMS and 16.96% energy consumption with the ZOS in 1 year. By applying the ZMS, the devices are in the soft-off power state, which consumes the least energy when they are not in operation. Without the help of the ZigBee modules in the ZOS system, the device could immediately go to the sleeping power state after the operations, and we note that the sleeping power state is also a low energy cost state as compared with the idle power state.

Of these four day types, an OECN solution is less effective at the weekend. We gain 10.58% with the ZMS and 7.28% with the ZOS. For day type 3, however, OECN solutions are quite effective for energy saving: 28% (ZMS) and 21% (ZOS). On day type 3, the devices change their power states more often than on day type 2. This is why OECN solutions are more effective on one day type than on another.

As we explained in the Section 4.3.2.2, the more frequently the device changes its state from active to inactive, the more energy is saved. That is why, on day type 3, we might have a large energy gain by using OECN solutions. Meanwhile, delay performance increases when the energy gain increases. The more frequently the device changes state, the more delays are accumulated from inactive to active state.

4.3.3.2.2 Home Network Delay

The OECN solutions are effective for energy saving. However, the energy gain is not free, and the OECN solutions pay by the delay to have the energy gain. As explained in the simulation setup, the daily delay is the sum of the delays for each home network device. This cumulative delay per day of the home network is 4.8 minutes when using the ZMS. This is acceptable for many users. For other users, the ZOS could be an opportune trade-off, where the daily home network delay is only 0.48 minutes. Note that by applying ZMS, devices are capable of entering the lowest power state, which needs longer time to return back to the working power state. That is why ZMS has a higher delay as compared to the other two solutions. ZOS and self-controlled solutions have similar delay for the different day types. However, using the self-controlled solution, the user needs to turn on/turn

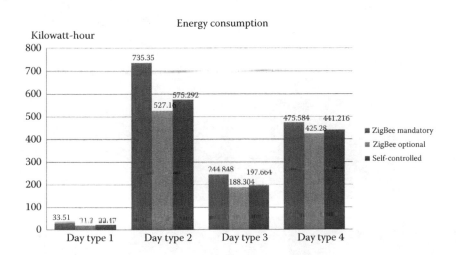

FIGURE 4.14 Home network energy consumption results.

off the user device manually, which is not favorable for the quality of the user's experience. From this point of view, the additional delay of ZOS and ZMS is favorably compensated by the automatic OECN management.

The greater the amount of energy gained, the more delay there is. As shown in Figure 4.15 for day type 2, the daily delay of all home network devices is 1.447 minutes (ZMS) and 0.146 minutes (ZOS). For day type 3, the OECN solutions are effective for energy saving but we also have an additional delay with the OECN solutions. Every time energy is gained from transitions, the OECN solutions put devices in a low or ultra-low power consumption state, which requires a longer time to wake up.

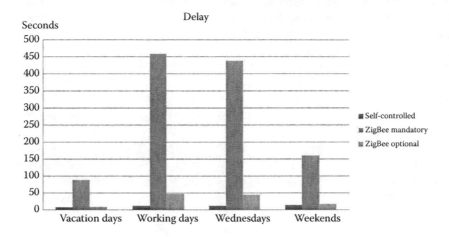

FIGURE 4.15   Home network delay results.

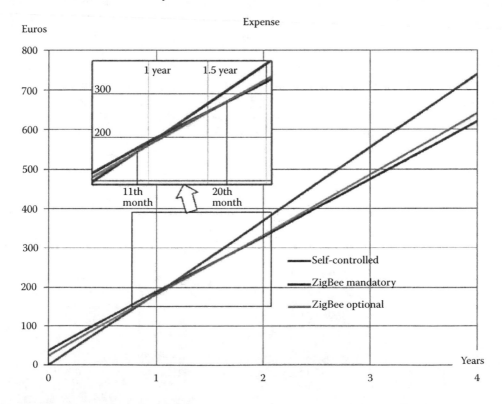

FIGURE 4.16   Home network cost results.

4.3.3.2.3   Cost

Reducing electricity bills can be a powerful motivation for home network users. At first, extra money must be spent on ZigBee modules. In Figure 4.16, this family uses six ZigBee modules for ZMS and four ZigBee modules for ZOS. However, after 1 year, the total cost for the OECN solutions are less than the cost for the self-controlled solution. Comparing the two OECN solutions, the ZMS with six ZigBee modules consumes less energy than the ZOS after 1.5 years in our use case. For users who want to have a short-term benefit in 1 year, they should choose the ZOS solution that is the most profitable solution from the 11th month to the 20th month. Users who want to have a long-term profit should choose the ZMS solution that is the most profitable energy-saving solution after the 20th month.

4.4   CONCLUSION

In this chapter, we have proposed energy-saving solutions based on an OECN. This proposition aims to reduce the energy consumption of the home network devices and thus reduce the environmental impacts caused by the carbon footprint from electricity generation. The proposed OECN provides two efficient energy-saving solutions for home network devices. Compared to the two OECN solutions, the device self-controlled energy solution is the solution that saves the least energy. The ZMS, which is based on a complete OECN, is more efficient in terms of energy saving, but it has a relatively high delay compared to the two other solutions. The ZOS, which is based on a partial OECN, is the second most efficient energy-saving solution. This solution also has a delay that is slightly greater than the device self-controlled energy solution but less than the ZMS.

Our ZMS is proven to be an effective energy-saving solution. In addition, the ZOS is proven to be a good trade-off of energy saving and delay. This trade-off depends first and foremost on the selection of non-ZigBee devices and on the time each device takes to change state.

The future challenge will lie in combining ZMS and ZOS solutions for a certain amount of time, depending on our knowledge of usage or user behavior. In the second phase, we plan to achieve better energy gain with minimum delay to assure better quality of user experience.

AUTHOR BIOGRAPHIES

Han Yan received her BS degree from Beijing University of Posts and Telecommunications, China, in 2008 and MS and engineering degrees from Telecom Bretagne, France, in 2010 and 2011, respectively. She has one year of industrial experience at Alcatel-Lucent during her gap year in 2009. Since 2011, she has been a PhD student involved in the entity Home Device Middleware Development Integration (HDMI) in France Telecom and the entity Advanced Technology in NETworking (ATNET) in IRISA. The topic of her thesis is "Collaboration of home networked devices for maximum energy efficiency." Her research interest is low-cost communication protocols, wireless sensor network, energy saving, and home network.

Cédric Gueguen received his MS degree in computer science from the University of Paris VI (UPMC) in France in 2006. He received his PhD degree in computer science from the UPMC in 2010 while working in the Networks and Performance Analysis (NPA) group at the Laboratory of Paris 6 (LIP6). In 2010, he was an assistant professor (ATER) at the University Paris-Est Marne-la-Vallée (UPEMLV) in the Protocols Architecture & Software for Networks (PASNET) team and was a member of the Gaspard Monge Computer Science Laboratory (LIGM).

He is currently working as an associate professor (maître de conférences) at the University of Rennes 1 and has been a member of the IRISA laboratory since September 2011. His research interests lie in the fields of wireless networking, scheduling algorithms, fairness issues, quality of service (QoS), and quality of experience (QoE).

Bernard Cousin, since 1992, has been a professor of computer science at the University of Rennes 1 in France. In 1987, he received his PhD degree in computer science from the University of Paris 6. Between 1985 and 1992, he has been successively an assistant professor at University of Paris 6 and an associate professor at University of Bordeaux 1.

Currently, he is at the head of a research group on advanced networking. He is a member of IRISA (a CNRS–University joint research laboratory located at Rennes in France). He has coauthored more than one hundred papers published in international journals and conferences.

His research interests include next-generation Internet, green networking, all-optical networks, dependable networking, high-speed networks, traffic engineering, multicast routing, network QoS management, network security, and multimedia distributed applications.

Jean-Paul Vuichard received his master's degree at Ecole nationale supérieure des Mines de Saint-Etienne and his PhD degree at University Rennes 1 in 1988. Then he worked as a network and embedded system engineer successively at MEMSOFT from 1988 to 1990 and Copernic from 1991 to 1993. Between 1994 and 2001, he worked as a network expert and system engineer at Capgemini.

Currently, he is the head of a research and development unit that is in charge of the home device middleware development and integration at France Telecom.

His research interests include green hardware and software recommendation, home devices and services API for B2B and B2B2C, tools and framework for device development, new user interface method, and framework and enablers in home devices.

Gil Mardon has always been passionate about technology development. He started his engineering career at SILICOM for 6 years between 1991 and 1998, and then he worked as a responsible engineer of integration simulation tools at Bell Labs Lucent Technologies from 1998 to 1999. He then became a project manager at Teamlog for 2 years. From 2001 to 2007, he worked as a software engineer and then became a software architect at France Telecom from 2007 to 2012. Currently, he is responsible for software development at DIWEL. His work interests are embedded systems, green networking, satellite transmission, and audio/video transmission.

REFERENCES

1. B. Rose. Home networks: A standards perspective. *IEEE Communications Magazine*, vol. 39, no. 12, pp. 78–85, Dec. 2001.
2. T. Wiedmann and J. Minx. A definition of "carbon footprint," CC Pertsova, *Ecological Economics Research Trends*, vol. 2, pp. 55–65, 2007.
3. The American Coalition for Clean Coal Electricity (ACCCE). Energy cost impacts on American families, 2001–2012, Tech. Rep., 2012.
4. L. Benini, R. Bogliolo, and G. D. Micheli. A survey of design techniques for system-level dynamic power management. *IEEE Transactions on VLSI Systems*, vol. 8, pp. 299–316, 2000.
5. Q. Qiu and M. Pedram. Dynamic power management based on continuous-time Markov decision processes. In *Proceedings of the 36th Annual ACM/IEEE Design Automation Conference*, ACM, pp. 555–561, 1999.
6. E.-Y. Chung, L. Benini, A. Bogiolo, and G. De Micheli. Dynamic power management for non-stationary service requests. In *Proceedings of the Conference on Design, Automation and Test in Europe*, ACM, pp. 77–81, 1999.
7. M. Gupta and S. Singh. Dynamic Ethernet link shutdown for energy conservation on Ethernet links. In *Proceedings of the International Conference on Communications*, pp. 6156–6161, 2007.
8. B. Khargharia, S. Hariri, and M. S. Yousif. An adaptive interleaving technique for memory performance-per-watt management. *IEEE Transactions on Parallel Distribution System*, vol. 20, no. 7, pp. 1011–1022, Jul. 2009.
9. X. Fan, C. Ellis, and A. Lebeck. Memory controller policies for DRAM power management. In *Proceedings of the 2001 International Symposium on Low Power Electronics and Design*, ACM, pp. 129–134, 2001.
10. Y.-K. Jeong, I. Han, and K.-R. Park. A network level power management for home network devices. *IEEE Transactions on Consumer Electronics*, vol. 54, no. 2, pp. 487–493, May 2008.
11. Use case scenarios white paper, DLNA [online]. Available at http://www.techrepublic.com/whitepapers/use-case-scenarios/160310. Last accessed: January 21, 2013.
12. Advance configuration and power interface specification [online]. Available at http://www.acpi.info/. Last accessed: January 21, 2013.
13. A true system-on-chip solution for 2.4-ghz IEEE 802.15.4 and ZigBee applications, Texas Instruments [online]. Available at http://www.ti.com/lit/ds/symlink/cc2430.pdf.2011. Last accessed: January 21, 2013.
14. UPnP low power architecture Version 1.0 [online]. Available at http://upnp.org/specs/lp/UPnP-lp-LPArchitecture-v1.pdf. Last accessed: January 21, 2013.

5 Intelligent Future Wireless Networks for Energy Efficiency
Overall Analysis and Standardization Activities

Adrian Kliks, Arturas Medeisis, Yoram Haddad,
Moshe Timothy Masonta, Luca de Nardis, and Oliver Holland

CONTENTS

5.1   INTRODUCTION

It has been estimated that a reasonable amount of worldwide energy is consumed by the information and communications technology (ICT) industry. Arnold et al. [1] indicate that this value has reached 2% of the worldwide energy consumption. Moreover, as stated in Ref. [2], around 0.5% of the total energy is used for wireless communications (equivalent to 650 TWh), and most of which (around 90%) is consumed in the outdoor cellular network, mainly by the base stations (BSs). One can assume that this percentage contribution to the total energy consumption will increase along with the advent of 4G wireless technologies and in general with the continuous tendency to increase total data traffic in the current and future communication networks. This increase in data traffic requires an improvement in the utilization of available resources. In general, this problem can be solved in at least three ways, but each of them entails higher energy consumption. First, one can go closer to the Shannon capacity limit, which means more complex systems with more processing energy consumption. Second, in some cases, the transmit power can be increased, and third, the density of frequency reuse can be enhanced, which generally results in an increase in network entities. Thus, in order to lower energy consumption while satisfying future demands, work on green communication and networking is being undertaken by the ICT community. Various initiatives have been started in the last couple of years, emphasizing the need for energy-efficient communication and networking, both wired and wireless. Moreover, energy efficiency can also be supported by the international standardization bodies by proposing energy-efficient solutions in their recommendations and documents.

Given that there are extremely significant links between the different forms of energy consumption and resulting solutions, energy-consumption reduction should be addressed in a holistic fashion. This requires care that the application of energy-efficient solutions at one stage will not lead to energy increases at another stage of the user data processing chain; indeed, all of the algorithms among layers of the open systems interconnection (OSI) model must be analyzed jointly. There are various examples that prove the necessity of such an approach. As one example, the number of retransmissions (at medium access control (MAC) and network layers) affects energy consumption at the physical (PHY) layer, and the application of (PHY layer) network coding can lead to better energy utilization in terms of reducing necessary retransmissions. However, such coding may be computationally expensive, therefore transferring the energy consumption from the radio transmission stage to the processing stage.

On the other hand, energy utilization must also be smartly managed within each layer. This observation is drawn from the fact that very often algorithms devoted to one separate layer do not take into account the impact of preceding or succeeding procedures. One glaring example of this is the relationship between the peak-to-average power ratio (PAPR), adaptive bit and power loading, and reduction of out-of-band (OOB) emission, where such aspects, although at the same layer, are highly intertwined and often contradict each other in terms of energy consumption. Such an observation is particularly crucial in the light of cross-layer solutions and future communication systems, such as intelligent or cognitive radio (CR) systems and networks. These solutions can lead to better energy utilization from a communication point of view (e.g., better opportunistic networking requiring fewer active network elements, selection of procedures and algorithms that fit best to the current need of the user, or application of a machine learning algorithm that assumes the usage of the gained knowledge in the future).

This chapter addresses a number of issues related to standardization and regulatory policies aiming at promoting energy-efficient communications (both wired and wireless) and networking, highlighting the need for a synergistic approach. It encompasses the analysis of various solutions supporting sophisticated energy management within each layer of the OSI stack model in the context of wireless networking, as well as the overview of cross-layer energy-efficient applications. In every part of this chapter, adequate references to the corresponding standardization activities on the European and global level have been identified. Finally, the steps toward implementation of energy-efficient networking solutions by means of CR technologies are presented.

The rest of the chapter is arranged as follows. First, the role of separate OSI layers on energy consumption is discussed, with particular emphasis on the holistic view of all the processing done in these layers. Then, the mutual dependencies between the neighboring layers are analyzed from the perspective of energy saving. This layer-centric and cross-layer discussion is followed by the presentation of the current challenges and opportunities of the future wireless systems. Finally, in that light, ongoing standardization activities are presented.

5.2 PER-LAYER ANALYSIS

In this section, a holistic view on the energy consumption in each layer will be provided. First, the impact of the PHY layer algorithms on the overall energy consumption of the wireless systems will be analyzed. Next, dynamic spectrum access (DSA) and radio resource management, as well as network layers, will be described. In each case, the guidelines for the next wireless systems will be identified.

5.2.1 PHY LAYER ANALYSIS

Typically, standards defining the wireless system concentrate on the lower layers of the OSI model, mainly focusing on the PHY and MAC, including protocols and network architecture. Algorithms proposed by the standardization bodies should ensure reliable data transmission with an accepted level of quality of service (QoS). On the other hand, mobile devices are powered by batteries, and from the perspective of end-user satisfaction, the duration between recharges of these batteries should be as long as possible. This creates a trade-off between the reliability and the simplicity of usage of mobile devices. In this section, we try to identify and briefly characterize various aspects of the PHY layer processing that influences this trade-off.

5.2.1.1 General Considerations

It is widely understood that most of the total power used within the whole wireless communication system is consumed by the BSs. This is the power consumed by air-conditioning, data transmission (transmit power), baseband processing, and various inefficiencies, including power amplifier (PA), power conversion, and feeder cable losses. One can observe that the reduction of power consumption due to the application of better PHY layer algorithms (such as decoding procedures or equalizers) measured per one device is relatively small compared with the power consumed by the BSs for data transmission, once all the various losses in the data transmission chain are taken into account. Moreover, better PHY algorithms are often more complicated, thus requiring more power for execution. However, it has to be highlighted that the application of particular routines in the PHY layer has either direct or more often indirect but still significant impact on the higher

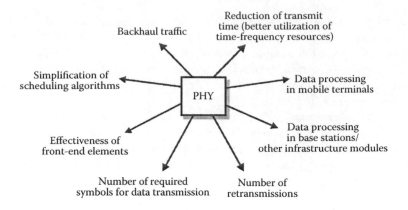

FIGURE 5.1   Impact of the PHY layer on the energy consumption in the whole communications system.

layers and finally on the total power consumption used in the system. As an example, worse (less complicated) equalization procedure will result in higher retransmission rate, since more frames will be under acceptable QoS levels. In other words, worse (less precise) equalization functions will cause an unexpected increase of total transmit power because of the higher number of retransmitted frames or symbols. Moreover, "better" adaptive algorithms can allow for more efficient utilization of the current channel realization, thus increasing the effective number of transmitted user bits per second and, in consequence, reducing the number of required transmit frames and simplifying the scheduling algorithms implemented on the BSs. Summarizing, the influence of the low-level OSI layers on the total power consumption seems to be significant, but the holistic view on the problem of power dissipation is required. For simplicity, the potential impact of the PHY layer on the total power consumption has been illustrated in Figure 5.1.

5.2.1.2   Energy Efficiency of the Power Amplifier

5.2.1.2.1   The Problem of PAPR and OOB Radiation

PA is one of the key elements of the wireless terminal from the energy-consumption perspective. It is responsible for amplifying the signal power to the predefined level, which is usually limited by national or international regulations, and hence it consumes a huge amount of the energy delivered to the wireless terminals. Various types of PAs can be defined (class A, B, AB, C, D, etc.). They classify the devices inter alia according to the topology and ways of signal processing and, in consequence, to the power efficiency η, which is the figure of merit of the PA. Power efficiency is defined as the ratio of alternating current output power (i.e., the power delivered to the load) to direct current input power (i.e., supply power). In the case of wireless terminals, it is the ratio of the radio frequency (RF) output power to the delivered supply power. In the ideal case, power efficiency is equal to 1; in practice, however, the theoretical limits can be defined for each of the classes of PAs. As an example, for class A, the maximum power efficiency is equal to 25%; for class B, 78.5%; and for class C, 99%. However, not all classes are suitable for mobile terminals or BSs owing to the characteristics of work. In general, in order to guarantee the highest power efficiency of the PA, its so-called operating point should be as close as possible to the compression region. However, when it falls into that region (or in the saturation region), nonlinear effects will appear at the output of the PA, causing significant signal degradation and OOB radiation [3,4]. Unwanted power emission outside the assumed, nominal band causes distortion to other users and degrades their signal. In consequence, the users observing such high interference could be forced to apply more sophisticated decoding algorithms or to retransmit the corrupted data (packets). In both cases, the increase of energy consumption by the victim user is observed. In order to avoid these phenomena, the operating point of the given PA should be backed off by some value X that depends on the variation of the amplitude of the time-domain signal. Unfortunately, high back-off values reduces the energy efficiency of PA, which again contradicts our assumption of optimization of the delivered power. Thus, other solutions should be prompted; that is, the signals that are characterized by the low variation of their envelope should be preferred. Usually, the PAPR ρ is used to express the variation of the signal; it is the ratio between the peak power of the time-domain signal and its mean power [3]:

$$\rho = \frac{P_{max}}{P_{mean}}. \tag{5.1}$$

The so-called single-carrier systems (e.g., WCDMA, UMTS) are characterized by very low PAPR values and thus can utilize the PA more efficiently. However, in the recent wireless communications systems, such as LTE, IEEE 802.11, IEEE 802.16, DVB-T2, or DVB-H, the multicarrier transmission is considered; moreover, multitone modulations are used in future generation systems (e.g., LTE-A). It is widely known, however, that multicarrier signals imply a high PAPR value that increases as the number of utilized subcarriers grows. It means that the efficient techniques for PAPR minimization should be implemented, as it is done, for example, in DVB-T2 standard, where the application of active constellation extension and tone reservation methods is proposed [5–7]. Significant reduction of PAPR leads to a reduction in the required back-off value and increases the power efficiency of the PA. Although the value of the required back-off will be reduced owing to the application of the PAPR reduction method, there is no sense to shift the operating point by the value corresponding to the maximum PAPR value. The reason for this is the very low probability of occurrence of the highest-amplitude samples in the time-domain signal. Thus, it is wiser to further increase the energy efficiency of the PA despite the very occasional clipping of the transmit signal. This implies that OOB emissions of future intelligent multicarrier systems will not be totally cancelled. Such an observation is particularly important since it is expected that such intelligent and cognitive wireless systems will be characterized by the high utilization of available resources including frequency spectrum, which will be characterized by reduced distance between neighboring signals in the frequency domain. This observation implies that the level of induced interference to neighboring users will be very challenging.

In conclusion, the presence of OOB emission can, in some sense, reduce the energy efficiency of the transmitter since a part of energy is wasted for undesired transmission. Moreover, it increases the level of interferences observed by other users [4], leading to the application of more sophisticated detection algorithms or causing packet/data retransmissions and, in consequence, higher energy consumption. Finally, high interference power reduces the effective range of the cells, since the maximum transmit power of the BS or mobile terminal is always upper bounded; thus, the required minimum value of the signal-to-noise ratio (SNR) cannot be guaranteed to the edge users. Clearly, it is not possible to eliminate the OOB leakage of the transmitted signal; however, electromagnetic compatibility should be maintained, and the minimization of the OOB emission should be forced by the standards. However, sophisticated and effective algorithms for OOB reduction consume a large amount of energy and trade-off strategies should be identified and incorporated in the standardization process. The need for such trade-offs is again highlighted by the fact that the allowance of very rare clipping can greatly improve the energy efficiency of the transmitter in question through more lenient PA design.

5.2.1.3 Pilot Signals and Training Sequences

The need for transmission of specific, predefined, and intentionally inserted data in each frame or symbol is nowadays unquestionable. These sets of steering data, that is, training sequences or pilot signals (TS/PS), are used for many purposes, such as channel estimation and correction, synchronization, and so on [3,8]. However, application of such sequences reduces the effective rate of the user (sometimes called goodput, or user-data rate) and thus degrades the power efficiency interpreted as an amount of power required to send one bit of the user data. If the length of the training sequence or the number of inserted pilots is high, the user-data rate will decrease, thus increasing the overall energy consumption. On the other hand, the minimum amount of steering data (training sequence bits or pilot tones) exists and cannot be exceeded if the required QoS should be kept on the constant level. Such an observation leads us to the suggestion that future wireless systems should adapt the amount of TS/PS sets to the current channel realization. However, such an approach requires adaptive modification of the estimation and decoding process, since different amounts of user data and TS/PS sets will be received in each frame, and this will increase the complexity of the receiver.

5.2.1.4 Ability for Adaptation

In the previous subsection, the need for adaptive pilots and training sequences has been discussed. This can be understood, however, in a much broader sense. In general, the ability of the adaptation of particular transmit-signal parameters should improve the overall energy efficiency, since either the received signal will be more robust to the channel influence or more user data will be transmitted in one channel instant [3,8]. Of course, the price paid for such an opportunity is an increase in transceiver complexity and an increase in the so-called backhaul traffic. The more adaptive the system, the more complicated the transceiver and higher background traffic. There is trade-off between the gain due to the system adaptability and the energy needed to provide those features, energy consumed by more complicated devices, and higher backhaul rate. The application of signal adaptation techniques is rather limited in the current wireless system; usually, only the coding rate and modulation format are changed from one superframe to another by proper selection of one of the predefined modulation and coding schemes. The next wireless systems should gain much more from signal adaptation; however, a further increase in backhaul traffic would be challenging and energy-consumption issues will start to play a significant role.

5.2.1.5 Cyclic Prefix

As mentioned in Section 5.2.1.2, the multicarrier transmission is perceived as a good candidate for the next wireless communications systems. However, the currently used orthogonal frequency division multiplexing (OFDM) signaling is inefficient from a spectrum and energy efficiency perspective. It is mainly due to the fact that the OFDM systems possess their great properties thanks to the very strict requirements put on the signal structure, such as precisely defined distance between adjacent subcarriers that ensures orthogonality or the presence of the cyclic prefix for resolving the problem of high intersymbol interference. Unfortunately, the duration of the cyclic prefix varies in the range from one-sixteenth to one-quarter of the payload size of one OFDM symbol. It means that up to 20% of the total power (and transmission time) is in some sense wasted. Further wireless systems should concentrate on application of other solutions, such us filter bank-based multicarrier signals (e.g., filter bank multicarrier [FBMC] technique [9]) or even generalized multicarrier signals (e.g., generalized multicarrier technique [10,11]). The architecture of the transceiver is of course higher (at least additional polyphase filters are required [12]) compared to the conventional OFDM-based devices, but there is no need for cyclic-prefix transmission, and the distances between pulses on time–frequency plane can be defined theoretically freely. In such an approach, much less requirements are put on the signal structure as it is in OFDM systems.

5.2.1.6 Spectrum Sensing—PHY Layer Perspective

It can be assumed that the future wireless systems will depend strongly on the accuracy of the neighborhood recognition [13,14]. The deeper the knowledge of the terminal about the environment (adjacent users, other wireless systems, etc.), the better the connection quality, since the terminal will be able to adapt its parameters to the current situation. The process of gathering all of the information about the vicinity can be called sensing—as it is done in the CR area. One has to observe that the ability of the end terminal to sense the particular features of the neighborhood is feasible only when some period will be devoted for that purpose, that is, explicitly for sensing and not for

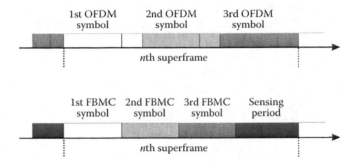

FIGURE 5.2 Comparison of conventional OFDM system with the cognitive FBMC one, where the dedicated time for spectrum sensing has been assigned.

transmission of either user or backhaul data. During the sensing time, the wireless device has to perform often highly complicated operations in order to recognize the vicinity on the acceptable accuracy level. Moreover, the wireless device is assumed to be mobile; thus, the sensing operation has to be repeated in a cyclic manner. The first issue is to define the ratio between the sensing and transmission time—the higher the ratio, the less effective usage of the supplied power. From the PHY layer point of view, other issues have to be considered as well, such as the required value of the sampling frequency and the accuracy of the collected data. One can observe that the higher the sampling rate, the higher the power consumption. It can be suggested that the period used for transmission of the cyclic prefix can be now devoted for spectrum sensing and neighborhood recognition, as it is illustrated in Figure 5.2.

5.2.1.7 Architecture

The final stage of the signal processing of the wireless device is realized by the so-called front-end, where the signal is converted between the digital and analogue forms, amplified, analogously filtered, and transmitted/received by the antenna. Various hardware architectures of the front-end have been defined and also implemented so far, for instance, the classical division between the homodyne and heterodyne classes of wireless transceivers. Recently, some proposals for dedicated white-space terminal front-end have been published [15]. The energetically effective wireless front-end is currently a vivid research area all over the world. Let us shortly analyze the particular components of the device front-end from the energy efficiency perspective. However, before we go into more detail, let us assume that the basis for our analysis will be either multistandard terminals (i.e., terminals that are compatible with many standards) or cognitive terminals (i.e., terminals that are able to choose the operating frequency band and transmission standard). Such an assumption, which is valid when referring to future wireless systems, has some significant consequences that can be seen as major requirements put on the particular elements of the wireless front-end. Thus, following that approach, one can characterize the front-end modules as below:

- Analogue-to-digital converters and digital-to-analogue converters—these modules are responsible for changing the form of the transmit/receive signal. These converters are usually both very fast (a speed expressed nowadays in mega samples per second) and very accurate (around 10–18 bits resolution), in order to ensure the acceptable level of QoS. The wider the scope of compatible systems, the higher the requirements put on the converters, and thus the higher the energy consumption.
- Mixers and local oscillators—these elements are responsible directly for multiplying the data signal by the carrier and shifting the desired signal to an appropriate spectrum band; the problem of these two modules is related to the range of served frequencies; it is technologically hard do develop a fully tunable mixer and local oscillator over a wide range of frequencies.

- Filters—the aim of the application of the (typically) analogue filters is to minimize the OOB emission, and thus not to harm other, neighboring users. However, parametrically steered filters, and thus fully adaptive filters, usually consume much energy.
- PAs—as described at the beginning of this subsection, highly efficient PAs can be treated as key factors influencing the power dissipation by the terminal front-end.
- Aerials—finally, the energy efficiency of the antennas should also be maximized; however the ratio between the radiated power to the delivered power will not be constant if the antenna is wideband or tunable.

5.2.2 DYNAMIC SPECTRUM ACCESS AND RADIO RESOURCE MANAGEMENT

Dynamic spectrum access (DSA) is a term used to describe a transient utilization of frequency holes, defined as radio spectrum resource (i.e., frequency and physical space intersection) not used at a given time, by frequency agile adaptive radiocommunications systems. It is obvious that in order to utilize spectrum holes, the prerequisite of DSA operation is the ability to discover the patterns and real-time information on spectrum utilization by other systems competing for access to spectrum. Often, the DSA will be considered to enable technological solution in cases where a new system wishes to gain access to spectrum already utilized by some incumbent, that is, primary system. Thus, the term *opportunistic spectrum access* emerged and is often used interchangeably with DSA. In this chapter, however, we shall use the more generic term DSA to highlight the fact that we speak not only of opportunistic competitive access by secondary systems but also of primary systems themselves that may use DSA principles to drive the most efficient utilization of spectrum licensed to them.

If one were to ascribe DSA operations to a specific layer in the OSI networking model, they clearly fall within the premises of the MAC layer. However, it should be kept in mind that DSA may be transgressing the confines of a particular network plane in cases when it is made to coordinate operation of multiple networks or network planes. The case in hand is the operation of modern cellular systems, which today are made through amalgamation of multiple networks—widely referred to as 2G, 3G, and 4G. Each of these has distinctive networking structures and is originally based on completely different routing and switching principles (voice circuit switched vs. packet switched). All of them however converge in the radio access plane where network nodes interact with suitably multiband/multistandard user terminals—mobile phones. Moreover, the growing functionality and processing power of the latter devices allow today seamless converging services from completely different networking technologies, such as 2G/3G/4G use on a broader scale with hot spot-based wireless local area networks (WLANs) implemented by using ubiquitous IEEE 802.11 technology or the wide area multicasting by means of terrestrial digital video broadcasting (DVB-T) TV technology.

Therefore, DSA takes the service differentiation aspect of MAC protocol to a completely new level in that it allows not only differentiation between different service streams but also selective association with different networking planes, different frequency bands, or even completely different networks. This led to the definition of a complementary concept of spectrum mobility [16]. This concept envisages the possibility for network nodes to switch between available frequency bands in order to seek various networking optimization functions:

- Most optimal frequency band—noting that different bands offer changing trade-off between coverage (owing to different radio wave propagation features) and capacity (owing to different bandwidths)
- Increased throughput and protocol stack adaptation—as the different coverage areas would mean different operational requirements in terms of SNR values, frequency of hand-offs, and so on
- Improving self-coexistence and synchronization of multiple cells that need to reuse the same frequency channels at decreasingly short distances from each other

However, other than these networking efficiency functions, spectrum mobility also opens up an important corollary field by contributing to energy efficiency.

In general, optimization of the MAC-level operations may relate to different aspects of energy efficiency, at different planes of the network structure:

- Radio access network (RAN) nodes—radio base station (RBS)
- User terminals—mobile phones

As previously mentioned, today most RBSs are made of a combination of radio transceivers that work in different bands and use different radio transmission standards (2G to 4G). Spectrum mobility allows the moving of active subscribers to only one or some of all accessible frequency bands, with the "parking band" typically being the lowest frequency band (e.g., 900 MHz) of the available complement, in order to ensure the largest possible coverage with the same output power. Once all subscriber terminals are accommodated in that "parking band," powering off the vacated RBS transmitters in higher bands (e.g., 1800 MHz, 2100 MHz) is allowed, as illustrated in Figure 5.3.

In the RAN plane, the operation of MAC and DSA may have significant impact on power consumption of RBS. Taking an example of a mobile operator in a large developed country, such as Vodafone in the United Kingdom, the network operation would be supported by around 8000 RBS sites and would annually consume around 259 GWh of electricity, some 86% of the total energy attributed to the company [17], whereas in each individual traditional RBS, the RF PAs (including feeder loss compensation) would consume approximately 1200 W or 50%–80% of the total RBS power [18]. Naturally, any savings on this scale would have huge benefits for the company in terms of reduced energy bills, not to mention the implicit environmental benefits. An example provided in Figure 5.4 shows results of simulations [19] of a power-saving benefit, expressed in terms of "energy reduction gain," if RBS transceivers for unused frequency bands are dynamically powered off. This shows the potential energy savings on the order of 10%–60% of maximum power consumed by RBS during peak loads.

In the user terminals, albeit dealing with much lesser power consumption levels, the impact of MAC/DSA operation would nevertheless be equally (if not more) important. In the terminals,

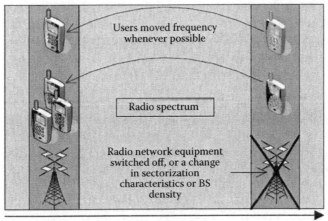

FIGURE 5.3 Reallocating traffic between bands to enable part of RBS transceivers to be switched off. (From Vodafone UK Corporate Responsibility Report, numbers based on energy consumption data for 2010/2011, online: http://www.vodafone.com/content/index/uk_corporate_responsibility/greener/carbon_energy/network_ efficiency.html, accessed Sept. 2012.)

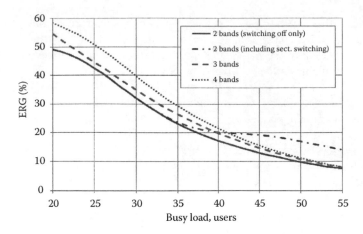

FIGURE 5.4 Power saving against busy hour load for RBS transceivers powering down solutions (streaming video traffic). (From Vodafone UK Corporate Responsibility Report, numbers based on energy consumption data for 2010/2011, online: http://www.vodafone.com/content/index/uk_corporate_responsibility/greener/carbon_energy/network_efficiency.html, accessed Sept. 2012.)

energy efficiency translates into battery life and this is one of the crucial elements in the perception of overall service quality by users.

It is important to keep in mind that the practice of energy saving has been around even before the dawn of cognitive networking. One notable example is the sleep mode envisaged in the IEEE 802.11 standardization for WLANs [20]. But the cognitive networking features such as DSA and spectrum mobility concepts allow taking energy saving and more generally energy efficiency to completely new levels of sophistication.

At the time of writing, there are no standards defining generic MAC protocols for DSA operations in wireless networking, especially those that relate to energy efficiency. As one aspect of its work, IEEE DySPAN-SC, through the IEEE 1900.4 standards working group, develops architectural blocks for cognitive radio networks (CRNs), on the basis of controller-based network configuration philosophy [21]. For example, DSA aspects might be addressed in the network reconfiguration manager and radio-access network reconfiguration controller (RRC). These try to merge the traditional approaches of controller-based telecommunications networks with the new functionalities and extended flexibility offered by software-defined radio (SDR).

IEEE 1900.4 deals with architecture for optimized spectrum utilization including studying of network heterogeneity aspects. Its standards may directly influence DSA functioning and its impact on energy efficiency. Another important standard that has been already produced by IEEE is IEEE 802.22 [22], which defines the DSA-based usage of the TV bands in the UHF range (470–862 MHz), utilizing the so-called TV white spaces, that is, gaps in coverage of terrestrial TV stations. The IEEE 802.22 MAC can be seen as similar to that of IEEE 802.16 (the standards widely known under its commercial trademark of "WiMAX") in that it utilizes combinations of time division multiplexing frames with OFDM-based partitioning of channel bandwidth into a number of subcarriers in order to organize transmissions to different users under tight control by a BS. Some notable differences however exist, such as special data transmission packets and novel protocol features (such as detection of incumbents, synchronization, and self-coexistence), which are foreseen to enable operation in DSA conditions with sufficient QoS, while avoiding interference to primary users—TV receivers.

In particular, the IEEE 802.22 reference architecture envisions a function of the spectrum manager as a part of MAC layer implementation in the BS. The spectrum manager is responsible for taking key decisions related to efficient spectrum access on a cell level. This function is complemented by the spectrum automation feature implemented in the MAC layer on the user equipment side. Although normally spectrum automation acts under instructions from the spectrum manager of a BS, in some cases, it may also perform autonomous operations such as spectrum sensing in idle mode or during device initialization.

IEEE 802.22 MAC standardization is a good example of a practical approach where the standard describes the various DSA-specific fields and packets as well as some decision making; however, the latter is not exhaustive. Sufficient freedom in defining DSA-related decision making is left to be deployment specific, in order to allow the possibility of complying with any specific local regulations pertaining to DSA spectrum use. Hence, only some very generic functions are made obligatory, such as spectrum sensing and incumbent database support.

The incumbent database support is arguably becoming the most important general feature for defining DSA-based wireless systems as it embraces a broad variety of envisaged shared spectrum use cases, most notably the TV white spaces and the licensed shared access (LSA) concept. The essence of the LSA concept, promoted by the cellular network industry, is to avail of the possibility of using multiple frequency bands for more optimal spectrum access by cellular networks by sharing with well-predictable incumbent users. Of these, the aforementioned TV broadcast stations are but just one example, the others possibly including radar stations, satellite Earth stations, event-related spectrum use pockets (wireless cameras and wireless microphones), and so on.

Another important corollary effect on RBS energy saving is achieved through the fact that SDR technologies allow general reduction of RBS hardware complement. The traditional multiband/multistandard RBSs were practically implemented by physically collocating on site a full complement of different transceivers, each designed for operation in a particular band under a particular RAN standard (see Figure 5.5).

Thus, a lot of energy was wasted to maintain separate RF PAs and compensate for power splitters and multiple long feeder lines to mast-top antennas. This has been dramatically changed by the introduction of reconfigurable SDR technologies in RBS, whereby the entire complement of different supported standards and frequency bands is contained in a single software-configured system (see Figure 5.6).

Although the new architecture may still require different antennas as well as different RF PAs owing to band-specific RF design and filtering, these final stage amplifiers may now be confined into outdoor expansion units mounted near respective antennas and thus minimizing power losses for lengthy feeder transmission. Moreover, this structure allows completely avoiding RF cable losses when baseband signal to outdoor units is transmitted by fiber-optic cables.

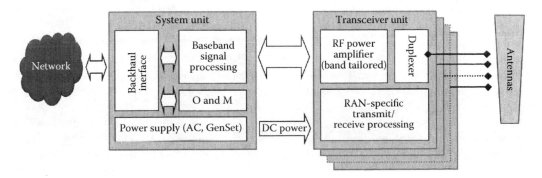

FIGURE 5.5 RBS architecture with separate radio transceiver units. (Adapted from ETSI TR 102 681. Reconfigurable Radio Systems (RRS); Radio Base Station (RBS) Software Defined Radio (SDR) status, implementations and costs aspects, including future possibilities, 2009-06.)

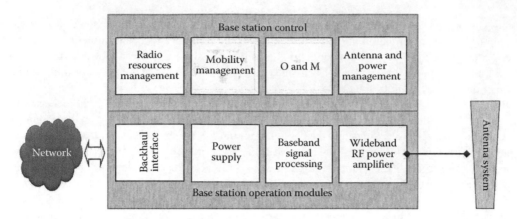

FIGURE 5.6 Reconfigurable RBS architecture. (Adapted from ETSI TR 102 681. Reconfigurable Radio Systems (RRS); Radio Base Station (RBS) Software Defined Radio (SDR) status, implementations and costs aspects, including future possibilities, 2009-06.)

Such RBS architectures are very well suited to exploit the advantages of DSA and spectrum mobility as the software-based operation allows seamless reconfigurability of operations and maximum power savings. Standardization efforts should attempt to incorporate or account for such technologies.

In conclusion, there appears to be a clear need for standards defining generic MAC protocols for DSA operations in wireless networking, especially those that would relate to energy efficiency. The only MAC definitions that deal with DSA aspects today may be found in technology and band-specific implementations, such as IEEE 802.11k that describes spectrum management aspects for operation in unlicensed bands or the IEEE 802.22 standard defining wireless networks in UHF TV bands. It may be therefore envisaged that the ultimate and logical outcome of the developments described here would be to develop standardized elements of a generalized CR MAC, which could become a genetic basis for developing specialized versions of MACs for different networking technologies [24].

5.2.3 Network Layer

Energy efficiency in routing protocols and algorithms is a long-standing research topic in the field of wireless networks. The issue of energy efficiency particularly gained wide attention from the research community in the context of mobile ad hoc networks (MANETs) and wireless sensor networks (WSNs).

In both scenarios, the possibility of extending network lifetime by optimizing the use of the limited energy reserve available in battery-powered devices led to significant research efforts since the mid-1990s, with pioneering results published in seminal works toward the end of the decade as in Refs. [25,26], where different options for introducing energy awareness in the routing function are explored.

In general, the most common approach to improve energy efficiency in the routing process is to adopt an energy-aware routing metric, possibly combined with energy-efficient procedures.

Potential approaches in the definition of energy-aware routing metrics may vary from simple, straightforward minimization of transmitted power, leading to lower emissions (typically at the price of a higher number of hops compared to traditional shortest path routing metrics) and to more advanced solutions that take into account several aspects in the operation of a wireless node and of the wireless network as a whole. Variables that may be taken into account in the design of an energy-aware routing strategy include the following:

- Transmission, reception, and processing energy consumption at the PHY layer—a correct modeling of the relative weight of each of the above three tasks in terms of energy consumption is instrumental in the design of an efficient energy-aware routing cost function. As an example, the knowledge that processing is a much cheaper task in terms of energy consumption than transmitting or receiving would lead to the definition of a routing solution that heavily relies on processing (e.g., compression or network coding) to reduce the number of bits and in turn of packets to be sent in the air. On the other hand, a mismatch between the model adopted in the routing cost function and the actual energy consumption budget in wireless devices composing the network may lead to severe degradation in energy efficiency. Figure 5.7 shows an example of such degradation by presenting the energy consumption in a simple scenario where processing energy cost is neglected. Assuming that the ratio between the energy consumed per bit in transmitting E_{RX} and the energy consumed per bit in receiving, E_{TX} is equal to $E_{RX}/E_{TX} = \alpha$, the figure shows how the energy consumption varies as a function of the mismatch between the actual value of α and its estimated value α^* used in the routing cost function.

  It is worth noting that several proposals in the literature only focused on some of the above tasks (e.g., in Ref. [27], the role of reception in energy consumption is not taken into account).

- MAC performance—a correct estimation of the performance of the MAC is also a key aspect in the deployment of energy-efficient routing solutions, as packet retransmissions owing to congestion in medium access may easily account for a significant portion of overall energy consumption. As a consequence, selecting MAC (and congestion)-aware routes may lead to a major energy efficiency boost. Including a fairly detailed model of MAC performance may thus be a smart move toward the achievement of higher energy efficiency.

The specific form taken by an energy-efficient routing cost function will also depend on the goals set in the design of such cost function. Performance goals typically deemed relevant include the following:

- Fairness in energy consumption across network wireless nodes—straightforward energy consumption minimization may lead to significant discrepancies in the rate of energy consumption between wireless nodes, with nodes on energy-efficient routes being depleted at a much faster rate than other nodes. Such unbalanced consumption may lead to undesirable side effects, such as early shutdown of such nodes owing to energy depletion and correspondingly a decrease in network connectivity and increased risk of network partitions. As a consequence, most solutions for energy-efficient routing include some measure of the expected lifetime of a node, expressed for example with its residual energy, with nodes low on energy becoming less and less appealing in the selection of new routes. This approach is, for example, adopted in Refs. [26,27], ensuring fair energy consumption across network nodes.

- Minimization of emissions—although transmitted power minimization will not lead to optimal energy consumption, as discussed previously, this may become relevant in scenarios where reducing the impact of emission on other nodes or other networks is a high priority goal. This might be the case for example of CRNs operating under coexistence constraints with collocated systems.

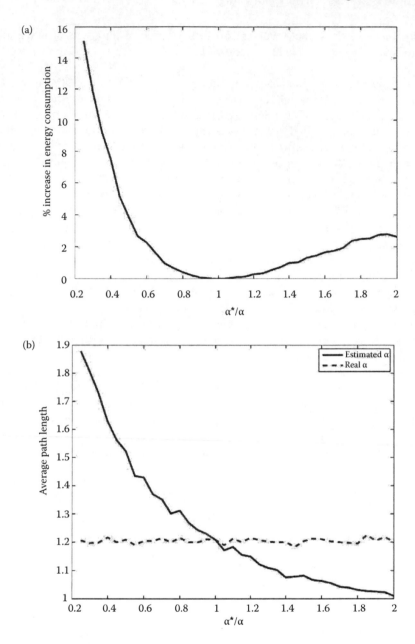

FIGURE 5.7 Impact of mismatch between estimated energy budget at physical layer and actual budget on energy efficiency. (a) Increase in energy consumption as a result of the use of estimated α^* compared to the case where the exact value of α is used in the routing cost function as a function of the ratio α^*/α. (b) Comparison between the average path length of routes determined with exact α versus estimated α^* as a function of the ratio α^*/α.

Designing an energy-aware routing metric is only one of the ways to achieve higher energy efficiency. Introducing energy awareness in routing procedures is in fact another key step toward the above goal. As an example, avoiding or mitigating the use of network-wide control packet broadcasts may significantly increase the energy efficiency and thus network lifetime. Procedures relying on flooding should thus be avoided as much as possible or modified so as to ensure that the number of packet broadcasts is kept as low as possible. This applies to both data packets, in case connection-less (datagram) routing is considered, and control packets, when connection-oriented routing protocols like Dynamic source routing are taken into account. Piggybacking control information in data transmissions may also lead to significant energy savings, by avoiding as much as possible to include overheads related to dedicated control packets.

In general, the design of an energy-efficient routing solution should take advantage of all potential information available about the target network. This includes, for example, expected network density and size, as well as traffic flows: for example, knowing if the traffic is going to be evenly shared between network nodes or rather directed toward a small set of nodes, typically referred to as sinks, may dramatically affect the characteristics required from the routing algorithm for it to be the optimal solution from an energy efficiency point of view. As a consequence, different application scenarios may lead to completely different choices in the definition of routing cost functions and procedure, as it will be detailed later in this section when dealing with standardization efforts.

As an additional example, knowledge about the position of network terminals may lead to much higher energy efficiency; several position-based energy-aware routing solutions were indeed proposed in the literature; see, for example, the work in Ref. [28]. The impact of position information accuracy on actual energy efficiency was investigated as well (e.g., in Ref. [29]).

The following subsection addresses another potential source of additional information and capabilities, which is the availability of more than one radio access technology (RAT) in the same device.

5.2.3.1 Energy-Efficient Routing in Multi-RAT Networks

Notably, most of the efforts devoted to the design of energy-efficient routing were devoted so far to scenarios assuming the availability of a single radio interface.

This is in most cases no longer the case, as modern radio terminals typically sport a plethora of RATs, each characterized by different transmission ranges and energy consumption levels. Taking into account the presence of multiple radio interfaces in the selection of the most energy-efficient path may indeed open the way to a new set of solutions, beyond what is achievable within the boundary of a single RAT. It should be noted however that taking advantage of such feature requires the development of new capabilities and functionalities within a radio device, as well as the design of protocols capable of dealing with the potential inconsistencies caused by the use of different radio interfaces across the same path.

A first functionality that would need to be introduced is a cross-RAT routing module, capable of collecting the information related to all available radio interfaces, integrating them with models of energy consumption for each of such interfaces and eventually selecting the best end-to-end path, either through a distributed selection procedure or by a centralized network entity. Figure 5.8 presents the module and its position in the protocol stack, as the collecting point of status information from each available RAT and the requirements set by traffic flows coming from the above layers.

It is interesting to note that the problem of multi-RAT routing may be further complicated by potential interactions between different RATs. It is in fact rather common for two or more RATs active in a radio terminal to operate in the same frequency range (WiFi and Bluetooth operating both in the 2.4 GHz band is a fitting example), with the side effect of having the decisions of the routing module taken for a specific RAT to potentially cause a feedback effect on other RATs.

In addition, as previously mentioned, particular care would need to be taken in order to ensure proper behavior of routing procedures when multiple RATs are used: well-known issues like hidden/exposed terminal problems and packet overhearing might easily be amplified by the use of different RATs on the links composing a multihop path. A preliminary analysis of such issues is proposed in Ref. [30] focusing on CR devices.

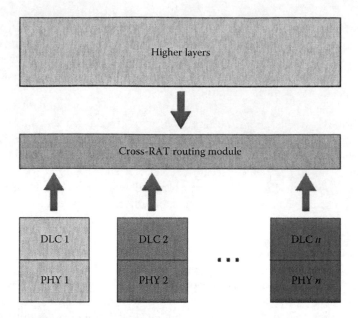

FIGURE 5.8   Representation of a cross-RAT routing module.

5.2.3.2   Standardization Efforts Related to Energy-Efficient Routing

Most standardization activities in the field of routing and higher layer protocols nowadays take place within the boundary of the Internet Engineering Task Force (IETF). Traditionally, the IETF MANET working group was responsible in the past years to propose and standardize routing protocols in MANETs, leading to the release of a wide set of both proactive and reactive routing protocols [31]. More recently, however, the increased general interest toward application scenarios typical of WSNs led to the creation of a second working group specifically dealing with routing issues in such networks. The IETF Routing Over Low power and Lossy networks (IETF ROLL) [32] working group started work in February 2008 with the specific goal of defining routing protocols capable of taking advantage of the peculiar characteristics of sensor networks, such as high number of nodes, low or no mobility, and directed traffic toward sinks, in order to achieve robust and power-efficient routing. At the same time, the group aims at defining a routing protocol capable of dealing with multiple different link layers with different constraints in terms of energy availability and frame length.

In the above context, the algorithm currently being refined within the IETF ROLL working group, labeled Routing Protocol on Lossy networks (RPL), relies indeed on the high directivity in traffic exchanged in sensor networks in order to achieve high energy efficiency, by building a hierarchy and a corresponding gradient related to the distance of a node from its reference sinks. Topologies generated by the RPL protocol are in fact typically based on directed acyclic graphs having the root in one of the sinks. The protocol also includes specific provisions to prevent unnecessary energy consumption, for example, by avoiding periodic exchanges to keep links alive, relying rather on reactive mechanisms that refresh links and local connectivity information when required for a data exchange [33].

5.3   CROSS-LAYER APPROACH

Following the per-layer analysis presented in the previous section, the holistic view on the cross-layer interactions will be discussed below. Such an approach seems to be particularly important, since the overall efficiency, including energy efficiency, of the current and future wireless communications systems strongly depend on the quality and amount of information exchanged between neighboring, not necessarily adjacent, layers. We will start with the PHY/MAC relations, extended to network/transport layer interactions in the next part of the chapter. Finally, the overall cross-layer analysis is presented.

5.3.1 PHY/MAC Layer Relations

Intelligent future wireless networks are expected to have the capability to adapt to a wide range of multiple protocol standards and across different OSI layers for collaboration between incompatible systems. For instance, a CR-based system is expected to have a very dynamic PHY layer adaptation for spectrum sensing, selection from a wide range of operating frequencies, rapid adjustment of modulation waveforms, and adaptive power control [34]. The flexibility and dynamic adaptation of intelligent networks introduce significant implications on the design of communication protocols for efficient resource management and QoS guarantees. It is therefore important for future intelligent systems to support the cross-layer design of algorithms and protocols that adapt to changes in physical link quality, radio interference, network topology, and traffic demand [34]. The idea behind cross-layer approach is to make the OSI protocol stack responsive to variations in the underlying network conditions in order to maintain an optimal operating point in wireless networks [35]. Cross-layer design between layers will allow runtime information sharing through direct communication, for example, between PHY and MAC layers [35]. Runtime information sharing between the lower layers (PHY and MAC) is crucial in CR systems, which are expected to dynamically adapt their operating parameters in a possible wide range of functionalities and values.

For instance, the interaction of the PHY and MAC layer will ensure that spectrum sensing and resource management are jointly optimized to improve the performance and energy efficiency [36] of the CRN. Figure 5.9 shows the cross-layer design approach of a typical intelligent radio system (e.g., CR) in an ad hoc network setup. Spectrum sensing function normally occurs at the PHY layer; however, it also spans into the MAC layer. Unlike in cellular networks, spectrum sensing in ad hoc CR networks is scheduled and controlled by end users (there is no central entity). Thus, each user terminal can decide when to start the sensing function, for how long, and when to stop sending in order to start transmitting [37]. DSA is performed at the MAC layer, but it relies on incumbent detection (i.e., spectrum sensing or geo-location database) and the current spectrum condition. Resource management in intelligent radio systems is also a cross-layer problem since it occurs in both the PHY and the MAC layer.

The interaction of the PHY and MAC layer (i.e., cross-layer design) can be optimized to address energy efficiency in future intelligent networks. Such optimization can also be extended to all the OSI layers since energy consumption of the wireless systems affects all layers [36]. Therefore, the CR

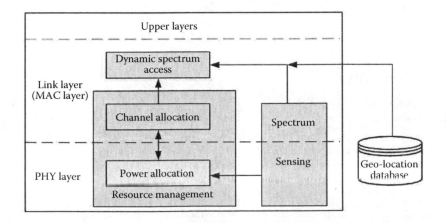

FIGURE 5.9   CR system PHY/MAC cross-layer design.

system's capability to sense the spectrum, to learn, and to adapt to the operating environment relies on direct interaction and runtime information sharing between the layers. It is therefore important for future intelligent networks standards to support the cross-layer design approach for efficient utilization of radio resources, especially the energy. Such standardization efforts should provide guidelines on how the interfaces between two (e.g., PHY and MAC) layers are to be designed and on the information flow. Simple guidelines may include the decision to use the geo-location database or spectrum sensing data, the amount of sensing time and how should it be shared at runtime (between the PHY and MAC layers), coordinating the channel coding and modulation process, and adjustments of parameters. However, care should be taken to ensure that cross-layer design does not introduce additional energy consumption, but instead makes the future intelligent networks more energy efficient.

5.3.2 PHY/MAC AND NETWORK/TRANSPORT LAYER RELATIONS

The analysis carried out in Section 5.2.3 highlighted how most of the solutions proposed in the past for introducing energy awareness in the network layer must rely on information on the status and capabilities of lower layers. A direct relation between PHY and network layer, in particular, will be fundamental in deploying energy-efficient network solutions under several aspects:

- Estimation of energy consumption related to PHY functions (transmission/reception) as already shown in Section 5.2.3
- Introduction of sensing information in routing, for example, in the case of underlay cognitive networks required to coexist with other systems, where information on the presence and characteristics of such systems could be integrated in the routing decision (see, e.g., Ref. [38])

Interaction between MAC and network will be relevant as well, as it enables the routing module to take informed decisions on the basis of the current status of the MAC domain belonged to by each node eligible to be selected in the end-to-end path. Congestion status, measured by average length of packet queues, as well as average number of packet collisions and back-off events in medium access will be valuable inputs in the route selection process. The interaction would be even more important in the multi-RAT scenario addressed in Section 5.2.3, where reliable and updated information on the status of each RAT is paramount in order to maximize routing efficiency and thus minimize energy consumption.

Interaction between higher layers, for example, transport layer, and PHY/MAC can be envisioned as well as a means of improving efficiency in the operation of a wireless network. A study of the literature reveals several performance issues related to lack of interaction between transport and lower layers. A well-known example is the unnecessary adjustment of contention windows in transmission control protocol (TCP) in reaction to packet loss owing to bad channel conditions [39], which could be avoided by enabling TCP to obtain information about channel conditions. The introduction of such direct relation would, however, benefit the network also in terms of energy efficiency, by leading to a lower number of transmissions and thus lower energy consumption.

5.3.3 CROSS-LAYER APPROACH

Finally, considering the cross-layer point or view, it has already been alluded to earlier in this chapter that optimal consideration of energy consumption and energy-saving options needs to take a holistic viewpoint, and that viewpoint also encompasses the joint consideration of all layers in the process. Nowhere is this more apparent than in the application and transport layer characteristics and the effects on energy-saving opportunities at lower layers, often taken advantage of through appropriate planning into the future facilitated by learning mechanisms in intelligent mobile/wireless networks.

Wherever there are elements that on the surface would appear unpredictable, but under the surface do yield predictable patterns that cannot be generalized to all deployment scenarios, machine learning mechanisms can be used to understand and predict those underlying patterns. Such scenarios are highly prevalent where humans are involved and therefore also apply to their data and mobile/wireless usage patterns. Machine learning in the context of cross-layer solutions to energy saving can learn about variations in users' data patterns and behavior and dynamically map those to the most energy-efficient solutions at lower layers, such as a simpler PHY or lower duty-cycle MAC, or even an opportunistic (or delay tolerant) routing at the network layer. Moreover, the use of energy-saving modes for radio network equipment in conjunction with dynamic radio resource planning reconfigurations can be planned for on the basis of learned usage and traffic variations, in conjunction with learned knowledge about the mobility patterns of users.

The incorporation of such considerations into standards should encompass the possibility for those standards to support a range of reconfiguration options, for example, in terms of resource planning and different PHY and MAC options that are coordinated appropriately with other systems in the radio proximity. Moreover, it should incorporate decision-making entities that are able to monitor the situation in terms of traffic variation, QoS, and other aspects, and utilize that monitored information to learn and plan for the future, also encompassing the ability to direct reconfiguration form/transport policies for making such reconfigurations to appropriate entities. Such characteristics can be found in the IEEE 1900.4 architecture [40], as supported by elements of IEEE 1900.5 [41], for example.

5.4   ENERGY EFFICIENCY OF INTELLIGENT RADIO SYSTEMS: CURRENT AND FUTURE RESEARCH

Following the discussion on the influence of particular layers as well as of the cross-layer relations on the total energy consumption of the wireless communications systems, let us now present the wide possibilities offered by the intelligent radios and networks and highlight the challenges related with the introduction of such smart solutions. Not only should future intelligent radio solve the problem of low utilization of available frequency resources, but one of its main goals should also be the lowering of the overall energy consumption in the system, starting from user terminals, through BSs and relays, finishing at specific entities of the core network.

5.4.1   CHALLENGES

Future intelligent networks using cognitive technologies are expected to provide improved user experiences such as higher throughput, acceptable QoS, and better quality of experience. For instance, an intelligent radio system may be expected to integrate with the smart home devices where it will be able to turn "on" the geyser once it is within the vicinity of the user's home or to read the weather forecast in order to advice the user on what to wear. To achieve these demands, such user terminals need to have high processing power and faster signal processing chips. This in turn consumes additional energy of the wireless networks and systems.

5.4.1.1   RF Front-End and Signal Processing

Despite more than 10 years of existence, with active research and development, there is still no clear path for practical implementation of CR systems in the real environment. One of the challenges yet to be addressed is finding a practical technique for performing wideband spectrum sensing by CR terminals. This is mainly due to the high expectation that CR systems should be able to reconfigure themselves and process signal to any kind of communication technologies [42]. This will require the design of linear and high-speed reconfigurable components in the RF front-end, baseband processing and data processing of the transceiver, as shown in Figure 5.10. This problem has already been mentioned in Section 5.2.1 but there is a need to mention this issue here, since the design of energy-efficient and cheap cognitive front-end still seems to be the bottleneck of the current and future wireless systems.

For DSA operation, CR systems should be capable of detecting exiting signals (both strong and weak) within an operational spectrum range. This requires a very high sensitive RF front-end with

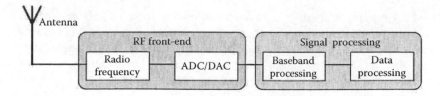

FIGURE 5.10 Simple CR transceiver.

a great level of selectivity. To improve the sensitivity of the RF front-end, several techniques, such as the use of multiple antennas, are proposed [43]. Regardless of the technique proposed to improve RF front-end sensitivity, it will introduce either new/additional components or high-speed processing (e.g., high-speed A/D converters). The raw data (received signal on spectrum sensing) gathered at the RF front-end should be processed in order to detect the presence of licensed users (through signal measurements). Such signal processing action required at this level should happen very fast to enable decision making. Thus, high-speed digital signal processors have to be used. High-speed processing consumes more energy and results into lots of heat dissipated from the chips. A challenge is to perform all the processing required by the CR systems in the most energy-efficient way. Cooling of electronic components within small and portable terminals is another interesting topic for energy efficiency in communications networks.

5.4.1.2 Future Wireless Broadband in Developing Countries

Wireless broadband provision in rural areas, especially in the developing countries, is one of the challenges experienced by the cellular network operators. It is estimated that over 1 billion people around the world still lack access to telecommunications services, and the majority of these people are in the developing countries [44]. Despite generating the low average revenue per user of $3/month to cellular operators, rural areas are also affected by lack of reliable electricity supply, lack of technical skills, and large amount of dispersed settlements. The lack of electricity grid or reliable electricity supply in some African countries means that alternative energy sources, such as diesel generators and solar systems, are used to power the remotely located BSs. From the consumer side, powering and recharging their user terminals (e.g., laptops and cell phones) remain a nightmare. In most cases, alternative energy sources (such as solar) are used to power and recharge user terminals. Recharging of user terminals (such as mobile phones) becomes expensive to consumers since one has to either travel to the nearest town for recharging or pay the local entrepreneur for recharging using solar systems. It is therefore important to ensure that future intelligent radio systems deployed in such areas are energy efficient (e.g., prolonged battery life of user terminals), low cost, self-healing, and remotely reconfigurable.

The cognitive system-based standard should consider the wireless systems deployed not only in urban areas (i.e., in areas where electricity grid exists) but also in remote rural areas where, for the lack of electricity supply, there are alternative energy sources (such as diesel-powered generators and solar systems). The standard should provide guidelines to the wireless systems vendors to develop intelligent wireless systems capable of learning their external operating environments and of reconfiguring and adapting. For instance, the BS should be able to know whether it is powered through the electricity grid or an alternative source. This will enable it to adapt its operations on the basis of the power source. Through the feedback mechanism from the power supply and the external environment, it can switch off some energy-consuming components to prolong the operating hours before the next fueling or battery recharge.

In wireless networks, providing connectivity to the cell edge users is another challenge also from an energy efficiency perspective. Especially in dispersed population and mountainous areas (with hills and valleys), network coverage planning and optimization become a challenge. WiFi offloading has been proposed to address cell edge users (especially in densely populated areas), but different mechanisms are required for sparsely populated remote areas. The recent IEEE 802.22 CR standard is one of the solutions to the rural areas. Television white spaces (TVWS) offer favorable propagation characteristics (longer distance coverage) and are the most suitable spectrum for rural or low-density areas [45]. The TVWS-based networks promise to be more energy efficient and cost effective since they require fewer cells (BSs) for a larger coverage area (more villages can be covered from a single BS within a radius of over 70 km). The TVWS can also be used for long-distance point-to-point links, thereby providing backhauling for remotely located networks.

In summary, users and operators from different regions (i.e., developed and developing) have different challenges and needs when it comes to accessing and providing intelligent wireless networks. For instance, the developed regions have sufficient supply of electricity, while most of the developing regions have insufficient supply of electricity. It is worth mentioning that, because of lack of electricity supply, some cellular operators in the developing regions rely on diesel power for their remote BSs [44]. Despite these differences, the deployment of energy-efficient wireless networks is a common goal shared by all the regions around the globe. However, it is therefore important for future communication standards to consider the deployment region for specific technology in order to address the energy efficiency requirements of a specific region (e.g., where there is lack of reliable electricity supply vs. regions with sufficient electricity supply).

5.4.1.3 Spectrum Sensing and Geo-Location Database Deployments

In licensed RF spectrum bands, incumbent detection for DSA-based networks introduces additional issues to both operators and consumers. As mentioned in previous subsections, spectrum sensing and geo-location database are the commonly preferred mechanisms for incumbent detection in DSA-based networks. We are aware of other mechanisms, such as beaconing and the use of pilot signals, but in this chapter, we focus mainly on spectrum sensing and geo-location database. Both these mechanisms introduce additional energy consumption into the communications network. For instance, the number and size of geo-location databases coupled with their high processing power will definitely increase the electricity bill of the operators or whoever owns them.

However, the decision to use either spectrum sensing or geo-location database depends on the deployment area and technology. For instance, in developing countries, particularly in the rural areas, the RF utilization is very low [46] when compared to urban areas. As a result, there may be no need to perform regular incumbent detection (spectrum sensing), but a geo-location database can be used at a low update rate (e.g., per week or more). This will lead to more energy-efficient deployment of intelligent wireless networks in rural areas than urban areas.

5.4.2 OPPORTUNITIES

With the increasing spread of intelligent mobile devices such as smart phones, a huge range of new possibilities is being opened up. However, this also introduces a challenge to the network operator, which is now requested to serve more and more "greedy" bandwidth users. It is estimated that the Internet protocol (IP) traffic will increase threefold over the next 5 years [47]. The idea of improving the capacity of a wireless link is not obvious since we are getting closer to the Shannon capacity limit [48]. A possible solution is to increase the number of BSs deployed so that the distance between the user device and its serving BS is reduced. However, a high density gives rise to several issues such as lack of available channels and thus increased co-channel interference and also increased energy consumption. One possibility considered a few years ago is to deploy small cellular BSs in small office and home office environments. This mini "base station" is referred to as "femtocell." A femtocell is a tiny box, similar in aspect and size to a wireless router but that supports cellular protocols such as UMTS and 4G (LTE-A). A femtocell is a plug-and-play device installed by the end user without intervention or prior localization planning by the cellular operator. The backhauling of the traffic to the core cellular network is done via classical Internet connection already available at the user home such as ADSL. Therefore, it has to fulfill self-organized properties especially for the radio resource management. There exist two kinds of femtocells, where each type provides a significant amount of energy saved, which will be discussed in detail in the following subsections.

5.4.2.1 Indoor Femtocell

The concept of indoor femtocell is at the origin of this technology. At the beginning, the main motivation was to enable users to experience better coverage when indoors as well as to increase the signal to interference plus noise ratio (SINR), which leads to an increased throughput [49]. But it appears to also be a very good energy-efficient solution for several reasons, both from the user and the cellular operator points of view. From the user side, when a femtocell is installed in the home, the received signal is clearly strengthened, which means that the device will not have to work hard (and consume more energy) to be able to decode the downlink signal from the BS. If we consider the uplink communication, this is even more emphasized since a large part of the energy consumption of the user device is expensed when transmitting to the BS. Thanks to the close location to the femtocell, the uplink transmitted power can now be meaningfully reduced, especially for these indoor users that otherwise would have to transmit at a high power to overcome the high path loss because of, for example, building penetration loss. This will also lead to the device's longer battery life. But this is also relevant for outdoor users in proximity of an indoor femtocell. In fact, there are several possibilities in allowing external users to connect to an indoor femtocell depending on the policy decided by the user who owns the femtocell. In the open access [50] scheme, the femtocell owner gives access to everyone (assuming they subscribed to the same cellular operator), whereas in the closed subscriber group scheme, only authorized users can connect through the femtocell. Therefore, in the former scheme, an outdoor user who experiences poor coverage from the macrocell BS could switch to the nearest femtocell and save energy in both downlink and uplink as previously explained. It is worth mentioning that in this way, we also deal with the major health concerns associated with cellular transmissions, since we also reduce the electromagnetic radiation near the head of the users [51]. It is also important to mention that a femtocell can also include WiFi technology and therefore does not require the user to hold two different boxes and hence double his energy consumption. From the operator point of view, this is also a good energy-saving potential since indoor users are known to be the most difficult to cover because of penetration loss as mentioned before. Thanks to a femtocell, energy saving occurs both in uplink, when processing the extremely low signal from the user, and in downlink, since there is no longer a need to cover these users and transmit close to the maximum authorized transmitted power.

Even though the main motivation was to improve indoor coverage, it has been quickly identified that the femtocell is in fact also a great potential solution for coverage of outdoor users. In fact, the concept of smaller-cell-than-macrocell coverage is not new. At the end of the 1980s, the microcell size was introduced [52]. Then, a concept of hierarchical networks or multitier (also called multilayer) cellular systems rapidly appeared to be a logical extension of the cellular system. It was proposed to shrink cell size to be able to support more users. This eventually gave rise also to the picocell size, where in this latter case the classical scenario is for airport or mall coverage. However, there is a main difference between femtocell technology and old techniques such as the picocell and microcell. These latter methods behave in fact almost exactly like a macrocell BS but just with a smaller coverage owing to a reduced transmitted power. However, all the cooling system, databases, and other common core components of a cellular system are still required since backhauling for these methods is done in the classical way, whereas the femtocell is a domestic small box that does not require all these additional components. In this way, the energy saved by the femtocell is far more important than what can be saved by simply reducing the transmission power of the macrocell to a microcell.

There are two main schemes that can be considered when one wishes to use an outdoor femtocell (denoted OF in the following) to save energy, namely, femto-gridding and femto-bordering [53].

5.4.2.2 Femto-Bordering

In femto-bordering, the cellular operator deploys OF in the edge of the macrocell. This translates into energy saving for both users and operators as detailed previously for the indoor femtocell.

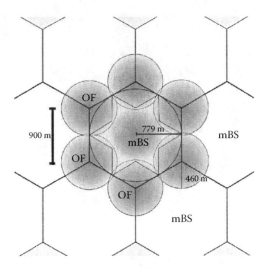

FIGURE 5.11 A macrocell surrounded by six outdoor femtocells. (From Y. Haddad and Y. Mirsky, Power efficient femtocell distribution strategies, In *Proceedings of 19th IEEE International Conference on Software, Telecommunications and Computer Networks (IEEE SoftCOM)*, Split, Croatia, Sept. 2011.)

Outdoor users located at the edge of the cell are known to be the most complicated to cover not only because of the high path loss but also because of high interference since the users located at the edge are in fact almost equidistant to multiple-macrocell BSs. If we enable an outdoor user to transmit instead to a nearby OF, this will allow the user to transmit at a lower power in uplink and allow the operator to reduce the transmitted power of the overlaying macrocell BS since it no longer has to cover them. In Figure 5.11, we show an example of a single-macrocell BS with six OFs deployed across the edge of the macrocell.

5.4.2.3 Femto-Gridding

Pushing the idea of femto-bordering one step further, we can consider the femto-gridding scheme. The idea comes first from the observation that in some hours of the day, there is an underutilization of the BS capacity. For instance, a BS located in a densely populated business district can be highly solicited during the day but is almost never used during the night when everyone is generally out of the office. Therefore, we can imagine completely turning off the macrocell BS during this off-peak period and replacing it with multiple femtocells. In this way, we make a grid-like tessellation of the area with femtocells. In Figure 5.12, we show two illustrations of femto-gridding where deployment (referred to as independent deployment in the left-hand illustration) is done without respect to the exact border of the original overlaying macrocell; in the right-hand illustration (where deployment is referred to as dependent deployment), the radius of the OF is planned with consideration of the macrocell.

Some numerical computations [53] reported an energy saving of approximately 37% per cell if the femto-bordering is used. In the femto-gridding case, this is even more emphasized with approximately 91% of energy saving compared to regular macrocell coverage. (Note that these numbers are valid under specific circumstances and additional details have to be taken into account to get a more accurate evaluation. A short discussion about this latter point is provided further in this section.) The energy savings as a result of the deployment of indoor femtocell and reduced energy consumption from the macrocell BS is estimated to be $0.50 per consumer of the cellular operator [54].

In view of all of the above, although femtocells might offer savings in terms of measures such as sum transmission power per area capacity, it is noted that the net from-the-socket power-saving potential of femtocells often depends on the ability to use power-saving modes for those femtocells as well as aspects such as the nature of the hierarchical RAN planning. This is due to the high

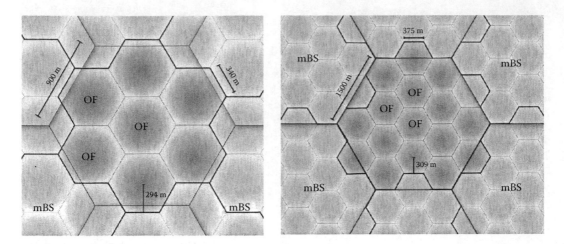

FIGURE 5.12 Two possible femto-gridding tessellations of an original macrocell. (From Y. Haddad and Y. Mirsky, Power efficient femtocell distribution strategies, In *Proceedings of 19th IEEE International Conference on Software, Telecommunications and Computer Networks (IEEE SoftCOM)*, Split, Croatia, Sept. 2011.)

necessary density of such components should they be used as the primary means of capacity provision, in tandem with the "latent" power consumption of them almost irrespective of measures such as transmission power. Statistically, femtocells are more amenable to such power-saving modes than other BS types, as the number of connected users is more likely to be low or zero.

5.4.2.4 WiFi Off-Loading

In the same spirit of femtocells, a similar opportunity for cellular operator energy saving relies on the WiFi technology. Since we assume that more and more user handheld devices support WiFi connection in addition to the regular cellular (3G, 4G) connection, the cellular operator can simply switch the user to a close WiFi hot spot when available. This is referred to as the off-loading technique and is already used by several operators to relieve the burden on macrocells [55]. It is important to distinguish this from the previous femtocell solution where we stay within the same cellular technology, and therefore, it does not require a very involved handheld device. In the off-loading technique, we consider a CR-enabled device that can switch from a technology to another one and perform what is called vertical hand-off. A second difference is that the femtocell traffic path goes through the cellular core network with dedicated capacity, whereas WiFi traffic goes through the regular Internet path, which sometimes entails long delays and does not guarantee QoS. Nevertheless, in both cases, a significant amount of energy saved is possible [56] for users together with cellular operators.

5.4.2.5 Ad Hoc Networks

It is quite interesting that when we want to send a message or establish a voice call session with a closely located user (e.g., within WiFi range), the packets still have to go through the BS and vice versa. A more efficient possibility could be the use of infrastructureless communication also referred to as ad hoc mode, where users communicate directly to each other. In this way, the gain is twofold. First, the macrocell BS is not used, and second, the required transmission power is reduced since the other user is closer. Some case studies have already been reported [57] with even the support of multi-hop when the users can be connected through a third user in the middle.

5.4.2.6 Mesh Networks

In the ad hoc network approach just presented above, we required the source and destination users to be within the same range of wireless communication or at most required some intermediate users to route the packet. Sometimes, this is not even possible, for example, due to the extremely large distance between two neighboring devices. In this case, transmitting the packet over the classical infrastructure mode through the BS may still be required. However, even in this case, we can prevent the packets from going through the core network to reach, perhaps, the close destination. For instance, assume that the source and destination users are attached to adjacent cells covered by two neighboring BSs. It could be enough to have the packet from the destination be forwarded by the serving BS to the destination BS and then to the destination users. This kind of network where BSs communicate directly without going through switching centers is referred to as "mesh networks" [58]. In this case, there is a significant amount of energy saved on the component of the core networks.

5.4.2.7 Content Caching Cellular BS

At the beginning of the cellular networks around the mid-1980s, the only service supported was voice communication. But in the last decade, this no longer true since people mostly prefer to communicate through short e-mails or "chatting" on Facebook. Thus, the focus is now on classical data services generally intended for desktop or laptop. In the regular wired network, it is well known that the content requested by the users is stored in Internet service provider (ISP) router by means of a technique known as "web caching." It reduces the traffic over the core Internet network and therefore reduces energy consumption. It could be a great opportunity for cellular operators to export this technique to cellular network [59]. Therefore, similarly to the case of mesh networks, we could also in this case save a great amount of energy preventing the BS from accessing the core network all the time to fulfill the requests of the users given that the data are already cached at the BS level. Also, cooperation between BSs through mesh networks could increase this potential.

5.4.2.8 Software-Defined Networks

From an operator perspective, it is now assumed that in the short term, we will require access to QoSoD (quality of service on demand)-based networks. These networks exhibit appealing properties such as the possibility to add or remove a service dynamically on the basis of availability, service level agreement of the customer, and so on. For this purpose, we need a complete paradigm shift on the way we manage the network. These future networks are referred to as "software-defined networks" [60] where there is a decoupling of the control and data forwarding plane. In fact, the control plane will play the same role as the operating system in the context of computers and is therefore referred to as "network operating system" in the network context. In this framework, there will be controllers that hold a global view of the network topology with real-time critical parameters such as which channel is currently used and where. Such a clear view will enable the controller to optimize the use of wireless resources and perform efficient load balancing between cellular BSs (as well as many other interesting capabilities). From an energy efficiency point of view, finding the shortest and fastest path for cellular traffic could allow the operator to completely turn off some of the BSs at some point of the day. The radio planning task, which is generally quite empirical, becomes more exact; for instance, a better control of the transmission power can be reached. This field is still in its infancy but has already been seriously considered by major manufacturers of routers such as Cisco and is also under experimental deployment at some academic laboratories [61].

5.4.2.9 Guideline toward Standardization

Some of the opportunities mentioned above do not need a complete rethinking of the network but sometimes just requires a "good willing" of the operators. Therefore, standardization bodies could help in the process of coordination and implementation of some simple rules to reach a "greener" world.

First, we propose to reduce the possibility for the operator to deploy a new macrocell BS. Instead, operators interested in extending their coverage should use outdoor femtocells at the edge of the cell.

Second, we propose to define a threshold of minimum users that justify the use of a macrocell BS at a given place and time. If the number of users currently active are below the threshold, the operator should turn off the macrocell BS and off-load the traffic to OF.

Third, a proxy server should be collocated with every BS so that there will be no useless request to the server located at a long distance from the BS.

Fourth, manufacturers and vendors of huge cellular equipment, such as BS and switching center, should be required to provide full compatibility to software-defined network capabilities so that the operator could implement intelligent and efficient policies to manage their network.

Fifth, manufacturers and vendors of handheld devices such as smart phones should be requested to implement intelligent algorithms into the devices so that ad hoc communication takes place whenever possible instead of using the energy-greedy operator's infrastructure.

5.4.2.10 Some Open Issues

At the beginning of Section 5.5.2, we mentioned that if the femtocell is considered in place of macrocell BSs, for instance, if we turn off completely the macrocell at night, this leads to a large amount of energy saved. We brought some results that came from very specific examples. However, for this idea to be feasible, we need a more general approach. Given as input the deployment of a set of cells with some typical patterns of traffic load in each cell, what would be the most efficient way to save energy? Femto-gridding or femto-bordering? How much femtocell would be needed? A general algorithm that would be able to deal with this kind of optimization problem would be welcome.

Another challenging issue that is yet to be solved concerns specifically the femto-bordering proposal. In this case, the macrocell is still on. Thus, the efficiency of using femtocells to relieve the burden on the cell edge is less obvious. However, we showed that in some cases, it can still lead to a significant amount of energy saved. The problem is that throughout the course of the study, we did not take into account the additional devices required for the femtocell to run. For instance, we omitted the energy consumption of the operator's gateway router that multiplexes the traffic from the femtocell (coming from a regular broadband access link like ADSL) into the cellular dedicated backbone. One cannot just account for the energy consumption for such a device to each femtocell deployed since this is a shared device for several femtocells.

This gave rise to a larger challenging issue, which is how to quantify the carbon footprint of shared devices such as routers. For instance, how can we assess the additional expenses (in terms of energy) that a new Internet connection bought by a user contributed to the overall carbon bill of the ISPs. We can simply account for the power used by the ISP router for each new customer but this would be erroneous since the same ISP router also serves other customers. In summary, further research is needed on the impact of the core network shared device on the endpoint (end user) carbon footprint bill.

5.4.3 CURRENT STANDARDIZATION ACTIVITIES FOR INTELLIGENT RADIO

Current standardization activities in intelligent radio are few and far between; however, standardization activities defining the elements of radio flexibility and opportunistic spectrum usage are well underway and in some cases complete.

5.4.3.1 Standards of the Institute of Electrical and Electronics Engineers

There are a large number of standards of strong interest being worked on within the Institute of Electrical and Electronics Engineers (IEEE) [62]. The key standards committees of interest within the IEEE are the IEEE 802 LAN/MAN Standards Committee (IEEE LMSC) [63] and the IEEE Dynamic Spectrum Access Networks Standards Committee (IEEE DySPAN-SC) [64].

5.4.3.2 Standards of the IEEE 802 LAN/MAN Standards Committee

Considerable effort on standardization in areas of relevance to spectrum usage flexibility is progressing within the scope of the IEEE 802 LAN/MAN Standards Committee [63]. These include the 802.22 wireless regional area network standard [65], which was the first such standard that was worked on in the area of "CR" applied to TV White Space, 802.11af [66], an amendment to WLAN standards for TV white space, 802.15.4m [67], covering the application of wireless personal area networks (WPANs) in TV white space, and 802.19 [68], handling coexistence, and specifically TV white space coexistence within 802.19.1.

5.4.3.2.1 IEEE 802.22

IEEE 802.22 was published relatively recently in July 2011. This standard aims to take advantage of the negative correlation between population density and the availability of TV channels for opportunistic use, thereby providing a viable means to provide broadband access to customers in sparsely populated areas to which high capacity wired broadband connectivity may not be a financially viable option. Such a standard is particularly applicable to areas such as rural United States, for example, where there are large sparsely populated areas interspersed by major communications hubs (cities) and towns [69].

IEEE 802.22 assumes a hierarchical architecture, that is, the definition of a network of communicating BSs that are able to transmit/receive data in a point-to-multipoint fashion to/from "consumer premises equipment (CPE)," located on the end users' homes. Moreover, in line with recent regulatory changes, 802.22 BSs refer to a geo-location database for information on local channel availability, and CPEs can only transmit upon being "enabled" by the BS through receiving a signal from a BS and associating with it.

In addition to ensuring that the interference is not caused by the primary system (which, of course, is a regulatory requirement), the standard also implements a number of self-coordination aspects. In such regard, it borrows heavily from standards such as 802.16 [70] in terms of organizing channel usage among cells (i.e., cellular frequency reuse) or even contention among cells if they need to exist on the same channel (e.g., as a result of there being a limited number of available TV channels in the area to opportunistically use). Such coordination is achieved through messages enclosed in "coexistence beacon protocol" bursts.

Although IEEE 802.22 is published, along with IEEE 802.22.1, which is a standard within the working group developing a beaconing protocol to assist coexistence [71], work in IEEE 802.22 is ongoing in the specification of a recommended practice for deployment of IEEE 802.22 devices (IEEE 802.22.2 [72]), and two amendments for which project authorization requests (PARs) have recently been approved: IEEE 802.22a, "Management and Control Plane Interfaces and Procedures and enhancement to the Management Information Base (MIB)" [73], and IEEE 802.22b, "Enhancement for Broadband Services and Monitoring Applications" [74].

5.4.3.2.2 IEEE 802.15.4m

IEEE 802.15.4m [67] is a very recently proposed amendment to the 802.15.4 (ZigBee) standard, developing a PHY for operation in TV white space. The PAR for work on 802.15.4m was approved, allowing work to begin on the amendment, only in September 2011 [75]. The amendment aims to achieve data rates of between 40 kbps and 2 Mbps, minimizing infrastructure requirements; the standard also specifies an intention to adapt the 802.15.4 MAC as necessary to be compatible with the created PHY layer. A reading of the PAR suggests the standard aims to serve applications such as machine-to-machine (M2M) communications, in addition to more conventional WPAN applications. Much of the early work on the standard has considered aspects such as requirements for the standard; hence, there is little detail to report thus far on the precise specification resulting from the standard amendment.

5.4.3.2.3 IEEE 802.11af

The IEEE 802.11af task group has been working on amending the IEEE 802.11-2007 standard for operation in TV white space. The PAR for IEEE 802.11af was approved in January 2010 [76]. The task group has undergone its first internal "Letter Ballot" in January 2011 and has since then been addressing comments received in that ballot. Minutes of the recent Jacksonville meeting of 802.11 indicate that a second attempt at Letter Ballot is intended to be made in March 2012, and the current project timeline is indicating progression to Sponsor Ballot in November 2012 and completion of the standard in mid- to late 2013. However, this depends entirely on the comments that are received in the second Letter Ballot and Sponsor Ballot and the time it takes the task group to address them.

The exact specification of IEEE 802.11 standards (and amendments) is closed until publication, except to participants in the associated meetings. However, it is clear that much of the work in 802.11af is related to interaction with the geo-location database, the satisfaction of other requirements as per regulations in various realms such as transmission power management for devices, and implementing the requirements for operation in TV white space (e.g., channel bandwidths, centre frequencies, OFDM tuning, etc.). Some interesting further additions do apply in the case of 802.11af, such as support for the developing 802.11 high-throughput PHY, which of course is being handled in close cooperation with the IEEE 802.11ac task group, which aims to achieve this task in the sub-6 GHz band. Detail on the developing IEEE 802.11af amendment standard is, however, very tentative and likely to change in forthcoming Letter and Sponsor Ballots. One relevant observation, however, is that, as alluded to above, an element of the standard (perhaps the PHY/MAC) looks to be developed from down-clocked versions of IEEE 802.11ac.

At the time of writing, IEEE 802.11af is still at the internal Letter Ballot stage, with the time to publication looking to be at least another year.

5.4.3.3 Standards of the IEEE Dynamic Spectrum Access Networks Standards Committee

The IEEE DySPAN-SC [77] manages the 1900 series standards on the topic of DSA and DySPAN. These standards are more generic and in many cases facilitate or incorporate the possibility of intelligent systems.

The current IEEE 1900 working groups are as follows:

- IEEE 1900.1: "Definitions and Concepts for Dynamic Spectrum Access: Terminology Relating to Emerging Wireless Networks, System Functionality, and Spectrum Management" [78]
- IEEE 1900.2: "Recommended Practice for the Analysis of In-Band and Adjacent Band Interference and Coexistence Between Radio Systems" [79]
- IEEE 1900.3: "Recommended Practice for Conformance Evaluation of Software Defined Radio (SDR) Software Modules" (disbanded) [80]
- IEEE 1900.4: "Architectural Building Blocks Enabling Network-Device Distributed Decision Making for Optimized Radio Resource Usage in Heterogeneous Wireless Access Networks" [81]
- IEEE 1900.5: "Policy Language and Policy Architectures for Managing Cognitive Radio for Dynamic Spectrum Access Applications" [82]
- IEEE 1900.6: "Spectrum Sensing Interfaces and Data Structures for Dynamic Spectrum Access and other Advanced Radio Communication Systems" [83]
- IEEE 1900.7: "Radio Interface for White Space Dynamic Spectrum Access Radio Systems Supporting Fixed and Mobile Operation" [84]

Figure 5.13 illustrates the current active working groups within DySPAN-SC.

5.4.3.3.1   IEEE 1900.1

IEEE 1900.1 was instantiated in March 2005 under the realization that many of the terms used in the fields of spectrum management, policy-defined radio, adaptive radio, software-defined radio, reconfigurable radio and networks, and related technologies do not have precise definitions or have multiple definitions. IEEE 1900.1 facilitates development of such technologies by clarifying the terminology and how these technologies relate to each other. The original standard was published

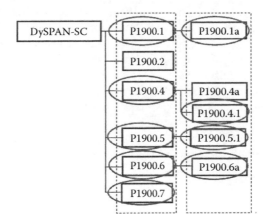

FIGURE 5.13 Working groups and standards within IEEE DySPAN-SC, with the currently active ones circled. (Heavily adapted from IEEE DySPAN-SC end of year report 2011, online: https://mentor.ieee.org/dyspan-sc/dcn/11/sc-11-0033-00-MTNG-year-end-report.pdf (requires IEEE Standards Association membership), accessed: March 2012.) Note that the IEEE 1900.2 standard has been published and the working group is not currently meeting.

in 2008, although the standards work has recently been reinitiated to incorporate new terms and definitions that have emerged since the publication of the original standard. This reinitiated work results in the finalization of the IEEE 1900.1a amendment at the moment of writing this deliverable.

5.4.3.3.2 IEEE 1900.6

The IEEE 1900.6 standards work was initiated in September 2008, and the standard was published in late April 2011. IEEE 1900.6 defines the information exchange between spectrum sensors and their clients for advanced radio communications systems or, more specifically, DSA systems. This is applicable to both cooperative/collaborative sensing scenarios and other scenarios where the intelligence that makes spectrum access and other decisions is at a different location from the spectrum sensors. The logical interface and supporting data structures in 1900.6 are deliberately defined abstractly, without constraining the sensing technology, client design, or data link between the sensor and client. Of course, the facilitation of sensing technologies through IEEE 1900.6 assists many spectrum coexistence techniques such as CR that utilize distantly obtained spectrum information. IEEE 1900.6 continues in the form of work on an amendment looking at adding "Procedures, Protocols, and Data Archive Enhanced Interfaces."

5.4.3.3.3 IEEE 1900.7

The IEEE 1900.7 standards work was initiated in September 2011, after a relatively long study period. IEEE 1900.7 work is concerned with defining a PHY/MAC for white space access. The aim is for this PHY/MAC to be generic, applicable to a range of possible white space bands and use cases, not constrained in the way that the radio interfaces of other TV white space standards are. Initial work in IEEE 1900.7 has considered use cases, spectrum bands for operation, and various generic requirements among other aspects.

5.4.3.4 The Weightless Standard for M2M over White Space

A relatively new standards effort is the Weightless standard for M2M communication in TV white space [86]. The requirements for such communications are low cost, excellent coverage, ultra-low power operation, and secure and guaranteed message delivery [86]. The inaugural meeting of this standards group, the Weightless SIG, was held September 30, 2011. As the initiators inform, "Weightless is not 1G, 2G, 3G or even 4G—it is ZERO G" [86]. One of the observations made by the researchers from the Weightless SIG is that the nature of spectrum access of various devices and the characteristic of the machine traffic are so different from human traffic that the use of the existing standard is not optimal. The main directions that can be identified within this document can be summarized as follows: (a) first, the single-carrier technique will be used with variable spreading factors in order to allow the flexibility of the data rates, (b) TDD scheme will be applied, (c) the need for relatively long frame duration up to 2 s, (d) application of the frequency hopping technique, and finally (e) broadband downlink (6–8 MHz) and narrowband uplink channels.

5.4.3.5 ETSI Reconfigurable Radio Systems

The ETSI Reconfigurable Radio Systems (ETSI-RRS) Technical Committee [87,88], which was created in January 2008, comprises four working groups, namely:

- WG1 System Aspects
- WG2 Equipment Architecture
- WG3 Cognitive Management and Control
- WG4 Public Safety

The scope of this technical committee is on standardization activities related to RRS encompassing system solutions related to SDR and CR; the committee's activities include the definition of RRS requirements, identifying gaps where existing standards do not fulfill those requirements, and proposing solutions to fill those gaps. This committee also aims at more generic forms of standards, much like IEEE DySPAN-SC, which could be applicable to intelligent radio systems or could support them. However, much of its work is "pre-standardization," for example, identifying issues or illustrating topics that could be standardized.

ETSI-RRS has produced several technical reports (TRs), for example, about the cognitive pilot channel [89,90] and the functional architecture for the management and control of RRS in a heterogeneous wireless radio environment [91].

Since September 2009, the mandate of ETSI-RRS has been extended so that technical specifications can also be produced.

The objective of a currently ongoing relevant work item in ETSI-RRS is to undergo a feasibility study on control channels for CR systems (DTR/RRS-03008 [TR 102 684]) [92]. The scope of this work item is "to identify and study communication mechanisms: (1) for the coexistence and coordination of different cognitive radio networks and nodes, operating in unlicensed bands like the ISM band or as secondary users in TV white spaces; (2) for the management of Opportunistic Networks, operating in the same bands as mentioned above." The latest draft of the relevant TR addresses various implementation options and presents several detailed solutions that are either system independent (such as an IP-based provision of cognitive information) or system dependent (such as the introduction of new MAC/RRC messages). More specifically, various options are outlined, including solutions based on IEEE 1900.4, Diameter, ANDSF, Distributed Agents, 3GPP RRC, IEEE 802.21, IEEE 802.11, WiMedia, and Bluetooth for implementation of the previously mentioned concepts. It is expected that a hybrid solution will finally be adopted building on a combination of several of those techniques.

5.5 CONCLUSIONS

Although efficiency is widely understood and has long been one of the key goals in the development of wireless communication systems, the pressing need to improve energy efficiency has been increasingly emphasized in recent years both at the academic and industrial levels. Year-by-year increases in data traffic along with the challenging requirements of end users (individuals and also companies) regarding QoS levels are only some reasons for the growth of overall energy consumption by the ICT industry. Moreover, methods to provide projecting data capacity requirement increases, such as increases in frequency reuse, imply a spiraling of energy consumption going into the future.

In this chapter, we have set out to demonstrate the need for a holistic view on the energy consumption of the whole wireless communications system. We started by discussing how the particular algorithms and solutions dedicated to each of the layers of the protocol stack interact with each other and how they might improve or otherwise degrade energy saving because of this interaction. Some guidelines for future standards have also been proposed, on the basis of such observations. It has been observed that new wireless communications standards should, first, consider solutions that will not contradict the effect of other procedures applied in the system (within the same layer but also at other layers) and, second, concentrate on overall energy efficiency of the whole system.

Having in mind how each layer contributes to energy consumption, a brief analysis of the cross-layer interactions has been presented. It has been shown that current wireless systems strongly rely on the information exchanged between neighboring layers and this trend will even further increase in the future. Cross-layer approaches are usually characterized by higher flexibility and efficiency than layer-based solutions, but require, in general, more energy to function properly. Thus, particular attention has to be put on the energy efficiency of such sophisticated, fully flexible, and adaptive solutions.

The conclusions summarized above have also been the basis for the analysis of opportunities and challenges of the promising techniques that will most likely be applied in future wireless systems. In this chapter, such an analysis has been made from the energy efficiency perspective, where it has to be highlighted that future smart and intelligent systems based on currently developed solutions have big potential for further optimization of energy consumption. Such an observation has also been linked to the last part of this chapter, where ongoing standardization activities have been discussed. Through this link, it is evident that "resource efficiency," understood mainly as spectrum resource as well as energy, is the main direction of action. Taking this into account, we can conclude that significant effort will be put on making the next wireless communications systems more energy efficient, whereby there is a strong interrelation between energy efficiency and spectrum efficiency, and in some sense, this observation is already supported by the standardization bodies. On the basis of the presented analysis, a clear message should be announced among the whole community: lowering the energy consumption of ICT is an urgent need, and the means to achieve such reduction can be realized.

ACKNOWLEDGMENTS

This work was supported by COST Action IC0902—"Cognitive Radio and Networking for Cooperative Coexistence of Heterogeneous Wireless Networks" (http://newyork.ing.uniroma1.it/IC0902/), by COST Action IC0905—TERRA – "Techno-Economic Regulatory Framework for Radio spectrum Access for Cognitive Radio/Software Defined Radio" (http://www.cost-terra.org), and by the Network of Excellence ICT-ACROPOLIS (http://ict-acropolis.eu/).

AUTHOR BIOGRAPHIES

Adrian Kliks received his MSc and PhD degree in telecommunication from Poznan University of Technology in 2005 and 2011, respectively. Since 2005, he has been employed at the Institute of Electronics and Telecommunications, and since 2007, he has been with the Chair of Wireless Communication in the Faculty of Electronics and Telecommunication. He was hired first as a senior researcher (2007–2009) and then as an assistant (2009–2011), and in October 2011, he became an assistant professor.

His research interests cover the wide spectrum of wireless communications. In particular, he is interested in multicarrier (both orthogonal and non-orthogonal) systems; in the area of software-defined, adaptive and cognitive radios; in hardware implementation on DSP/FPGA; and in radio planning. From 2007 to 2008, he was a member of the PhD self-government at the Faculty of Electronic and Telecommunication and the chairman of the PhD Self-Government at the Poznan University of Technology. He was the originator of organization and a co-chairman of the organizing committee at the First International Interdisciplinary Conference of Young Scientists InterTECH, which was held last June 17–18, 2008. He was also involved in the preparation of the national conferences in the area of wireless communications, as well as of the European Wireless conference held in Poznan in April 2012. Adrian Kliks is a member of the IEEE for 8 years, and since 2012, he has been a senior member. He was (and still is) involved in industrial and international projects (like ICT-URANUS, NoE NEWCOMM++, COGEU, ACROPOLIS, COST Action IC-0902, COST-Terra, and Newcom#), where he also acted (acts) as the task leader. During the course of these projects, he was awarded with international exchange grants (realized in University of Pisa and CTTC in Barcelona). He also acted as a reviewer for various journals (e.g., *IEEE Transactions on Wireless Communications Wireless Personal Communications*, *Wireless Networks*, *EURASIP Journal on Wireless Communications and Networking*) and conference papers (published at IEEE VTC, European Wireless, IEEE Globecom, etc.). Adrian Kliks was a TPC member in various international conferences (like IEEE Globecom, IEEE PIMRC, IEEE ICC, and IEEE WCNC). He organized various special sessions at international conferences (like ISWCS, CrownCom, WSA, and EUSIPCO). Currently, he participates actively in working groups established for the definition of IEEE 1900.x standards on cognitive radio.

Arturas Medeisis has been working in the field of radio spectrum management for nearly 20 years. He started with the national regulatory authority of Lithuania and then moved to the European Radiocommunications Office (formerly ERO, now ECO), where he served as spectrum management expert and later as deputy director.

Since 2007, Arturas has held the position of associate professor at the Telecommunications Engineering Department of the Vilnius Gediminas Technical University. Since 2010, he has served as the chairman of the management committee of the COST Action IC0905 TERRA (COST-TERRA), which is a multidisciplinary European research forum, focused on coordinating techno-economic studies to assist the development of a harmonized regulatory framework to facilitate the advancement and commercial deployment of cognitive radio/software-defined radio. Arturas actively participates in various spectrum management-related activities and missions of ITU BDT in the capacity of field expert.

Yoram Haddad received his BSc, Engineer diploma, and MSc (radiocommunications) from SUPELEC in 2004 and 2005, and his PhD in computer science and networks from Telecom ParisTech in 2010. From January 2011 to October 2012, he has been a post-doctoral research associate at the Ben-Gurion University in Beer-Sheva, Israel.

Since 2010, he has been a tenure-track senior lecturer (assistant professor) at the Jerusalem College of Technology (JCT) in Jerusalem, Israel. He recently co-founded the Mobile Network and Security Research Group at JCT. Yoram is a member of the management committee of COST-TERRA IC0905 and COST-Intellicis IC0806. He served as TPC and reviewer for several conferences such as IEEE VTC and IEEE PIRMC.

Yoram's main research interests are in the area of wireless networks and algorithms for networks. He is interested in energy-efficient wireless deployment, femtocell, modeling of wireless networks, security for wireless networks, wireless application to intelligent transportation systems, and, more recently, wireless software-defined networks.

Moshe Timothy Masonta is a senior researcher at the Council for Scientific and Industrial Research: Meraka Unit, in Pretoria, and a doctorate candidate in electrical engineering (telecommunications technology) at Tshwane University of Technology (TUT), Pretoria, South Africa. He received a BTech (Honours) degree in electrical engineering (2005) and a masters in technology (MTech) degree (2008), both from TUT. He also holds an MSc in electronic engineering degree (2010) from Ecole Supérieure d'Ingénieurs en Electrotechnique et Electronique de Paris, France. From September to December 2011, he was a visiting student at the University of California,

Santa Barbara. He is a member of the European Cooperation on Science and Technology (COST) Action IC0905 TERRA's Special Interest Group on energy efficiency. His research interests are in dynamic spectrum access and management, cognitive radio systems, television white space spectrum, spectrum regulations, and energy efficiency in wireless networks.

Luca de Nardis received both his Laurea degree and his PhD from Sapienza University of Rome in 2001 and 2005, respectively. Since December 2008, he has been an assistant professor at the DIET Department of Sapienza University of Rome. In 2007, he was a post-doctoral fellow at the EECS Department at the University of California at Berkeley.

He authored or co-authored over 60 publications in international peer-reviewed journals and conferences and served as a member of the technical program committee of over 30 international IEEE conferences.

He is currently involved in two European COST Actions on Cognitive Radio: IC0902, where he leads the Working Group on cross-layer design and cognitive engine, and IC0905, for which he is a national delegate for Italy and leads the Working Group on application scenarios. He also participates in the FP7 ACROPOLIS NoE, where he leads the WorkPackage on Neighbourhood and Network Awareness.

Oliver Holland is a research fellow at King's College London, serving as project manager and deputy coordinator of the ICT-ACROPOLIS Network of Excellence (www.ict-acropolis.eu). Oliver is a leadership member of IEEE DySPAN-SC, is a member and vice-chair of IEEE 1900.1 and IEEE 1900.7, and a member of IEEE 1900.6. Oliver was a technical editor of IEEE 1900.4 and is technical editor of the developing IEEE 1900.1a and 1900.6a standards. He is a management committee member of COST IC0902 and IC0905 "TERRA," holding various leadership positions within these COST actions. Oliver has served on the TPC of all major conferences in the area of mobile and wireless communications, has served as session chair and panelist at numerous conferences covering green radio and cognitive radio, among other topics, and frequently serves as reviewer for various prestigious international conferences and journals. Oliver has assumed leadership positions in numerous international workshops, conferences, and journals. Among many examples, he was guest editor of the special issue "Achievements and the Road Ahead: The First Decade of Cognitive Radio," appearing in *IEEE Transactions on Vehicular Technology (TVT)*, was co-chair of the "Cognitive Radio and Cooperative Communications" track of IEEE VTC 2010-Fall, was tutorials co-chair of IEEE CCNC 2012, was TPC co-chair of ISWCS 2012, was co-chair of the "Transmission Technologies" track of IEEE VTC 2013-Spring, was tutorials co-chair of ACM MSWiM 2012, is co-chair of the "Wireless Access" track at IEEE VTC 2013-Fall, and is tutorials co-chair of IEEE PIMRC 2013. Oliver is an associate editor of *IEEE TVT* and the *SAIEE Africa Research Journal*. He is an officer of the IEEE Technical Committee on Cognitive Networks (TCCN), serving as liaison between TCCN and COST IC0905 "TERRA" and between TCCN and DySPAN-SC, and he is chair of the UKRI Chapter of the IEEE VTS. Oliver has authored over 100 publications; according to Google Scholar, his publications have been cited more than 500 times.

REFERENCES

1. O. Arnold, F. Richter, G. Fettweis and O. Blume, Power consumption modeling of different base station types in heterogeneous cellular networks, In *Future Network and Mobile Summit 2010 Conference Proceedings*, Florence, Italy, pp. 1 8, June 16–18, 2010.
2. W. Guo and T. O'Farrell, Reducing energy consumption of wireless communications, online: http://de2011.computing.dundee.ac.uk/wp-content/uploads/2011/11/Reducing-Energy-Consumption-of-Wireless-Communications.pdf, accessed Oct. 30, 2012
3. A. Goldsmith, *Wireless Communication*. Cambridge University Press, New York, 2005.
4. P. Kryszkiewicz, H. Bogucka and A.M. Wyglinski, Protection of primary users in dynamically varying radio environment: Practical solutions and challenges, *EURASIP Journal on Wireless Communications and Networking*, vol. 2012, no. 23, January 2012, doi:10.1186/1687-1499-2012-23.

5. B.S. Krongold and D.L. Jones, PAR reduction in OFDM via active constellation extension, *IEEE Transactions on Broadcasting*, vol. 49 no. 3, pp. 258–268, Sept. 2003.

6. H. Bogucka, Directions and recent advances in PAPR reduction methods, In *IEEE International Symposium on Signal Processing and Information Technology*, pp. 821–827, Aug. 2006.

7. ETSI EN 302 755 V1.1.1 (2009-09). ETSI Digital Video Broadcasting (DVB); frame structure channel coding and modulation for a second generation digital terrestrial television broadcasting system (DVB-T2), Sept. 2009.

8. J. Proakis and M. Salehi, *Digital Communications*. McGraw-Hill Higher Education, Singapore, 2008.

9. F. Bader and M. Shaat, Pilot pattern adaptation and channel estimation in MIMO WiMAX-like FBMC system, In *6th International Conference on Wireless and Mobile Communications (ICWMC)*, pp. 111–116, 2010.

10. S. Pagadarai, A. Kliks, H. Bogucka and A.M. Wyglinski, On non-contiguous multicarrier waveforms for spectrally opportunistic cognitive radio systems, In *International Waveform Diversity and Design Conference (WDD)*, pp. 177–181, 2010.

11. H. Bolcskei, F. Hlawatsch and H.G. Feichtinger, Equivalence of DFT filter banks and Gabor expansions, *SPIE*, vol. 2569, pp. 128–139, 1995.

12. P.P. Vaidyanathan, *Multirate Systems and Filters Banks*. PTR Prentice-Hall, Englewood Cliffs, 1993.

13. S. Haykin, Cognitive radio: Brain-empowered wireless communications, *IEEE Journal on Selected Areas in Communications*, vol. 23, no. 2, pp. 201–220, Feb. 2005.

14. Y. Zeng, Y.-C. Liang, A.T. Hoang and R. Zhang, A review on spectrum sensing for cognitive radio: Challenges and solutions, *EURASIP Journal on Advances in Signal Processing*, ID 381465, 2010.

15. R.A. Elliot, M.A. Enderwitz, F. Darbari, L.H. Crockett, S. Weiss and R.W. Stewart, Efficient TV white space filter bank transceiver, In *Proceedings of the 20th European Signal Processing Conference (EUSIPCO), 2012*, pp. 1079–1083, Aug. 27–31, 2012.

16. E. Hossain, D. Niyato and Z. Han, *Dynamic Spectrum Access and Management in Cognitive Radio Networks*, Cambridge University Press, Cambridge, UK, p. 58, 2009.

17. Vodafone UK Corporate Responsibility Report, numbers based on energy consumption data for 2010/2011, online: http://www.vodafone.com/content/index/uk_corporate_responsibility/greener/carbon_energy/network_efficiency.html, accessed Sept. 2012.

18. L.M. Correia, D. Zeller, O. Blume, D. Ferling, Y. Jading, I. Gódor, G. Auer and L. Van der Perre, Challenges and enabling technologies for energy aware mobile radio networks, *IEEE Communications Magazine*, vol. 48, no. 11, pp. 66–72, Nov. 2010.

19. O. Holland, O. Cabral, F. Velez, A. Aijaz, P. Pangalos and A.H. Aghvami, Opportunistic load and spectrum management for mobile communications energy efficiency, In *IEEE 22nd International Symposium on Personal Indoor and Mobile Radio Communications (PIMRC)*, pp. 666–670, 2011.

20. IEEE 802.11, Wireless LAN Media Access Control (MAC) and physical layer (PHY) specifications, 1999.

21. M. Sherman, A.N. Mody, R. Martinez, C. Rodriguez and R. Reddy, IEEE standards supporting cognitive radio and networks, dynamic spectrum access and coexistence, *IEEE Communications Magazine*, vol. 46, no. 7, pp. 72–79, 2008.

22. IEEE 802.22, Standard for cognitive wireless regional area networks (RAN) for operation in TV bands, 2011.

23. ETSI TR 102 681. Reconfigurable Radio Systems (RRS); Radio Base Station (RBS) Software Defined Radio (SDR) status, implementations and costs aspects, including future possibilities, 2009-06.

24. A.M. Wyglinski, M. Nekovee and Y.T. Hou, Eds., *Cognitive Radio Communications and Networks: Principles and Practice*, Elsevier, USA, 2010.

25. V. Rodoplu and T.H.-Y. Meng, Minimum energy mobile wireless networks, *IEEE Journal on Selected Areas in Communications*, vol. 17, no. 8, pp. 1333–1344, Aug. 1999.

26. S. Singh, M. Woo and C.S. Raghavendra, *Power-Aware Routing in Mobile Ad Hoc Networks*, ACM/IEEE MOBICOM, Dallas, Texas, pp. 25–30, Oct. 1998.

27. J.-H. Chang and L. Tassiulas, Routing for maximum system lifetime in wireless ad-hoc networks, In *37th Annual Allerton Conference on Communication, Control, and Computing*, Sept. 1999.

28. I. Stojmenovic and X. Lin, Power-aware localized routing in wireless networks, *IEEE Transactions on Parallel and Distributed Systems*, vol. 12, no. 10, pp. 1–12, Oct. 2001.

29. L. De Nardis, D. Domenicali and M.-G. Di Benedetto, Performance and energy efficiency of position-based routing in IEEE 802.15.4a low data rate wireless personal data networks, In *IEEE International Conference on UWB 2007 (ICUWB2007)*, Singapore, pp. 264–269, Sept. 24–26, 2007.

30. O. Holland, A. Georgakopoulos, V. Stavroulaki, K. Tsagkaris, P. Demestichas, L. De Nardis, M.-G. Di Benedetto and H. Aghvami, Comparison of in-band and out-of-band common control channels for cognitive radio, In *ACROPOLIS 2nd Annual Workshop on "Advanced Coexistence Technologies for Radio Resource Usage Optimisation" and Industry Panel*, Brussels, Belgium, June 27–28, 2012.
31. IETF MANET Working Group webpage, online: http://datatracker.ietf.org/wg/manet/charter/, accessed Jan. 2013.
32. IETF ROLL Working Group webpage, online: http://datatracker.ietf.org/wg/roll/charter/, accessed Jan. 2013.
33. IETF RPL routing protocol draft, online: http://tools.ietf.org/html/draft-ietf-roll-rpl-19, accessed Jan. 2013.
34. D. Raychaudhuria, X. Jinga, I. Seskara, K. Lea and J.B. Evans, Cognitive radio technology: From distributed spectrum coordination to adaptive network collaboration, *Elsevier Pervasive and Mobile Computing*, vol. 4, pp. 278–302, 2008.
35. V. Srivastava and M. Motani, Cross-layer design: A survey and the road ahead, *IEEE Communications Magazine*, vol. 43, no. 12, pp. 112–119, Dec. 2005.
36. G. Miao, N. Himayat, Y. Li and A. Swami, Cross-layer optimization for energy-efficient wireless communications: A survey, *Wiley Wireless Communications and Mobile Computing,* vol. 9, no. 4, pp. 529–542, April 2009.
37. I.F. Akyildiz, W.-Y. Lee and K.R. Chowdhury, CHRANS: Cognitive radio ad hoc networks, *Elsevier Journal on Ad Hoc Networks,* vol. 7, pp. 810–826, 2009.
38. L. De Nardis, M.-G. Di Benedetto, A. Akhtar and O. Holland, Combination of DOA and beamforming in position-based routing for underlay cognitive wireless networks, In *7th International Conference on Cognitive Radio Oriented Wireless Networks and Communications (CROWNCOM 2012)*, Stockholm, Sweden, June 18–20, 2012.
39. D.-J. Deng, R.-S. Cheng, H.-J. Chang, H.-T. Lin and R.-S. Chang, A cross-layer congestion and contention window control scheme for TCP performance improvement in wireless LANs, *Springer Telecommunications Systems*, vol. 42, nos. 1–2, pp. 17–27, 2009.
40. IEEE Std 1900.4™-2009, IEEE standard for architectural building blocks enabling network-device distributed decision making for optimized radio resource usage in heterogeneous wireless access networks.
41. IEEE Std 1900.5™-2011, IEEE standard for policy language requirements and system architectures for dynamic spectrum access systems.
42. F.K. Jondral, Software defined radio—Basics and evolution to cognitive radio, *EURASIP Journal on Wireless Communications and Networking*, vol. 3, pp. 275–283, 2005.
43. D. Cabric and R.W. Brodersen, Physical layer design issues unique to cognitive radio systems, In *16th IEEE International Symposium on Personal Indoor and Mobile Radio Communications, (PIMRC 2005)*, September 2005.
44. A. Fehske, G. Fettweis, J. Malmodin and G. Biczok, The global footprint of mobile communications: The ecological and economic perspective, *IEEEE Communications Magazine*, vol. 49, no. 8, pp. 55–62, 2011.
45. M. Fitch, M. Nekovee, S. Kawade, K. Briggs and R. Mackenzie, Wireless services provision in TV white space with cognitive radio technology: A telecom operator's perspective and experience, *IEEE Communications Magazine*, vol. 49 no. 3, pp. 64–73, 2011.
46. M.T. Masonta, D. Johnson and M. Mzyece, The white space opportunity in Southern Africa: Measurements with Meraka cognitive radio platform, In *Springer Lecture Notes of the Institute for Computer Sciences, Social Informatics and Telecommunications Engineering*, R. Popescu-Zeletin et al., Eds., vol. 92, pp. 64–73, Feb. 2012.
47. Cisco, Visual Networking Index (VNI): Forecast and methodology, 2011–2016, white paper, May 2012, online: http://www.cisco.com/en/US/solutions/collateral/ns341/ns525/ns537/ns705/ns827/white_paper_c11-481360_ns827_Networking_Solutions_White_Paper.html, accessed: Jan. 2013
48. P. Jacobs, Tech-on interview by H. Yomogita, online: http://techon.nikkeibp.co.jp/english/NEWS EN/20080905/157548/, Sept. 5 2008, accessed Jan. 2013.
49. Y. Haddad and D. Porrat, Femtocell: Opportunities and challenges of the home cellular base station for the 3G, In *Proceedings of International Conference on Wireless Applications and Computing*, Algarve, Portugal, June 2009.
50. S. Saunders, Ed., S. Carlaw, A. Giustina, R. Rai Bhat, V. Srinivasa Rao, R. Siegberg et al., *Femtocells: Opportunities and Challenges for Business and Technology*, Wiley, UK, 2009.
51. J. Zhang and G. De la Roche, *Femtocells: Technologies and Deployment*, Wiley, UK, 2010.
52. X. Lagrange, Multitier cell design, *IEEE Communications Magazine*, vol. 35, no. 8, pp. 60–64, Aug. 1997.

53. Y. Haddad and Y. Mirsky, Power efficient femtocell distribution strategies, In *Proceedings of 19th IEEE International Conference on Software, Telecommunications and Computer Networks (IEEE SoftCOM)*, Split, Croatia, Sept. 2011.

54. V. Chandrasekhar, J.G. Andrews and A. Gatherer, Femtocell networks: A survey, *IEEE Communications Magazine*, vol. 46, no. 9, pp. 59–67, Sept. 2008.

55. M. Hamblen, Verizon offers free Wi-Fi to consumer customers, expands on Boingo relationship offering free Wi-Fi to business users, July 2009, online: http://www.computerworld.com/s/article/9135949/ Verizon_offers_free_Wi_Fi_to_consumer_customers, accessed Jan. 2013.

56. K. Lee, Mobile data offloading: How much can WiFi deliver?, In *Proceedings of ACM CoNEXT*, Philadelphia, USA, 2010.

57. P. Gardner-Stephen, J. Lakeman, R. Challans, C. Wallis, A. Stulman and Y. Haddad, MeshMS: Ad hoc data transfer within mesh network, *International Journal of Communications, Network and System Sciences*, vol. 5 no. 8, pp. 496–504, 2012.

58. Y. Al-Hazmi, H. de Meer, K. A. Hummel, H. Meyer, M. Meo and D. Remondo, Energy efficient wireless mesh infrastructures, *IEEE Network*, vol. 25, no. 2, pp. 32–38, March–April 2011.

59. J.Z. Wang, Z. Du and P.K. Srimani, Cooperative proxy caching for wireless base stations, *Mobile Information Systems*, vol. 3, no. 1, pp. 1–18, Jan. 2007.

60. N. McKeown, T. Anderson, H. Balakrishnan, G. Parulkar, L. Peterson, J. Rexford, S. Shenker and J. Turner, OpenFlow: Enabling innovation in campus networks, *SIGCOMM Computer Communication Review*, vol. 38, no. 2, pp. 69–74, March 2008.

61. K.-K. Yap, M. Kobayashi, D. Underhill, S. Seetharaman, P. Kazemian and N. McKeown, The Stanford OpenRoads deployment, In *4th ACM International Workshop on Experimental Evaluation and Characterization (WINTECH '09)*, ACM, New York, USA, pp. 59–66, Sept. 2009.

62. The Institute of Electrical and Electronics Engineers (IEEE), online: www.ieee.org, accessed March 2012

63. IEEE 802 LAN/MAN Standards Committee, online: http://www.ieee802.org, accessed March 2012.

64. IEEE DySPAN Standards Committee, online: http://www.dyspan-sc.org, accessed March 2012.

65. IEEE 802.22 Working Group on Wireless Regional Area Networks, online: http://www.ieee802.org/22, accessed March 2012.

66. IEEE 802.11 Task Group af (TGaf), Wireless LAN in the TV white space, online: http://www.ieee802. org/11/Reports/tgaf_update.htm, accessed March 2012.

67. IEEE 802.15 Task Group 4m (TG4m), TV white space amendment to 802.15.4, online: http://www. ieee802.org/15/pub/TG4m.html, accessed March 2012.

68. IEEE 802.19 Wireless Coexistence Working Group, online: http://www.ieee802.org/19, accessed March 2012.

69. K. Harrison, S.M. Mishra and A. Sahai, How much white-space capacity is there?, In *IEEE DySPAN 2010*, Singapore, April 2010.

70. IEEE 802.16 Broadband Wireless Access Standards Working Group, online: http://www.ieee802.org/16, accessed March 2012.

71. IEEE 802.22.1 Project Authorization Request (PAR), online: http://www.ieee802.org/22/802-22-1_ PAR_Approval.pdf, accessed March 2012.

72. IEEE 802.22.2 Project Authorization Request (PAR), online: http://www.ieee802.org/22/802-22-2_ PAR_Approval.pdf, accessed March 2012.

73. IEEE 802.22a Project Authorization Request (PAR), online: http://www.ieee802.org/22/P802_22a_ PAR_Approved.pdf, accessed March 2012.

74. IEEE 802.22b Project Authorization Request (PAR), online: http://www.ieee802.org/22/P802_22b_ PAR_Approved.pdf, accessed March 2012.

75. IEEE 802.15.4m Project Authorization Request (PAR), online: https://mentor.ieee.org/802.15/dcn/11/15-11-0643-00-004m-tg4m-par.pdf, accessed March 2012.

76. IEEE 802.11af Project Authorization Request (PAR), online: https://development.standards.ieee.org/ P684500033/par (requires IEEE Standards Association Membership), accessed March 2012.

77. IEEE Dynamic Spectrum Access Networks Standards Committee (IEEE DySPAN-SC), online: http:// www.dyspan-sc.org, accessed March 2012.

78. IEEE 1900.1 Working Group on Definitions and Concepts for Dynamic Spectrum Access: Terminology Relating to Emerging Wireless Networks, System Functionality, and Spectrum Management, online: http://grouper.ieee.org/groups/dyspan/1/index.htm, accessed March 2012.

79. IEEE 1900.2 Working Group on Recommended Practice for the Analysis of In-Band and Adjacent Band Interference and Coexistence between Radio Systems, online: http://grouper.ieee.org/groups/dyspan/2/ index.htm, accessed March 2012.

80. IEEE 1900.3 Working Group on Recommended Practice for Conformance Evaluation of Software Defined Radio (SDR) Software Modules, online: http://grouper.ieee.org/groups/dyspan/3/index.htm, accessed March 2012.
81. IEEE 1900.4 Working Group on Architectural Building Blocks Enabling Network-Device Distributed Decision Making for Optimized Radio Resource Usage in Heterogeneous Wireless Access Networks, online: http://grouper.ieee.org/groups/dyspan/4/index.htm, accessed March 2012.
82. IEEE 1900.5 Working Group on Policy Language and Policy Architectures for Managing Cognitive Radio for Dynamic Spectrum Access Applications, online: http://grouper.ieee.org/groups/dyspan/5/index.htm, accessed March 2012.
83. IEEE 1900.6 Working Group on Spectrum Sensing Interfaces and Data Structures for Dynamic Spectrum Access and Other Advanced Radio Communication Systems, online: http://grouper.ieee.org/groups/dyspan/6/index.htm, accessed March 2012.
84. IEEE 1900.7 White Space Radio Working Group, online: http://grouper.ieee.org/groups/dyspan/7/index.htm, accessed March 2012.
85. IEEE DySPAN-SC end of year report 2011, online: https://mentor.ieee.org/dyspan-sc/dcn/11/sc-11-0033-00-MTNG-year-end-report.pdf (requires IEEE Standards Association membership), accessed March 2012.
86. Weightless standard for machine-to-machine (M2M) communication over white space, online: http://www.weightless.org, accessed March 2012.
87. ETSI Reconfigurable Radio Systems (ETSI-RRS), online: http://www.etsi.org/website/technologies/RRS.aspx, accessed March 2012.
88. M. Mueck, A. Piipponen, K. Kalliojärvi, G. Dimitrakopoulos, K. Tsagkaris, P. Demestichas, F. Casadevall, J. Pérez-Romero, O. Sallent, G. Baldini, S. Filin, H. Harada, M. Debbah, T. Haustein, J. Gebert, B. Deschamps, P. Bender, M. Street, S. Kandeepan, J. Lota and A. Hayar, ETSI reconfigurable radio systems: Status and future directions on software defined radio and cognitive radio standards, *IEEE Communications Magazine*, vol. 48, pp. 78–86, Sept. 2010.
89. ETSI-RRS TR 102.683, v1.1.1, Reconfigurable radio systems (RRS); cognitive pilot channel (CPC), 2009.
90. ETSI-RRS TR 102.802, v1.1.1., Reconfigurable radio systems (RRS); cognitive radio system concepts, 2010.
91. ETSI-RRS TR 102.802, v1.1.1., Reconfigurable radio systems (RRS); functional architecture (FA) for the management and control of reconfigurable radio systems, 2009.
92. Draft ETSI-RRS TR 102 684, Feasibility study on control channels for cognitive radio systems, 2011.

6 Green Network Security for Ad Hoc and Sensor Networks

M. Bala Krishna

CONTENTS

6.1 INTRODUCTION

Wireless networks have proven to be the foremost, affordable, and reliable technology effectively used in real-time indoor and outdoor applications. Energy efficiency, reliability, and security are the primary factors considered in the design of green communications. The emerging technology of machine-to-machine communication [1] across various intelligent devices shares information and performs cooperative tasks without any human interaction. Green wireless networking [2] for WPAN, WLAN, or WMAN is based on network reconfiguration and redeployment. This improves the energy efficiency and increases the network life cycle. Recent trends in wireless networks are attributed with accuracy, quality of service, and interoperability across various functional devices like base station (BS), source nodes, mobile nodes, data gathering nodes, cluster head nodes, and the sink nodes. Due to the constraints in energy and memory resources, the energy-efficient protocols are the primary concern in the design of wireless networks. Wireless networks operate in hostile environments that are vulnerable to security threats. Hence, the security protocols based on energy-efficient methodologies are used in wireless networks. Energy and security management protocols [3] are primarily used in localization and routing to establish a secure communication channel in wireless networks. The security protocols are attributed with authenticity, integrity, and confidentiality [4,5] between the nodes in the network. Secure-path routing (SPR) [6] aims to design the network topology with minimum route cost and decrease the rate of compromised nodes in the network. The routing cost is a function of latency, packet transmission rate, and number of hops between the source node and the sink node.

Robust network security can be achieved by (i) the exclusion of compromised nodes from accessing the data in current transaction, (ii) repudiation of new nodes from accessing the previous data, and (iii) minimizing frequent topology changes in the network [7]. Authenticated encryption (AE) [8] and network accountability [9] are based on factors like (i) integrity and authenticity of data, (ii) reliability and confidentiality of user transaction, and (iii) time period of secure interaction. Wireless mesh networks (WMNs) [10] achieve maximum connectivity and improve the performance by providing multiple alternate routes for reliable communication in the network. The network deployment and resource management for green mesh networks [11] are based on APs that are driven by renewable energy resources.

This chapter is organized as follows. Section 6.2 explains the issues and challenges of energy-efficient security protocols for wireless ad hoc and sensor networks. Section 6.3 elucidates the proposed architecture for Green Network Security (GNS) in wireless networks. The layered GNS architecture is an integrated approach of energy-efficient design and security design for wireless networks. Energy-efficient design layer is based on localization, routing, and data aggregation protocols. Security design layer is based on key management, secure data aggregation, and secure routing protocols. Section 6.4 enumerates the green network metrics for the node and the network with respect to energy and security attributes. Section 6.5 explains the components of GNS for wireless ad hoc and sensor networks. This section explains energy-efficient and security protocols based on localization, routing, and key management for ad hoc and sensor networks. Section 6.6 explains the proposed design for GNS in sensor networks based on energy-efficient and secure key management. Simulation results indicate that the integrated approach of energy-efficient and security design improves the network performance as compared to the distinctive designs of energy-efficient and security protocols. Section 6.7 concludes the chapter and highlights the future directions for GNS in wireless networks.

6.2   ISSUES AND CHALLENGES

The primary tasks of energy-efficient security protocols are given as follows:

1. Usage of alternate route paths for multiple sessions in the network. For longer sessions, the route paths are periodically authenticated to establish the secure communication channel in the network.
2. The node compromise level is evaluated for each session. If the compromise level is more than the threshold level, then the route paths and secure keys are modified.

The challenges and improvisations of robust energy-efficient security protocols are as follows:

1. To establish a secure network, the compromised nodes and failure nodes must be identified and filtered from the communication channel. Periodic and event-based modification of group-wise keys and individual keys ensures a secure network. Further, centralized secure network management systems are used to monitor and update the network configurations to ensure the safety and integrity of the network.
2. To sustain the degree of integrity in vulnerable route paths. Authorized access points (AAPs) are used to scan and detect the compromised and defective access points in the network.
3. To maintain the required levels of assurance with the end user.

4. For highly sensitive and critical areas, additional infrastructure is used to establish the secure communication channel.
5. The channel is prone to signal loss owing to interference and hidden terminal problems (the intruder can interfere with the channel but cannot be detected). The main challenge is to minimize the signal-to-noise ratio and bit error rate in the communication channel.
6. End-to-end delays and time synchronization are the two primary issues to be addressed by energy-efficient security protocols for wireless networks.

6.3 ARCHITECTURE OF GNS FOR WIRELESS NETWORKS

The architecture of GNS comprises an energy-efficient design layer and a security design layer. The main components of GNS for wireless networks are as follows:

1. Energy-efficient design
   a. Localization
   b. Routing
   c. Data aggregation
2. Energy-efficient security design
   a. Secure key management
   b. Mobile beacon-based key management

Figure 6.1 illustrates the proposed architecture of GNS for wireless networks based on the integrated approach of energy-efficient design and security design. In the lower layer of GNS architecture, the energy-efficient design is based on localization, routing, and data aggregation protocols.

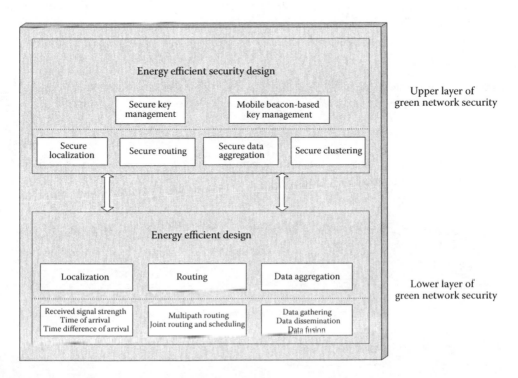

FIGURE 6.1 Architecture of GNS for wireless networks.

In the upper layer of GNS architecture, the energy-efficient security design is based on secure key management and security attributes of localization, and routing and data aggregation.

6.4 GREEN NETWORK METRICS

6.4.1 ENERGY METRICS

Transmission energy—The energy consumed by the node during data transmission. This is a function of data packet size, current node transaction energy, and the adjacent node distance. Node transmission energy E_{Tx} based on the energy coefficient e is given as:

$$E_{Tx} = DPktSize * E_{CurrTransaction} + DPktSize * e * Dist^2$$

Reception energy—The energy consumed by the node during data reception. This is a function of data packet size and current node transaction energy. Node reception energy E_{Rx} is given as:

$$E_{Rx} = DPktSize * E_{CurrTransaction}$$

Energy efficiency—The data is aggregated by the CH node or data aggregation node to save the node energy and establish energy-efficient route paths in the network.

Average energy consumed per session—The total energy consumed by the transmitting nodes and receiving nodes in the network per session.

Adaptability—Dynamic network traffic conditions and heterogeneous nodes affect the performance of energy-efficient security protocols. Hence, adaptable protocols with varying node and network attributes are designed for wireless networks.

Energy consumption [3] can be reduced by varying the CPU clock speed, optimizing the sleep schedules, and varying the transmission and reception cycles in the network. Security protocols use random querying techniques to identify the intruders in the network.

6.4.2 SECURITY METRICS

Authentication—The BS authenticates the node based on valid ID and the secure key. The authentication process for the $Node_i$ is given as:

$$Node_i = Authenticate(Node_i, Key_{public/private}, MAC).$$

Confidentiality—The group members interact with the confidential neighboring nodes in the network. The compromised nodes are traced using the message exchange and excluded from the network. The transaction between two confidential nodes, $Node_i$ and $Node_j$, is given as:

$$
\begin{aligned}
Transaction(Node_i, Node_j) &= confidential && \text{if } Key_i \equiv Key_j \\
&= nonconfidential && \text{if } Key_i \neq Key_j \text{ and} \\
& && Key_i \vee Key_j \Rightarrow Key_{Intruder}
\end{aligned}
$$

Integrity—Authenticated nodes exchange the data with authorized nodes using message authentication code. The integrity between $Node_i$ and $Node_j$ is given as:

$$Integrity(Node_i, Node_j) = \begin{cases} True & if\ Transaction(Node_i, Node_j) = confidential \\ False & if\ Transaction(Node_i, Node_j) \neq confidential \end{cases}$$

Control access—The network access is confined to the authorized nodes with control access rights. The member nodes with valid MACs are allowed to access the channel for valid transactions. The control access scheme between $Node_i$ and $Node_j$ is given as:

$$C_{access}(Node_i, Node_j) = \begin{cases} permitted & if\ Integrity(Node_i, Node_j) = True \\ not\ permitted & if\ Integrity(Node_i, Node_j) = False \end{cases}$$

Availability—The data is available to the intended recipients with authorized keys that are verified by the BS. The data availability between $Node_i$ and $Node_j$ is given as:

$$Data(Node_i, Node_j) = \begin{cases} available & if\ C_{access}(Node_i, Node_j) = permitted \\ not\ available & if\ C_{access}(Node_i, Node_j) = not\ permitted \end{cases}$$

Secure energy-efficient—The nodes interact with each other based on valid key exchange. Secure energy efficient protocols minimize the key size and frequent key modifications in the network. The symmetric keys consume less energy and are used in energy-efficient security protocols.

6.5 GNS COMPONENTS FOR WIRELESS NETWORKS

Wireless networks like IEEE 802.11 (WLANs) use mobile stations (MSs) and access points (APs) to update the network configurations. The security protocols use cryptographic methods to encrypt the message and hash functions to authenticate and verify message integrity. The secure keys in the route paths are periodically modified to establish the secure channel in the network. Advanced power management (APM) and Advanced Configuration and Power Interface (ACPI) techniques are used to monitor the energy consumption levels in security protocols. Secure verification techniques like Common Electronic Purse Specifications (CEPS) and Secure Electronic Transaction (SET) [9] are used to address E-commerce transactions in ad hoc networks. Purchase Security Application Module (PSAM) [9] installed in point of sale (POS) devices are used to improve accountability in bank transactions. Elliptic curve cryptography (ECC) with smaller key size reduces the size of message buffers and minimizes the complexity in security protocols.

6.5.1 COMPONENTS OF GNS FOR AD HOC NETWORKS

Wireless ad hoc networks are infrastructureless networks based on dynamic network topology and multihop routing techniques. Group management schemes provide access rights to exchange the information with authentic users and avoid network attacks [12]. Malicious nodes [13] compromise the consistency and confidentiality of the network and alter the route paths. COllaborative REputation (CORE) and Cooperation Of Nodes Fairness In Dynamic Ad hoc NeTworks (CONFIDANT) [13] are based on degree of trust levels to forward data packets to the neighboring nodes. These protocols measure the degree of reputation, forward the data packets to the neighboring nodes, and resolve the issues related to secure route discovery and network management. Figure 6.2 classifies the

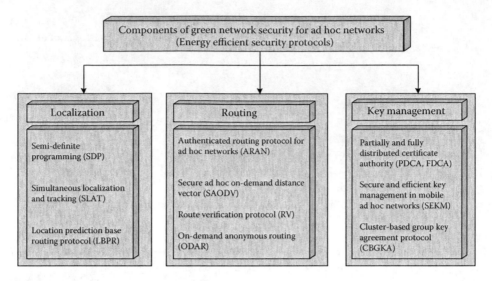

FIGURE 6.2   GNS components for ad hoc networks.

components of Green Network Security for ad hoc networks based on localization, routing, and key management techniques.

6.5.1.1   Energy-Efficient Localization

Localization techniques are used to estimate the physical position of the node in the network. Triangulation, trilateration, and multilateration techniques are used to determine the node positions in ad hoc and sensor networks. Secure localization verifies the consistency of location parameters and estimates the precise node position of intruders in the network. For single-hop distance, the node locations are estimated using the distance, signal strength (received signal strength indicator [RSSI]) and angle measurements (time of arrival [ToA], time difference of arrival [TDoA]). For multihop distance, the node locations are estimated using the DV-distance and DV-hop techniques.

6.5.1.1.1   Energy-Efficient Secure Localization

Since the energy source in wireless devices are irreplaceable, energy-efficient localization schemes are crucial in the design of ad hoc networks. Bounded degree and plane topology [14] methods are used to estimate the power stretch factors in the network. This protocol reduces the energy consumption by restricting the message length. Location-based group-key management [15] supports lightweight design attributes and provides solutions to mobile ad hoc networks that are susceptible to security extortions in hostile environments.

6.5.1.1.2   Simultaneous Localization and Tracking

Simultaneous Localization and Tracking (SLAT) [16] based on energy-efficient attributes uses the probabilistic estimation technique based on Bayesian method. This protocol operates on sparse inverse covariant matrices and applies the linear and angular laws to estimate the positions of unknown node. This method applies Laplacian and Gaussian distribution methods to minimize the complexity in Bayesian filters and reduces the error rate in localization. The average node positions are calculated to improve the node tracking accuracy in the network. This protocol is suitable for small-scale and indoor networks.

In mobile ad hoc networks, the cryptographic methods use the trust-based broadcast schemes to forward the data packets to the intermediate nodes. Secure source anonymous message authentication scheme (SAMAS) [17] addresses this issue by using the trusted intermediate nodes to forward the messages in the network. The communicating nodes are further classified as normal and super nodes. The normal node communicates with the neighboring nodes in the current MANET and the super node communicates with the surrounding nodes of the adjoining MANET and act as the secured data forwarding node.

6.5.1.1.3 Location Prediction-Based Routing Protocol

Location Prediction-Based Routing protocol (LBPR) [18] aims to address flooding in route discovery and preserve the route information at the sink node. The location information of the destination node is distributed along the route discovery paths. This protocol predicts the minimum-hop route and reduces the cost of route discovery. LPBR is further extended to nonreceiver-aware multicast LPBR (NR-MLPBR) and receiver-aware multicast LPBR (R-MLPBR). The NR-MLPBR algorithm minimizes the number of hop counts while R-MLPBR assumes that the receiver nodes have the knowledge of neighboring receiving nodes in the multicast group.

6.5.1.2 Energy-Efficient Secure Routing

6.5.1.2.1 Authenticated Routing Protocol for Ad Hoc Networks

Authenticated Routing Protocol for Ad Hoc Networks (ARAN) [19] is a low, energy-efficient protocol that verifies the digital signatures assigned to the initiator node, packet forwarding node, and the destination node. This protocol uses cryptographic certificates for route establishment. The authentication process confirms the node identity and the nonrepudiation process confirms the transactions performed by the authenticated nodes. The exchange of Route Discovery Packet identifier and REPly packet identifier between the source node and destination node confirms the route path in the network. The main limitation of this protocol is the increase in control overhead and cost of packet delivery owing to the authenticated certification process for each session.

6.5.1.2.2 Secure Ad Hoc Routing Protocol

Secure ad hoc on-demand distance vector (SAODV) routing protocol [20,21] identifies the default message initiators and tracks the malicious nodes. The integrity of the routing paths is verified using RREQ and RREP messages. This protocol uses (i) authenticated digital signatures to estimate the number of additional hops traversed by the data packets along the compromised route paths and (ii) hash functions to identify the modified hop count by the intruders in the network. For hop count within the defined range, AODV and SAODV protocols [22] consume less energy. The energy consumption in SAODV increases with the increase in hop count and data packet size. Secure dynamic source routing (SDSR) [23] assigns the sessions keys to the source node, destination node and the authenticated intermediate nodes based on the secure route paths. The delay in the control packet is minimized based on the threshold node energy.

6.5.1.2.3 Route Verification Protocol

Route verification protocol (RV) [22] detects the intruders that attack the network from radio coverage of a wide area network and disables the nodes along the existing routing paths. This protocol supports multigrade monitoring (MGM) technique to overhear and detect the intruder nodes. MGM saves the energy resources in the network.

6.5.1.2.4 On-Demand Anonymous Routing

On-Demand Anonymous Routing (ODAR) [24] maintains the secure route path and periodically updates the routing information. The identities of all the member nodes along the routing path are acknowledged by the source node and the sink node. This protocol uses filter keys to safeguard the privacy of intermediate nodes based on node identity and location information. The filter key is composed of m bits and k hash functions. The filter keys are attributed with packet transmissions based on the parameters like node identification, packet length, and sequence number. Since the header fields of the previous transactions are retained, the filters avoid loops along the multiple route paths. While transmitting the data, the source node is provided with the pseudonym identity of the destination node, and the identities of intermediate nodes are kept confidential.

6.5.1.3 Energy-Efficient Secure Key Management

Wireless ad hoc networks operate in hostile environments and are vulnerable to intruders in the network. Hence, the key management techniques are used to secure the nodes in the communication channel. In dynamic ad hoc networks, the certificate authority (CA) assigns secure keys to the nodes in the network. The following protocols explain energy-efficient secure key management protocols in ad hoc networks. Table 6.1 highlights the energy-efficient security protocols in ad hoc networks.

6.5.1.3.1 Partially Distributed Certificate Authority

Partially distributed certificate authority (PDCA) [25], a public key management scheme with k server nodes and n ad hoc nodes distributes the keys in the time interval $[t_1-t_m]$ with varying network attributes. Fully distributed certificate authority (FDCA) [26] is a public key management scheme that extends PDCA to all the server nodes in the network. Each node in the network is guaranteed with a minimum of k immediate neighbor nodes to enables the secure paths along multiple routes in the network.

6.5.1.3.2 Secure and Efficient Key Management in Mobile Ad Hoc Networks

Secure and Efficient Key Management (SEKM) [27] in dynamic ad hoc networks reduces the complexity of computational overhead. This technique uses public key infrastructure (PKI) based on CA scheme for the servers and group members. The secret shareholding nodes with private keys act as the server node in multicast group and issue partial certificates to the forwarding nodes. The protocol functions in five phases: (i) server group formation, (ii) group maintenance, (iii) member upgrading, (iv) certificate modification, and (v) addressing the upgraded servers. This protocol is used in mesh-based network, where the server node distributes the keys in the network.

6.5.1.3.3 Cluster-Based Group Key Agreement Protocol

Cluster-based group key agreement protocol (CBGKA) [28], a group key energy-efficient protocol, partitions the cluster into three parts, which are further subdivided into branches. Each node is connected to a minimum of two nodes. The clusters use a tripartite key, which is shared between the three members of the cluster. This protocol has two variations: (i) the unauthenticated version, where the nodes agree with the usage of same elliptic curve, base point parameters, and one hash function, and (ii) the authenticated version, where the nodes have same parameters as the

---

TABLE 6.1

Energy-Efficient Security Protocols in Ad Hoc Networks

Design Feature	Functionality	Applied Protocols
Localization	To estimate the accurate node positions and secure the node positions of end users.	SLAT, SDP, matching algorithms (M2), LBPR
Routing	To establish the energy-efficient route paths and secured route paths in the network. To conceal the route path information from the intruders.	ARAN, SAODV, SDSR, SPAAR, ODAR
Key management	To implement energy-efficient key distribution and key management schemes controlled by the BS or CH node. Key management techniques are based on symmetric key and asymmetric key methods.	PDCA, SEKM, CBGKA

unauthenticated version and an additional hash function, key generation center (KGC). This protocol considerably reduces the overhead complexity as compared to SEKM.

6.5.2 COMPONENTS OF GNS FOR SENSOR NETWORKS

Wireless sensor networks (WSNs) are infrastructureless networks composed of self-organized nodes with constraints in computing, memory, and energy resources. In sensor networks, security is established using a trusted server and key distribution protocols. Key management schemes use the key revocation technique and disables the identity of compromised nodes to establish the secure channel in the network [29]. In mobile sensor networks, Group Selection Protocol (GSP) improves the estimate accuracy of node distribution and node localization. This approach can be used in secure node localization to establish the authentic route paths in the network.

Figure 6.3 classifies the GNS components for sensor networks. Localization, routing, data aggregation, and key management techniques contribute to energy-efficient secure paths in the network. In resilient sensor networks, the location-based energy-efficient key management protocol [30] broadcasts the re-keying process to establish a robust secure network.

6.5.2.1 Energy-Efficient Secure Localization

6.5.2.1.1 Semi-Definite Programming

Semi-Definite Programming (SDP) [31] is a position-based estimation method in which the non-convex node distances are transformed into linear constraints with relaxation in energy and secure metrics. This method is used for unknown and inaccurate node coordinates. The accuracy of node positions is achieved by converting the quadratic constraints into the linear constraints. This method is used when the anchor nodes are not placed within a given area. The sensor nodes are deployed at different points in the geographical area surrounded by the position known anchor nodes and the random error is calculated based on radio range R. Distributed localization technique uses the linear and angular measurements to achieve robustness and scalability in the network.

6.5.2.1.2 Secure Location Verification with Hidden and Mobile BSs

The BS verifies the location of sensor nodes in the network. Covert base stations (CBSs) [32] are formed by the hidden and mobile stations in the network. CBSs prevent the sensor nodes from

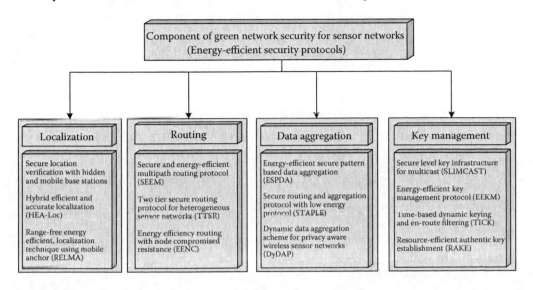

FIGURE 6.3 GNS components for sensor networks.

altering the location positions in the network. CBS systems use ultrasonic and radio frequency waves to estimate the received signal strength of the node. The location verification is based on secure key exchange and GPS attributes of the node. The degree of security is based on precise position estimates of mobile anchor nodes and CH nodes in the network. This protocol requires the installation of efficient mobile infrastructure and GPS system in the network.

6.5.2.1.3 Hybrid Efficient and Accurate Localization

Hybrid Efficient and Accurate Localization (HEA-Loc) [33] is a distributed range-free localization technique to locate the sensor node in small-scale networks. The locator nodes are equipped with multidirectional antennas and transmission power that is more than the sensor nodes. This protocol uses the extended Kalman filter (EKF) [34] to estimate the node positions based on primary anchor nodes and direct neighbor nodes. The proximity-distance map (PDM) [35] algorithm recursively estimates the location of sensor nodes in the network.

6.5.2.1.4 Range-Free Energy-Efficient Localization Using Mobile Anchor

Range-free Energy-efficient Localization using Mobile Anchor (RELMA) [36] uses minimum number of anchor nodes to estimate the node positions. The protocol divides the network into p zones with estimated hop degree within each zone and estimated hop distance from the BS. Each member node broadcasts the message with node identity and node status (anchor node or normal node) and estimates the number of neighbor nodes within each cluster. The sensor node forwards the identity of neighbor nodes to the CH, which further node position. The adjacency node list is updated and the neighbor set of each mobile node estimates the overlapping cluster regions. This protocol is suitable for small-scale sensor networks.

In unattended and hostile environments, the intruders disrupt the network services by manipulating the node positions in the network. Location-aided key distribution [37] schemes based on GPS attributes are incorporated in the CH node to detect the intruders in the network. This protocol considerably reduces the computational overhead as compared to SDP and HEA-Loc protocols. This protocol successfully prevents the wormhole attack, but does not address the issue of intruder attacks when broadcasting the neighborhood keys in the network.

6.5.2.2 Energy-Efficient Secure Routing

In energy-efficient secure routing protocols, the route paths with minimum energy and secure authentication between the nodes is used to establish the connectivity in the network. A malicious node consumes the node energy, increases the packet size, and manipulates the route path in the network. Hence, in secure multipath routing protocols, the attributes like confidentiality, nonrepudiation, authenticity, and integrity are considered for the transaction of data packets between the source and destination nodes.

6.5.2.2.1 Secure and Energy-Efficient Multipath Routing Protocol

Secure and Energy-Efficient Multipath routing protocol (SEEM) [38] resolves the network alterations caused by the malicious nodes. The BS controls the transaction and chooses the route path. The protocol assumes the predominance of the BS by selecting the optimal route path from existing multiple paths in the network. Thus, the BS is trusted to control the overall performance of the network. This approach considerably reduces the communication overhead and increases the network life cycle.

6.5.2.2.2 Two-Tier Secure Routing Protocol for Heterogeneous Sensor Networks

Two-Tier Secure Routing Protocol for Heterogeneous Sensor Networks (TTSR) [39] considers the varying functional parameters of the sensor node and network traffic. The sensor network is divided into high-end sensor nodes (H-sensor nodes) and the low-end sensor nodes (L-sensor nodes). Intra-cluster routing establishes the secure route paths for low-end sensors and inter-cluster routing establishes the secure route paths for high-end sensors. This protocol minimizes the packet modifications by the intruders in the network.

6.5.2.2.3 Energy Efficiency Routing with Node Compromised Resistance

Energy Efficiency Routing with Node Compromised Resistance (EENC) [40] is an ant colony optimization (ACO)-based technique that supports node reliability and minimizes the energy consumption in the network. The trust level and energy efficiency of the route path are evaluated by the current transmitting node and selects the forwarding node based on minimum cost towards the destination. The packet forwarding technique is adaptive to the varying network traffic conditions. Based on the pheromone cost and hop count, the next hop neighbor node is selected. The transmitting node eliminates the compromised nodes, with the degree of trust less than the threshold value. The ACO technique applies the reinforcement method in which the ants act as self-organized nodes and choose the routing path based on the pheromone cost and pheromone strength toward the sink node. The compromised nodes indicate discrepancies in the node energies and trust values along the route paths and are thus discarded by the forwarding nodes.

6.5.2.3 Energy-Efficient Secure Data Aggregation

The data gathering node (DGN) collects the sensor data, aggregates the relevant information, and forwards it to the CH node, which further forwards the aggregated data to the BS. Data gathering protocols eliminate redundant packets along the route paths and save the node energy. Secure data aggregation protocols support authenticity and confidentiality of the nodes in the network.

Figure 6.4 illustrates the energy-efficient routing based on secure data aggregation. The energy is saved using multihop routing and efficient data aggregation techniques. Since the CHs are far away from the sensor nodes, the data aggregation node (DAN) acts as a weighted relay node to (i) collect the data from neighboring sensor nodes, (ii) aggregate the relevant data, and (iii) forward the aggregated data to the CH node, which finally forwards the data to the BS. The energy-efficient secure route paths are selected based on the node energy and key verification process

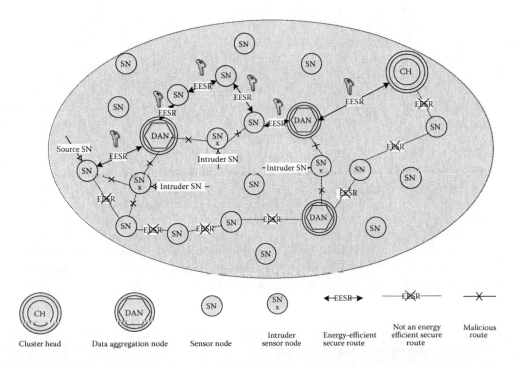

FIGURE 6.4 Energy-efficient routing based on secure data aggregation in sensor networks.

between the pair of nodes. In state-based energy-efficient secure routing, the previous path history and periodic authentication establish the route paths in the network. In Figure 6.4, the intruders and energy-consuming paths are depicted as crossed paths "X" and the data transactions are disabled along these paths. Alternate energy-efficient secure route paths are formed in the neighborhood of intruder and energy-consuming paths to forward the data packets towards the destination.

6.5.2.3.1 Energy-Efficient Secure Pattern-Based Data Aggregation

Energy-Efficient Secure Pattern-Based Data Aggregation (ESPDA) [41] eliminates the redundant data packets based on pattern codes. The data packets are converted to distinctive encrypted form and transmitted to the CH, which further aggregates the encrypted data packets and establishes a secure route path in the network. The CH nodes apply a different key pattern, broadcast the code, and encrypt the data packets. The data packets with repeated codes are eliminated by the CH nodes. The key pattern codes generated by the CH nodes are symmetric and save the node energy. These symmetric keys are generated based on (i) the unique ID of the sensor node, (ii) the private key ($K_{private}$) assigned to the sensor node, and (iii) the public key (K_{public}) common to all the sensor nodes in the network. The CH node selects the sensor node based on acceptable node energy and valid key exchange. This protocol reduces the number of redundant data transmissions at the CH node and the BS. The main limitation of this protocol is that it does not address the hidden node problem.

6.5.2.3.2 Secure Routing and Aggregation Protocol with Low Energy Protocol

Secure Routing and Aggregation Protocol with Low Energy (STAPLE) [42] uses an incremental approach to identify the neighboring nodes and finalize the routing paths toward the destination node. The number of multipath routes created by the authentic nodes is reduced to decrease the communication cost and save the node energy. To ensure the secure data aggregation in the network, the CH and BS periodically eliminate the compromised nodes and tampered data packets in the network. The multilevel route paths are established using the valid key exchange between the one-hop neighbor nodes toward the sink node. Hierarchy Message Authentication Code (HMAC) verifies the valid key exchange with the neighborhood nodes based on (i) previous-hop state, (ii) current-hop state, and (iii) next-hop state of the network. The unauthenticated data packets are filtered and discarded by the intermediate nodes of the network. The main limitation of this protocol is that it does not support multiple route paths in varying network traffic conditions.

6.5.2.3.3 Dynamic Data Aggregation Scheme for Privacy Aware Sensor Networks

Dynamic data aggregation scheme for privacy aware wireless sensor networks (DyDAP) [43] is a time-based protocol that addresses the confidentiality and integrity of the data packets based on the aggregation function. DyDAP supports dynamic data aggregation with variations in traffic conditions, number of active nodes, and size of data packets. DyDAP uses source nodes (data collectors to sense the information), intermediate nodes (to forward and processing the data), and controller nodes (to monitor the validity of intermediate nodes). The working mechanism of DyDAP is given as follows: (i) the sensor nodes with valid key integrity are enabled to exchange the data with neighboring nodes; (ii) based on the valid node ID and key exchange, the node privacy is maintained in the network; (iii) the data packets are aggregated from the neighborhood nodes using secure keys; and (iv) the integrity of the compromised nodes is verified in the network. For multilevel security in the network, the data packet is first encrypted by the node, then aggregated and encrypted by the DAN or CH node, and finally transmitted to the BS. This protocol is suitable for small-scale sensor networks.

6.5.2.4 Energy-Efficient Secure Key Management

6.5.2.4.1 Secure Level Key Infrastructure for MultiCAST and Group Communication Protocol

Secure Level key Infrastructure for MultiCAST and group communication protocol (SLIMCAST) [44] addresses the cost of key maintenance based on addition or deletion of sensor nodes in the network. The sensor nodes in the multicast route use a different set of keys to transmit and receive the data packets. Based on the hop count, SLIMCAST subdivides the multicast routing trees into hierarchical subtrees. Common keys are used within each subtree to encrypt and decrypt the data packets. The intruder nodes are detected by the multicast group node using key authentication. The keys for a given hierarchical level are modified with respect to the rate of intrusions in the corresponding multicast group. This protocol considerably reduces the energy consumption and route cost in the network.

6.5.2.4.2 Energy-Efficient Key Management Protocol

Energy-Efficient Key Management protocol (EEKM) [30] is a lightweight broadcast protocol that uses the key revocation and key modification techniques for the compromised nodes in the network. Because the frequent rekeying technique increases the complexity of encryption and decryption process, EEKM uses dynamic key composition to reduce complexity and save node energy. The sensor network is divided into virtual groups. BS broadcasts the message and the CH node generates group-wise keys and pair-wise keys in the network. The sensor nodes use private keys that are synchronized with the master key of the BS. The pair-wise keys of the adjacent nodes are verified and the compromised nodes are eliminated from the network. EEKM estimates the average energy consumed during the keying assignment and rekeying process. The new nodes added to the virtual group are reassigned the unused IDs and pair-wise keys to minimize the key management process. The main limitation of this protocol is, it does not address the issue of large-scale attacks in the network.

Energy-efficient hybrid key management protocol (EHKM) [45] uses the predeployed keys to sustain the level of trust between the sensor nodes, and dynamically adapt new keys to improve the security in the network. This protocol operates in three modes: (i) low security mode, which uses static keys; (ii) high security mode, which uses dynamic keys, and (iii) hybrid security mode, which is a combination of the above modes. EHKM uses group-wise keys and subnetwork-wise keys in the protocol design. The group-wise key management is used for noncritical messages and the subnetwork-wise key management is used for dynamic subnetworks. EHKM assigns a group-wise key (K1) for group communication between the nodes, a pair-wise key (K2) for adjacent nodes interaction, and a unique key (K3) for each sensor node in the network. The subnetwork key is assigned to the CH node, which further generates the partial keys for the sensor nodes. The main advantage of this protocol is that it is applied for large-scale sensor networks.

Table 6.2 highlights the components of GNS for sensor networks based on functionality and applied protocols.

6.5.2.4.3 Time-Based Dynamic Keying and En-Route Filtering Protocol

Time-Based Dynamic Keying and En-Route Filtering (TICK) protocol [46] reduces the transmission cost by enabling the sensor nodes to use the local time attributes with the dynamic keys to encrypt and decrypt the message. The main components of TICK are (i) time based key management—to create and update the keys, (ii) crypto—to encrypt and decrypt the message, and (iii) filtering–forwarding (FFWD) module—to filter the secure data packets and forward them toward the destination node. This protocol successfully filters the malicious nodes based on local time variations and authentic key verification. This protocol reduces the control overhead and saves the node energy.

6.5.2.4.4 Resource-Efficient Authentic Key Establishment

Resource-Efficient Authentic Key Establishment (RAKE) [47] used in heterogeneous WSNs saves the node energy and key storage space. RAKE for static and mobile sensor nodes support resilience against the network attacks. RAKE uses symmetric hierarchical keys, where the number of keys in the hierarchy depends on the size of cluster. The working mechanism of RAKE is divided into two phases: (i) the symmetric keys are generated by the BS and distributed to the CH and the sensor nodes. Single key is stored between the two adjacent nodes to save the storage space; (ii) the secure

TABLE 6.2

Energy-Efficient Security Protocols in Sensor Networks

Design Feature	Functionality	Applied Protocols
Localization	To estimate the node positions with accuracy and precision, address the queries, and reveal the node positions to authorized users only. To minimize the cost and control overhead.	HEA-Loc, RELMA, secure location verification with hidden and mobile base stations, location privacy and resilience in WSNs, practical and secure localization and key distribution for WSNs
Routing	To establish the energy-efficient routing paths in the network and secure the routing paths controlled by BS and CH	SEEM, TTSR, EENC
Data aggregation	To aggregate the data received from authorized neighboring nodes and forward it to the CH node	ESPDA, STAPLE, DyDAP
Key management	To implement energy-efficient key management techniques for valid data transactions in the network. The keys are revoked for the compromised nodes and the keys are updated as the rate of intrusions increase in the network.	SLIMCAST, EEKM, EEHKM, TICK, RAKE

keys of the sensor nodes (lower level) are registered with the CHs and BS (higher level). This registration process saves the storage space and optimally uses the keys in the network.

6.6 DESIGN OF GNS FOR SENSOR NETWORKS

The proposed design for GNS in sensor networks is divided into two phases: (i) to generate the energy-efficient route paths in the network and (ii) to generate the secure route paths in the network. In the first phase, sensor nodes are grouped into clusters. The CH manages and controls the network operations within each cluster. The sensor nodes are distributed evenly within each cluster to support the energy-efficient design. To maintain the optimum connectivity, the sensor nodes with energy more than the threshold level are selected as the active nodes. The route paths are formed based on the network topology and the number of active sensor nodes per session. In the second phase, secure keys are distributed to the sensor nodes.

The BS manages the keys across various CHs and further the CH node manages the keys across various sensor nodes within the cluster. Symmetric key cryptography methods are used in the ad hoc and sensor networks to save the node energy and minimize the complexity of security protocol. The keys are updated based on the rate of intrusions in the network.

Figure 6.5 illustrates energy-efficient secure routing based on authentic key management. The energy consumption is reduced by using multihop and secure key management techniques. CH verifies the pair-wise keys between the adjacent nodes in the network and finalizes the route path between the nodes. The route paths are periodically monitored based on node energy and secure route authentication. In Figure 6.5, the dotted line between the sensor node SN1 and the cluster head CH indicates energy-efficient secure path in the network. The crossed route paths marked as "X" between the nodes indicate (i) the intrusions of malicious nodes or (ii) the energy-consuming paths in the network. The data transactions are not permitted along these paths.

Figure 6.6 illustrates the sequence of flow to form a GNS-based energy-efficient secure route path in the sensor network. In the first phase, energy-efficient route paths are established based on cluster formation and threshold energy. In the second phase, secure route paths are established using centralized or distributed methods based on symmetric key management techniques. The algorithms for each phase are given as follows:

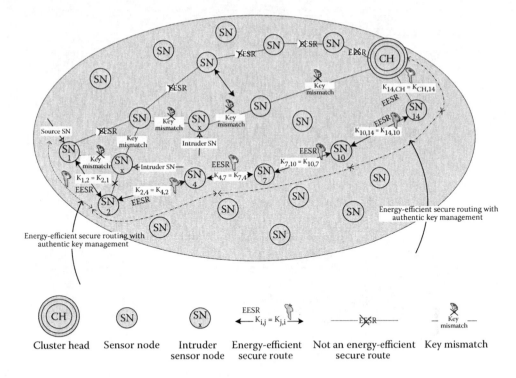

FIGURE 6.5 Design of GNS for sensor networks.

Algorithm Phase I (GNS: Energy-Efficient Paths)

1. Form clusters in the network.
2. Distribute the nodes evenly in the network to:
   a. Balance the energy in the route paths.
   b. Support the multihop and fault tolerant paths in the network.
3. The route paths are updated based on the node threshold energy.
4. Form the network topology based on node distance and maximum neighborhood set.
5. Periodically, update the route paths in the network.
6. Repeat Step 3 and Step 4 to form energy-efficient paths.
7. Finalize the energy-efficient paths in the network.

Algorithm Phase II (GNS: Energy-Efficient Secure Paths)

1. Select centralized or distributed key management technique.
2. BS generates the keys and distributes it to the CH nodes and sensor nodes.
3. For energy-efficient secure paths use symmetric key management.
4. Establish pair-wise keys between the nodes in the network.
5. Update the keys based on the threshold energy and rate of intrusions in the network
6. Repeat Step 4 and Step 5 to form energy-efficient secure paths.
7. Finalize the energy-efficient secure paths in the network.

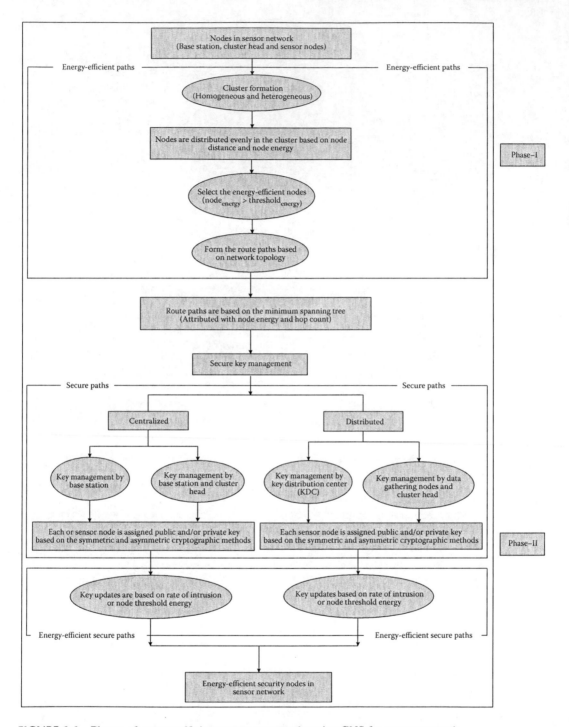

FIGURE 6.6 Phases of energy-efficient secure route paths using GNS for sensor network.

6.6.1 SIMULATION RESULTS

Simulations were performed using NS2 and MATLAB® with varying energy and secure keys. Simulation parameters for the GNS for sensor networks are as follows: sensor nodes: 100, area: 900 m × 900 m, mobility model: static, traffic: CBR, number of sources: 10–15 varying, packet size: 128 bytes, transmission power: 1.8 W, reception power: 0.8 W, node range: 180 m, and pause time: 20 s.

Figure 6.7 indicates the energy comparisons of GNS-based Energy-Efficient Secure Path (EESP) with Energy-Efficient Path (EEP) protocol and Secure Path (SP) protocol. As the number of data packets increase, the GNS-based energy-efficient secure routing consumes less energy as compared to EEP and SP protocols. Because of the periodic route verification process, the network load across various nodes in the proposed GNS-based EESP is relatively more as compared to EEP and SP protocols.

The energy gradients for the proposed GNS-based EESP are relatively less as compared to the energy gradients of EEP and SP protocols. The energy consumption range is given as follows: GNS-based EESP: 3.2 to 4.5 mJ, EEP: 4.5 to 6.3 mJ, and SP: 5 to 7.5 mJ. GNS-based EESP periodically verifies the node energies and selects the route path with optimum node energy. In EEP and SP protocols, the energy gradients rapidly vary with more number of data packets in the network.

Figure 6.8 indicate the throughput success ratio comparisons of GNS-based EESP with EEP and SP protocols. As the number of sensor nodes increase, the throughput rate of GNS-based EESP increases as compared to EEP routing and SP routing. The increase in throughput rate is due to the elimination of route paths with (i) the node energy less than the threshold energy and (ii) the intruded nodes in the network. Simulation results indicate that the network load across various route paths in GNS-based EESP is relatively more as compared to the EESP and SP protocols. The symmetric key management technique in GNS-EESP saves the node energy and increases the throughput rate. The optimum throughput is achieved when the route paths are stabilized. The throughput success ratio in EEP and SP protocols is less due to varying packet size and complexity in group-wise key management approach.

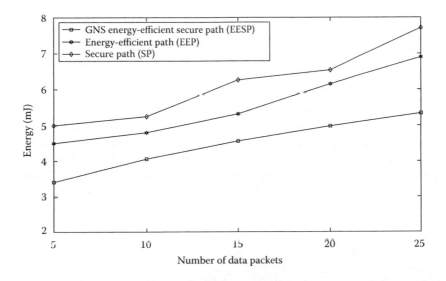

FIGURE 6.7 Energy comparisons of GNS-based EESP with EEP and SP protocols with varying data packets.

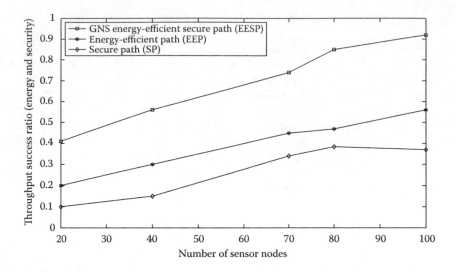

FIGURE 6.8 Throughput success ratio of GNS-based EESP with EEP and SP protocols with varying number of sensor nodes.

6.7 CONCLUSIONS AND FUTURE DIRECTIONS

This chapter explains the significance of GNS for wireless ad hoc and sensor networks. The design metrics for energy and security are elucidated briefly. Energy-efficient security protocols based on localization, routing, and key management for ad hoc and sensor networks are explained in detail. The proposed architecture of GNS for sensor networks is an integrated framework of energy-efficient design and security design. The primary attributes considered in the design are threshold energy and rate of intrusion. GNS improves the throughput rate and saves the node energy. This work can be further extended to include node scheduling and load balancing along the route paths, to improve the throughput rate and increase the network life cycle.

AUTHOR BIOGRAPHY

M. Bala Krishna received his bachelor of engineering (BE) degree in computer engineering from the Delhi Institute of Technology (presently Netaji Subhash Institute of Technology), University of Delhi, Delhi, India, and his master of technology (MTech) degree in information technology from the University School of Information Technology (presently University School of Information and Communication Technology), GGS Indraprastha University, Delhi, India. He had earlier worked as a senior research associate and project associate at the Indian Institute Technology, Delhi, India, in the areas of digital systems and embedded systems. He had also worked in the projects related to communication networks. He is presently working as assistant professor in the University School of Information and Communication Technology, GGS Indraprastha University, Delhi, India. His teaching and research areas include computer networks, wireless networking and communication, mobile computing, and embedded system design. He has publications in international conferences and journals. His current research areas include wireless ad hoc and sensor networks, mobile and ubiquitous computing, and security in wireless networks.

ABBREVIATIONS

AAPs	Authorized access points
ACPI	Advanced configuration power interface
AE	Authenticated encryption
APM	Advanced power management
APs	Access points
ARAN	Authenticated routing protocol ad hoc networks
BS	Base station
CA	Certification Authority
CBGKA	Cluster-Based Grouped Key Agreement protocol
CBSs	Covert base stations
CEPS	Common Electronic Purse Specifications
CH	Cluster head
CONFIDANT	Cooperation Of Nodes: Fairness In Dynamic Ad hoc NeTworks
CORE	COllaborative REputation
DAN	Data aggregation node
DGN	Data gathering node
DyDAP	Dynamic data aggregation scheme for privacy aware wireless sensor networks
ECC	Elliptic curve cryptography
EEKM	Energy-Efficient Key Management protocol
EENC	Energy Efficiency Routing with Node Compromised Resistance
EEP	Energy-Efficient Path
EESP	Energy-Efficient Secure Path
EKF	Extended Kalman filter
ESPDA	Energy-Efficient Secure Pattern-Based Data Aggregation
FFWD	Filtering–forwarding
GNS	Green Network Security
GSP	Group Selection Protocol
HEA-Loc	Hybrid efficient and accurate localization
HMAC	Hierarchy Message Authentication Code
KGC	Key generation center
LBPR	Location Prediction-Based Routing protocol
MGM	Multigrade monitoring
MSs	Mobile stations
ODAR	On-Demand Anonymous Routing
PDCA, FDCA	Partially and fully distributed certificate authority
PDM	Proximity-distance map
PKI	Public key infrastructure
POS	Point of sale
PSAM	Purchase Security Application Module
RAKE	Resource-Efficient Authentic Key Establishment
RELMA	Range-free Energy-efficient, Localization technique using Mobile Anchor
RSSI	Received signal strength indicator
RV	Route verification protocol
SAMAS	Secure source anonymous message authentication scheme
SAODV	Secure Ad hoc On-demand Distance Vector

SDP	Semi-Definite Programming
SDSR	Secure Dynamic Source Routing
SEEM	Secure and Energy-Efficient Multipath Routing protocol
SEKM	Secure and Efficient Key Management in mobile ad hoc networks
SET	Secure Electronic Transaction
SLAT	Simultaneous Localization and Tracking
SLIMCAST	Secure Level key Infrastructure for MultiCAST
SN	Sensor node
SP	Secure path
SPR	Secure-path routing
STAPLE	Secure Routing and Aggregation Protocol with Low Energy protocol
TDoA	Time difference of arrival
TICK	Time-Based Dynamic Keying and En-Route Filtering
ToA	Time of arrival
TTSR	Two-Tier Secure Routing Protocol for Heterogeneous Sensor Networks
WMNs	Wireless mesh networks
WSNs	Wireless sensor networks

REFERENCES

1. Rongxing L, Xu L, Xiaohui L, Xuemin S and Xiaodong L, GRS: The green, reliability, and security of emerging machine to machine communications, *IEEE Communications Magazine*, April 2011, 49(4), pp. 28–35.
2. Pablo S, Antonio De La O, Paul P, Vincenzo M and Albert B, Greening wireless communications: Status and future directions, *Elsevier Journal of Computer Communications*, 1 August 2012, 35(14), pp. 1651–1661.
3. Phongsak P and Prashant K, On a framework for energy-efficient security protocols in wireless networks, *Elsevier Journal of Computer Communications*, 1 November 2004, 27(17), pp. 1716–1729.
4. Erdal C and Chunming R, Chapter 2. Wireless ad hoc, sensor and mesh networks, In *Security in Wireless Ad Hoc and Sensor Networks*, Erdal C and Chunming R (Eds.), John Wiley & Sons Ltd, Great Britain by CPI Antony Rowe, Chippenham, England, 2009, pp. 9–28.
5. Anjum N and Salil K, Chapter 1. Authentication and confidentiality in wireless ad hoc networks, In *Security in Ad Hoc and Sensor Networks, Computer and Network Security,* Vol. 3, Beyah R, McNair R and Corbett C (Eds.), World Scientific Publishing Co. Pte. Ltd, Singapore, 2010, pp. 3–28.
6. Huzaifa A N, Jitender S D and Eric D M, Proactive mitigation of impact of wormholes and sinkholes on routing security in energy-efficient wireless sensor networks, *Springer Journal of Wireless Networks*, May 2009, 15(4), pp. 431–441.
7. Xiangqian C, Kia M, Kang Y and Niki P, Sensor network security: A survey, *IEEE Communications Surveys and Tutorials*, Second Quarter 2009, 11(2), pp. 52–73.
8. Jongdeog L, Krasimira K and Sang H S, The price of security in wireless sensor networks, *Elsevier Journal of Computer Networks*, 3 December 2010, 54(17), pp. 2967–2978.
9. Yang X, Accountability for wireless LANs, ad hoc networks, and wireless mesh networks, *IEEE Communications Magazine*, April 2008, 46(4), pp. 116–126.
10. Sahibzada A M, Shahbaz K, Hamed Al-R and Kumarendra S, Meshed high data rate personal area networks, *IEEE Communications Surveys*, First Quarter 2008, 10(1), pp. 58–69.
11. Lin X C, Vincent Poor H, Yongkang L, Tom H L, Xuemin S and Jon W M, Dimensioning network deployment and resource management in green mesh networks, *IEEE Wireless Communications Magazine*, October 2011, 18(5), pp. 58–65.
12. Sakarindr P and Ansari N, Security services in group communications over wireless infrastructure, mobile ad hoc and wireless sensor networks, *IEEE Wireless Communications Magazine*, October 2007, 14(5), pp. 8–20.
13. Laniepce S L, Chapter 9, Security for ad hoc routing and forwarding, In *Wireless Ad Hoc and Sensor Networks*, Labiod H (Ed.), ISTE Ltd, John Wiley & Sons, New York, 2008, pp. 195–223.

14. Wen Zhan S, Yu W, Xiang-Yang L and Ophir F, Localized algorithms for energy efficient topology in wireless ad hoc networks, *Springer Journal of Mobile Networks and Applications*, 2005, 10(6), pp. 911–923.

15. Depeng L and Srinivas S, An efficient group key establishment in location-aided mobile ad hoc networks, In *Proceedings of ACM Second International Workshop on Performance Evaluation of Wireless Ad Hoc, Sensor, and Ubiquitous Networks (PE-WASUN)*, Montreal, Quebec, Canada, 10–13 October 2005, pp. 57–64.

16. Christopher T, Ali R and Jonathan B, Simultaneous localization, calibration, and tracking in an ad hoc sensor network, In *Proceedings of ACM Fifth International Conference on Information Processing in Sensor Networks (IPSN)*, Nashville, Tennessee, USA, 19–21 April 2006, pp. 27–33.

17. Jian R, Yun L and Tongtong L, Providing source privacy in mobile ad hoc networks, In *Proceedings of IEEE Sixth International Conference on Mobile Adhoc and Sensor Systems (MASS)*, Macau (S.A.R.), China, 12–15 October 2009, pp. 332–341.

18. Natarajan M, A location prediction based routing protocol and its extensions for multicast and multipath routing in mobile ad hoc networks, *Elsevier Journal of Ad Hoc Networks*, September 2011, 9(7), pp. 1104–1126.

19. Sanzgiri K, Dahill B, Levine B N, Shields C and Belding-Royer E M, A secure routing protocol for ad hoc networks In *Proceedings of IEEE International Conference on Network Protocols (ICNP'02)*, 12–15 November 2002, pp. 78–87.

20. Zapata M G, Secure ad hoc on-demand distance vector (SAODV) routing, *ACM SIGMOBILE Mobile Computing and Communications Review*, 3 July 2002, 6(3), pp. 106–107.

21. Papadimitratos P and Haas Z J, Secure on-demand distance vector routing in ad hoc networks, In *Proceedings of IEEE Symposium on Advances in Wired and Wireless Communication*, Princeton, NJ, USA, 18–19 April 2005, pp. 168–171.

22. Creti M T, Beaman M, Bagchi S, Zhiyuan L and Yung-Hsiang L, Multigrade security monitoring for ad-hoc wireless networks, In *Proceedings of IEEE Sixth International Conference on Mobile Adhoc and Sensor Systems (MASS)*, Macau (S.A.R.), China, 12–15 October 2009, pp. 342–352.

23. Kargl F, Geiss A, Schlott S and Weber M, Secure dynamic source routing, In *Proceedings of IEEE Thirty Eighth Annual Hawaii International Conference on System Sciences (HICSS)*, Hawaii, USA, 3–6 January 2005, pp. 1–10.

24. Lichun B, Rex C and Denh S, Chapter 5. On-demand anonymous routing, In *Security in Ad hoc and Sensor Networks, Computer and Network Security–Vol. 3*, Beyah R, McNair J and Corbett C (Eds.), World Scientific Publishing Co. Pte. Ltd, 5 Toh Tuck Link, Singapore, 2010, pp. 137–157.

25. Panagiotis P and Zygmunt J H, Chapter 31. Securing mobile ad hoc networks, In *The Handbook of Ad hoc Wireless Networks*, Ilyas M (Ed.), CRC Press LLC, Boca Raton, FL, January 2003, pp. 31:1–31:17.

26. Ming C W, Yang X and Xu S, Chapter 9. Security issues in ad hoc networks, In *Security in Sensor Networks*, Xiao Y (Ed.), Auerbach Publications, Taylor & Francis Group, LLC, Boca Raton, FL, 2007, pp. 215–236.

27. Bing W, Jie W, Eduardo B F, Mohammad I and Spyros M, Secure and efficient key management in mobile ad hoc networks, *Elsevier Journal of Network and Computer Applications*, 3 August 2007, 30(3), pp. 937–954.

28. Elisavet K, Efficient cluster-based group key agreement protocols for wireless ad hoc networks, *Elsevier Journal of Network and Computer Applications*, January 2011, 34(1), pp. 384–393.

29. Han K, Kim K, Park J and Shon T, Efficient sensor node authentication in third generation-wireless sensor networks integrated networks, *IET Journal of Communications: Special Issue on Distributed Intelligence and Data Fusion for Sensor Systems*, 12 August 2011, 5(12), pp. 1744–1754.

30. Kwang-Jin P, Jongwan K, Chong-Sun H and Ui-Sung S, An energy-efficient key management protocol for large-scale wireless sensor networks, In *Proceedings of IEEE International Conference on Multimedia and Ubiquitous Engineering (MUE)*, Seoul, South Korea, 26–28 April 2007, pp. 201–206.

31. Pratik B and Yinyu Y, Semidefinite programming for ad hoc wireless sensor network localization, In *Proceedings of ACM Third International Symposium on Information Processing in Sensor Networks (IPSN)*, Berkeley, California, USA, 26–27 April 2004, pp. 46–54.

32. Srdjan C, Kasper B R, Mario C and Mani S, Secure location verification with hidden and mobile base stations, *IEEE Transactions on Mobile Computing*, April 2008, 7(4), pp. 470–483.

33. Yuanyuan H, King-Shan L and Yik-Chung W, HEA-Loc: A robust localization algorithm for sensor networks of diversified topologies, In *Proceedings of IEEE International Wireless Communications and Networking Conference (WCNC)*, Sydney, NSW, Australia, 18–21 April 2010, pp. 1–6.

34. Dan S, *Optimal State Estimation: Kalman, H Infinity, and Nonlinear Approaches*, John Wiley & Sons, Hoboken, NJ, 2006.

35. Hyuk L and Hou J C, Localization for anisotropic sensor networks, In *Proceedings of IEEE Twenty Fourth Annual Joint Conference of the Computer and Communications Societies (INFOCOM)*, Miami, Florida, USA, 13–17 March 2005, vol. 1, pp. 138–149.

36. Karim L, Nasser N and El Salti T, RELMA: A range free localization approach using mobile anchor node for wireless sensor networks, In *Proceedings of IEEE Conference on Global Telecommunications (GLOBECOM)*, Miami, Florida, USA, 6–10 December 2010, pp. 1–5.

37. Qi M, John A S and Radu S, Practical and secure localization and key distribution for wireless sensor networks, *Elsevier Journal of Ad Hoc Networks*, August 2012, 10(6), pp. 946–961.

38. Nidal N and Yunfeng C, SEEM: Secure and energy-efficient multipath routing protocol for wireless sensor networks, *Elsevier Journal of Computer Communications*, 10 September 2007, 30(11–12), pp. 2401–2412.

39. Xiaojiang D, Mohsen G, Yang X and Hsiao-Hwa C, Two tier secure routing protocol for heterogeneous sensor networks, *IEEE Transactions on Wireless Communications*, September 2007, 6(9), pp. 3395–3401.

40. Kai L, Chin-Feng L, Xingang L and Xin G, Energy efficiency routing with node compromised resistance in wireless sensor networks, *Springer Journal of Mobile Network Applications*, February 2012, 17(1), pp. 75–89.

41. Hasan C, Suat O, Prashant N, Devasenapathy M and Ozgur Sanli H, Energy-efficient secure pattern based data aggregation for wireless sensor networks, *Elsevier Journal of Computer Communications*, 20 February 2006, 29(4), pp. 446–455.

42. Nike G, Ruichuan C, Zhuhua C, Jianbin H and Zhong C, A secure routing and aggregation protocol with low energy cost for sensor networks, In *Proceedings of IEEE International Symposium on Information Engineering and Electronic Commerce (IEEC)*, Ternopil, 16–17 May 2009, pp. 79–84.

43. Sabrina S, Luigi Alfredo G, Gennaro B and Coen-Porisini A, DyDAP: A dynamic data aggregation scheme for privacy aware wireless sensor networks, *Elsevier Journal of Systems and Software*, January 2012, 85(1), pp. 152–166.

44. Jyh-How H, Buckingham J and Han R, A level key infrastructure for secure and efficient group communication in wireless sensor networks, In *Proceedings of IEEE First International Conference on Security and Privacy for Emerging Areas in Communications Networks (SECURECOMM)*, Athens, Greece, 5–9 September 2005, pp. 1–12.

45. Landstra T, Zawodniok M and Jagannathan S, Energy-efficient hybrid key management protocol for wireless sensor networks, In *Proceedings of IEEE Thirty Second International Conference on Local Computer Networks (LCN)*, Dublin, Ireland, 15–18 October 2007, pp. 1009–1016.

46. Uluagac A S, Beyah R A and Copeland J A, TIme-Based DynamiC Keying and En-Route Filtering (TICK) for wireless sensor networks, In *Proceedings of IEEE Global Telecommunications Conference (GLOBECOM)*, Miami, FL, USA, 6–10 December 2010.

47. Qi S, Ning Z, Madjid M and Kashif K, Resource-efficient authentic key establishment in heterogeneous wireless sensor networks, *Elsevier Journal of Parallel and Distributed Computing*, 2 February 2013, 73(2), pp. 235–249.

Section II

Cellular Networks

7 Interplay between Cooperative Device-to-Device Communications and Green LTE Cellular Networks

Elias Yaacoub, Hakim Ghazzai,
Mohamed-Slim Alouini, and Adnan Abu-Dayya

CONTENTS

7.1 INTRODUCTION

Traditional efforts for saving energy in cellular networks focus on reducing the transmit powers of base stations (BSs) and mobile devices. However, when a BS is in its working mode, studies show that more than 50% of the energy consumed is due to circuit processing, air conditioning, and other factors [1]. In addition to the growing awareness of energy-efficient wireless networks and their environmental benefits, electricity bills have become a significant cost factor for mobile operators [2]. This trend in reducing energy consumption and ensuring green wireless networks is supported by political and national initiatives, which are beginning to put requirements on lowering the CO_2 emissions. For example, the European Commission research project EARTH concentrates on energy efficiency in radio access networks with the goal of finding solutions and concepts that can reduce energy consumption of mobile broadband systems by 50% [3]. An effective approach to save energy in wireless networks is to completely turn off selected BSs when the traffic load becomes light (e.g., see Refs. [4–7]).

On the other hand, the demands on mobile data traffic are growing exponentially, with web browsing, audio streaming, and video dominating the traffic on the wireless web [8]. Therefore, efficient techniques that allow the wireless networks to cope with the increasing demands while still providing the desired quality of service (QoS) and quality of experience (QoE) to mobile users should be investigated. In many practical scenarios, mobile users might be interested in the same content: this could happen when several users are subscribed in a live news service, attending a live webinar, watching a live sports match on their portable devices, and so on. For example, during Olympic games, confined areas in the city will host thousands of fans with similar interests, wanting to watch the latest results of their favorite sport, and hence all the traffic will fall on the same cell's BS. In such scenarios, when several of these users are in the same cell, the load on the BS increases significantly if the BS is sending the content to each user individually, that is, via unicasting. On the other hand, multicasting requires that the BS transmits at the lowest rate among the subscribed users, which could lead to a large delay and sometimes unacceptable QoS, in addition to being unfair to the other users having better channel conditions with the BS [9]. Cooperation between mobile terminals (MTs) via peer-to-peer (P2P) communications is a possible solution that can help in mitigating this problem. Consequently, the content of common interest could be sent once by the BS to each group of cooperating MTs over long-range (LR) wireless links, and these MTs could exchange the content via cooperation on high-rate, short-range (SR) wireless links. In addition to freeing bandwidth at the BS and increasing network throughput [10,11], SR collaboration between MTs leads to a reduced energy consumption [12,13]. In fact, higher rates can be achieved over SR communications between MTs that are relatively close from each other in a single cooperating cluster. This leads to shorter transmission and reception times and hence less energy consumption from the batteries of the MTs.

It would be simpler for practical purposes if both LR and SR communications can be performed over the same technology. This would also facilitate the cross-layer operation of the content distribution operation between the wireless interfaces of the same device. For example, this would be useful when receiving video packets on the LR from the BS, transferring them to the SR interface, and then forwarding them to neighboring MTs on the physical layer while maintaining real-time display of the video on the application layers of the MTs. Such an option is under investigation in long-term evolution-advanced (LTE-A), and it is referred to as device-to-device (D2D) communication.

In fact, D2D communication has received some research attention in the literature [14–16] as part of LTE-A. D2D enables linking an MT to another MT directly using the cellular spectrum. This could allow large amounts of data (e.g., multimedia) to be transferred from one MT to another over short distances and using a direct connection. This data exchange occurs over the SR without the need to use the cellular network itself, thus leading to off-loading some traffic from the network.

It would be interesting to study the interplay between green communications at the BS level and green communications at the MT level. Although most of the energy consumption occurs at BSs, SR collaboration between MTs can significantly reduce the number of LR links with the BS. Thus, for example, instead of having 10 LR connections to BS A, SR collaboration could lead to having two links, with the other eight MTs receiving the data via SR collaboration. Moving the connections of two MTs from BS A to BS B and putting BS A in sleep mode is obviously much easier than moving all the 10 links without compromising QoS and overloading BS B and other neighboring BSs.

Hence, in this chapter, we present an efficient approach for MT clustering, an energy-efficient approach for putting BSs in sleep mode, and describe an approach that combines the two methods in order to ensure green communications for both the users' MTs and the operator's BSs. The proposed techniques are investigated in the framework of orthogonal frequency division multiple access (OFDMA)-based state-of-the-art LTE cellular networks, while taking resource allocation and intercell interference into account.

The chapter is organized as follows. Related previous work in the literature is described in Section 7.2. The system model is presented in Section 7.3. Section 7.4 presents the cooperative energy-efficient D2D method. Section 7.5 describes the green approach for LTE cellular networks. Section 7.6 presents the combined approach that jointly uses the two previous methods. Simulation results are presented and analyzed in Section 7.7. Interesting directions for future research are outlined in Section 7.8. Finally, conclusions are drawn in Section 7.9.

7.2 BACKGROUND/RELATED WORK

This section presents an overview of background work related to cooperative D2D communications and energy-efficient green cellular network.

7.2.1 Overview of Cooperative P2P Communications

Wireless P2P networks with collaboration are studied in Ref. [17], where peers receive a video description on the LR and distribute it on the SR via broadcast. The broadcast scope can be determined statically or dynamically. In static broadcast scope, each description has a time-to-live that is decremented at each broadcast. With dynamic broadcast scope, peers exchange information about their available descriptions. A peer decides to broadcast one of the descriptions it has if a significant number of its direct neighbors do not have that description. The metrics of interest in Ref. [17] are bandwidth consumption, delay, and fairness between peers in pulling descriptions from the server. Energy consumption is not considered explicitly but is affected by the video distribution delay. The concept of resource sharing in order to adapt video contents in a distributed manner to address peer mobility and heterogeneity in a community network is presented in Ref. [18]. The idle computing power of the peers is utilized to adapt videos in addition to bandwidth sharing. In the approach of Iqbal and Shirmohammadi [18], participating peers are responsible for video adaptation and distribution. The limited energy capabilities of handheld devices are taken into consideration by allowing these devices to receive only but not to retransmit. Multiple parents are considered for each node in order to handle the volatility of node arrival and departure. The metrics of interest in Ref. [18] are the bandwidth and the CPU resources. In Ref. [19], it is argued that tree-based cooperative live streaming is more stable than mesh-based methods in constructing and maintaining an efficient topology in mobile environments, although the opposite might be true in wired networks. The proposed algorithm uses hop count distance to estimate the distance between the media source and the client peer destination. A peer requests a move to a higher level in the tree topology on the basis of hop count distance. When peers keep close to the source, they obtain more energy-efficient and delay-efficient performance. Thus, if the peer's hop count distance is larger than the peer's parent, the algorithm of Kim and Chong [19] exchanges the two nodes to keep a short distance between the data source and the client peer. P2P streaming in high-mobility environments is considered in Ref. [20], where the example considered is in a moving train. Peers are connected to 3G/HSDPA networks and they exchange video chunks on an SR wireless local area network (WLAN) in ad hoc mode. An efficient protocol is presented where one of the peers acts as controller and coordinates the download of video chunks among peers.

Cooperative P2P Repair (CPR) is introduced in Ref. [21] and shown to be effective in improving video quality. With CPR, peers within a single cooperating group listening to the same wireless wide area network (WWAN) video broadcast/multicast and connected to each other via ad hoc WLAN exchange received WWAN packets locally via WLAN to repair WWAN losses. Network coding is used for forward error correction in order to improve the WLAN P2P repair approach.

The strength of CPR in correcting channel errors is exploited in Ref. [22] in order to perform joint source/channel coding. This allows more bits to be dedicated for source coding, which enhances the video quality. The structured network coding approach of Liu et al. [21] is applied in Ref. [23] for H.264 SVC where layered coding using two layers is assumed. Multiple network interfaces on one device are exploited. H.264 two-layered video is transmitted over an LR WWAN. Peers receiving different subsets of WWAN broadcast/multicast packets can perform CPR over an SR WLAN, by exchanging received WWAN packets with their local WLAN peers. This improves the transmission success from a WWAN broadcast/multicast source to a CPR cooperative cluster. In a first time epoch, peers receive a group of pictures (GOP) on the LR. In another epoch, they perform CPR on the SR while receiving the next GOP on the LR. The problem is formulated as an optimization problem aiming for the minimization of the overall distortion in the cluster. Because of the exponential complexity of the problem, an efficient suboptimal approach is presented.

The work in Refs. [24,25] investigates cooperative content distribution between MTs by subdividing the problem into LR and SR transmissions. An energy consumption model at the MTs is presented, although no formulation into an energy minimization problem is adopted. Instead, simple collaborative techniques that lead to energy efficiency are proposed. TDMA (time division multiple access) is assumed over both the LR and SR technologies. All MTs are assumed to be within a single cooperating cluster. The energy consumption is computed in Refs. [24,25] for several investigated scenarios. Different combinations of LR/SR technologies are studied in the simulation results. Bluetooth is shown to be more energy efficient than WLAN on the SR in Ref. [24].

These works assume that MTs communicate over different technologies on the LR (e.g., LTE, 3G/HSPA, and GPRS) and SR (e.g., Bluetooth and WiFi). Energy consumption over the various wireless interfaces of a given mobile device is investigated in Ref. [26]. Mahmud et al. [26] studied the energy consumption of wireless interface cards by measuring the amount of current that is drawn from the card from the host terminal. Then, a power-saving algorithm is used to predict the total energy consumption through each attached interface and choose the optimal one in order to make these attached wireless cards more energy efficient.

It would be simpler for practical purposes if both LR and SR communications can be performed over the same technology. This is under ongoing study in LTE-A under the name of D2D communication. D2D communication in LTE-A has been investigated in Refs. [14–16]. The main challenge with D2D communication is to keep the interference to the primary cellular network at tolerable levels [16]. In Ref. [14], the D2D communication is investigated as a network underlying the LTE-A cellular network. Mechanisms for D2D communication session setup and management are proposed. The D2D communication can operate in multiple modes. It can underlie the cellular transmission, or the cellular network can assign dedicated resources to the D2D terminals, or they can reuse the same resources used by the cellular network [15]. On the basis of the results of the single-cell studies, a mode selection procedure for a multicell scenario is proposed in Ref. [15]. The problem of radio resource allocation to the D2D communications is formulated as a mixed integer nonlinear programming problem in Ref. [16]. Because of the difficulty of solving this problem, a greedy heuristic algorithm that reduces the interference to the primary cellular network by utilizing channel gain information is proposed. A proprietary D2D communication technology, Qualcomm's "FlashlinQ," is presented [27,28]. This technology enables automatic discovery and communication between devices without using the cellular infrastructure. Hence, it builds an SR wireless network between FlashLinQ-enabled devices, allowing those devices to share content. FlashLinQ can perform SR content sharing while still being simultaneously connected to a cellular network on the LR. Furthermore, FlashLinQ allows all pairs that coexist to communicate simultaneously.

7.2.2 OVERVIEW OF GREEN CELLULAR COMMUNICATIONS

Cellular operators are increasingly concerned about energy efficiency in cellular network. In addition to maintaining their profitability, the purpose is to also reduce negative environmental effects.

Thus, standardization authorities and network operators are being motivated by the energy efficiency objective to explore advanced technologies leading to improved network infrastructures [29].

Since BSs consume a large portion of the total energy consumed in a cellular system, an efficient technique to save energy in wireless networks is to turn off selected BSs when the traffic load becomes light, or, in other words, put them in sleep mode. BS sleeping strategy has been investigated extensively in the literature (e.g., Refs. [4–6]). Switching off BSs leads to considerable energy savings, since traffic in peak hours can be as much as 10 times higher than the traffic in off-peak periods in the same area [2]. In Ref. [7], a cell zooming approach is proposed to adaptively adjust the cell sizes according to traffic load, user requirements, and channel conditions in order to reduce energy consumption. A solution based on traffic load balancing is proposed in Ref. [30], where the power ratio is introduced. It consists of the ratio between the dynamic and fixed power parts of BS power consumption. This power ratio is then used to determine the relationship between the optimal number of active BSs and the traffic load. As expected, the results of the study by Xiang et al. [30] show that less BSs need to be switched on at light traffic loads whereas more BSs are needed during high load periods.

Other techniques focus on switching off certain components instead of switching off the whole BS. For example, in Ref. [31], an algorithm named Green Antenna Switching (GAS) is introduced. It consists of switching off one of the multiple-input multiple-output (MIMO) antennas along with its corresponding power amplifiers, while taking into account the traffic intensity and topology handled by each BS. Different traffic loads and service classes are investigated using an LTE system level simulator. Guaranteed savings are obtained with GAS compared to the traditional approach (keeping the MIMO antennas always on) without introducing excessive performance degradation. In Ref. [32], the objective of the proposed technique is to shut down unnecessary network elements while meeting traffic demands with the remaining network elements. Hence, energy-efficient multi-path routing is proposed in Ref. [32] and shown to lead to reduced energy consumption compared to single/shortest path routing.

Key enhancements that are shaping the design of next-generation 4G cellular systems, such as LTE-A, include reducing effective cell sizes by using combinations of microcells [33], distributed antenna systems [34], relays [35,36], and indoor femtocells [37,38]. These are complemented by advanced interference coordination/mitigation techniques, heterogeneous fractional frequency reuse patterns, and cooperative multipoint transmission/reception techniques. These heterogeneous network deployments can be used to achieve the goal of green cellular communications [29]. Studies in this direction have been presented in Refs. [39,40] for example.

To maintain a smooth network operation guaranteeing user satisfaction, BSs that are turned off should know when to automatically switch on based on local decisions, external trigger, or pre-configuration. Hence, there is a need for coordination between BSs in order to configure the network automatically and dynamically on the basis of the current traffic situation, which leads to a self-organizing network (SON) case [41]. The benefits of SONs expand beyond designing green networks. In fact, they can be used for efficient operation and maintenance [42–45] and for reducing CAPEX (capital expenditures) [46] and OPEX (operational expenditures) [47] by eliminating many on-site operations and reducing the workload for site survey and analysis of network performances, respectively.

7.2.3 Chapter Novelty

Most of the previous works consider energy efficiency either at the MT level, thus investigating cooperative P2P networks, or at the cellular network level, thus investigating BS power-saving techniques. The novelty of the topic treated in this chapter is that it investigates the interplay between these two concepts using the same technology on the SR and LR. Hence, P2P communications occur using LTE-A D2D communications, and the collaboration between MTs on the SR helps reduce the number of LR links with the LTE BSs, which facilitates the process of switching off some BSs through the implementation of green cellular communication techniques.

7.3 SYSTEM MODEL

We consider a cellular network of N_{BS} BSs. Content of common interest is to be delivered from the BS to K requesting MTs distributed throughout the area of interest. The MTs are interested in the same content. They communicate with the BSs using LTE or with neighboring MTs using D2D communications. MTs form cooperating clusters for the purpose of energy minimization during cooperative content distribution. Within each cooperating cluster, the content is delivered on the LR to a single MT, the cluster head (CH), which in turn multicasts the content to the other MTs in that cluster using SR collaboration. Figure 7.1 shows the scenario considered with two cells as example.

At a given fading realization, each MT receives the content from a single source, which could be either a BS or another MT. Receiving parts of the content from different sources is suboptimal in terms of energy minimization. In fact, the energy minimization problem for content distribution in a single cluster of cooperating MTs is formulated and solved in Ref. [48]. It was shown that the optimal solution in a single cluster is to send all data to a single MT, the CH, and that CH should distribute the content to all other MTs cooperating with it.

We assume that MTs can form coalitions where the energy consumption in the coalition is lower than the sum of the individual energy consumptions of the coalition members. We use the term *node* to refer to either an MT or a BS. The N_{BS} BSs are numbered from node n_1 to node $n_{N_{BS}}$. Having K MTs in the system, they are numbered from node $n_{N_{BS}+1}$ to node $n_{N_{BS}+K}$. We denote by C_i the coalition of nodes forming a single cooperative cluster with n_i as CH communicating on the LR with a BS on behalf of all the cluster members.

7.3.1 RATE CALCULATIONS AND CHANNEL MODEL

OFDMA is the access scheme for the downlink (DL) of LTE systems. In LTE, the available spectrum is divided into resource blocks (RBs) consisting of 12 adjacent subcarriers. The assignment

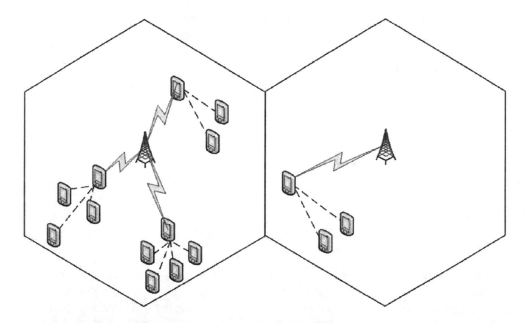

FIGURE 7.1   Multicell system model with clustering and D2D communications.

of a single RB takes place every 1 ms, agreed to be the duration of one transmission time interval (TTI) or the duration of two 0.5 ms slots [49,50]. Given the transmit power $P_{t,kj}^{(x)}$ that n_k is using in order to transmit to n_j over subcarrier x, the channel gain $H_{kj}^{(x)}$ of the channel between n_k and n_j over subcarrier x, the signal-to-interference plus noise ratio (SINR) $\gamma_{kj}^{(x)}$ on the link between n_k and n_j over subcarrier x, and the target bit error rate P_e, the bit rates on the link between any two nodes n_k and n_j over subcarrier x can be calculated as follows:

$$R_{kj}^{(x)} = W^{(x)} \log_2\left(1+\beta\gamma_{kj}^{(x)}\right) \tag{7.1}$$

where $W^{(x)}$ is the passband bandwidth of the subcarrier, and β is called the SNR gap. It indicates the difference between the SNR needed to achieve a certain data transmission rate for a practical M-QAM system and the theoretical Shannon limit [51]. It is given by $\beta = -1.5/\ln(5P_e)$. Having subcarriers of similar bandwidth, we can write $W^{(x)} = B_{sub}$. The subcarrier bandwidth can be expressed as

$$B_{sub} = \frac{B}{N_{sub}}, \tag{7.2}$$

with B being the total usable bandwidth and N_{sub} being the total number of subcarriers.

The channel gain $H_{kj}^{(x)}$ is expressed as

$$H_{kj,dB}^{(x)} = (-\kappa - \upsilon \log_{10} d_{kj}) - \xi_{kj} + 10\log_{10} F_{kj}^{(x)}. \tag{7.3}$$

In Equation 7.3, the first factor captures propagation loss, with d_{kj} being the distance between nodes n_k and n_j, κ being the pathloss constant, and υ being the path loss exponent. The second factor, ξ_{kj}, captures log-normal shadowing with a standard deviation σ_ξ, whereas the last factor, $F_{kj}^{(x)}$, corresponds to Rayleigh fading (generally considered with a Rayleigh parameter a such that $E[a^2] = 1$). A block fading model is considered, where the fading remains constant over the subcarriers of an RB for a duration T_{dec}. After T_{dec}, the channel decorrelates and a new realization occurs for another T_{dec}, and so on. The fading is considered to be independent identically distributed (iid) over RBs.

Assuming that the transmission is taking place from node n_k to node n_j over subcarrier x, the SINR on the link between n_k and n_j over subcarrier x is given by

$$\gamma_{kj}^{(x)} = \frac{P_{t,kj}^{(x)} H_{kj}^{(x)}}{I_j^{(x)} + \left(\sigma_j^{(x)}\right)^2}, \tag{7.4}$$

where $\left(\sigma_j^{(x)}\right)^2$ is the thermal noise power measured at the receiver of n_j and $I_j^{(x)}$ is the interference on subcarrier x measured at the receiver of node n_j. The expression of the interference is given by

$$I_{kj}^{(x)} = \sum_{l \neq k,d; \, l=1}^{N_{BS}+K} \sum_{m=N_{BS}+1}^{N_{BS}+K} \alpha_{l,m}^{(x)} P_{t,lm}^{(x)} H_{lj}^{(x)}. \tag{7.5}$$

The content distribution starts at the BSs, and the data are received by the MTs. Hence, the interference is measured at the MTs. Consequently, the transmitter n_k could be either a BS or an MT (which explains the first summation from 1 to $N_{BS} + K$ in Equation 7.5), but the receiver n_j is an MT (which explains the second summation from $N_{BS} + 1$ to $N_{BS} + K$ in Equation 7.5). The binary indicator variable $\alpha_{l,m}^{(x)}$ is set to 1 if subcarrier x is used for communication on the link between nodes n_l and n_m, and it is set to zero otherwise.

In LTE, the smallest allocation unit is an RB. Denoting by $l_{sub}(k, j)$ the set of consecutive subcarriers forming the RB allocated for communication on the link between nodes n_k and n_j, then the rate over the link between n_k and n_j is given by

$$R_{kj} = \sum_{x \in l_{sub}(k,j)} B_{sub} \log_2\left(1 + \beta\gamma_{kj}^{(x)}\right) \tag{7.6}$$

since the rate achievable over an RB is the sum of the rates on the subcarriers that form that RB.

With D2D communications, we consider that the CH multicasts the data to the members of its cluster over a single RB. On the LR, the BS transmits individually the content to each CH (or each MT in the case without D2D communications) after performing appropriate resource allocation (as described in Section 7.5). The BS transmit power is assumed to be divided equally over all subcarriers:

$$P_{t,kj}^{(x)} = \alpha_{k,j}^{(x)} \frac{P_{t,k}^{max}}{N_{sub}} \text{ for } k = 1, \ldots, N_{BS}. \tag{7.7}$$

It should be noted that for transmissions starting from the BSs, in each cell, an LTE RB, and hence the subcarriers constituting that RB, can be allocated to a single user at a given TTI. In other words, there is an exclusivity of subcarrier allocations to avoid intracell interference. Hence, in each cell with n_k as BS, we have

$$\sum_{j=N_{BS}+1}^{N_{BS}+K} \alpha_{k,j}^{(x)} \leq 1 \forall k = 1, \ldots, N_{BS}; \ \forall x = 1, \ldots, N_{sub}. \tag{7.8}$$

7.4 ENERGY-EFFICIENT CLUSTERING APPROACH FOR D2D COMMUNICATIONS

In this section, we zoom into a single cell and describe an energy-efficient cluster formation method. Within each cluster, D2D communications are adopted to perform cooperative content distribution. In Section 7.5, a multicell network is considered, and a green communication approach is applied for the BSs without assuming D2D communications. Both methods are combined in Section 7.6.

7.4.1 SINGLE-CELL SYSTEM MODEL WITH D2D COMMUNICATIONS

Content of common interest is to be delivered from the BS to requesting MTs distributed throughout the cell area of the BS. The MTs are interested in the same content. They communicate with the BS using LTE or with neighboring MTs using D2D communications. MTs form cooperating clusters for the purpose of energy minimization during cooperative content distribution. Within each cooperating cluster, the content is delivered on the LR to a single MT, the CH, which in turn multicasts

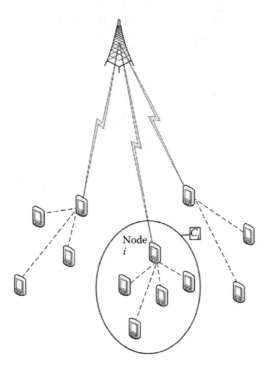

FIGURE 7.2   Single-cell system model with clustering and D2D communications.

the content to the other MTs in that cluster using SR collaboration. Figure 7.2 shows the scenario considered.

At a given fading realization, each MT receives the video content from a single source, which could be either the BS or another MT. We assume that MTs form coalitions where the energy consumption in the coalition is lower than the sum of the individual energy consumptions of the coalition members. In a coalition cluster C_i, n_i is the CH communicating on the LR with the BS on behalf of all the cluster members.

7.4.2   ENERGY CALCULATIONS

The time needed to transmit a content of size S_T bits on a link between nodes n_k and n_j having an achievable rate R_{kj} bps is given by S_T/R_{kj}. Denoting the power drained from the battery of node n_j to receive the data from node n_k by $P_{\mathrm{Rx},kj}$, then the energy consumed by n_j to receive the data from n_k is given by $S_T P_{\mathrm{Rx},kj}/R_{kj}$. Similarly, denoting by $P_{\mathrm{Tx},k}$ the power drained by the battery of n_k to transmit the data via multicasting, then the energy consumed by n_k to transmit the content to n_j is given by $S_T P_{\mathrm{Tx},k}/R_{kj}$. It should be noted that $P_{\mathrm{Tx},k}$ can be expressed as

$$P_{\mathrm{Tx},k} = P_{\mathrm{Tx}_{\mathrm{ref}},k} + P_{t,k}, \tag{7.9}$$

where $P_{\mathrm{Tx}_{\mathrm{ref}},k}$ corresponds to the power consumed by the circuitry of node n_k during transmission on the communication interface and $P_{t,k}$ corresponds to the power transmitted over the air interface by n_k.

Denoting by E_{C_k} the energy consumed by the MTs that are members of cluster C_k with node n_k as CH, and assuming, without loss of generality, that we are considering cell l with n_l as BS in this section, then the energy consumed in C_k is given by

$$E_{C_k} = \frac{S_T P_{Rx,lk}}{R_{lk}} + \frac{S_T P_{Tx,k}}{\min\limits_{i \neq k; n_i \in C_k} R_{ki}} + \sum_{j \neq k; n_j \in C_k} \frac{S_T P_{Rx,kj}}{\min\limits_{i \neq k; n_i \in C_k} R_{ki}} \qquad (7.10)$$

where the first term corresponds to the energy consumed by node n_k to receive the data from the BS on the LR cellular link, the second term corresponds to the energy consumed by node n_k to transmit the data to the other nodes in its cluster on the SR via D2D communications, and the last term corresponds to the energy consumed by the nodes to receive their data from node n_k on the SR. To avoid multiple transmissions among nodes of the same cluster, MT k transmits using multicasting. Therefore, the second term does not involve any summation over the MTs, conversely to the third term. In addition, with SR multicasting, $R_{kj} = \min_{i \neq k; n_i \in C_k} R_{ki}$, since transmission should take place at the minimum achievable rate in the cluster in order to guarantee that all MTs in the cluster receive the desired multimedia information. The LTE resource allocation on the LR is described in Section 7.5.

When all MTs have similar characteristics in terms of power consumption, then we have $P_{Rx,lk} = P_{Rx,LR}$ $\forall k$ for the power drained from the batteries of the MTs during reception on the LR from the BS, $P_{Rx,kj} = P_{Rx,SR}$ $\forall k, j$ for the power drained from the batteries of the MTs during reception on the SR from the CHs, and $P_{Tx,k} = P_{Tx,SR}$ $\forall k$ for the power drained from the batteries of the CHs during multicasting on the SR to the other MTs. In this case, Equation 7.10 can be simplified as follows:

$$E_{C_k} = S_T \left(\frac{P_{Rx,LR}}{R_{lk}} + \frac{P_{Tx,SR} + (|C_k| - 1)P_{Rx,SR}}{\min\limits_{i \neq k; n_i \in C_k} R_{ki}} \right), \qquad (7.11)$$

where $|\cdot|$ denotes set cardinality.

7.4.3 Cooperative Cluster Formation Method

The presented method consists of an initialization phase and a coalition formation phase.

Initialization phase: At the start of the proposed method, all the nodes n_k are directly connected to the BS via LR LTE links; that is, each cluster consists of a single node such that $C_k = \{n_k\}$ and $|C_k| = 1$. This is equivalent to the scenario without D2D collaboration. All clusters are in the search space $S = \{k; C_k \neq \varnothing\}$.

Coalition formation phase:

- Find the cluster having the highest energy consumption per node: $k = \arg\max_{i \in S} E_{C_i} / |C_i|$.
- Coalition candidate search: Find the cluster C_j that leads to the lowest energy consumption when merged with cluster C_k: $j = \arg\min_{i \neq k} E_{C_i \cup C_k}$.
- Coalition formation: Form a coalition between the members of clusters C_j and C_k if the following condition is verified: $E_{C_j \cup C_k} \leq E_{C_j} + E_{C_k}$. This condition indicates that a coalition between two clusters is formed only if it is more energy efficient than having the two clusters operate independently.
- If the merger condition is satisfied, set $C'_j = C_j \cup C_k$. The new cluster has n_j as CH since it has the lowest energy consumption on the link with the BS. Thus, it is the CH of the coalition cluster.

- Update the clusters by setting $C_j = C'_j$ and $C_k = \varnothing$. If the merger condition is not satisfied, keep clusters C_k and C_j separate since this scenario turned out to be more energy efficient than collaboration. In both cases, remove k from the search space: $S = S\backslash\{k\}$.
- Repeat the process until no improvement can be made, that is, until $S = \varnothing$. This means that $\forall k,j$ such that $C_k \neq \varnothing$ and $C_j \neq \varnothing$, we have $E_{C_j \cup C_k} > E_{C_j} + E_{C_k j}$.

7.5   GREEN APPROACH FOR LTE CELLULAR NETWORKS

In this section, an LTE network with multiple cells is considered. MTs are assumed to communicate directly with their respective BSs without SR D2D communications. An energy-efficient approach that puts redundant BSs into sleep mode, thus leading to green LTE cellular communications, is described. In Section 7.6, this approach is combined with D2D communications to lead to better performance in terms of energy efficiency for both the MTs and the BSs.

7.5.1   MULTICELL SYSTEM MODEL FOR GREEN LTE NETWORKS

We consider a geographical area of interest where an LTE network is deployed. The user distribution is considered to be uniform in the area of interest. The area is subdivided into cells of equal size, with a BS placed at the center of each cell. Consequently, the BS distribution in the area is also uniform. The proposed method presented in Section 7.5.4 will be used to switch off certain BSs according to a certain network utility metric. This leads to energy savings and contributes in achieving green communications in the deployed LTE network.

An example is shown in Figure 7.3. One of the cells is lightly loaded with only one user connected to the BS, whereas the other BS has a higher load with three users connected to it, as shown

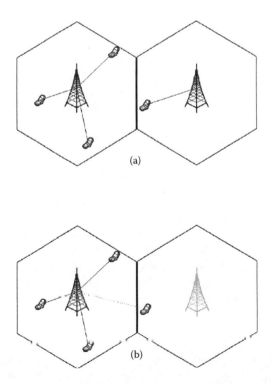

FIGURE 7.3   Green communications example.

in Figure 7.3a. If the user in the second cell can be connected to the BS of the first cell without compromising its QoE, then the BS of the second cell can be switched off, which leads to green energy-efficient communications, as shown in Figure 7.3b.

7.5.2 ADMISSION CONTROL AND RESOURCE ALLOCATION

We assume that the subcarriers constituting a single RB are subjected to the same fading and hence the channel gain on the subcarriers of a single RB is considered to be the same. In addition, the fading is assumed to be iid across RBs. We allocate one DL RB for each user.

Hence, when a user n_j joins the network, it is associated with cell l^* and the DL RB for which the subcarriers x^* satisfy

$$(x^*, l^*) = \arg\max_{(x,l)} \left(1 - \sum_{k=N_{BS}+1; k \neq j}^{N_{BS}+K} \alpha_{l,k}^{(x)} \right) H_{lj}^{(x)}. \tag{7.12}$$

In Equation 7.12, the first term in the multiplication indicates that the search is on the RBs that are not yet allocated to other users by BS n_l.

7.5.3 UTILITY CALCULATION

After having users in the network associated with their respective BSs, and after performing resource allocation as described in Section 7.5.2, a utility function for each BS can now be computed. In this chapter, the utility considered depends on the total energy consumption to distribute the content: the energy consumed by the MTs (drained from their batteries) and the cost of energy consumed at the BS during the distribution of the content of common interest. Thus, the utility of BS n_l is expressed as follows:

$$U_l = \left(\sum_{k; \exists x \text{ such that } \alpha_{l,k}^{(x)}=1} E_{C_k} \right) + P(E_l), \tag{7.13}$$

where U_l is the utility of cell l, E_l is the energy consumed by BS n_l during the content distribution process in cell l, and $P(E_l)$ is the price of the energy consumed E_l. In the first term of Equation 7.13, the summation is over the clusters having CHs connected to BS n_l. This is translated by having an RB allocated in cell l to the CH n_k, which leads to having subcarriers x such that $\alpha_{l,k}^{(x)} = 1$.

It should be noted that the utility in Equation 7.13 is selected as such because it satisfies the description presented above. However, other utility functions can be used. In addition, in some situations, different metrics might be used to assess the performance of the network. In this case, different utility functions should be derived. The approach presented in Section 7.5.4 is independent from the utility selected and can be implemented with any utility. Selecting the appropriate utility suitable for each implementation scenario, in addition to analyzing and comparing the performance with different utilities, is left as a topic for future research.

7.5.4 GREEN COMMUNICATION METHOD

In this section, we present a utility minimizing green communication method. Although the proposed method is utility independent, it will be used with the utility of Section 7.5.3 in this chapter.

Hence, its purpose is to minimize energy consumption in the network, for both BSs and MTs. The steps of the approach can be described as follows:

- **Step 1:** Sort the BSs in decreasing order of utility. After this step, BS $n_m = 1$ would be the one having the worst performance and BS $n_m = N_{BS}$ would be the one having the best utility (least energy consumption). Compute the total network utility $U_{tot} = \sum_{l=1}^{N_{BS}} U_l$.
- **Step 2:** Start from BS $m = 1$.
- **Step 3:** For each user n_j served by BS n_m, find the best BS that is "on" other than BS n_m, and that can serve n_j; that is,

$$(x^*, l^*) = \arg \max_{(x, l \neq m)} \left(1 - \sum_{k=N_{BS}+1; k \neq j}^{N_{BS}+K} \alpha_{l,k}^{(x)} \right) H_{lj}^{(x)}. \tag{7.14}$$

Then, calculate the new interference levels in the network after moving user n_j from cell m to cell l^*, in addition to the new achievable rates and consumed energies for all users.
- **Step 4:** After moving the users in cell m and computing the consumed energies in the previous step, compute the new network utility: $U_{tot}^{new} = \sum_{l=1}^{N_{BS}} U_l^{new}$.
- **Step 5:** If $U_{tot}^{new} \leq U_{tot}$, accept the changes made, switch BS n_m off, and set $U_{tot} = U_{tot}^{new}$. If, on the other hand, $U_{tot}^{new} > U_{tot}$, reject the change and keep BS n_m on.
- **Step 6:** Increment m and go back to Step 3.
- **Step 7:** Repeat Steps 3 to 6 until the steps are implemented with the last BS $m = N_{BS}$.
- **Step 8:** Repeat Steps 1 to 7 on the BSs that are still on until no improvement in the network can be made (i.e., no change leads to a utility decrease).

Hence, the above approach tries to ensure energy efficiency in the network by switching off the BSs having a utility that is too high. It can be shown that the algorithm has a worst-case complexity of $O\left(JN_{RB}KN_{BS}^2 \right)$, with J being the number of iterations in Step 8. The actual complexity is less owing to switching BSs off as the algorithm progresses (and hence excluding them from the search) and to the fact that a user does not "hear" the pilot signals of all BSs. Consequently, only the BSs with which it has a relatively good link quality will be included in the search.

7.6 JOINT APPROACH: INTERPLAY BETWEEN D2D COLLABORATION AND GREEN CELLULAR NETWORKS

In this section, we describe an approach that couples between collaborative D2D communications on the SR and green cellular communications on the LR. This approach consists of joining the methods described in Sections 7.4 and 7.5 into a single energy-efficient approach. In fact, D2D communications consist of reducing the number of LR links, thus reducing the number of cellular connections to the BS in each cell. Afterward, the reduced number of LR LTE links makes it easier to implement the green communications algorithm, since a reduced number of links should be transferred to other cells when a given BS is put into sleep mode.

The joint approach can be described as follows:

- **Step 1:** When a user n_j joins the network, it is associated with cell l^* and the DL RD for which the subcarriers x^* satisfy (Equation 7.12).
- **Step 2:** Calculate the rates achievable by each user on the LR cellular links in its corresponding cell.

- **Step 3:** Implement the cluster formation method described in Section 7.4.3. The outcome of the implementation of the clustering method is a set of collaborative clusters, each having a CH connected to the BS on the LR LTE links and communicating with its peers in the cluster via SR D2D communications.
- **Step 4:** Implement the green communications method described in Section 7.5.4 considering only the links between the CHs and the BSs. With a lot of LR connections replaced by SR D2D communications in Step 3, a reduced number of LR connections will be considered.

An example of the joint approach is shown in Figure 7.4. In fact, Figure 7.4a shows the implementation of the clustering approach for D2D communications, described in Section 7.4, independently in each cell. Figure 7.4b shows the implementation of the green cellular communication method presented in Section 7.5, on the MTs that act as CHs. Hence, with one CH in the right cell of Figure 7.4, it was possible to put that cell into sleep mode and move the CH to the other cell. However, without the D2D approach, it might be difficult to switch off the cell to the right of Figure 7.4 since it might not be possible to move the three MTs in that cell to the neighboring cell without significant degradation in performance.

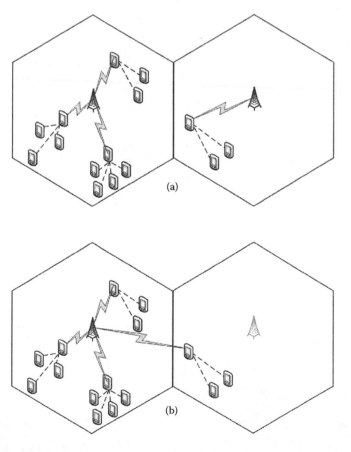

(a)

(b)

FIGURE 7.4   System model.

7.7 RESULTS AND DISCUSSION

This section presents the simulation results by comparing different case study scenarios. The scenarios investigated are listed in Table 7.1. Case 1 corresponds to the traditional case, without green communications or D2D communications. Case 2 corresponds to green communications without D2D collaboration, that is, the scenario of Section 7.5. Case 3 corresponds to D2D collaboration without green communications, that is, the scenario of Section 7.4. Finally, Case 4 corresponds to the approach of Section 7.6, including both green communications and D2D collaboration.

We consider an LTE coverage area with uniform user distribution. BSs are uniformly placed in the area according to the cell radius. An LTE bandwidth of 10 MHz is considered. It is subdivided into 50 RBs of 12 subcarriers each [50,52]. The simulation parameters are shown in Table 7.2. LTE parameters are obtained from Refs. [50,52], and channel parameters are obtained from Ref. [53]. All BSs are considered to have the same transmit power and the same operation cost. In addition, mobile devices are considered to have the same maximal transmit power.

Figure 7.5 shows the number of active BSs. Without implementing the green communications approach, all BSs are active. When green communications are implemented (Cases 2 and 4), the number of active BSs is reduced, until the number of users exceeds 200. This leads to energy savings in lightly loaded networks. The best results are achieved with Case 4, when green communications

TABLE 7.1
The Different Investigated Scenarios

	Green Communications	D2D Communications
Case 1	No	No
Case 2	Yes	No
Case 3	No	Yes
Case 4	Yes	Yes

TABLE 7.2
Simulation Parameters

Parameter	Value
κ	-128.1 dB
σ_ξ (dB)	8 dB
Area size	5×5 km^2
B	10 MHz
N_{RB}	50
$P_{Tx,SR}$	1.425 joules/s
S_T	1 Mbit
Max. BS transmit power	10 W
υ	3.76
Rayleigh parameter a	$E[a^2] = 1$
Inter-BS distance	1 km
B_{sub}	15 kHz
$P_{Rx,LR}$	1.8 joules/s
$P_{Rx,SR}$	0.925 joules/s
$P(\bar{E}_l)$	35 cents/kWh
Max. MT transmit power	0.125 W

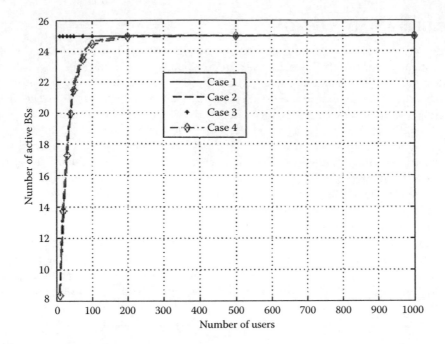

FIGURE 7.5   Number of active BSs versus the number of users in the network.

are combined with D2D communications. This scenario leads to the lowest number of active BSs and hence the largest energy savings in the cellular LTE network.

Figure 7.6 shows the energy consumed by the MTs in the different investigated scenarios. The scenarios with D2D communications (Cases 3 and 4) lead to significant savings compared to the other scenarios (Cases 1 and 2), where SR collaboration is not adopted. Figure 7.7 shows the total content distribution delay, that is, the time needed to distribute the content of common interest to all

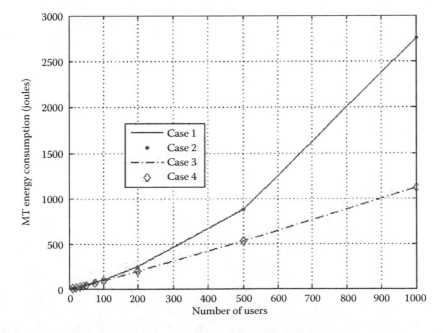

FIGURE 7.6   Total mobile energy consumption versus the number of users in the network.

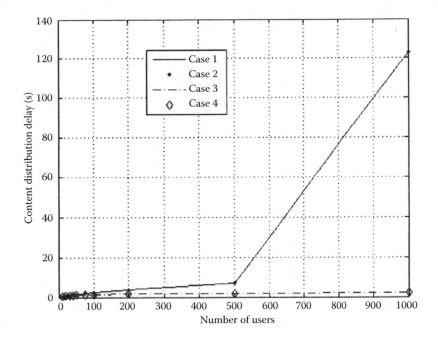

FIGURE 7.7 Total delay for content distribution versus the number of users in the network.

MTs in the network. Clearly, significantly shorter times are needed with collaborative D2D communications (Cases 3 and 4), which leads to better user experience in downloading the content of interest.

Analyzing Figures 7.5 through 7.7, it can be seen that the combination of green communications with D2D collaboration (Case 4) leads to the best results. In fact, Case 4 has the best results in all three figures. Although Case 3 has comparable results in terms of MT energy consumption (Figure 7.6) and content distribution speed (Figure 7.7), it performs worse in terms of BS energy consumption, as shown in Figure 7.5.

Consequently, D2D collaboration allows communication at faster rates as a result of the proximity between MTs. Hence, shorter delays are needed to distribute the content, especially that CHs are selected as MTs having good channel conditions on the LR. Therefore, the wireless interface of the MTs is open for a shorter time to receive the content and thus less energy is drained from the MTs batteries. When coupled with green communications, the reduced number of LR links as a result of clustering allows to put more BSs into sleep mode, which leads to more energy savings at the BSs.

7.8 FUTURE RESEARCH DIRECTIONS

This section describes interesting future research directions that complement the work presented in this chapter. Some of the most important topics for future work include the following:

- Investigating joint uplink and downlink performance: The work in this chapter was focused on the downlink direction, from the BS to the MTs. To be able to implement the proposed techniques in practice, the uplink direction should be considered in conjunction with the downlink. In fact, a BS put in sleep mode will be in this situation for both uplink and downlink transmissions. The challenge in the uplink is that each MT is transmitting different information to the BS. Consequently, the scenario of the content of common interest is not applicable. Hence, a CH needs to aggregate the data of all the MTs in its cluster before transmitting the data to the BS. The same concept applies in the downlink where the MTs are interested in different content. Generalizing the results presented in this chapter in this direction is an obvious subject for further research.

- Utility functions: Beyond the utility function used in Equation 7.13, different utility functions can be investigated and compared. For example, weighted utilities can be considered, where different weights can be given to the BS and MT energy consumptions. Detailed models for energy consumption at the BSs or MTs can also be derived, validated, and incorporated in these novel utility metrics.

- D2D communications in heterogeneous macrocell/small cell networks: The proliferation of small cells (microcells, picocells, femtocells, etc.) is expected to increase in the coming years [54]. However, the challenge is that the density of small cells and their operation affect the overall interference variation in the network and, thus, should affect the allocation and configuration of macrocell sites [55]. In Ref. [56], this problem is treated by investigating macrocell–femtocell cooperation, where a femtocell user may act as a relay for macrocell users, and in return each cooperative macrocell user grants the femtocell user a fraction of its superframe. A multi-tier network composed of macrocells and small cells controlled by the same operator is studied in Ref. [57], where it was shown that the operator can manipulate the system loads by tuning the pricing and the bandwidth allocation policy between macrocells and small cells. It would be interesting to investigate the collaboration between macrocells and small cells/femtocells for the purpose of energy efficiency. Preliminary results in this direction were obtained in Ref. [40]. A more challenging research topic is to include the impact of D2D communications on energy efficiency in heterogeneous macrocell/small cell networks.

- Analyzing QoS/QoE performance during D2D/Green LTE operation: In this chapter, content distribution delay was tracked during the simulations as a QoS metric. An interesting future research direction is to include advanced QoS/QoE performance in addition to content distribution delay. For example, video QoE during real-time video streaming of a content of common interest can be investigated. The impact of clustering and D2D communications along with BS on/off switching on the video quality experienced by mobile users can be analyzed and assessed. Furthermore, QoE can be included as a metric in the optimization problem or it can be imbedded in certain utility functions. The objective would be to reach a trade-off between energy efficiency (for BSs and MTs) and the QoS/QoE experienced by mobile users.

- Trust, security, and social aspects: Content distribution techniques for the purpose of energy efficiency or enhanced QoS should be customized to take into account the social aspects of human behavior during content distribution. In fact, people are unlikely to exchange data with complete strangers, even if the content is of common interest. For example, a group of friends sitting in a cafe might exchange content collaboratively, but they might not exchange data with people sitting on the next table, although proximity constraints and link quality might justify this exchange. Hence, adding a social network dimension to collaborative content distribution is an interesting direction for further investigation.

- Revenue aspects: The mobile network operators generate more revenue if the content is completely distributed on the LR (generally, D2D communication on SR is expected to be free, because it occurs directly between MTs), since more users will be downloading the content from the operator's network. On the other hand, SR collaboration frees more LR bandwidth, which leads to reduced user blocking rates, and the freed bandwidth can be used to accommodate more content distribution requests (thus leading to more revenue) or to switch off a BS (thus leading to power cost savings). Consequently, a trade-off should be made between increasing an operator's revenues and freeing the LR wireless resources. A possible solution to this problem is to have the operator enable user subscription in a cooperative content distribution framework. In this case, a mobile application could be downloaded on the subscribers' mobiles, and it would enable D2D collaboration in a joint LR/SR content distribution approach. The subscription fee should be selected such that subscribers would still save (some) money while saving battery energy, and the operator would still make revenue from subscription fees. Deriving an optimization framework to find the best trade-off for setting the prices is consequently an interesting topic worthy of further study.

7.9 CONCLUSION

In this chapter, D2D communications between MTs were investigated jointly with green cellular communication techniques. Reduced energy consumption was achieved for MTs by collaborative D2D communications, whereas energy savings were achieved in the cellular network by implementing the green communication approach. Thus, the combined method described in this chapter led to energy savings for both the mobile users and the mobile network operator.

ACKNOWLEDGMENT

This work was made possible by NPRP grant # 09-180-2-078 from the Qatar National Research Fund (a member of the Qatar Foundation). The statements made herein are solely the responsibility of the authors.

AUTHOR BIOGRAPHIES

Elias Yaacoub received his BE degree in electrical engineering from the Lebanese University in 2002, his ME degree in computer and communications engineering from the American University of Beirut (AUB) in 2005, and his PhD degree in electrical and computer engineering from AUB in 2010. He worked as a research assistant in the AUB from 2004 to 2005 and in the Munich University of Technology in spring 2005. From 2005 to 2007, he worked as a telecommunications engineer with Dar Al-Handasah, Shair and Partners. Since November 2010, he has been a research scientist at the QUWIC (QU Wireless Innovations Center), rebranded to QMIC (Qatar Mobility Innovations Center) in July 2012. His research interests include wireless communications, antenna theory, sensor networks, energy efficiency in wireless networks, video streaming over wireless networks, and bioinformatics.

Hakim Ghazzai is currently working toward his PhD degree in electrical engineering at King Abdullah University of Science and Technology in Saudi Arabia. He received his Diplome d'Ingenieur in Telecommunication Engineering with highest distinction from the Ecole Superieure des Communications de Tunis (SUP'COM), Tunisia, in 2010 and his masters degree in high-rate transmission systems from the same institute in 2011. His general research interests are at the intersection of wireless networks, green communications, and optimization.

Mohamed-Slim Alouini was born in Tunis, Tunisia. He received the Diplome d'Ingenieur from the Ecole Nationale Supérieure des Telecommunications (TELECOM Paris Tech) and the Diplome d'Etudes Approfondies (D.E.A.) in electronics from the Université Pierre et Marie Curie in Paris, both in 1993. He received his MSEE degree from the Georgia Institute of Technology in the United States in 1995 and a doctorate in electrical engineering from California Institute of Technology in 1998. He also received a Habilitation degree from the Université Pierre et Marie Curie in 2003. Dr. Alouini started his academic career at the Department of Electrical and Computer Engineering of the University of Minnesota, in the United States in 1998. In 2005, he joined the Electrical and Computer Engineering program at Texas A&M University at Qatar. He has been a professor of electrical engineering at King Abdullah University of Science and Technology since 2009. His research interests include statistical characterization and modeling of fading channels, performance analysis of diversity combining techniques, MIMO (multiple input–multiple output), and multi-hop/

cooperative communications systems, capacity and outage analysis of multiuser wireless systems subject to interference or jamming, cognitive radio systems, and design and performance evaluation of multi-resolution, hierarchical, and adaptive modulation schemes. Dr. Alouini has published several papers on the above subjects, and he is coauthor of the textbook *Digital Communication over Fading Channels* published by Wiley Interscience. He is a Fellow of the Institute of Electrical and Electronics Engineers (IEEE), a member of the Thomson ISI Web of Knowledge list of Highly Cited Researchers, and a co-recipient of best paper awards in eight IEEE conferences (including ICC, GLOBECOM, VTC, and PIMRC).

Adnan Abu-Dayya received his PhD in electrical engineering from Queens University, Canada, in 1992. He then worked as a manager at Nortel Networks in Canada in the advanced technology group and as a senior consultant at the Communications Research Center in Ottawa, Canada. Before moving to Qatar in March 2007, he worked for 10 years at AT&T Wireless in Seattle, USA, where he served in a number of senior management positions covering product innovations and emerging technologies, systems engineering, and product realization. He was also responsible for developing and licensing the extensive patent portfolio of AT&T Wireless. From April 2007 to December 2008, he was the chairman of the Electrical Engineering Department at Qatar University (QU). He was appointed as the executive director of the QUWIC (QU Wireless Innovations Center) in December 2008 (QUWIC was rebranded to QMIC [Qatar Mobility Innovations Center] in July 2012). Dr. Abu-Dayya has more than 20 years of international experience in the areas of wireless/telecom R&D, innovations, business development, and services delivery. He has many issued patents and more than 50 publications in the field of wireless communications.

REFERENCES

1. J.T. Louhi. Energy efficiency of modern cellular base stations. In *International Telecommunications Energy Conference (INTELEC)*, September 2007.
2. L.M. Correia, D. Zeller, O. Blume, D. Ferling, Y. Jading, I.G. Gunther Auer, and L. Van der Perre. Challenges and enabling technologies for energy aware mobile radio networks. *IEEE Communications Magazine*, 48(11):66–72, November 2010.
3. EARTH. Energy Aware Radio and neTwork tecHnologies. EU Funded Research Project FP7-ICT-2009-4-24733-EARTH, January 2010–June 2012, http://www.ict-earth.eu, last accessed: January 3, 2013.
4. M.A. Marsan, L. Chiaraviglio, D. Ciullo, and M. Meo. Optimal energy savings in cellular access networks. In *IEEE ICC Workshops 2009*, June 2009.
5. D. Ciullo, L. Chiaraviglio, M. Meo, and M.A. Marsan. Energy-aware UMTS access networks. In *International Symposium on Wireless Personal Multimedia Communications (WPMC'08)*, September 2008.
6. S. Zhou, J. Gong, Z. Yang, Z. Niu, and P. Yang. Green mobile access network with dynamic base station energy saving. In *MobiCom 2009*, September 2009.
7. Z. Niu, Y. Wu, J. Gong, and Z. Yang. Cell zooming for cost-efficient green cellular networks. *IEEE Communications Magazine*, 48(11):74–79, November 2010.
8. Alcatel-Lucent. Mobile data demand drives all-IP strategy. Alcatel-Lucent Business E-Zine, http://www2.alcatel-lucent.com/enrich/en/v5i1/mobile-data-demand-drives-all-ip-strategy/, last accessed: Jan. 3, 2013, February 2011.
9. J. Liu, W. Chen, Z. Cao, and K.B. Letaief. Dynamic power and sub-carrier allocation for OFDMA-based wireless multicast systems. In *IEEE ICC 2008*, May 2008.
10. L. Popova, T. Herpel, and W. Koch. Improving downlink UMTS capacity by exploiting direct mobile-to-mobile data transfer. In *Proceedings of International Symposium on Modeling and Optimization in Mobile, Ad Hoc and Wireless Networks*, April 2007.
11. M. Peng and W. Wang. Investigation of cooperative relay node selection in heterogeneous wireless communication systems. In *Proceedings of IEEE International Conference on Communications*, May 2008.
12. F. Fitzek and M. Katz. *Cooperation in Wireless Networks: Principles and Applications*. Springer, Dordrecht, The Netherlands, 2006.
13. Q. Zhang, F.H.P. Fitzek, and V.B. Iversen. Design and performance evaluation of cooperative retransmission scheme for reliable multicast services in cellular controlled P2P networks. In *Proceedings of IEEE International Symposium on Personal, Indoor and Mobile Radio Communications*, September 2007.

14. K. Doppler, M. Rinne, C. Wijting, C.B. Ribeiro, and K. Hugl. Device-to-device communication as an underlay to LTE-advanced networks. *IEEE Communications Magazine*, 47(12):42–49, December 2009.
15. K. Doppler, C.-H. Yu, C.B. Ribeiro, and P. Jänis. Mode selection for device-to-device communication underlaying an LTE-advanced network. In *IEEE WCNC 2010*, April 2010.
16. M. Zulhasnine, C. Huang, and A. Srinivasan. Efficient resource allocation for device-to-device communication underlaying LTE network. In *6th IEEE International Conference on Wireless and Mobile Computing, Networking and Communications (WiMob 2010)*, October 2010.
17. M. Leung and S. Chan. Broadcast-based peer-to-peer collaborative video streaming among mobiles. *IEEE Transactions on Broadcasting*, 53(1):350–361, March 2007.
18. R. Iqbal and S. Shirmohammadi. A cooperative video adaptation and streaming scheme for mobile and heterogeneous devices in a community network. In *IEEE International Conference on Multimedia and Expo (ICME'09)*, pp. 1768–1771, 2009.
19. Y. Kim and I. Chong. An adaptive algorithm for live streaming using tree-based peer-to-peer networking in mobile environments. In *International Conference on Ubiquitous and Future Networks (ICUFN'11)*, pp. 191–195, June 2011.
20. M. Stiemerling and S. Kiesel. Cooperative P2P video streaming for mobile peers. In *International Conference on Computer Communications and Networks (ICCCN'10)*, pp. 1–7, August 2010.
21. X. Liu, G. Cheung, and C.-N. Chuah. Structured network coding and cooperative wireless ad-hoc peer-to-peer repair for WWAN video broadcast. *IEEE Transactions on Multimedia*, 11(4):730–741, June 2009.
22. X. Liu, G. Cheung, and C.-N. Chuah. Joint source/channel coding of WWAN multicast video for a cooperative peer-to-peer collective using structured network coding. In *IEEE International Workshop on Multimedia Signal Processing (MMSP'09)*, pp. 1–6, 2009.
23. X. Liu, G. Cheung, C.-N. Chuah, and Y. Ji. Bit allocation of WWAN scalable H.264 video multicast for heterogeneous cooperative peer-to-peer collective. In *IEEE International Conference on Acoustics Speech and Signal Processing (ICASSP'10)*, pp. 5574–5577, 2010.
24. F. Albiero, M. Katz, and F. Fitzek. Energy-efficient cooperative techniques for multimedia services over future wireless networks. In *IEEE ICC'08*, pp. 2006–2011, 2008.
25. F. Albiero, J. Vehkaperä, M. Katz, and F. Fitzek. Overall performance assessment of energy-aware cooperative techniques exploiting multiple description and scalable video coding schemes. In *Communication Networks and Services Research Conference (CNSR'08)*, pp. 18–24, 2008.
26. K. Mahmud, M. Inoue, H. Murakami, M. Hasegawa, and H. Morikawa. Energy consumption measurement of wireless interfaces in multi-service user terminals for heterogeneous wireless networks. *IEICE Transactions on Communications*, E88-B(3):1097–1110, March 2005.
27. X. Wu, S. Tavildar, S. Shakkottai, T. Richardson, J. Li, R. Laroia, and A. Jovicic. FlashLinQ: A synchronous distributed scheduler for peer-to-peer ad hoc networks. In *48th Annual Allerton Conference on Communication, Control, and Computing*, 2010.
28. M.S. Corson, R. Laroia, J. Li, V. Park, T. Richardson, and G. Tsirtsis. Toward proximity-aware internetworking. *IEEE Wireless Communications*, 17(6):26–33, December 2010.
29. Z. Hasan, H. Boostanimehr, and V.K. Bhargava. Green cellular networks: A survey, some research issues and challenges. *IEEE Communications Surveys and Tutorials*, 13(4):524–540, Fourth Quarter 2011.
30. L. Xiang, F. Pantisano, R. Verdone, X. Ge, and M. Chen. Adaptive traffic load-balancing for green cellular networks. In *IEEE PIMRC 2011*, September 2011.
31. P. Pace. Green antenna switching to improve energy saving in LTE networks. In *IEEE Online Conference on Green Communications (GreenCom 2012)*, September 2012.
32. Y.O. Lee and A.L. Narasimha Reddy. Multipath routing for reducing network energy. In *IEEE Online Conference on Green Communications (GreenCom 2012)*, September 2012.
33. I. Chih-Lin, L.J. Greenstein, and R.D. Gitlin. A microcell/macrocell cellular architecture for low- and high-mobility wireless users. *IEEE Journal on Selected Areas in Communications*, 11(6):885–891, 1993.
34. A. Saleh, A. Rustako, and R. Roman. Distributed antennas for indoor radio communications. *IEEE Transactions on Communications*, 35(12):1245–1251, 1987.
35. O. Oyman, N. Laneman, and S. Sandhu. Multihop relaying for broadband wireless mesh networks: From theory to practice. *IEEE Communications Magazine*, 45(11):116–122, 2007.
36. S. Peters and R. Heath. The future of WiMAX: Multihop relaying with IEEE 802.16j. *IEEE Communications Magazine*, 47(1):104–111, 2009.
37. V. Chandrasekhar, J.G. Andrews, and A. Gatherer. Femtocell networks: A survey. *IEEE Communications Magazine*, 46(9):59–67, 2008.

38. J. Hoydis and M. Debbah. Green, cost-effective, flexible, small cell networks. *IEEE ComSoc MMTC E-Letter Special Issue on "Multimedia Over Femto Cells"*, 5(5):23–26, 2010.
39. D. Chee, M.S. Kang, H. Lee, and B.C. Jung. A study on the green cellular network with femtocells. In *Third International Conference on Ubiquitous and Future Networks (ICUFN 2011)*, pp. 235–240, 2011.
40. E. Yaacoub. Green Communications in LTE networks with environmentally friendly small cell base stations. In *IEEE Online Conference on Green Communications (GreenCom 2012)*, September 2012.
41. NEC Corporation. Self Organizing Network: NEC's proposals for next-generation radio network management. White Paper, February 2009.
42. 3rd Generation Partnership Project (3GPP). 3GPP TS 32.500 Telecommunication management; Self-Organizing Networks (SON); concepts and requirements (Rel. 8), 2008.
43. 3rd Generation Partnership Project (3GPP). 3GPP TS 32.501 Telecommunication management; self-configuration of network elements; concepts and Integration Reference Point (IRP) requirements (Rel. 8), 2008.
44. 3rd Generation Partnership Project (3GPP). 3GPP TS 32.502 Telecommunication management; self-configuration of network elements Integration Reference Point (IRP); Information Service (IS) (Rel. 8), 2008.
45. 3rd Generation Partnership Project (3GPP). 3GPP TS 32.511 Telecommunication management; Automatic Neighbour Relation (ANR) management; concepts and requirements (Rel-8), 2008.
46. J.M. Celentano. Carrier capital expenditures. *IEEE Communications Magazine*, 46(7):82–88, July 2008.
47. S. Shen. *How to Cut Mobile Network Costs to Serve Emerging Markets?* Gartner Inc., London, UK, November 2005.
48. E. Yaacoub, L. Al-Kanj, Z. Dawy, S. Sharafeddine, F. Filali, and A. Abu-Dayya. A utility minimization approach for energy-aware cooperative content distribution with fairness constraints. *Transactions on Emerging Telecommunications Technologies (ETT)*, 23(4):378–392, June 2012.
49. T. Lunttila, J. Lindholm, K. Pajukoski, E. Tiirola, and A. Toskala. EUTRAN uplink performance. In *International Symposium on Wireless Pervasive Computing (ISWPC) 2007*, February 2007.
50. 3rd Generation Partnership Project (3GPP). 3GPP TS 36.211 3GPP TSG RAN Evolved Universal Terrestrial Radio Access (E-UTRA) physical channels and modulation, version 8.3.0, Release 8, 2008.
51. X. Qiu and K. Chawla. On the performance of adaptive modulation in cellular systems. *IEEE Transactions on Communications*, 47(6):884–895, June 1999.
52. 3rd Generation Partnership Project (3GPP). 3GPP TS 36.213 3GPP TSG RAN Evolved Universal Terrestrial Radio Access (E-UTRA) physical layer procedures, version 8.3.0, Release 8, 2008.
53. 3rd Generation Partnership Project (3GPP). 3GPP TR 25.814 3GPP TSG RAN Physical layer aspects for evolved UTRA, v7.1.0, 2006.
54. J.G. Andrews, H. Claussen, M. Dohler, S. Rangan, and M.C. Reed. Femtocells: Past, present, and future. *IEEE Journal on Selected Areas in Communications*, 30(3):497–508, April 2012.
55. V. Chandrasekhar, M. Kountouris, and J.G. Andrews. Coverage in multi-antenna two-tier networks. *IEEE Transactions on Wireless Communications*, 8(10):5314–5327, October 2009.
56. F. Pantisano, M. Bennis, W. Saad, and M. Debbah. Spectrum leasing as an incentive towards uplink macrocell and femtocell cooperation. *IEEE Journal on Selected Areas in Communications*, 30(3):617–630, April 2012.
57. C. Gussen, V. Belmega, and M. Debbah. Pricing and bandwidth allocation problems in wireless multi-tier networks. In *Proceedings of Asilomar Conference on Signals Systems and Computers*, November 2011.

8 Interference-Limited Fixed Relaying-Aided Macrocellular CDMA Networks

Álvaro Ricieri Castro e Souza and Taufik Abrão

CONTENTS

8.1 INTRODUCTION

In interference-limited networks, such as code division multiple access (CDMA) networks, the energy efficiency (EE) is limited by the interference power level (Buzzi and Poor 2008; Meshkati et al. 2005; Souza et al. 2012a), since this forces all users to increase the transmission power in order to maintain their quality of service (QoS) quantified for example as the minimum acceptable information rate or the maximum tolerable symbol/frame error rate (SER/FER). This problem is worst at the cell edges, given the higher distance to the base station (BS), which further limits the EE. When the number of users increases, which directly affects the interference power level, the percentage of users transmitting at their maximum power increases, resulting in a reduction of the achievable EE, even when employing advanced multiuser detection (MuD) techniques (Meshkati et al. 2005; Souza et al. 2012b), which mitigate the UL interference.

In order to improve the user or overall EE, mainly for the cell-edge users, one of the most promising techniques is cooperative communications. This approach, which is part of the most recent 4G standards (Loa et al. 2010), uses relay nodes to retransmit the information from the mobile user terminals (MTs) to the BS. The relay stations (RSs) are strategically placed into the cell, aiming to reduce the path loss term and, depending on the implementation technique, even reduce the multiple access interference (MAI), while providing spatial diversity with the addition of a second uncorrelated signal source at the destination detection side. With these two features added, EE could be increased, resulting in precious resources savings despite adding extra components and complexity to the network.

In this chapter, an energy-efficient relaying design for macrocellular systems and networks is proposed through the deployment of a number of fixed relay stations (FRSs) strategically located inside the covered area. Initially, important concepts such as area power consumption (APC) are reviewed. The proposed scheme presents several improvements compared to the conventional schemes. Relay selection is performed for the uplink (UL) and downlink (DL) based on the simple minimum distance criterion, but considering different RS–BS distance placement.

Besides, traffic load conditions of the UL and DL should be considered to enable to adapt to the heterogeneous and asymmetric traffic services in the next-generation wireless communications. However, because of the problem complexity in this work, only the UL fixed-relay minimum distance-based selection and positioning has been considered, but including the optimal number of RS definition and proposition of power allocation policies for RSs and mobile terminals (MTs). The relay selection results are illustrated in terms of the network geometry for a range of potential relaying locations.

Another important concept explored in this work is the energy-efficient cooperation regions; the optimal relaying location is determined for cooperative cellular systems with asymmetric traffic. It is worth noting that the MT–RS and RS–BS channels have a different influence over relay selection decision in order to attain the optimal energy-efficient scheme. This way, since the RSs (as well as BSs) have a fixed power consumption term, which is related to the signal processing circuits, channel estimation, cooling, and so on, which effectively affect the overall EE of the network, in this chapter, the optimal number of relays regarding the power cost of new equipment (RS) installation has also been investigated. In other words, we investigate the point when the resource saving obtained with relay deployment is supplanted by the circuit power cost generated by these new equipment.

After determining the best fixed-relay placement and quantity, a distributed power control algorithm with minimum rate constraint problem in macrocell DS/CDMA networks is developed to allocate power to the users according to the defined topology and system parameters. Indeed, we use the concept of game theory (Fudenberg and Tirole 1991), which now is well established for this kind of optimization (Buzzi and Poor 2008; Meshkati et al. 2005; Miao et al. 2011; Sampaio et al. 2012; Zappone et al. 2011). Since there is great interest in providing distributed solutions, and the games are played between MT and each one of the RS and the BS, the suitable approach is to use non-coalitional games, which are able to describe the selfish and distributed nature of this type of resource allocation (RA) optimization problem. Hence, the RA problem is analyzed under various interference density configurations, and its reliability is studied in terms of solution existence and uniqueness.

Finally, in order to corroborate the effectiveness of the proposed method and framework, extensive numerical results have been examined in realistic interference scenarios taking into account several figures of merit, such as outage probability, convergence rate, average sum power, average sum rate, and average EE, all obtained by the conventional and cooperative RA approaches.

The rest of this chapter is organized as follows. We first briefly revisit the main relay techniques from the perspective of 4G standards in Section 8.2. The system model and the multichannel interference-aware power control problem formulation are described in Section 8.3. An energy-efficient model is introduced in Section 8.4. We then discuss an energy-efficient power optimization non-cooperative game and study the existence and uniqueness of the equilibrium for this game in Section 8.5. Numerical results supporting the proposed approach are discussed in Section 8.6. Finally, we conclude this chapter in Section 8.7 and provide future research trends in Section 8.8.

8.2   RELAY TECHNIQUES AND 4G STANDARDS

RSs can be implemented (a) with dedicated equipment that are part of the cell infrastructure, called FRSs, and (b) by enabling the MT to retransmit signals from other users, exploring random positioning and quantity, called mobile relay stations (MRSs). Using FRS has the advantage of installing equipment in strategic areas, in order to increase data rate or even coverage, reducing the size of the main cell using equipment with lower costs than a complete BS, the capacity of more antennas, and the energy supply. On the other hand, the MRS has the advantage of higher diversity, given the number of MTs into the cell. As pointed out by Nosratinia and Hedayat (2011), there is no impediment to using the mobile devices as RSs, but limitations in size, power, and processing capacity and mainly in sharing the resources make this approach less practical than the FRSs.

Regarding the retransmission scheme, there are also two possibilities: the non-regenerative and the regenerative protocols. The first class refers to protocols that only retransmit an amplified version of the received signal, without decoding it, whose primary example is the Amplify and Forward (AF) protocol. Although simple, the main drawback of this protocol is noise amplification and interference propagation, since no interference cancelation technique is applied. Even with this weakness, AF achieves a higher diversity order than the regenerative approach (Laneman 2002). Furthermore, the main distinction in regenerative protocols is with regard to the estimation of the transmitted data; hence, in applying data estimation/detection/decoding procedure(s), a system based on regenerative relays could remove noise and interference on the retransmitted signal, with the cost of increased complexity. The main drawbacks of this protocol are the detection process and an increase in service delay, which means respectively that if the transmitted symbols are detected wrongly at the relay, this error is propagated, and the detection/decoding extra processing step is both time and energy consuming.

The choice of which protocol to use depends on the adopted system. For instance, the two systems proposed for 4G networks (WiMAX 802.16m and LTE-Advanced release 11) deploy adaptive modulation and coding (AMC) to adjust system parameters according to the instantaneous channel conditions. Since the RS–BS path introduces extra background noise and fading, it is impossible for the AF protocol to change the AMC status during the relaying operation (Kim et al. 2011) in order to allocate the necessary data rate to satisfy the supported users. Besides, there are interference cancelation techniques that cannot be applied when using the AF protocol. Hence, the most recent standards for WiMAX and LTE-A are able to adopt only DF relays schemes (Loa et al. 2010). In the LTE-A standard, there are three types of relay implementations, depending on which layer the information relaying occurs (Kim et al. 2011):

- Layer 1 relay: this relay uses the AF protocol, since the information at the radio-frequency level is retransmitted.
- Layer 2 relay: the information is relayed at the PHY layer using the DF protocol.

- Layer 3 relay: the information is relayed at the MAC layer, again, deploying the DF protocol. Relaying at this level makes it possible for the RS to control its own cell, acting as a simplified BS, with fewer capacities.

Only the last two have been considered as part of the LTE-A standard implementation, currently recognized as Type 1 (layer 3) and Type 2 (layer 2).

As discussed for layer 2 and layer 3 relays, there are two methods of cell control when using RSs for LTE-A and WiMAX 802.16j: non-transparent relay and transparent relay modes (Liebl et al. 2011; Shen et al. 2008). In the transparent relay mode, an MT associated to an RS is located within the coverage of the BS. Control signaling from the BS can directly reach the MTs, while data traffic is relayed via RS. In this relay mode, control signaling and data traffic are separated. Transparent relay only supports centralized scheduling. BS coordinates and allocates the radio resources to MTs and RSs within the cell by distributing control information and arbitrating access requests. RS only has the functionality of forwarding UL traffic to BS and vice versa. Transparent relay is dedicated for throughput enhancement, where MTs are located within the coverage of BS DL control channel. DL control information is always transmitted with the most robust modulation scheme, which guarantees the highest possible probability of correctly decoding the information. Conversely, in non-transparent relay mode, all data and control signaling transmissions between BS and MT are relayed. The non-transparent RS can operate in both centralized and distributed scheduling. The non-transparent relay mode has the capability of extending the coverage.

If the RS simultaneously receives and transmits at the same frequency, it can generate self-interference. Hence, it is necessary to separate the MT–RS and RS–BS links, which is made by one of the following techniques, deploying time, frequency, or spatial diversity (Loa et al. 2010):

- In-band relaying: RSs use the same RF channel for MT–RS and RS–BS links, sharing the resources between the two links. This can be made using time division duplex (TDD), frequency division duplex, or spatial separation between the MT–RS and RS–BS antennas.
- Out-of-band relaying: RSs deploy different RF channels (or the same channel, without resource sharing) to MT–RS and RS–BS links.

Out-of-band relaying reduces the interference that could be generated by the RSs, at the cost of an extra RF channel. Considering again the LTE-A standard, there are three possible implementations for Type 1 relay: (a) out-of-band relaying with a second RF channel, (b) in-band relaying with TDD (half-duplex in-band), and (c) spatial separation (full-duplex in-band). Note that for the Type 2 relay, there is only the TDD in-band relaying configuration. For the WiMAX 802.16m standard, the IEEE Standard for Local and Metropolitan Area Networks Part 16: Air Interface for Broadband Wireless Access Systems Amendment 3: Advanced Air Interface 2011 discusses in-band relaying using TDD and the out-of-band relaying.*

8.3 SYSTEM MODEL FOR RELAY-BASED COMMUNICATION

8.3.1 RELAY TOPOLOGY

In this chapter, the UL of a synchronous DS-CDMA network has been considered, which relies on pseudo-noise spreading codes (PN) of length N for supporting K users uniformly distributed in a macrocell having a radius of r_O (km). The FRSs implement the AF protocol and are placed uniformly in a circle, with a radius r_1 from the BS. The set of all RSs is denoted by $\mathcal{R} = \{R_1, R_2, \ldots, R_i, \ldots, R_{N_{RS}}\}$, where N_{RS} is the number of RSs. We consider the scenario using two different carrier frequencies

---

* In the WiMAX standard, the in-band relaying is referred to as time-division transmit and receive (TTR) relaying and the out-of-band relaying is termed as simultaneous transmit and receive (STR) relaying.

and non-overlapping spectrum, where the MT–RS and MT–BS links operate in the F_1 band, while the RS–BS links are hosted by the F_2 band. This way, all MTs use the band F_1 to communicate with the BS or with the RS, while the RSs use the band F_2 to communicate with the BS, avoiding extra interference on the MT–BS links. Another advantage of this approach is that the original infrastructure may be maintained, including the traditional 120° sectorization. However, the relaying band reduces the spectral efficiency (SE). Hence, we consider N_{RS} sectors at the BS antenna operating in the F_2 band in order to avoid interference between the RSs. This can be carried out without affecting the F_1 band equipment. Figure 8.1 shows the fixed relay-based macrocell topology.

Herein, the out-of-band relaying strategy is adopted in order to reduce the MAI level. Since the EE of CDMA systems is severely limited by the interference power level, and considering that the interference generated by the extra transmissions from the RS to BS in the F_2 band does not affect the original macrocell's MAI level, the network performance results tend to be better than that achieved with the aid of in-band relaying (Miao et al. 2011; Souza et al. 2012b). As we consider the AF protocol, the cell control is made in a transparent way, since the AF protocol does not support the non-transparent relaying.

We also assume that the FRSs are positioned at high elevation, such as lampposts and at the top of buildings, in order to achieve a line-of-sight (LoS) communication and hence to reduce the path-loss attenuation (Nourizadeh et al. 2006; Yamao et al. 2002). One of the major problems of the power allocation of DS-CDMA networks is the near–far effect, since users that are within (but near) the circle formed by relays, trying to allocate power to transmit to the BS, substantially increase the interference power level, if the RS antenna is omni-directional (Badruddin and Negi 2004). Hence, we adopt directional antennas for the RSs, while the polygon within the macrocell of Figure 8.1 has been formed by the idealized coverage area of the directional antennas of the RSs. Furthermore, in order to cover all the area outside the circle formed by the RSs, some parts of the two-hop area are covered by two relays. This polygon has $n = 2N_{RS}$ sides, and the antenna's coverage angle is defined by the external angle of the polygon:

$$\angle_{RS} = 2\pi - \frac{(n-2)\pi}{n} = \frac{N_{RS}+1}{N_{RS}}\pi \tag{8.1}$$

where n is the number of sides of the regular polygon.

Since the FRSs are installed on high elevations, it is reasonable to adopt the Rician fading model in order to describe the LoS in the RS–BS links, while the MT–RS and MT–BS links are subjected

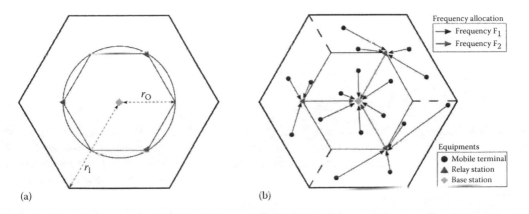

FIGURE 8.1 Cell structure with (a) cell radius r_1 and r_0 and (b) carrier frequency bands F_1 and F_2 for the fixed relay-assisted system model considered in this work.

to Rayleigh fading (Chen et al. 2010; Ding et al. 2011). In Rician–Rayleigh fading conditions, k_R is the Rician k-factor* defined as the ratio of the LoS power component to the sum power of the NLoS multipath components. Practical values for k can be found in European Telecommunications Standards Institute (2007), Senarath et al. (2007), and Kyösti et al. (2008).

8.3.2 SINR in Fixed-Relaying CDMA

In order to describe the signal-to-interference plus noise ratio (SINR) expressions in the UL of our fixed-RS CDMA system, it is necessary to define some user sets. The first of these sets refers to users that rely on a single hop (1H users), which are covered directly by the BS's jth directional antenna, denoted by the set $K_{D,j}$. The second set refers to the MTs that use relays to transmit their signals and is within the coverage area of the BS's jth sector, denoted by $K_{R,j} = K_{BS,j} \backslash K_{D,j}$, where $K_{BS,j}$ is the set of all users covered by the jth sector of the BS. Since these MTs also use the band F_1 to communicate with the RS, the interference generated by them must be taken into account. Hence, the equivalent base-band received signal within the jth BS sector is given by

$$
y_j^{1H} = \sum_{k \in K_{BS,j}} \sqrt{p_k} x_k h_k s_k + \eta_{BS_j} = \sum_{k \in K_{D,j}} \sqrt{p_k} x_k h_k s_k
$$
$$
+ \sum_{\ell \in K_{R,j}} \sqrt{p_\ell} x_\ell h_\ell s_\ell + \eta_{BS_j} \tag{8.2}
$$

where p_k is the power transmitted from the kth MT, x_k is the information symbol, and s_k refers to the spreading code that identifies the kth user of length N, with N being the processing gain; h_k identifies the complex channel gain, which includes shadowing, path loss, and the multiplicative fading channel effects; η_{BS_j} is the additive background noise (with Gaussian distribution with zero mean and variance σ^2) related to the receive antenna of BS's jth sector. Hereafter, without loss of generality, we have assumed $\mathbb{E}\left[|x_k|^2\right] = 1$.

The corresponding SINR of the kth user belonging to the set $K_{D,j}$ is given by

$$
\gamma_{k,j}^{1H} = \frac{p_k h_k^2 \left|d_k^T s_k\right|}{\displaystyle\sum_{\ell \in K_{BS,j}}^{\ell \neq k} p_\ell h_\ell^2 \left|d_k^T s_\ell\right| + \sigma_{BS,j}^2 \left\|d_k\right\|^2}
$$
$$
= \frac{p_k h_k^2 \left|d_k^T s_k\right|}{\underbrace{\displaystyle\sum_{\iota \in K_{D,j}}^{\iota \neq k} p_\iota h_\iota^2 \left|d_k^T s_\iota\right|}_{(I)} + \underbrace{\displaystyle\sum_{\ell \in K_{R,j}} p_\ell h_\ell^2 \left|d_k^T s_\ell\right|}_{(II)} + \sigma_{BS,j}^2 \left\|d_k\right\|^2}, \tag{8.3}
$$

where d_k is the detection filter (matched filter or multiuser filter) for the kth user at the BS receiver, $\sigma_{BS,j}^2$ is the noise variance at the BS receiver input and, $h_k^2 = |h_k|^2$ is the channel's power gain.

Note that term (I) of Equation 8.3 indicates the interference generated by single-hop users and term (II) of Equation 8.3 indicates the interference generated by two-hop users, that is, users relying on the RS covered by the jth BS antenna sector. To make notation in Equation 8.3 more compact, we can write

---

* It is widely recognized that the Rician k-factor significantly affects the outage performance, while the outage probability rapidly decreases, as k increases. More particularly, the fading conditions degrade to Rayleigh fading as $k \to 0$.

$$\gamma_{k,j}^{1H} = \gamma_k^{1H} = p_k \Gamma_{k,j}, \text{ with } \Gamma_{k,j} = \frac{h_k^2 \left| d_k^T s_k \right|^2}{\displaystyle\sum_{\ell \in K_{\text{BS},j}}^{\ell \neq k} p_\ell h_\ell^2 \left| d_k^T s_\ell \right|^2 + \sigma_{\text{BS},j}^2 \left\| d_k \right\|^2} = \frac{h_k^2 \left| d_k^T s_k \right|^2}{\mathfrak{T}_{k,j}},$$

(8.4)

where the term $\Gamma_{k,j}^{1H}$ refers to the channel gain normalized by the interference plus noise power levels, $\mathfrak{T}_{k,j}$. This notation will be used for other SINR expressions in this section.

Now, to describe the SINR expression for two-hop (2H) users, we must define other sets of users. The first set refers to the interest users of the ith relay R_i and identified by K_{R_i}; it is the set of users for which the ith relay is allocated. The second set refers to the users that are covered by the ith relay and denoted by $K_{R_i}^C$, in which, because of the overlapping in the adjacent RSs coverage, $K_{R_i} \subset K_{R_i}^C$ always holds. With these two sets, we can define the set of users that are covered by the ith relay but belong to another relay interest area, denoted by the index $K_{R_i}^{\text{int}} = K_{R_i} \backslash K_{R_i}^C$, as described in Figure 8.2.

On the basis of these user set definitions, the equivalent base-band received in-band DS-CDMA signal at the ith relay considering the UL direction can be written following Equation 8.2 as

$$y_{R_i} = \sum_{k \in K_{R_i}^C} \sqrt{p_k} x_k h_{k,i} s_k + \eta_{R_i} = \sum_{k \in K_{R_i}} \sqrt{p_k} x_k h_{k,i} s_k + \sum_{\ell \in K_{R_i}^{\text{int}}} \sqrt{p_\ell} x_\ell h_{\ell,i} s_\ell + \eta_{R_i}$$

(8.5)

and the corresponding SINR for the kth user belonging to K_{R_i} set at the R_i fixed-RS receiver is given by

$$\gamma_{k,R_i}^{2H} = \frac{p_k h_k^2 \left| d_k^T s_k \right|}{\displaystyle\sum_{\ell \in K_{R_i}^C}^{\ell \neq k} p_\ell h_{\ell,i}^2 \left| d_k^T s_\ell \right| + \sigma_{R_i}^2 \left\| d_k \right\|^2}$$

$$= \frac{p_k h_{k,i}^2 \left| d_k^T s_k \right|}{\underbrace{\displaystyle\sum_{\iota \in K_{R_i}}^{\iota \neq k} p_\iota h_{\iota,i}^2 \left| d_k^T s_\iota \right|}_{(I)} + \underbrace{\displaystyle\sum_{\ell \in K_{R_i}^{\text{int}}} p_\ell h_{\ell,i}^2 \left| d_k^T s_\ell \right| + \sigma_{R_i}^2 \left\| d_k \right\|^2}_{(II)}},$$

(8.6)

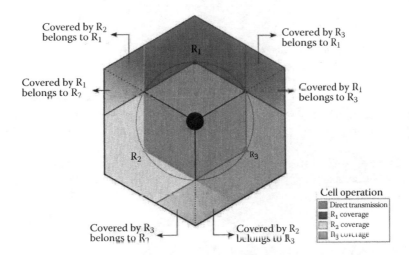

FIGURE 8.2 Macrocell coverage model, with 3 RSs and 120° BS antenna sectors, emphasizing the areas related to the $K_{R_1}^{\text{int}}, K_{R_2}^{\text{int}}, K_{R_3}^{\text{int}}$ user sets.

where term (I) refers to interference users on the ith relay, while term (II) holds for the MAI generated by the users' signals not only relayed by the qth adjacent relays but also covered by the ith relay. $h_{k,i}^2$ is the power channel gain defined by the link between the kth user and the ith relay; $\sigma_{R_i}^2$ is the additive noise variance at the input of the ith relay; and d_k is the detection filter applied to the kth user detection at the ith relay receiver if a regenerative protocol is used.

Since the AF protocol is implemented in the RSs, the SINR formula given in Equation 8.6 just refers to an approximation* and has been used to quantify the interference level at the ith RS, because of the signal amplification carried out at the RS (without detection operation).

Considering that no interference is added in the RS–BS path, the signal retransmitted by the ith RS and detected by the jth directive antenna on the BS is given by

$$
\begin{aligned}
y_{R_i} &= \frac{\sqrt{P_{R_i}}}{\sqrt{P_i}}\left[\sum_{k\in K_{R_i}^C}\sqrt{p_k}\,x_k h_{k,i} s_k + \eta_{R_i}\right] g_i + \eta_{BS,j} \\
&= \frac{\sqrt{P_{R_i}}}{\sqrt{P_i}}\left[\sum_{k\in K_{R_i}^C}\sqrt{p_k}\,x_k h_{k,i} s_k + \sum_{\ell\in K_{R_i}^{int}}\sqrt{p_\ell}\,x_\ell h_{\ell,i} s_\ell + \eta_{R_i}\right] g_i + \eta_{BS,j},
\end{aligned}
\tag{8.7}
$$

where g_i identifies the complex channel gain with LoS between RS–BS, which includes shadowing, path loss, and the multiplicative fading channel effects. The related SINR for the kth retransmitted user is given by

$$
\begin{aligned}
\gamma_{k,R_i}^{2H} &= \frac{P_{R_i} p_k h_{k,i}^2 g_i^2 \left|d_k^T s_k\right|}{P_i\sigma_{BS,j}^2\left\|d_k\right\|^2 + P_{R_i}g_i^2\sigma_{R_i}^2\left\|d_k\right\|^2 + P_{R_i}\sum_{\substack{\ell\in K_{R_i}^C \\ \ell\neq k}} p_\ell h_{\ell,i}^2 g_i^2 \left|d_k^T s_\ell\right|} \\[2mm]
&= \frac{P_{R_i} p_k h_{k,i}^2 g_i^2 \left|d_k^T s_k\right|}{\underbrace{P_i\sigma_{BS,j}^2\left\|d_k\right\|^2}_{(I)} + \underbrace{P_{R_i}g_i^2\sigma_{R_i}^2\left\|d_k\right\|^2}_{(II)} + \underbrace{P_{R_i}\sum_{\substack{\iota\in K_{R_i} \\ \iota\neq k}} p_\iota h_{\iota,i}^2 g_i^2 \left|d_k^T s_\iota\right|}_{(III)} + \underbrace{P_{R_i}\sum_{\ell\in K_{R_i}^{int}} p_\ell h_{\ell,i}^2 g_i^2 \left|d_k^T s_\ell\right|}_{(IV)}},
\end{aligned}
\tag{8.8}
$$

where $g_i^2 = \left|g_k\right|^2$ is the channel power gain for the RS–BS UL. Term (I) holds for the background noise at the jth directive BS antenna sector, term (II) refers to background noise from the ith relay, and term (III) stands for the MAI power level generated by the other users served by the interest ith relay. Finally, term (IV) describes the MAI power level from the users covered by the ith users but relayed by the adjacent relaying stations; besides, P_i refers to the signal normalization introduced by the ith RS; similarly, the ratio $\sqrt{P_{R_i}}/\sqrt{P_i}$ stands for the (amplitude) gain factor introduced by the ith RS.

Note that, in the absence of interference cancelation, the relaying process is merely a retransmission procedure. For compactness purpose, the SINR expression in Equation 8.8 can immediately be rewritten as

$$
\gamma_{k,R_i,j}^{2H} = \gamma_k^{2H} = p_k\,\Gamma_{k,j}^{R_i},
\tag{8.9}
$$

---

* This approximation is also an upper bound of the achieved SINR, given the extra background noise inserted by the RS–BS link.

where

$$\Gamma_{k,j}^{R_i} = \frac{P_{R_i} h_{k,i}^2 g_i^2 \left| \boldsymbol{d}_k^T \boldsymbol{s}_k \right|^2}{P_i \sigma_{BS,j}^2 \boldsymbol{d}_k^2 + P_{R_i} g_i^2 \sigma_{R_i}^2 \boldsymbol{d}_k^2 + P_{R_i} \displaystyle\sum_{\ell \in K_{R_i}^C}^{\ell \neq k} p_\ell h_{\ell,i}^2 g_i^2 \left| \boldsymbol{d}_k^T \boldsymbol{s}_\ell \right|}$$

$$= \frac{P_{R_i} h_{k,i}^2 g_i^2 \left| \boldsymbol{d}_k^T \boldsymbol{s}_k \right|^2}{\mathfrak{I}_{k,j}^{R_i}}. \tag{8.10}$$

In this context, and considering the discussion presented in Zappone et al. (2011), the term that normalizes the signal received by the ith relay can be defined as

$$\sqrt{P_i} = \sqrt{\mathbb{E}\left[y_{R_i}^2 \right]}, \tag{8.11}$$

where P_i is equivalent to the sum of the power of all users covered by the ith relay plus the background noise at the relay, with

$$P_i = \underbrace{\sum_{k \in K_{R_i}^C} p_k h_{k,i}^2}_{} + N\sigma_{R_i}^2 = \underbrace{\sum_{k \in K_{R_i}} p_k h_{k,i}^2}_{(I)} + \underbrace{\sum_{\ell \in K_{R_i}^{\text{int}}} p_\ell h_{\ell,i}^2}_{(II)} + \underbrace{N\sigma_{R_i}^2}_{(III)} \tag{8.12}$$

where term (I) refers to the users allocated to the ith relay, term (II) holds for users covered by the ith relay but allocated to another qth relay ($q \neq i$), and term (III) refers to the background noise that is amplified, which is independent of the number of users covered by the RS. This way, the retransmission power is defined as the necessary power level to maintain the allocated power at the MT–RS path, resulting in

$$P_{R_i} = \frac{\sqrt{\mathbb{E}\left[\left\| y_{R_i} \right\|^2 \right]}}{g_i^2} = \frac{P_i}{g_i^2} = \underbrace{\frac{\displaystyle\sum_{k \in K_{R_i}} p_k h_{k,i}^2}{g_i^2}}_{(I)} + \underbrace{\frac{\displaystyle\sum_{\ell \in K_{R_i}^{\text{int}}} p_\ell h_{\ell,i}^2}{g_i^2}}_{(II)} + \underbrace{\frac{N\sigma_{R_i}^2}{g_i^2}}_{(III)}, \tag{8.13}$$

where term (I) refers to relay transmission power consumption with interest users, term (II) refers to the power consumptions with interfering users covered by the ith RS, and term (III) refers to noise amplification, since the AF protocol is deployed.

Despite keeping the power level the same, this approach becomes inadequate if the kth user power (p_k) is just a few orders of magnitude greater than the additive white Gaussian noise (AWGN) power level at the BS. Since the AF protocol does not remove the interference and noise received at the first hop (MT–RS), the only way to have the same SINR level in the first and second hop (equaling Equation 8.6 to Equation 8.8) is with

$$\frac{\sigma_{BS,j}^2 P_i \left\| \boldsymbol{d}_k \right\|^2}{P_{R_i}} = 0.$$

As the noise power is, in general, not null, this condition becomes inappropriate. Despite this fact, in general, when the EE problem considers the operational cost (power that is consumed independent of the transmission power level), the optimal power level should be higher to compensate for this dissipated power (Betz and Poor 2008). Hence, this problem tends to not affect the achieved SINR under interest conditions (mainly under high system loading).

To compare the obtained results with the non-cooperative (nco) paradigm, we also define the system model for this case. Here, the frequency band F_2 and the RSs are not considered; as a result, all users communicate directly with the BS. Considering the jth 120° sector of the BS antenna, we can define the set K_j^{nco} as the users' group covered by this jth directive BS antenna, whose signal is given by

$$y_j^{nco} = \sum_{k \in K_j^{nco}} \sqrt{p_k} \, x_k h_k s_k + \eta_{BS_j}, \tag{8.14}$$

and the SINR for the kth user belonging to K_j^{nco} set is equivalent to

$$\gamma_{k,j}^{nco} = \frac{p_k h_k^2 \left| d_k^T s_k \right|}{\displaystyle\sum_{\ell \in K_j^{nco}}^{\ell \neq k} p_\ell h_\ell^2 \left| d_k^T s_\ell \right| + \sigma_{BS,j}^2 \left\| d_k \right\|^2}. \tag{8.15}$$

In the same way, to simplify the notation, we rewrite Equation 8.15 as

$$\gamma_{k,j}^{nco} = \gamma_k^{nco} = p_k \, \Gamma_k^{nco}, \text{ with } \Gamma_k^{nco} = \frac{h_k^2 \left| d_k^T s_k \right|^2}{\displaystyle\sum_{\ell \in K_j^{nco}}^{\ell \neq k} p_\ell h_\ell^2 \left| d_k^T s_\ell \right|^2 + \sigma_{BS,j}^2 \left\| d_k \right\|^2} = \frac{h_k^2 \left| d_k^T s_k \right|^2}{\mathfrak{I}_{k,j}^{nco}} \tag{8.16}$$

8.3.3 RS PLACEMENT

There were some issues that can help us understand the effects of relay position and quantity into the obtained EE. The first one refers to power consumption. Adding more relays reduces the average distance between the MTs and the nearest relay, which reduces the path-loss term and can incur lower power consumption. But as all relays have a fixed power cost and need to amplify the received signal, there probably exists a point when the power costs of one extra relay outperforms the power saving obtained for the MTs, which can translate into EE reduction.

Another drawback of increasing the number of relays is the increase of interfering users' density in $K_{R_i}^C$. As the relays are placed uniformly around a circle and the coverage angle is strictly decreasing when the number of relays increase, the minimum coverage angle is obtained taking $N_{RS} \rightarrow \infty$ in Equation 8.1, obtaining $\angle_{min,RS} = \pi$. On the other hand, since all relays divide equally the two-hop area, the number of users in the interest set K_{R_i} tends to be reduced while increasing the number of RS. This way, we see that the coverage area (consequently, the number of users, since we assume uniform distribution) of the ith relay tends to stabilize while the number of interest users tends to be constantly reduced with relay adding. Since we use the AF protocol, it is impossible to eliminate or even reduce this interference, and then the EE tends to be reduced. This effect can be seen with Figures 8.3 and 8.4, where the area covered only by the correspondent relay is constantly reduced when increasing the number of relays.

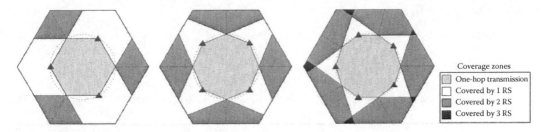

FIGURE 8.3   Coverage areas considering 3, 4, and 5 RSs.

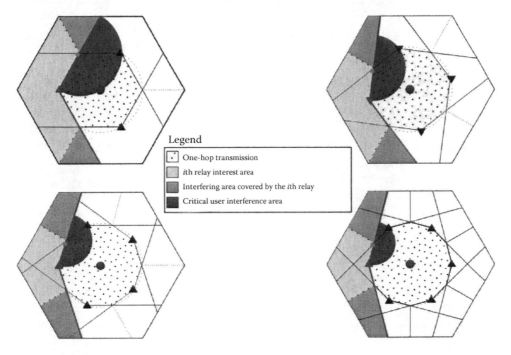

FIGURE 8.4   Behavior of the interference area for the ith relay and critical user interference area for $N_{RS} =$ 3, 4, 5, and 6. For all cells, $r_1 = 0.5\ r_0$.

Regarding the RS distance placement, if we choose to place the relays near the cell edge, there can be subserved users inside the one-hop polygon, and the relay benefits will not be totally exploited. On the other hand, if the relays are placed near the BS, the cell-edge user's problem may continue and the interference generated by the two-hop users can be more intense at the BS, and again the benefits will be wasted. Some insights regarding this problem have been reported in the literature. For instance, in Badruddin and Negi (2004), the best radius to install the RSs has been found to be approximately 64% of the macrocell's radius, which divides the macrocell area into two equal areas, that is, one for direct transmission and the other for two-hop users.

In order to demonstrate that the high number of relays is not the only problem in determining the placement, Figure 8.4 shows one possible effect of few RSs, meaning higher interference from two-hop users. When considering a hypothetical user (dot in Figure 8.4) located at the edge of the two-hop area, in the nearest point to the BS, the lower the number of relays, the higher distance between this hypothetical user to the RS, which implies higher transmission power in order to compensate the path-loss term. As one can conclude from the geometry of Figure 8.4 for the case when $N_{RS} =$ 3, the power received at the BS is almost the same as that received by the ith RS allocated to serve this user, considering $r_1 = 0.5\ r_0$.

8.3.4 Detection Techniques

There are two detection methods in DS-CDMA systems—single-user detection (SuD) and MuD. The main difference between then is the treatment of the interfering users' signals; while the first one considers those signals as MAI, the second one uses that information to mitigate the interference, resulting in higher SINR at the cost of higher computational complexity.

The SuD technique is performed by the matched filter-based (MF) detector. There are K paths, each one detecting only one user without sharing information, which is equivalent to treating the interfering signals as MAI. The MF detector consists of a matched filter bank where the filter associated to the kth path (d_k) is simply defined by the kth user spreading code (s_k). This filter is applied to the received signal y, and then the resulting signal is integrated and sampled and finally a hard estimation is made. This way, the correlation between the codes and the system loading defines the amount of MAI for each user. As the orthogonality condition between the codes is hard to achieve (mainly because of the user's signal asynchronism), the MF detection is highly limited by the MAI level.

In order to reduce the MAI effects, Verdú (1984) developed the concept of MuD. In this technique, the information of the interfering users is used to mitigate (or even eliminate) the MAI, reducing the necessary power to achieve a target SINR γ^*. The optimum MuD (OMuD) developed by Verdú compares the received signal with any possible combination of symbols for all users and chooses the combination that minimizes the Euclidian distance for the received signal. As the complexity is of the order $\mathcal{O}(2^K)$, the idea is to use sub-optimum MuD detectors, with polynomial complexity and suboptimal results compared to OMuD. There are several classes of suboptimal MuD detectors: linear, non-linear, or heuristic based. As the detection methods based on linear MuD (LMuD) provides closed-form expressions for the achieved SINR (which makes it possible to obtain analytical solutions), they are suitable for power allocation problems.

The LMuD strategy is to apply linear transformations to the MF soft estimation (the information before hard estimation) in order to decouple de MAI. Some of the most popular LMuD techniques are the minimum mean-square error (MMSE), zero-forcing (ZF), and decorrelator (DEC). The most efficient LMuD is the MMSE detector, as it considers the spreading codes and the amplitudes of the interfering signals and background noise in the filter calculation, as shown in Equation 8.17. This way, MAI decoupling does not result in background noise amplification, as occurs with DEC and ZF detectors.

$$D_{\mathrm{MMSE}} = [R + \sigma^2 A^{-2}]^{-1}, \tag{8.17}$$

where $R \in \mathbb{R}^{K \times K}$ is the spreading codes correlation matrix, $R = S^T S$, with $S = \left[s_1^T, s_2^T, \ldots, s_k^T, \ldots, s_K^T \right]$ being the spreading codes $N \times N$ matrix, and $A \in \mathbb{R}^{K \times K}$ is the received signal amplitude diagonal matrix.

One drawback to adopting the MMSE in distributed problems is the necessity of transmitting the amplitude matrix to all users or recalculating the filter matrix at each step of the power allocation (if iterative algorithms are used) or each network configuration (if there is mobility). This way, one possible approach is to adopt the DEC detector, which only considers the correlation and spreading codes matrices. This way, the overhead of control messages can be reduced at the cost of slight performance degradation (for higher SINR). The DEC filter matrix is given by

$$D_{\mathrm{DEC}} = [d_1, d_2, \ldots, d_k, \ldots, d_K] = S(S^T S)^{-1} = S R^{-1}. \tag{8.18}$$

8.4 RA: ENERGY EFFICIENCY MODEL

The RA problem proposed in this work involves the EE maximization with relay placement (position and quantity) and power optimization, under channel fading effect and random MT users' location. Taking into account all these conditions is highly complex, given that the relay placement radius is continuous and for each position we should test the impact of more RSs for each considered distance. Hence, we separate the problem into two sub-problems:

- Determine the best relay placement in which the network achieves the lower power consumption, given the performance metrics.
- Maximize the EE for the topology determined with the first problem.

8.4.1   ENERGY-EFFICIENT COOPERATION REGIONS

In Yang et al. (2010), the definition of the best placement of relays is considered for a single-user scenario with some relays, one of which is selected to help the transmission in both UL and DL, with asymmetric traffic. Despite some simplification, the metric defined to choose the best placement of relays will be deployed in this work. In Figure 3 of Yang et al. (2010), the authors defined regions of energy saving (ε_{sav}), varying the position in both axes, where the percentage energy saving is expressed by

$$\varepsilon_{sav} = \frac{\varepsilon_{nco} - \varepsilon_{coo}}{\varepsilon_{nco}} \times 100 \ [\%], \tag{8.19}$$

where ε_{coo} is the energy consumed when the cooperative mode is used and ε_{nco} is the energy consumed with the one-hop system, discarding the effects of Rayleigh fading.

As the problem proposed in Yang et al. (2010) does not involve EE maximization as proposed in this chapter, the specification in that work is a system with a fixed rate R, where the relays are deployed to reduce the energy consumption per information bit, but not necessarily achieving the best EE operational point as the concern herein. Hence, we can define the percentage of EE improvement metric regarding the non-cooperative scheme as

$$\Delta\xi_{\%} = \frac{\xi_{nco} - \xi_{coo}}{\xi_{nco}} \times 100 \ [\%]. \tag{8.20}$$

In order to obtain the best relay positioning, which includes location and number of deployed retransmission stations, we will consider the APC metric, for a given SE. The model presented in Xu and Qiu (2010) reduces the complexity for estimating the best relay positioning, since only the path-loss term has been considered, while taking the average for the small-scale fading term.

8.4.1.1   SE for Cooperative and Non-Cooperative Cases

The SE can be defined by the total rate divided by the available bandwidth; in CDMA systems, it is equivalent to the sum of the SE for each user:

$$\zeta \underset{\text{def}}{=} \frac{1}{w_c} \sum_{k=1}^{K} r_k(\gamma_k) = \sum_{k=1}^{K} \zeta_k(\gamma_k) \ \left[\frac{bps}{Hz}\right], \tag{8.21}$$

where r_k is the achieved rate and w_c is the cell total bandwidth. If we consider the Shannon capacity equation (Shannon 1948), we can rewrite Equation 8.21 as

$$\zeta \leq \frac{1}{w_c} \sum_{k=1}^{K} w_k \log_2(1+\gamma_k), \tag{8.22}$$

where w_k is the bandwidth allocated for the kth user. For simplicity, hereafter we assume that the attainable rate $r(\gamma) = w \log_2 (1 + \gamma)$ in practice can be achieved, after deploying coding, among other things.

Since the average SE ($\overline{\zeta}$) is of great interest, and the only variable component in Equation 8.21 is the data rate r_k, we can immediately define

$$\overline{\zeta} = \frac{1}{w_c} \sum_{k=1}^{K} \overline{r_k(\gamma_k)}, \qquad (8.23)$$

with $\overline{r_k(\gamma_k)} = \mathbb{E}\{r_k(\gamma_k)\}$. Thus, to obtain the average data rate, we simply use the definition of statistical expectation (Alouini and Goldsmith 1999):

$$\overline{r_k(\gamma_k)} = \int_0^\infty r_k f_{\gamma_k}(\gamma_k) d\gamma_k = \int_0^\infty w_k \log_2(1+\gamma_k) f_{\gamma_k}(\gamma_k) d\gamma_k, \qquad (8.24)$$

where f_{γ_k} is the probability density function of the SINR γ_k.

Assuming all user experiments have the same conditions and have the same probabilities for fading and random location, the expected data rate $\overline{r(\gamma)}$ can be considered the same for all users, which results in

$$\overline{\zeta} = \frac{K\overline{r}}{w_c}, \qquad (8.25)$$

where $\overline{r} = \overline{r(\gamma)} = \mathbb{E}\{r_k(\gamma_k)\}$.

As can be seen in Equations 8.3 and 8.8, the kth user SINR depends strongly on the random locations of the interest user and the other $K - 1$ users (if no BS antenna sector is used), owing to the path-loss and fading terms. One possible way to reduce computation consists in taking the average value of fading and then only considering the path-loss term, but the K terms of path-loss calculations remain.

8.4.1.2 Area Power Consumption

The concept of APC for relay-assisted networks is discussed in Xu and Qiu (2010) and is related to the total power consumed, including power wasted in idle mode and circuitry power, to maintain a target SE (ζ^*) in a cell with area S_c:

$$APC(\zeta^*) = \frac{P_T}{S_c} \qquad (8.26)$$

In Xu and Qiu (2010), the APC for DL with TDD in-band relaying and round-robin scheduling (no multiple-access interference) has been considered. Besides, the authors carried out a study of optimum cell radius for one-hop transmissions, best relay placement, and relay density impact, considering a ring area placement. Since the analyzed system is half-duplex, the power wasted while the RS and BS are idle is also computed. Hence, the half-duplex in-band systems divide time or frequency resources to retransmit, and the data rate is, in general, 50% of the minor rate achieved in the MT–RS or RS–BS path.

Different from Xu and Qiu (2010), in this chapter, since we are concerned with the UL of AF out-of-band scheme, that penalty does not exist,* and if we look to a pure minimum rate criterion, it is not necessary to double the target data rate to maintain the QoS metric. However, if the purpose is to maintain a target SE, the data rate in cooperative mode must be adjusted to compensate the extra bandwidth of the F_2 band. For example, if we consider the same bandwidth w for F_1 (MT–RS and MT–BS links) and F_2 (RS–BS link) bands, then $w_c = w$ for non-cooperative systems and $w_c = 2w$ for the cooperative case. This way, the SE for non-cooperative (nco) and cooperative (coo) modes is given respectively by

$$\zeta_{nco} = \frac{1}{w}\sum_{k=1}^{K} r_k^{nco} \text{ and } \zeta_{coo} = \frac{1}{2w}\sum_{k=1}^{K} r_k^{coo}. \tag{8.27}$$

Hence, in order to obtain the same SE in both modes, it is necessary to guarantee $\sum_{k=1}^{K} r_k^{nco} = 2\sum_{k=1}^{K} r_k^{coo}$ or $\overline{r^{coo}} = 2\overline{r^{nco}}$, which can be guaranteed with $r^{coo} = 2r^{nco}$ or $\zeta_k^{coo} = 2\zeta_k^{nco}$.

Under this statement, the APC for direct and cooperative transmission can be described as follows. For the first case, we have to allocate the vector $\boldsymbol{p}^{nco}$ [W], defined by

$$\boldsymbol{p}^{nco} = \left[p_1^{nco}, p_2^{nco}, \ldots, p_k^{nco}, \ldots, p_K^{nco} \right] \tag{8.28a}$$

$$\text{with } p_k^{nco} = \frac{\left(2^{r_k^{nco}/w} - 1\right)}{\ell_k \Gamma_k}, \forall k \in K, \tag{8.28b}$$

where Γ_k is given by Equation 8.16. Hence, the APC for the nco mode results is

$$\text{APC}_{nco} = \frac{1}{S_c}\left(K p_c + \varrho_{MT}\sum_{k=1}^{K} p_k^{nco} \right), \tag{8.29}$$

where p_c refers to the power circuitry consumption, described by Equation 8.60 in Appendix A; $\varrho_{MT} = \left(\dfrac{\text{PAPR}}{\rho_{MT}} - 1\right) > 1$ refers to the power amplifier (PA) inefficiency at the MTs, where PAPR is the peak-to-average power ratio and ϱ_{MT} is the PA efficiency of the MT. As we consider multimedia CDMA networks and make the hypothesis of sufficient data to be transmitted all the time, the transmission buffer never goes empty, and the idle power consumption has not been considered in this model.

In the proposed CDMA cooperative relaying mode, the power allocation must guarantee twice the minimum rate per service used in the direct (nco) mode. As a result, the transmitted power vector for the MTs and relay stations are written, respectively, as

$$\boldsymbol{p}^{nco} = \left[p_1^{coo}, p_2^{coo}, \ldots, p_k^{coo}, \ldots, p_K^{coo} \right] \tag{8.30a}$$

---

* As pointed out in Kim et al. (2011), the delay introduced by one RS with the amplify-and-forward scheme is less than a microsecond and can be ignored.

$$\mathbf{p}_{RS} = \left[P_{R_1}, P_{R_2}, \ldots, P_{R_i}, \ldots, P_{R_{N_{RS}}} \right] \tag{8.30b}$$

$$\text{with } p_k^{coo} = \frac{\left(2^{r_k^{coo}/w} - 1\right)}{\ell_k \Gamma_k} = \frac{\left(2^{2r_k^{nco}/w} - 1\right)}{\ell_k \Gamma_k}, \forall k \in K \tag{8.30c}$$

where Γ_k is given by Equation 8.4 or Equation 8.10 for one-hop users or two-hop users, respectively. As a result, the total APC in the UL cooperative relay mode can be formulated as

$$\begin{aligned}
\text{APC}_{coo} &= \frac{P_T^{UL}}{S_c} = \frac{P_{T_{RS}} + P_{T_{MT}}}{S_c} \\
&= \frac{1}{S_c} \left[N_{RS} P_{c,R_i} + \varrho_{RS} \sum_{i=1}^{N_{RS}} P_{R_i} + K p_c + \varrho_{MT} \sum_{k=1}^{K} p_k^{coo} \right],
\end{aligned} \tag{8.31}$$

where P_{c,R_i} is the power wasted with circuitry at each RS and $\varrho_{RS} = \left(\dfrac{\text{PAPR}}{\rho_{RS}} - 1 \right) > 1$ is the RS PA inefficiency. Discussion about relay circuit power consumption can be found in Appendix A.

In order to find the best number of FRSs N_{RS} for a given cell area and overall target SE, the following power consumption minimization problem for the coo mode can be formulated:

$$\arg\max_{r_1, N_{RS}} \frac{1}{S_c} \left[N_{RS} P_{c,R_i} + \varrho_{RS} \sum_{i=1}^{N_{RS}} P_{R_i} + K p_c + \varrho_{RS} \sum_{k=1}^{K} p_k^{coo} \right] \text{ for } \zeta = \zeta^* \tag{8.32a}$$

$$s.t. \quad \text{Pr}_{out}^{coo} \le \text{Pr}_{out}^{nco} \tag{8.32b}$$

$$P_{R_i} \le P_{R_{max}}; \quad p_k \le P_{max} \tag{8.32c}$$

where Pr_{out}^{nco} is the outage probability obtained in non-cooperative mode and Pr_{out}^{coo} is the outage probability for the cooperative mode. An outage event occurs when the target SE is not achieved; that is, $\gamma_k < \left(2^{r_k^{nco}/w} - 1\right)/\ell_k$ for non-cooperative users and $\gamma_k < \left(2^{2r_k^{nco}/w} - 1\right)/\ell_k$ for cooperative users.

8.4.2 BEST FRSs PLACEMENT

We can define the best RS placement as the topology network that minimizes the APC subject to the performance metrics constraints (the SE metric). Since the closed-form expressions for SE and APC are hard to obtain, even for TDMA or FDMA systems (Alouini and Goldsmith 1999; Xu and Qiu 2010), the Monte Carlo simulation approach has been used. The experiment developed herein follows the assumptions described in Alouini and Goldsmith (1999), but adapted to multiuser and single-cell scenarios, as described by Algorithm 1.

Algorithm 1: Relay placement with APC/SE metric

Initialization: $i \leftarrow 1$, It, ζ^*, r_O;
> **while** $i \leq It$
>> Generate random positions for the K users;
>> Evaluate MT–BS channel conditions;
>> Calculate the APC and $\mathrm{Pr}_{\mathrm{out}}$ for nco mode (Equation 8.29), given ζ^*;
>> **For** each relay placement (r_1 and N_{RS}):
>>> Evaluate channel conditions for MT–RS and RS–BS;
>>> Calculate the APC and $\mathrm{Pr}_{\mathrm{out}}$ for coo mode (Equation 8.31), given ζ^*;
>> **end**;
>> $i = i + 1$
> **end**;
> Choose r_1 and N_{RS} that minimize the APC maintaining $\mathrm{Pr}_{\mathrm{out}}^{\mathrm{coo}} \leq \mathrm{Pr}_{\mathrm{out}}^{\mathrm{nco}}$;

Output: r_1, N_{RS} (best fixed-RS placement)

8.4.3 EE Metric

The EE formulation for the kth user is given by

$$\xi_k = r_k(\gamma_k) \frac{L}{M} \frac{f(\gamma_k)}{P_{\mathrm{T},k}} \quad \forall k \in [1,\ldots,K], \quad \left[\frac{\mathrm{bit}}{\mathrm{joule}}\right], \tag{8.33}$$

where r_k is the transmit rate, including redundancy bits, L is the number of information bits per packet, M is the number of bits per packet, $f(\gamma_k)$ is the efficiency function, and $P_{\mathrm{T},k}$ is the total power consumption, including circuitry power. Note that the power consumption model is quite different at MTs regarding the FRS or BS model, since the power consumption requirements in MTs are remarkably different from those in the BS.

In the next subsections, we describe the EE and other related concepts, then we apply it to one- and two-hop users (including non-cooperative case), and finally discuss the other necessary parameters to apply the EE metric (Equation 8.33) in both cases. In order to analyze the EE of the system, our attention is on the overall EE.

8.4.4 EE Metric for Cooperative and Non-Cooperative Modes

8.4.4.1 Non-Cooperative and One-Hop (1H) Users

For the direct communication case, that is, nco mode, and users inside the one-hop area for cooperative case, we can use the parameters defined in Souza et al. (2012b), resulting in

$$r_k(\gamma_k) = w \log_2(1 + \iota_k \gamma_k), \quad \iota_k = \frac{-1.5}{\ln(5\mathrm{BER}_k)} \in [0,1], \tag{8.34}$$

where w is the system bandwidth and ι_k represents the SINR gap factor from the Shannon capacity that accounts for the deployment of practical modulation and coding schemes (Goldsmith 2005) and BER_k is the maximum tolerable bit error rate by the kth user. This gap measures the efficiency of the transmission scheme with respect to the best possible performance in AWGN ($\iota_k = 1$).

Furthermore, we need to define an efficiency function $f(\gamma_k)$ that approximates the probability of error-free packet reception in order to express the net amount of detected data (free of error). In general, the efficiency function can be expressed as $f(\gamma_k) = (1 - \text{BER}(\gamma_k))^M$, where M is the number of transmitted bits in a packet and BER is bit error rate. However, the exact expression for the efficiency function is hard to derive and depends on the channel and modulation type, modulation order, and so forth. For instance, in the absence of channel coding, and assuming that $\text{BER}(\gamma_k) \approx e^{-\gamma_k/2}$ is a good approximation for bit error rate under the medium- and high-SNR regime for BFSK, BPSK, and QPSK modulation formats, the efficiency function can be expressed by

$$f(\gamma_k) = (1 - \text{BER}(\gamma_k))^M \approx (1 - 2\text{BER}(\gamma_k))^M = \left(1 - e^{-\gamma_k/2}\right)^M \equiv \left(1 - e^{-\gamma_k}\right)^M. \tag{8.35}$$

This efficiency function approximation is adopted herein since it has a well-behaved shape, while presenting desirable properties at the limiting points $\gamma_k = 0$ and $\gamma_k = \infty$ and simultaneously holding the same shape as that of the original utility function $f(\gamma_k) = (1 - \text{BER}(\gamma_k))^M$ for wide digital modulation formats and orders (Goodman and Mandayan 2000).

Besides, the total power consumption at the kth MT is modeled as

$$P_{\text{T},k} = \varrho_{\text{MT}} p_k + p_c. \tag{8.36}$$

Hence, the EE function for the kth user in the nco mode can be defined as

$$\xi_k^{\text{nco}} = \frac{L}{M} \frac{r_k\left(\gamma_k^{\text{nco}}\right)\left(1 - e^{-\gamma_k^{\text{nco}}}\right)^M}{\varrho_{\text{MT}} p_k + p_c}, \tag{8.37}$$

with γ_k^{nco} given by Equation 8.15 and

$$\xi_k^{\text{1H}} = \frac{L}{M} \frac{r_k\left(\gamma_k^{\text{1H}}\right)\left(1 - e^{-\gamma_k^{\text{1H}}}\right)^M}{\varrho_{\text{MT}} p_k + p_c} \tag{8.38}$$

for users inside the one-hop area for the cooperative case, where γ_k^{1H} is given by Equation 8.3.

8.4.4.2 Two-Hop (2H) Users

For the two-hop case, we have to consider the SINR of the two-hop MT–RS–BS link. As we consider a full-duplex out-of-band, the achieved rate can be modeled as

$$r_k\left(\gamma_k^{\text{2H}}\right) = w \log_2\left(1 + \iota_k \gamma_k^{\text{2H}}\right). \tag{8.39}$$

Note that the bandwidth here is w and not 2w, since the only signal from the MT–RS–BS path is considered as the interest signal. Furthermore, as the SE ζ_k^{nco} is doubled to maintain the system SE metric and the system and there is no data rate penalty inserted by the RS (since the whole time slot is used by the kth user), the data rate achieved in cooperative mode is doubled.

As we consider the user side of the 2H communication, the power consumption model is the same as defined for the 1H users in Equation 8.36. Hence, the EE for the two-hop case is given by

$$\xi_k^{\text{2H}} = \frac{L}{M} \frac{r_k\left(\gamma_k^{\text{2H}}\right)\left(1 - e^{-\varsigma^{\text{2H}}\gamma_k^{\text{2H}}}\right)^M}{\varrho_{\text{MT}} p_k + p_c}, \tag{8.40}$$

where ς^{2H} stands for a penalty factor given by the two-hop transmission, indicating that a higher SINR is necessary to achieve the same efficiency obtained in one-hop transmissions.

To simplify notation, we define a generic expression for EE, given by

$$\xi_k^{\varpi} = \frac{L}{M} \frac{r_k\left(\gamma_k^{\varpi}\right)\left(1 - e^{-\varsigma\gamma_k^{\varpi}}\right)^M}{\varrho_{MT}p_k + p_c} \tag{8.41}$$

where ϖ identifies the non-cooperative (nco) mode and one-hop (1H) or two-hop (2H) users. For nco and 1H users, we have $\varsigma = 1$, while for 2H users, $0 < \varsigma^{2H} < 1$.

8.4.5 Overall EE for nco and coo Modes

8.4.5.1 Non-Cooperative Mode

Since there are no relay cost for the non-cooperative case, the overall system EE is simply the computation of all individual EE functions of Equation 8.37, defined as

$$\bar{\xi}_{nco} = \frac{L}{M} \frac{\displaystyle\sum_{k=1}^{K} r_k(\gamma_k)\left(1 - e^{-\gamma_k}\right)^M}{Kp_c + \varrho_{MT}\displaystyle\sum_{k=1}^{K} p_k}. \tag{8.42}$$

8.4.5.2 Cooperative Mode

Since we consider that the RS only transmits amplified copies of other users' signals (and thus does not generate any information), the total data rate is given by the sum of MTs data rate r_k transmitting in either coo submode 1H or 2H. Hence, the overall EE can be expressed by

$$\bar{\xi}_{coo} = \frac{L}{M} \frac{\displaystyle\sum_{k=1}^{K} r_k\left(\gamma_k^{\varpi}\right)\left(1 - e^{-\varsigma\gamma_k^{\varpi}}\right)^M}{\left(\overbrace{Kp_c + \varrho_{MT}\displaystyle\sum_{k=1}^{K} p_k}^{(I)} + \overbrace{\varrho_{RS}\displaystyle\sum_{i=1}^{N_{RS}}\displaystyle\sum_{\ell\in K_{R_i}} p_\ell h_{\ell,i}^2 \frac{P_{R_i}}{P_i}}^{(II)} + \underbrace{\varrho_{RS}\displaystyle\sum_{i=1}^{N_{RS}}\displaystyle\sum_{\ell\in K_{R_i}^{int}} p_\ell h_{\ell,i}^2 \frac{P_{R_i}}{P_i}}_{(III)} + \underbrace{\varrho_{RS}\displaystyle\sum_{i=1}^{N_{RS}} N\sigma_i^2 \frac{P_{R_i}}{P_i}}_{(IV)} + \underbrace{N_{RS}P_{c,R_i}}_{(V)} \right)}, \tag{8.43}$$

where index ϖ holds for coo submodes 1H and 2H, $\varsigma = 1$ holds for one-hop users, and $\varsigma = \varsigma^{2H}$ holds for two-hop users; term (I) indicates the power consumed by the MT–BS and MT–RS transmission, term (II) indicates the power consumed by the RS with the interest users belonging to the K_{R_i} set, term (III) indicates the power wasted with interfering users (since AF relays cannot remove these users), term (IV) is equal to the power wasted with noise amplification, and term (V) stands for the circuit power spent by the RS. A compact version of overall EE is presented as follows:

$$\bar{\xi}_{coo} = \frac{L}{M} \frac{\displaystyle\sum_{k=1}^{K} r_k\left(\gamma_k^{\varpi}\right)\left(1 - e^{-\varsigma\gamma_k^{\varpi}}\right)^M}{N_{RS}P_{c,R_i} + \varrho_{RS}\displaystyle\sum_{i=1}^{N_{RS}} P_{R_i} + Kp_c + \varrho_{MT} + \displaystyle\sum_{k=1}^{K} p_k}. \tag{8.44}$$

Hence, the trade-off between the increasing number of relays N_{RS} and the EE gain can be evaluated using Equation 8.43 or Equation 8.44. Increasing the number of relays possibly increases the achieved data rate or, equivalently, decreases the power to attain a specific data rate. However, since the relay-based network has a fixed power consumption given by P_{c,R_i}, under certain scenarios, the achieved performance gain could not compensate the necessary power to activate one extra RS. Furthermore, the impact of inserting more relays can be viewed, basically, as a trade-off between the power resource saving through term (I) of Equation 8.43 and the power consumption described by terms (II) to (V) of Equation 8.43.

8.5   DISTRIBUTED NON-COALITIONAL EE POWER OPTIMIZATION GAME

The power allocation games at the RSs and each BS sector are mutually independent; in other words, we have $N_{RS} + N_\delta$ non-coalitional games, where N_δ is the number of sectors for the F_1 band, and the objective is to allocate the necessary power to achieve EE as higher as possible.

The 2H non-coalitional game for the N_{RS} relays is described by

$$\mathcal{G}_i^{2H} = \left[\mathcal{K}_i, \ \{\mathcal{A}_{k,i}\}, \ \{u_{k,i}\} \right], \tag{8.45}$$

where the players' set is defined by $\mathcal{K}_i \equiv K_{R_i}$; the strategy set $\{\mathcal{A}_{k,i}\}$ is defined by the power levels of the users in the set K_{R_i}, and $\{u_{k,i}\}$ refers to the application of these strategies in the utility function ξ_k^{2H}, defined formally in Equation 8.40.

On the other hand, the power allocation game for users inside the one-hop area and directly covered by the jth BS sector is modeled by a 1H non-coalitional game:

$$\mathcal{G}_j^{1H} = \left[\mathcal{K}_j, \ \{\mathcal{A}_{k,j}\}, \ \{u_{k,j}\} \right], \tag{8.46}$$

where the set $\mathcal{K}_j$ is given by $K_{D,j}$, $\{\mathcal{A}_{k,j}\}$ is the set of available power levels to the users in the set $K_{D,j}$, and $\{u_{k,j}\}$ is created by applying the strategies defined by the utility function ξ_k^{1H} in Equation 8.38 for 1H users.

Besides the independent behavior between those games, it is possible to note that the players can interfere with other games; for instance, the two-hop users interfere in the one-hop signal, and one specific user can be covered by two different RSs. For simplicity, we consider this kind of power interference as part of the interference term $\mathfrak{I}_k$. Hence, the BS only needs to communicate the SINR estimation to obtain $\mathfrak{I}_k$ and, as a consequence, Γ_k in each MT in order for each MT to determine its optimum operation point.

For the non-cooperative case, we have only N_δ games, one for each sector, and since all users interfere only in its sector, the games are mutually independent. This way, the non-coalitional game for the users covered by the jth sector of BS in the non-cooperative mode is given by

$$\mathcal{G}_j^{nco} = \left[\mathcal{K}_j, \ \{\mathcal{A}_{k,j}\}, \ \{u_{k,j}\} \right], \tag{8.47}$$

where $\mathcal{K}_j \equiv K_j^{nco}$, $\{\mathcal{A}_{k,j}\}$ is the set of available power levels to the users in the set K_j^{nco}, and $\{u_{k,j}\}$ is created by applying the strategies defined by the utility function ξ_k^{nco} in Equation 8.37.

Let us consider the power allocation p_k; the respective power vector for the interfering users of kth user is defined by

$$\boldsymbol{p}_{-k} = [p_1, p_2, \ldots, p_{k-1}, p_{k+1}, \ldots, p_K]. \tag{8.48}$$

Hence, given the power allocation of all interfering users, $\boldsymbol{p}_{-k}$, the best response for the power allocation of the kth user can be expressed as

$$p_k^{\text{best}} = q_k(\boldsymbol{p}_{-k}) = \arg\max_{p_k} u_k(p_k, \boldsymbol{p}_{-k}), \tag{8.49}$$

where u_k is given by Equation 8.38 or Equation 8.40, and $q_k(\boldsymbol{p}_{-k})$ is called the kth best response function.

Given the definition of the EE utility function and the non-coalitional games for both cooperative and non-cooperative modes, we state, generically, the optimization problem at the kth MT as

$$\arg\max_{p_k} u_k = \arg\max_{p_k} \xi_k = \arg\max_{p_k} \frac{L}{M} \frac{\mathrm{w} \log_2\left(1 + \iota_k \gamma_k^\varpi\right)\left(1 - e^{-\varsigma \gamma_k^\varpi}\right)^M}{\varrho_{\mathrm{MT}} p_k + p_c} \tag{8.50a}$$

$$s.t. \quad 0 \le p_k \le P_{\max} \tag{8.50b}$$

$$0 \le P_{\mathrm{R}_i} \le P_{\mathrm{R}_i}^{\max}, \tag{8.50c}$$

where the constraint (Equation 8.50c) is not considered for the non-cooperative mode since there are no relays.

8.5.1 Numerical Solution

The channel gain normalized by the interference plus noise power level Γ_k, generically defined in Equations 8.4, 8.10, and 8.16, can be easily estimated by the MTs when they receive the estimated SINR from BS or RS, allowing the implementation of a distributed solution. The only centralized process is the SINR estimation and information exchanging carried out by the BS and fixed RSs. Note that once an interference power is added to the RS–BS link, the MT is able to estimate the SINR level provided that it has a reasonable estimation of the channel gain g_i and the AWGN power level $\sigma_{\mathrm{BS},j}^2$. As we can see from Equation 8.33, the EE expression depends on both γ_k and p_k. Since these two parameters are related, we can rewrite p_k as

$$p_k = \frac{\gamma_k}{\Gamma_k} = \gamma_k \tilde{\mathfrak{T}}_k, \tag{8.51}$$

where $\tilde{\mathfrak{T}}_k$ is the interference plus noise power level normalized by the channel gain and is directly proportional to MAI and other sources of interference. With this term, we can quantify the effects of the interference over the EE, simplifying the analysis. Using Equation 8.51, we can modify the optimization problem to use γ_k as the optimizing argument and to map the power range into the SINR range. Hence, the power consumption can be modeled as

$$P_{\mathrm{T},k} = \varrho_{\mathrm{MT}} \gamma_k \tilde{\mathfrak{T}}_k + p_c, \tag{8.52}$$

and the generic EE utility function can be modeled as

$$\xi_k^\varpi = \frac{L}{M} \frac{r_k\left(\gamma_k^\varpi\right)\left(1 - e^{-\varsigma \gamma_k^\varpi}\right)^M}{\varrho_{\mathrm{MT}} \gamma_k^\varpi \tilde{\mathfrak{T}}_k^\varpi + p_c}. \tag{8.53}$$

In order to find the optimal power p_k^*, we take the utility function first derivative regarding γ_k, which is equivalent to the derivative regarding p_k, and then obtain the points where $\dfrac{\partial \xi_k}{\partial \gamma_k} = 0$, with normalized MAI $\tilde{\mathfrak{T}}_k$ hold fixed. After that, these points are tested to verify which one maximizes ξ_k.

In order to guarantee that the utility function Equation 8.53 has only one maximizer, we introduce the concept of quasi-concavity, defined in the following (Boyd and Vandenberghe 2004; Rodriguez 2003):

Definition 1: (Quasi-concavity)

A function $z : \mathbb{R}^n \to \mathbb{R}$ that maps a convex set of n-dimensional vectors $\mathfrak{D}$ into a real number is quasi-concave if for any $x_1, x_2 \in \mathfrak{D}$, $x_1 \neq x_2$,

$$z(\lambda x_1 + (1 - \lambda)x_2) \geq \min \{z(x_1), z(x_2)\}, \tag{8.54}$$

where $\lambda \in (0, 1)$.

The function z is said to be strictly quasi-concave if

$$z(\lambda x_1 + (1 - \lambda)x_2) > \min \{z(x_1), z(x_2)\} \tag{8.55}$$

The proof of the strict quasi-concavity of the proposed utility function is described in Appendix C, and the result is described by the next lemma.

Lemma 1.1: The Utility Function Described in Equation 8.53 is Strict Quasi-Concave in γ_k

Proof: See Appendix C.

Since the utility function is strict quasi-concave, we can find the optimal SINR γ_{EE}^* taking the first derivative of the utility function (Equation 8.53) and then finding the point where $\dfrac{\partial \xi_k}{\partial p_k} = 0$ or $\dfrac{\partial \xi_k}{\partial \gamma_k} = 0$. The latter condition is equivalent to solve

$$M\varsigma e^{-\varsigma \gamma_k^{\varpi}} \log_2\left(1 + \iota_k \gamma_k^{\varpi}\right) + \frac{\iota_k\left(1 - e^{-\varsigma \gamma_k^{\varpi}}\right)}{\left(1 + \iota_k \gamma_k^{\varpi}\right)\ln 2} = \frac{\tilde{\mathfrak{T}}_k \varrho_{\text{MT}} \log_2\left(1 + \iota_k \gamma_k^{\varpi}\right)\left(1 - e^{-\varsigma \gamma_k^{\varpi}}\right)}{\varrho_{\text{MT}} \gamma_k^{\varpi} \tilde{\mathfrak{T}}_k^{\varpi} + p_c}, \tag{8.56}$$

for γ_k^{ϖ}. Figure 8.5 illustrates the characteristic of strict quasi-concavity for the EE utility function described in Equation 8.53 for one-hop and two-hop users, indicating the optimum points obtained when solving Equation 8.56 for both cases. Since the utility function for non-cooperative users is the same as that for one-hop users, this curve is omitted.

As one can see from Equation 8.56, the optimum SINR for EE $\left(\gamma_{k,\text{EE}}^*\right)$ depends on the adopted penalty value ς, as well as on the MAI power level, since the $\tilde{\mathfrak{T}}_k$ term appears at both equations. This behavior is due to the circuitry power consumption p_c contribution on the EE calculation: since

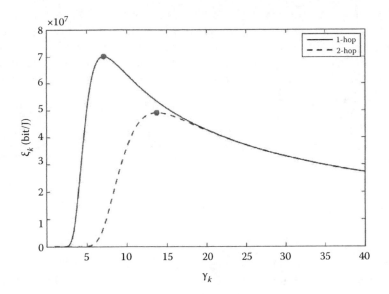

FIGURE 8.5 EE utility function ξ_k as a function of SINR for one-hop users with $\varsigma = 1$ and two-hop users with $\varsigma = 0.5$. The dot represents the optimum EE point found via Equation 8.56 for both cases.

all MTs waste a fixed amount of power to transmit, the allocated power is increased to compensate this cost (Betz and Poor 2008).

When $p_c \rightarrow 0$ or $\varrho_{MT} \gamma_k^{\varpi} \tilde{\mathfrak{T}}_k^{\varpi} \gg p_c$, the optimum SINR tends to converge to a fixed value, which depends on the following system parameters: packet size M, SINR gap ι, efficiency penalty ς, PA efficiency ϱ_{MT}, and circuit power p_c. Moreover, when the interference level is low, more power is allocated to compensate the fixed cost, increasing the optimum SINR. Figure 8.6 shows the optimum values $\gamma_{k,EE}^*$ for different levels of $\tilde{\mathfrak{T}}_k$. Note that $\gamma_{k,EE}^*$ tends to converge to a fixed point when $\tilde{\mathfrak{T}}_k \geq 10^{-2}$; that is, it achieves the values $\gamma_{k,EE}^* \approx 6.86$ for one-hop users and $\gamma_{k,EE}^* \approx 13.44$ for

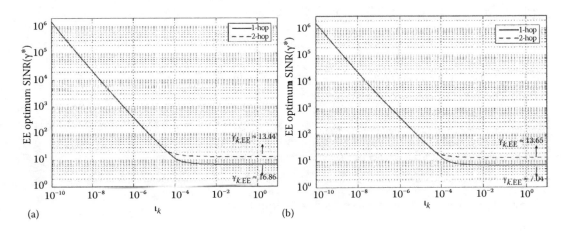

FIGURE 8.6 Optimum SINR for EE for one-hop and two-hop users, with (a) $\iota_k = 1$ and (b) $\iota_k = 0.283$.

two-hop users when ι_k is not considered ($\iota_k = 1$, Figure 8.6a) and $\gamma^*_{k,EE} \approx 7.04$ for one-hop users and $\gamma^*_{k,EE} \approx 13.65$ for two-hop users when $\iota_k = 0.283$ (Figure 8.6b). These values are the same values found when setting $p_c = 0$, and it is possible to conclude that ι_k has a minimum increasing impact in the optimum SINR value. The difference between one-hop and two-hop users is caused by the penalty inserted with ς for two-hop users. When the optimum SINR becomes higher, the difference tends to disappear since $\left(1 - e^{-\gamma_k}\right)^M \approx \left(1 - e^{-\varsigma\gamma_k}\right)^M$.

An important concept when using game theory is to verify if the proposed game achieves Nash equilibrium and if the achieved equilibrium is unique. The concept of Nash equilibrium is given by the following definition, and the existence of the Nash equilibrium of the games defined in Equations 8.45 and 8.46 is summarized in Theorem 1.1. ∎

Definition 2: (Nash equilibrium)

An equilibrium is said to be a Nash equilibrium if and only if any user cannot unilaterally improve their response by changing the optimum value (Fudenberg and Tirole 1991). In the context of the EE problem, this statement is equivalent to the fact that any user cannot improve their utility value by changing the optimum power for any other value:

$$u_k\left(p_k^*, \boldsymbol{p}_{-k}\right) \geq u_k(p_k, \boldsymbol{p}_{-k}), \quad \forall k. \tag{8.57}$$

Theorem 1.1:

The proposed game achieves at least one equilibrium $\boldsymbol{p}^* = \left[p_1^*, p_2^*, \ldots, p_k^*, \ldots, p_K^*\right]$, where p_k^* is defined by the following condition:

$$p_k^* = \min\left(\gamma^*_{k,EE} \tilde{\mathfrak{T}}_k, P_{\max}\right).$$

Furthermore, the achieved equilibrium $\boldsymbol{p}^*$ is also unique.

Proof: See Appendix C.

Another important point is the Pareto optimality, given in Definition 3. As discussed in Betz and Poor (2008), the Nash equilibrium obtained with the MF detector is not Pareto optimal, while for the DEC, the equilibrium is Pareto optimal. This result occurs because when DEC is used, the SINR does not depend on the power of the other users, while when the MF detector is used, the SINR is affected by other users; that is, the selfish behavior of non-coalitional games affect the EE, and the EE can be improved if all users decrease their transmission power (Betz and Poor 2008). Such a result can also be seen in Saraydar et al. (2002), where pricing techniques are considered, resulting in a Pareto-dominant technique over the non-pricing strategy. ∎

Definition 3: (Pareto dominance and Pareto optimal)

A vector $\hat{\boldsymbol{p}}$ is Pareto dominant over another vector $\boldsymbol{p}$ if $u_k(\hat{\boldsymbol{p}}) \geq u_k(\boldsymbol{p}) \ \forall k \in \mathcal{K}$ and for some $k \in \mathcal{K}$, $u_k(\hat{\boldsymbol{p}}) > u_k(\boldsymbol{p})$, where $\mathcal{K}$ is the set of players. A vector $\boldsymbol{p}^*$ is said to be Pareto optimal if for any vector $\boldsymbol{p} \neq \boldsymbol{p}^*$, $u_k(\hat{\boldsymbol{p}}) \geq u_k(\boldsymbol{p}) \ \forall k \in \mathcal{K}$ and for some $k \in \mathcal{K}$, $u_k(\boldsymbol{p}^*) > u_k(\boldsymbol{p})$.

8.5.2 Power Control Algorithm

The Verhulst-based distributed power control algorithm (Gross et al. 2011) has been deployed for allocating power to all users. The Verhulst mathematical model was first idealized to describe population dynamics on the basis of food and space limitation. In Gross et al. (2011), that model was adapted to single-rate DS/CDMA distributed power control using the following discrete iterative convergent equation:

$$p_k[n+1] = (1+\alpha)p_k[n] - \alpha \left[\frac{\gamma_k[n]}{\gamma_k^*} \right] p_k[n], \ k = 1,\dots,K, \tag{8.58}$$

where $p_k[n + 1]$ is the *kth* user power at the $n + 1$ iteration, bounded by $P_{min} \le p_k[n + 1] \le P_{max}$; the factor $\alpha \in [0, 1]$ is the Verhulst convergence factor, $\gamma_k[n]$ is the kth user's SINR at iteration n, and γ_k^* is the target SINR for the kth user.

For each user, the SINR that maximizes EE $\left(\gamma_{k,EE}^* \right)$ is calculated, which is used as the target SINR for the Verhulst-based PCA. The procedure is summarized by Algorithm 2.

Algorithm 2: EE with Verhulst-based PCA

Initialization: $i \leftarrow 1$, It, $p_k[0] = \sigma^2 \ \forall k$;

    while $i \le$ It

        For $k = 1{:}K$

            Obtain $\tilde{\mathfrak{T}}_k^{\varpi}$ via SINR measurement;

            Calculate $\gamma_{k,EE}^*$ by solving Equation 8.56;

            Find p_k^* iteratively using Verhulst-based PCA (Gross et al. 2011);

        end;

        $i = i + 1$

    end;

Output: $p_k^* \ \forall k$ (EE optimum solution)

8.6 NUMERICAL RESULTS

This section provides numerical evidences for the EE gains achieved with fixed RS-aided cooperative CDMA systems. Those numerical results were obtained as average values over 2000 realizations employing the Monte Carlo simulation method. Table 8.1 lists the main system and channel parameters deployed across this section. Details of path-loss models are presented in Appendix B. In the following, we will analyze the numerical values for EE, APC, and outage probability achieved with nco and coo modes under different system loading $\mathcal{L}$, and target SE ζ_k^*, as well.

8.6.1 APC and EE for MF Filter

Figure 8.7 shows APC, in watts per square kilometer, for cooperative and non-cooperative modes, considering MF detection, relative to the number of RSs. It can be concluded that, for low SE target, it is possible to achieve APC savings when using the cooperative mode. For example, when $\mathcal{L} = 1$ and $\zeta_k^* = 0.5$ bps/Hz, the APC gain is about 23.8% for $N_{RS} = 12$ and $r_1 = 0.575r_O$. Besides this interesting result, when the SE target is increased, the necessity of doubling ζ_k^* for each user decreases APC gains (or even makes it impossible to achieve any APC gain). When $\zeta_k^* = 1.25$ bps/Hz (Figure 8.7c), it is impossible to achieve any gain in APC because of the increase in MAI power level, and for $\mathcal{L} = 1$, no topology satisfies the constraints of the problem defined in Equation 8.32.

TABLE 8.1
Adopted Parameters for the Fixed Relay CDMA System, Channel, and Power Control Algorithm

Parameters	Adopted Values
Macrocell DS-CDMA System Parameters	
Noise power	-117 [dBm]
Spreading codes	Pseudo-noise (PN), $N = 256$
External cell radius	$r_O = 2$ [km]
Bands, $F_1 \cap F_2 = \varnothing$	F_1: MT–RS and MT–BS links, F_2: RS–BS links $w_c = 10^6$ [Hz]
SE target	$\zeta_k^* \in \{0.5, 0.9, 1.25\}$ [bps/Hz]
Carrier center frequency	$f_c = 5$ [GHz]
BS antenna height	$h_{BS} = 30$ [m]
FRSs	
No. of relaying stations	$N_{RS} \in [3, 20]$
Relay placement radius	$r_I \in [0.04, 0.9] \, r_O$ [km]
Maximum transmit power	$P_{R_i}^{max} = 20$ [dBW]
Circuit power	$P_{c,R_i} = 0.78$ [W][a]
PA inefficiency	$\varrho_{RS} = 3.3$
Antenna height	$h_{RS} = 12.5$ [m]
MTs	
System loading	$\mathcal{L} = K/N \in [0.5, 1]$
Spatial distribution	$\mathcal{U}(d_{min}, r_O), \; \mathcal{U}(0, 2\pi)$
Max. power per MT user	$P_{max} = 30$ [dBm]
PA inefficiency	$\varrho_{RS} = 2.5$
Circuit power	$p_c = 7$ [dBm]
Antenna height	$h_{MT} = 1.5$ [m]
Channel Gain (WINNER II Channel Models—Part I Channel Models 2008)	
MT–BS links	NLoS C2 scenario, $d_{min} = 35$ [m]
MT–RS links	Mixed LoS and NLoS C2 scenario, $d_{min} = 10$ [m]
RS–BS links	LoS B5a scenario, $d_{min} = 30$ [m]
Monte Carlo Simulation	
No. of realizations	2000 trials
Energy Efficiency	
Packet size	$M = 80$ [bits]
No. of information bits per packet	$L = 50$ [bits]
SINR gap	$\iota_k = 1$
Efficiency penalty	$\varsigma^{2H} = 0.5$

[a] The relay circuit power is equal to $P_{c,R_i} = 0.78$ W since each relay station consumes 0.39 W per channel (Dohler and Li 2010), and the RSs operate in both F_1 and F_2 bands.

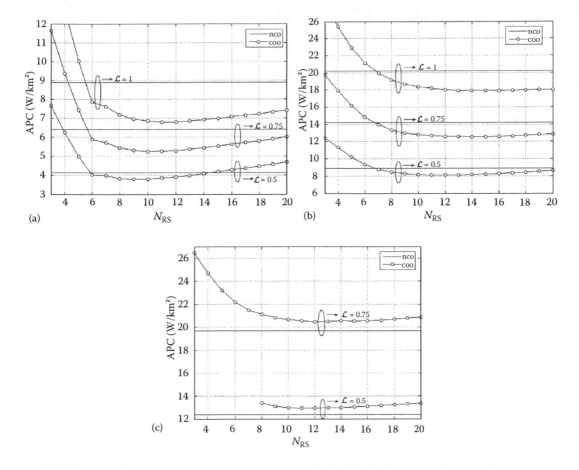

FIGURE 8.7 APC for coo and nco modes with MF detection relative to the number of RSs. $\mathcal{L} = [0.5,\ 0.75,\ 1]$, $N = 256$, $r_1 \in [0.04, 0.9]$ r_0, and $\xi_k^* \in$ [(a) 0.5, (b) 0.9, (c) 1.25] bps/Hz.

The optimized parameters and respective results are summarized in Table 8.2. Note that when system loading increases, the optimum placement radius r_1 is reduced and the number or RSs is increased, while when the SE target is increased, r_1 is reduced. The reduction of r_1 is explained by the one-hop users: as the only effect caused by the insertion of RSs is a possible reduction in the MAI power level, the necessity of doubling the SE target results in higher power consumption for one-hop users, demanding that more users are served by the RSs, which translates into placement radius reduction.

To analyze the impact of noise amplification and interfering area (i.e., interfering users) coverage into RS transmission power consumption, it is decomposed into three terms, amplified noise (Equation 8.13 (III)), interfering users' coverage (Equation 8.13 (II)), and interest users' coverage (Equation 8.13 (I)), described in Table 8.3. It is possible to note that noise amplification is the biggest cost to the RS transmission power consumption, corresponding to at least 57% of total power and becoming close to 95% when $\zeta_k^* = 0.5$ bps/Hz and $\mathcal{L} = 0.5$. When interference increases, with higher SE or system loading, the power consumption with MTs tends to grow, but the power wasted with interfering users also increases, degrading system performance. This way, some technique of interference cancelation or noise reduction is necessary to reduce APC.

Given the topologies that minimize APC, the EE maximization problem described by Algorithm 2 is applied, and the obtained results, considering global EE, are described in Table 8.4. Note that

TABLE 8.2

Optimized Parameters and Respective Results*

	$\zeta_k^* = 0.5$ bps/Hz					
	$\mathcal{L} = 0.5$		$\mathcal{L} = 0.75$		$\mathcal{L} = 1$	
Parameters	nco	coo	nco	coo	nco	coo
APC (saving)	4.122	3.802 (0.320)	6.410	5.259 (1.151)	8.904	6.791 (2.113)
Radius (relative)	–	1.15 (0.575 r_O)	–	1.15 (0.575 r_O)	–	1.15 (0.575 r_O)
No. of RSs	–	10	–	10	–	12
Outage probability	0	0	0	0	0.001	0

	$\zeta_k^* = 0.75$ bps/Hz					
	$\mathcal{L} = 0.5$		$\mathcal{L} = 0.75$		$\mathcal{L} = 1$	
Parameters	nco	coo	nco	coo	nco	coo
APC (saving)	8.913	8.114 (0.799)	12.223	12.530 (1.693)	20.209	17.867 (2.342)
Radius (relative)	–	1.10 (0.550 r_O)	–	1.05 (0.525 r_O)	–	1.00 (0.500 r_O)
No. of RSs	–	11	–	14	–	16
Outage probability	0.065	0.025	0.089	0.038	0.115	0.064

	$\zeta_k^* = 1.25$ bps/Hz					
	$\mathcal{L} = 0.5$		$\mathcal{L} = 0.75$		$\mathcal{L} = 1$	
Parameters	nco	coo	nco	coo	nco	coo
APC (saving)	12.388	12.928 (–0.54)	19.675	20.604 (–0.929)	27.858	–
Radius (relative)	–	0.90 (0.450 r_O)	–	0.85 (0.425 r_O)	–	–
No. of RSs	–	11	–	12	–	–
Outage probability	0.213	0.199	0.244	0.244	0.278	–

*Minimum APC (in parentheses, $APC_{nco} - APC_{coo}$) given in W/km². Optimum radius r_1 (in parentheses, relative distance to cell radius r_O) in km. Optimum Number of RSs N_{RS}^*, and outage probability given system loading $\mathcal{L}$ and SE target (ζ_k^*) for MF.

the gain in EE obtained with the cooperative mode is elevated when compared with the nco mode, which is near 290% when $\zeta_k^* = 0.9$ bps/Hz and $\mathcal{L} = 1$, and the outage probability is also significantly reduced. As the cooperative system uses two RF channels, we introduce the EE normalized by bandwidth, measured in [bits/J · Hz]. Even if the obtained EE is normalized, the cooperative mode obtained a minimum gain of 44% for $\zeta_k^* = 0.5$ bps/Hz and $\mathcal{L} = 0.5$ and a maximum gain of 94% when $\zeta_k^* = 0.9$ bps/Hz and $\mathcal{L} = 1$, reinforcing the benefits of the cooperative mode to increase EE for the whole system.

Comparing the outage probability in Tables 8.2 and 8.4, it is possible to note that it increases when the EE is maximized. This occurs given the higher operational point of EE, as the SINR that maximizes EE is bigger than the necessary SINR to achieve ζ_k^*. If the minimum SINR to achieve ζ_k^* was higher than that obtained with optimum EE, this result will be inverted.

As one can see from Table 8.4, the obtained EE is the same for the non-cooperative mode for any SE target ζ_k^*. This occurs because ζ_k^* is not considered at the EE maximization problem; that is, there are no SE constraint. For the cooperative mode, since the topologies obtained with APC minimization are different, the EE presents a small variation.

In order to measure the RS transmission power components after EE maximization, Table 8.5 brings the same analysis made in Table 8.3. As the EE optimum SINR is higher than what is necessary to achieve any SE target ζ_k^*, the relay power consumption with interest users is higher after

TABLE 8.3

Components of RS Power Consumption for Optimized Topologies Described in Table 8.2 with MF Filter

Parameters	$\mathcal{L} = 0.5$	$\mathcal{L} = 0.75$	$\mathcal{L} = 1$
	$\zeta_k^* = 0.5$ bps/Hz		
Total power [W]	4.953	5.076	6.166
Noise (%)	4.718 (95.239)	4.718 (92.944)	5.661 (91.817)
Interfering (%)	0.052 (1.057)	0.080 (1.569)	0.132 (2.141)
Interest (%)	0.183 (3.704)	0.278 (5.587)	0.373 (6.042)
	$\zeta_k^* = 0.9$ bps/Hz		
Parameters	$\mathcal{L} = 0.5$	$\mathcal{L} = 0.75$	$\mathcal{L} = 1$
Total power [W]	5.285	6.353	6.904
Noise (%)	4.675 (88.452)	5.334 (83.951)	5.435 (78.720)
Interfering (%)	0.159 (3.018)	0.361 (5.689)	0.623 (9.020)
Interest (%)	0.451 (8.530)	0.658 (10.360)	0.846 (12.260)
	$\zeta_k^* = 1.25$ bps/Hz		
Parameters	$\mathcal{L} = 0.5$	$\mathcal{L} = 0.75$	$\mathcal{L} = 1$
Total power [W]	3.775	4.082	–
Noise (%)	2.917 (77.274)	2.783 (68.162)	–
Interfering (%)	0.269 (7.137)	0.463 (11.353)	–
Interest (%)	0.588 (15.589)	0.836 (20.485)	–

TABLE 8.4

Energy Efficiency (EE) [bits/J], EE Normalized by Bandwidth [bits/J.Hz] and Outage Probability for Cooperative and Non-Cooperative Modes with MF Filter

	$\zeta_k^* = 0.5$ bps/Hz					
	$\mathcal{L} = 0.5$		$\mathcal{L} = 0.75$		$\mathcal{L} = 1$	
Parameters	nco	coo	nco	coo	nco	coo
Energy efficiency	3.759×10^5	1.084×10^6	2.886×10^5	9.243×10^5	2.297×10^5	8.312×10^5
Normalized EE	0.376	0.542	0.288	0.462	0.229	0.415
Outage probability	0.189	0.046	0.332	0.121	0.441	0.202
	$\zeta_k^* = 0.75$ bps/Hz					
	$\mathcal{L} = 0.5$		$\mathcal{L} = 0.75$		$\mathcal{L} = 1$	
Parameters	nco	coo	nco	coo	nco	coo
Energy efficiency	3.759×10^5	1.084×10^6	2.886×10^5	9.989×10^5	2.297×10^5	8.950×10^5
Normalized EE	0.376	0.558	0.288	0.499	0.229	0.447
Outage probability	0.354	0.169	0.469	0.252	0.557	0.324
	$\zeta_k^* = 1.25$ bps/Hz					
	$\mathcal{L} = 0.5$		$\mathcal{L} = 0.75$		$\mathcal{L} = 1$	
Parameters	nco	coo	nco	coo	nco	coo
Energy efficiency	3.759×10^5	1.084×10^6	2.886×10^5	1.007×10^6	2.297×10^5	–
Normalized EE	0.376	0.575	0.288	0.503	0.229	–
Outage probability	0.443	0.309	0.541	0.389	0.618	–

TABLE 8.5

Components of RS Power Consumption after EE Maximization with MF Filter

	$\zeta_k^* = 0.5$ bps/Hz		
Parameters	$\mathcal{L} = 0.5$	$\mathcal{L} = 0.75$	$\mathcal{L} = 1$
Total power [W]	10.670	15.088	20.974
Noise (%)	4.718 (44.214)	4.718 (31.269)	5.661 (26.993)
Interfering (%)	0.478 (4.485)	0.771 (5.110)	1.486 (7.087)
Interest (%)	5.474 (51.301)	9.599 (63.621)	13.827 (65.920)
	$\zeta_k^* = 0.9$ bps/Hz		
Parameters	$\mathcal{L} = 0.5$	$\mathcal{L} = 0.75$	$\mathcal{L} = 1$
Total power [W]	10.553	15.590	20.517
Noise (%)	4.675 (44.300)	5.334 (34.213)	5.435 (26.492)
Interfering (%)	0.565 (5.358)	1.307 (8.385)	2.256 (10.993)
Interest (%)	5.313 (50.342)	8.949 (57.402)	12.826 (62.515)
	$\zeta_k^* = 1.25$ bps/Hz		
Parameters	$\mathcal{L} = 0.5$	$\mathcal{L} = 0.75$	$\mathcal{L} = 1$
Total power [W]	7.421	9.491	–
Noise (%)	2.917 (39.310)	2.783 (29.314)	–
Interfering (%)	0.520 (7.005)	0.931 (9.817)	–
Interest (%)	3.984 (53.685)	5.777 (60.869)	–

EE maximization, consuming up to 60% of the total power with these users. However, the power wasted by the RSs with noise amplification and interfering users is still high, in the range of 38% to 50%, which implies EE reduction.

8.6.2 APC AND EE FOR DEC FILTER

Figure 8.7 and Table 8.6 show that for low SE target, the DEC filter presents marginal degradation in terms of APC when compared to MF detection. This behavior is caused by the low MAI power level in the system, since its cancellation results in a higher increment of the background noise power for DEC. This can be seen looking to the results of cooperative mode, when it is necessary to double ζ_k^*. However, for the interest cases of SE, that is, $\zeta_k^* = 0.9$ or $\zeta_k^* = 1.25$ bps/Hz, APC is reduced when deploying the cooperative mode and DEC, obtaining save of about 5.4 W/km^2 (22%) for $\zeta_k^* = 0.9$ bps/Hz and $\mathcal{L} = 1$. Given the robustness of DEC relative to MAI power level and the near–far effect, it is possible to obtain cooperative topologies that minimize APC even for higher ζ_k^* and $\mathcal{L}$, which was impossible when deploying MF detection.

The optimum topologies and their parameters about APC and outage probability are summarized in Table 8.6.

The components of RS transmission power consumption are shown in Table 8.7. As occurred to the MF detector, after APC minimization, the major component is the noise amplification term, responding for at least 66% of the total power consumption. Even when this term is discarded, the power wasted with interfering users is about 25% to 50% of the power spent with interest users, showing the impact of the multiple-covered areas in the two-hop area. In the best case ($\zeta_k^* = 1.25$ bps/Hz and $\mathcal{L} = 1$), the power consumed by the RSs with interest users is about 21% of the total power, demonstrating that the AF protocol is highly inefficient in the presence of interference/noise.

Looking to the EE performance when deploying the DEC detector, described in Table 8.8, one realizes that the gain is considerable: about 40% when $\zeta_k^* = 1.25$ bps/Hz and $\mathcal{L} = 0.5$ and about

TABLE 8.6

Optimized Topologies and Parameters*

	$\zeta_k^* = 0.5$ bps/Hz					
	$\mathcal{L} = 0.5$		$\mathcal{L} = 0.75$		$\mathcal{L} = 1$	
Parameters	nco	coo	nco	coo	nco	coo
APC (saving)	4.605	3.923 (0.682)	7.658	5.582 (2.076)	11.496	7.463 (4.033)
Radius (relative)	–	1.15 (0.575 r_O)	–	1.15 (0.575 r_O)	–	1.15 (0.575 r_O)
No. of RSs	–	9	–	11	–	12
Outage probability	0	0	5.989×10^{-5}	1.302×10^{-5}	0.001	0

	$\zeta_k^* = 0.75$ bps/Hz					
	$\mathcal{L} = 0.5$		$\mathcal{L} = 0.75$		$\mathcal{L} = 1$	
Parameters	nco	coo	nco	coo	nco	coo
APC (saving)	9.083	7.687 (1.396)	14.715	11.423 (3.292)	21.139	15.688 (5.451)
Radius (relative)	–	1.15 (0.575 r_O)	–	1.10 (0.550 r_O)	–	1.10 (0.550 r_O)
No. of RSs	–	11	–	13	–	14
Outage probability	0.072	0.013	0.103	0.023	0.135	0.023

	$\zeta_k^* = 1.25$ bps/Hz					
	$\mathcal{L} = 0.5$		$\mathcal{L} = 0.75$		$\mathcal{L} = 1$	
Parameters	nco	coo	nco	nco	coo	nco
APC (saving)	12.041	11.562 (0.479)	19.046	17.368 (1.678)	26.626	23.743 (2.883)
Radius (relative)	–	1.00 (0.500 r_O)	–	0.95 (0.475 r_O)	–	0.95 (0.475 r_O)
No. of RSs	–	11	–	12	–	14
Outage probability	0.198	0.133	0.225	0.160	0.252	0.164

*Minimum APC (in parentheses, $APC_{nco} - APC_{coo}$) given in W/km². Optimum radius r_1 (in parentheses, relative distance to cell radius r_O) in km. Optimum number of RSs N_{RS}^*, and outage probability given system loading $\mathcal{L}$ and SE target (n_k^*) for DEC.

85% for $\zeta_k^* = 0.5$ bps/Hz and $\mathcal{L} = 1$ when comparing the cooperative mode for both filters, and about 75% for $\mathcal{L} = 0.5$ and 150% for $\mathcal{L} = 1$ when comparing the non-cooperative mode of the detectors. When the non-cooperative mode with MF and the cooperative mode with DEC are compared (Figures 8.8 through 8.10), the gain becomes even higher, about 307% for $\zeta_k^* = 0.5$ bps/Hz and $\mathcal{L} = 0.5$ and 581% for $\zeta_k^* = 0.9$ bps/Hz and $\mathcal{L} = 1$. When comparing the normalized EE, as shown in Figure 8.10, the obtained gains varies from 103% for $\zeta_k^* = 0.5$ bps/Hz and $\mathcal{L} = 0.5$ to 241% for $\zeta_k^* = 0.9$ bps/Hz and $\mathcal{L} = 1$. In terms of outage probability, the combination of filter optimization and cooperative mode provides a considerable improvement when compared to MF. These results demonstrate the impact of cooperative mode and the necessity of filter optimization to mitigate MAI and improve QoS maintenance, which results in higher EE.

Finally, the analysis of the RS transmission power components after EE optimization, described in Table 8.9, leads to the same conclusions obtained when using MF detection. Besides the increase in percentage consumption with interest users when compared to Table 8.7, which is in the range of 45% to 59%, there is significant power wasting with noise amplification and interfering users. It is worth noting that this wasted power is even bigger (in percentage) with DEC. As this filter reduces the MAI power level, all users allocate less power to maximize EE, and given the fact that the noise term is constant, more power is wasted in terms of percentage.

TABLE 8.7

Components of RS Power Consumption for Optimized Topologies Described in Table 8.6 with DEC Filter

Parameters	$\zeta_k^* = 0.5$ bps/Hz $\mathcal{L} = 0.5$	$\mathcal{L} = 0.75$	$\mathcal{L} = 1$
Total power [W]	4.483	5.571	6.196
Noise (%)	4.246 (94.714)	5.189 (93.144)	5.661 (91.367)
Interfering (%)	0.048 (1.065)	0.093 (1.670)	0.140 (2.259)
Interest (%)	0.189 (4.221)	0.289 (5.186)	0.395 (6.374)
Parameters	$\zeta_k^* = 0.9$ bps/Hz $\mathcal{L} = 0.5$	$\mathcal{L} = 0.75$	$\mathcal{L} = 1$
Total power [W]	5.799	6.519	7.347
Noise (%)	5.189 (89.491)	5.525 (84.740)	5.950 (80.985)
Interfering (%)	0.147 (2.536)	0.306 (4.696)	0.458 (6.238)
Interest (%)	0.463 (7.973)	0.688 (10.564)	0.939 (12.777)
Parameters	$\zeta_k^* = 1.25$ bps/Hz $\mathcal{L} = 0.5$	$\mathcal{L} = 0.75$	$\mathcal{L} = 1$
Total power [W]	4.686	5.041	6.349
Noise (%)	3.737 (79.734)	3.614 (71.681)	4.216 (66.397)
Interfering (%)	0.267 (5.708)	0.463 (9.187)	0.798 (12.577)
Interest (%)	0.682 (14.558)	0.964 (19.132)	1.335 (21.026)

TABLE 8.8

Energy Efficiency (EE) [bits/J], EE Normalized by Bandwidth [bits/J.Hz], and Outage Probability for Cooperative and Non-Cooperative Modes with DEC Filter

	$\zeta_k^* = 0.5$ bps/Hz					
	$\mathcal{L} = 0.5$		$\mathcal{L} = 0.75$		$\mathcal{L} = 1$	
Parameters	nco	coo	nco	coo	nco	coo
Energy efficiency	6.630×10^5	1.532×10^6	6.303×10^5	1.561×10^6	5.949×10^5	1.536×10^6
Normalized EE	0.663	0.766	0.630	0.780	0.594	0.768
Outage probability	0	5.156×10^{-4}	4.687×10^{-5}	1.302×10^{-5}	0.001	0
	$\zeta_k^* = 0.75$ bps/Hz					
	$\mathcal{L} = 0.5$		$\mathcal{L} = 0.75$		$\mathcal{L} = 1$	
Parameters	nco	coo	nco	coo	nco	coo
Energy efficiency	6.630×10^5	1.579×10^6	6.303×10^5	1.595×10^6	5.949×10^5	1.565×10^6
Normalized EE	0.663	0.789	0.630	0.797	0.594	0.782
Outage probability	0.074	0.014	0.104	0.023	0.135	0.022
	$\zeta_k^* = 1.25$ bps/Hz					
	$\mathcal{L} = 0.5$		$\mathcal{L} = 0.75$		$\mathcal{L} = 1$	
Parameters	nco	coo	nco	coo	nco	coo
Energy efficiency	6.630×10^5	6.606×10^6	6.303×10^5	1.588×10^6	5.949×10^5	1.562×10^6
Normalized EE	0.663	0.803	0.630	0.794	0.594	0.781
Outage probability	0.198	0.133	0.224	0.161	0.253	0.165

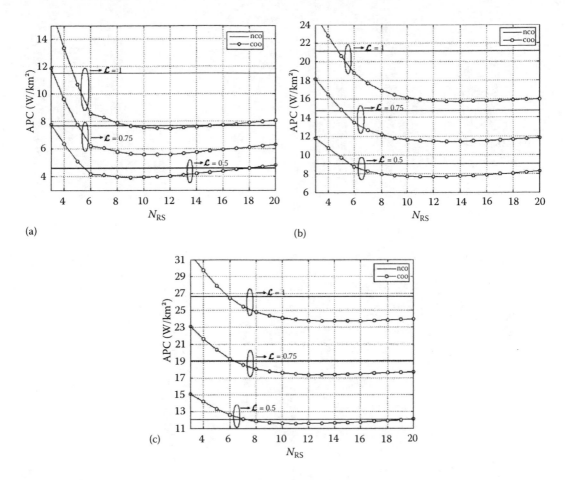

FIGURE 8.8   APC for coo and nco modes with DEC detection relative to the number of RSs. $\mathcal{L} = [0.5, \ 0.75, \ 1]$, $N = 256$, $r_I \in [0.04, 0.9] \ r_O$ and $\xi_k^* \in [$(a) 0.5, (b) 0.9, (c) 1.25$]$ bps/Hz.

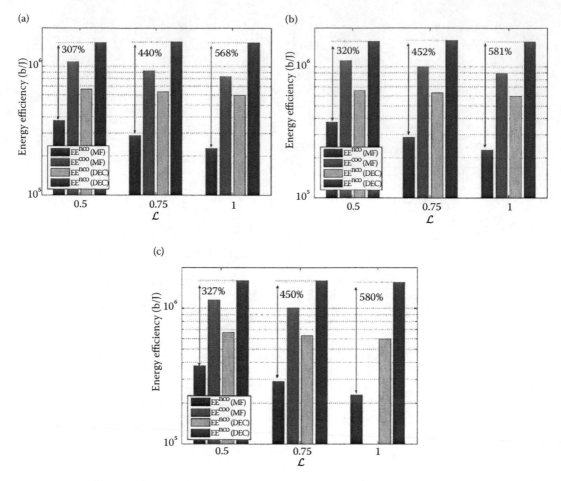

FIGURE 8.9 Energy efficiency for cooperative and non-cooperative modes, for MF and DEC filters. $\mathcal{L} = [0.5, 0.75, 1]$, $N = 256$, $r_1 \in [0.04, 0.9]$ r_O and $\xi_k^* \in$ [(a) 0.5, (b) 0.9, (c) 1.25] bps/Hz.

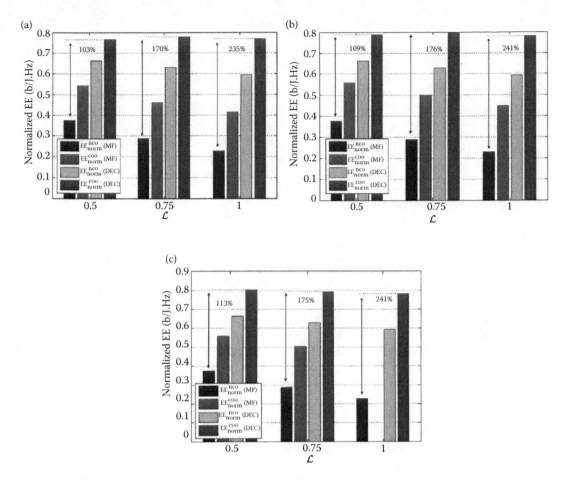

FIGURE 8.10 Normalized EE for cooperative and non-cooperative modes, for MF and DEC filters. $\mathcal{L} = [0.5, 0.75, 1]$, $N = 256$, $r_1 \in [0.04, 0.9]$ r_O and $\xi_k^* \in [$(a) 0.5, (b) 0.9, (c) 1.25] bps/Hz.

TABLE 8.9

Components of RS Power Consumption after EE Maximization with DEC Filter

Parameters	$\zeta_k^* = 0.5$ bps/Hz		
	$\mathcal{L} = 0.5$	$\mathcal{L} = 0.75$	$\mathcal{L} = 1$
Total power [W]	8.912	12.866	16.636
Noise (%)	4.246 (47.643)	5.189 (40.336)	5.661 (34.031)
Interfering (%)	0.353 (3.968)	0.741 (5.757)	1.157 (6.955)
Interest (%)	4.313 (48.389)	6.936 (53.907)	9.818 (59.014)

Parameters	$\zeta_k^* = 0.9$ bps/Hz		
	$\mathcal{L} = 0.5$	$\mathcal{L} = 0.75$	$\mathcal{L} = 1$
Total power [W]	10.372	13.765	17.634
Noise (%)	5.189 (50.036)	5.525 (40.136)	5.950 (33.741)
Interfering (%)	0.490 (4.716)	0.974 (7.075)	1.474 (8.358)
Interest (%)	4.693 (45.248)	7.266 (52.789)	10.210 (57.901)

Parameters	$\zeta_k^* = 1.25$ bps/Hz		
	$\mathcal{L} = 0.5$	$\mathcal{L} = 0.75$	$\mathcal{L} = 1$
Total power [W]	8.093	10.223	13.827
Noise (%)	3.737 (46.173)	3.614 (35.345)	4.216 (30.489)
Interfering (%)	0.481 (5.947)	0.819 (8.015)	1.411 (10.204)
Interest (%)	3.875 (47.880)	5.790 (56.640)	8.200 (59.307)

8.7 CONCLUSIONS

This chapter presents an out-of-band cooperative model for DS-CDMA systems and analyzes the problem of optimum placement of RSs and EE maximization. In terms of APC, the deployment of cooperative networks and filter optimization presented a maximum gain of 20% when compared with non-cooperative system and SuD, caused by the necessity of doubling the SE target criterion. When analyzing the EE of the proposed system, the combination of the coo mode and MuD presented high gains, mainly when system loading increases. Even when the extra RF channel bandwidth is considered, the obtained gains are at least 103%, achieving a maximum value of 241%.

Regarding the non-regenerative protocol, it is demonstrated that it wastes considerable power with non-interesting terms, such as interfering users and noise amplification. This way, it is necessary to investigate the gains that could be obtained with regenerative protocols, as for example Decode and Forward protocols. Besides this fact, combining the system model presented in this chapter with regenerative protocols could imply spectral resource waste: as the regenerative protocols must receive the whole package to decode, each one of the frequencies will be in idle state 50% of the time. In order to adopt regenerative protocols, the in-band relaying is the best choice.

Finally, splitting the placement/EE optimization problem into two sub-problems results in a suboptimal solution: as the SINR that maximizes EE is different for one-hop and two-hop users and the SINR is the same for these two class of users for the APC optimization problem, the topology defined in APC minimization could not be the optimum placement for EE maximization. This way, deploying an analytical solution or using techniques as heuristic algorithms to jointly solve placement and EE maximization tends to maximize EE.

8.8 FUTURE RESEARCH

- *Description of a system model to implement regenerative protocols:* As discussed through this chapter, the non-regenerative protocols waste a considerable part of transmission power with noise and interference. This way, adopting regenerative protocols could be a solution, but the out-of-band model considered here must be adapted. Adopting regenerative protocols approximates the system model to the 4G standards, presenting an interesting scenario.

- *Joint solution of the placement EE problem, analytically or via computational simulation:* Since the division of the RSs placement and EE maximization problem in two sub-problems results in a non-optimal solution, solving this problem jointly results in higher EE. This way, it is possible to develop a suboptimal solution (that could be based in heuristic algorithms or system simplifications) or even analytically solve the problem.
- *Impact of cooperative diversity in EE maximization:* For simplicity, only one RS is selected and the direct signal MT–BS is unconsidered. If more than only one relay could be selected and a better relay selection scheme is deployed, it is possible to take advantage of the cooperative diversity, which results in BER reduction and affects the efficiency function and probably results in EE improvement.
- *Impact of different implantation topologies:* Considering another placement than the circular adopted in this chapter could reduce the average distance between MTs and RSs and the multiple coverage areas, improving system performance.

APPENDIX A: CIRCUITRY POWER CONSUMPTION

In the PHY layer, a linear BS power consumption model is adopted (Arnold et al. 2010; Cui et al. 2005), consisting of two parts: static and dynamic power consumption parts. The first part describes the static power consumption, a power figure that is consumed already in an empty BS. Depending on the load situation, a dynamic power consumption part adds to the static power (Arnold et al. 2010). The static and dynamic parts of power consumption at the BS or FRS can be translated to the transmit power P and circuit power P_c.

The overall power consumption at the BS or FRS is given by

$$P_T = \varrho P + P_c, \tag{8.59}$$

where ϱ is a parameter related to the drain efficiency of the PA.

The circuitry consumption is determined by active circuit blocks, such as analog-to-digital converter (ADC), digital-to-analog converter (DAC), synthesizer (syn), mixer (MIX), low PA (LNA), intermediate frequency amplifier (ifa), and transmitter and receiver filters (filt, filr) (Cui et al. 2005). The circuitry power consumption can be decomposed as

$$P_c = 2P_{syn} + P_{mix} + P_{LNA} + P_{filt} + P_{filr} + P_{IFA} P_{ADC} \tag{8.60}$$

For the AF RS, the main subsystem blocks are as follows (Dohler and Li 2010):

- Receive bandpass filter
- Low-noise amplifier
- Synthesizer
- Bandpass filter
- Power amplifier
- Transmit bandpass filter
- Power supply

The last block can be discarded since the RSs do not operate with batteries.* When considering the regenerative protocols, for example, decode-and-forward (DF), other features are added to the RF equipment, such as

- Intermediate frequency stage
- ADC and DAC converter
- Digital signal processing

It is a common assumption that the power dissipation can be satisfactorily modeled as the sum of a *static term* and a *dynamic term,*

$$P_{chip} = V_{dd}I_{leak} + af_{ck}C_{ap}V_{dd}^2, [W]$$

where a is related to the effective fraction of gates switching, f_{ck} is the clock frequency, and C_{ap} is the circuit capacitance. If the frequency is dynamically scaled with the information rate r, it is reasonable to model the power dissipation as a linear function of the rate with a constant offset, as discussed in Schurgers et al. (2001). Thus, as pointed out above, the power dissipation in the RF front end during transmission is dominated by the MIX, syn, filter, and DAC and can be modeled as a rate-independent constant (Cui et al. 2005). On the basis of those assumptions, the circuit power can be modeled as

$$P_c(r) = \alpha + \beta r. \qquad (8.61)$$

Since the main concern in this chapter is relay placement and EE optimization, we can use an average value for circuit power and then discard the dynamic term. As discussed in Dohler and Li (2010), there were two main possibilities to implement the AF protocol: using analog or digital hardware. The advantage of digital equipment is the possibility of half-duplex relaying in time (since the received signal cannot be stored in an analog way). Since we use relaying with a separated frequency band and no digital signal processing is done, we can maintain the analog equipment just for simplicity.[†] As the AF protocol does not insert complexity costs, the circuit power consumption depends only on the receive/transmit RF chain. This way, we can adopt the average circuit power as $P_c = 780$ mW, since the RSs do not have its own data and use two separated bands. To corroborate the elimination of the dynamic part, the scenario created for the analog AF relaying in Dohler and Li (2010) shows that complexity and power consumption are independent of the number and type of relayed traffic, considering UMTS voice/HSPDA traffic.[‡]

APPENDIX B: PATH-LOSS MODELS

There are several models to build the path-loss model for our scenario. For example, there are documents from 3GPP (3GPP-ETSI TR125996, 2007), WiMAX (Senarath et al. 2007) and WINNER (Kyösti et al. 2008) that describes scenarios for urban/suburban macrocell/microcell, consider-

---

* In order to guarantee constant voltage, when the RSs use battery, a voltage stabilizer is needed and consumes power to maintain the voltage level.

† As pointed out by Dohler and Li (2010), the memory/storage increases 2.2 mW in the RF power consumption; thus, adopting this hardware has minimal impact in the final circuit power.

‡ The ADC and DCA power terms have been discarded, since the AF protocol does not use these blocks; they are probably inserted since the relay can transmit and receive its own data. The values in Table 5.9 (p. 350) of Dohler and Li (2010) represent the power consumption for both uplink and downlink operation, considering average PA and transmitted power. The values discussed in this paragraph are presented in Dohler and Li (2010, p. 347).

ing LoS and NLoS cases, reference distances and other parameters, such as shadowing and delay metrics. The following are the basic characteristics of the three links considered in this work:

- **MT–RS:** Urban or suburban scenario, where the MTs are at the street level and the FRSs are located in higher places, like rooftops or lampposts. As the FRS can be near the BS, the macrocell case is most appropriate.
- **MT–BS:** Urban or suburban macrocell, where the MTs are at the street level and the BS is in a higher place.
- **RS–BS:** LoS propagation, since the FRSs are installed in higher places.

On the basis of these basic properties, we first consider the path-loss scenarios described for 802.16j. Looking at Table 2 of Multi-Hop Relay System Evaluation Methodology (Channel Model and Performance Metric) (2007), the best choices are the Type-E model for MT–RS and MT–BS links and Type-H for RS–BS link. Both models are based on the COST 231 Walfish–Ikegami (WI) model and is given, basically, by

$$PL = L_0 + L_{rts} + L_{msd}, \tag{8.62}$$

where L_0 is the free-space path loss, L_{rts} is the rooftop to street diffraction, and L_{msd} is the multi-screen loss. As in the RS–BS link, both antennas are higher than the rooftop height and the rooftop to street diffraction is discarded.

Despite the scenario conformity, these two scenarios cannot be applied to our proposed scenario, since the maximum carrier frequency for the COST 231 WI is about 2 GHz. For $f_c = 5$ GHz (Multi-Hop Relay System Evaluation Methodology (Channel Model and Performance Metric) 2007), consider using as an alternative the WINNER model (WINNER II Channel Models—Part I Channel Models 2008), where the carrier frequency is in the range $2\text{GHz} < f_c < 6$ GHz.

Now looking at Table 2.1 of WINNER II Channel Models—Part I Channel Models (2008), the models chosen to the three links are C2 (MT–RS and MT–BS) and B5a (RS–BS). The C2 model is considered for urban macrocell scenarios, and the LoS and NLoS cases are given, respectively, by

$$PL_{LoS} = \begin{cases} 26\log_{10}(d) + 39 + 20\log_{10}(f_c/5), & 10\text{ m} < d < d'_{BP} \\ 40\log_{10}(d) + 13.47 - 14\log_{10}(h'_{rx}) - 14\log_{10}(h'_{tx}) + 6\log_{10}(f_c/5), & d'_{BP} < d < 5\text{ km} \end{cases} \tag{8.63}$$

$$PL_{NLoS} = [44.9 - 6.55\log_{10}(h_{rx})]\log_{10}(d) + 34.46 + 5.83\log_{10}(h_{rx}) \\ + 23\log_{10}(f_c/5), 50\text{ m} < d < 5\text{ km}, \tag{8.64}$$

where h_{tx} and h_{rx} are, respectively, the transmitter and receiver antenna height in meters, h'_{tx} and h'_{rx} are the transmitter and receiver effective antenna height, given by $h'_{tx} = h_{tx} - 1$ and $h'_{rx} = h_{rx} - 1$, d is the distance between receiver and transmitter in meters, f_c is the carrier frequency in megahertz, and d'_{BP} is the breakpoint distance in meters, reflecting the distance when the scatterers significantly affect the LoS signal, and is given by

$$d'_{BP} = \frac{4h'_{rx}h'_{tx}f_c}{c} \tag{8.65}$$

where c is the propagation velocity in free space.

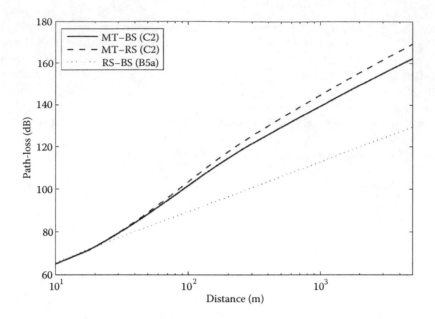

FIGURE 8.11   Path-loss models considered in this work for the three possible links.

It is difficult to determine when an MT has LoS communication without the analysis of the cell area constructions (like streets and buildings) and equipment placement. To adequately model this behavior, we use the LoS probability equation for C2 scenario, presented in Table 4.7 of WINNER II Channel Models—Part I Channel Models (2008) and given by

$$\text{Pr}_{\text{Los}} = \min(18/d, 1)(1 - e^{-d/63}) + (1 - e^{-d/63}) \tag{8.66}$$

and the final path-loss model for the C2 scenario is equivalent to

$$PL_{\text{C2}} = \text{Pr}_{\text{LoS}}PL_{\text{LoS}} + (1 - \text{Pr}_{\text{LoS}})\,PL_{\text{NLoS}}. \tag{8.67}$$

For the RS–BS link, we consider the B5a model only for the LoS case, given the fact that the FRSs are placed in order to guarantee LoS with the BS. The path-loss term is given by

$$PL_{\text{B5a}} = 23.5\log_{10} d + 42.5 + 20\log_{10}(f_c/5),\ 30\ \text{m} < d < 8\ \text{km}. \tag{8.68}$$

Figure 8.11 shows the path-loss term, in dB, for the three different links according to distance d.

APPENDIX C: PROOF OF PROPOSED THEOREMS AND LEMMAS

To simplify the proofs developed herein, we use the generic definition of the EE utility function, given in Equation 8.53 and repeated here for convenience:

$$\xi_k^{\varpi} = \frac{\ell_k \log_2\left(1 + \iota_k\gamma_k^{\varpi}\right)\left(1 - e^{-\varsigma\gamma_k^{\varpi}}\right)^M}{\varrho_{\text{MT}}\gamma_k^{\varpi}\tilde{\mathfrak{L}}_k^{\varpi} + p_c}$$

with

- One-hop and non-cooperative users: $\varsigma = 1$
- Two-hop users: $0 < \varsigma < 1$
- $\ell_k = \dfrac{L}{M}$

This is only used to avoid proving all theorems and lemmas for each utility function.

A. Proof of Lemma 1

In order to demonstrate that the utility function ξ_k^{ϖ} is strict quasi-concave regarding the SINR γ, and for consequence p, given the one-to-one mapping between transmission power and SINR, we first demonstrate that the numerator of the utility function, that is,

$$\text{num}_{EE} = w \log_2 (1 + \iota\gamma)(1 - e^{-\varsigma\gamma})^M$$

is S shaped. As demonstrated in Rodriguez (2003), if the utility function is a ratio $f(x)/x$ and $f(x)$ is sigmoidal, then $f(x)/x$ is strict quasi-concave regarding x. According to Rodriguez (2003), there are six conditions that must be demonstrated to confirm if one function $f(x)$ is sigmoidal:

C1. The domain of the function (x) is the non-negative part of the real line $[0, \infty]$.
C2. The image is given by the interval $[0, B]$, usually with $B = 1$ (without loss of generality, normalization is assumed).
C3. It is increasing in whole domain.
C4. The first derivative is continuous.
C5. Strictly convex in the interval $[0, x_i]$, $x_i \in x$.
C6. Possibly strictly concave in the interval $[x_i, L]$, $L > x_i$.

Condition C1 is obvious, since $\gamma > 0$.

At first, Condition C2 appears to be false, because $\lim\limits_{\gamma \to \infty} \text{num}_{EE} = \infty$, but since our scenario includes power limitation, it is obvious that $\lim\limits_{\gamma \to \gamma_{max}} \text{num}_{EE} = b$, where $b \in \mathbb{R}$ and γ_{max} is the maximum achievable SINR given P_{max}.

Condition C3 is trivial, since as $\left(1 - e^{-\varsigma\gamma_i}\right)^M > \left(1 - e^{-\varsigma\gamma_j}\right)^M$ and $\log_2 (1 + \iota\gamma_i) > \log_2 (1 + \iota\gamma_j)$ for $\gamma_i > \gamma_j$, it is obvious that

$$\ell w \log_2(1 + \iota\gamma_i)\left(1 - e^{-\varsigma\gamma_i}\right)^M > \ell w \log_2(1 + \iota\gamma_j)\left(1 - e^{-\varsigma\gamma_j}\right)^M.$$

Condition C4 is also true, since the first derivative of num_{EE}, given in Equation 8.69, has only one restriction to continuity: $1 + \iota\gamma > 0$. As $0 < \iota < 1$ and $\gamma \geq 0$, $1 + \iota\gamma \geq 1$ $\forall\gamma$ and the first derivative of num_{EE} is continuous in all domain.

$$\frac{\partial \text{num}_{EE}}{\partial \gamma} = \ell w \left[M\varsigma e^{-\varsigma\gamma}(1 - e^{-\varsigma\gamma})^{(M-1)} \log_2(1 + \iota\gamma) + \frac{\iota(1 - e^{-\varsigma\gamma})^M}{(1 + \iota\gamma)\ln 2} \right]. \tag{8.69}$$

In order to demonstrate Conditions C5 and C6, we took the second derivative of the EE numerator. If the inflection point γ_i exists, we try to demonstrate that in the interval $[0, \gamma_i]$, the numerator is strict convex, $\dfrac{\partial^2 \text{num}_{EE}}{\partial \gamma^2} > 0$, and in the interval $[\gamma_i, \gamma_{max}]$, it is strict concave, $\dfrac{\partial^2 \text{num}_{EE}}{\partial \gamma^2} < 0$; if the inflection point does not exist, it is necessary to demonstrate the strict convexity of num_{EE}. Taking the second derivative of num_{EE}, we obtain

$$\frac{\partial^2 num_{EE}}{\partial\gamma^2} = \ell w \left[\frac{(M-1)M\varsigma^2 e^{(-2\varsigma\gamma)}(1-e^{-\varsigma\gamma})^{(M-2)} \ln(1+\iota\gamma)}{\ln 2} \right.$$
$$\left. + \frac{M\varsigma^2 e^{(-\varsigma\gamma)}(1-e^{-\varsigma\gamma})^{(M-1)} \ln(1+\iota\gamma)}{\ln 2} + \frac{2M\iota\varsigma e^{(-\varsigma\gamma)}(1-e^{-\varsigma\gamma})^{(M-1)}}{(1+\iota\gamma)\ln 2} \right]. \quad (8.70)$$

Analyzing Equation 8.70, it is hard to conjecture if any inflection point exists or if the numerator is strict convex. In order to verify the existence of the inflection point, we took $\dfrac{\partial^2 num_{EE}}{\partial\gamma^2} = 0$ and try to find γ_i. After some simplifications, the inflection point can be obtained by solving

$$f_{inf}(\gamma) = (1+\iota\gamma)\ln(1+\iota\gamma)\varsigma(Me^{-\varsigma\gamma}-1) + 2\iota(1-e^{-\varsigma\gamma}) = 0. \quad (8.71)$$

Even with this simplification, it still hard to demonstrate the existence or the non-existence of γ_i, but some considerations can be made to demonstrate the existence of at least one inflection point. Note that $\lim_{\gamma\to 0^+} f_{inf}(\gamma) > 0$, while $\lim_{\gamma\to\infty} f_{inf}(\gamma) < 0$. As $f_{inf}(\gamma)$ is continuous, there must be at least one inflection point γ_i for which $f_{inf}(\gamma) = 0$. As the numerical solution of $f_{inf}(\gamma) = 0$ is hard to obtain when considering all parameters (M, ι and ς), we solve Equation 8.71 considering the system parameters defined in Table 8.1:

- Non-cooperative and one-hop users: $M = 80$, $\iota = 1$, and $\varsigma = 1$
- Two-hop users: $M = 80$, $\iota = 1$, and $\varsigma = 0.5$

The results are shown in Figure 8.12, and the inflection points obtained are as follows:

- Non-cooperative and one-hop users: $\gamma_i \approx 4.611$
- Two-hop users: $\gamma_i \approx 9.133$

To confirm these results, we plot $\dfrac{\partial^2 num_{EE}}{\partial\gamma^2}$ and the obtained inflection points γ_i for all users. As one can note from Figure 8.13, the inflection points obtained with Equation 8.71 correspond to the inflection points of num_{EE}; in the interval $[0, \gamma_i]$, we obtain $\dfrac{\partial^2 num_{EE}}{\partial\gamma^2} > 0$ (except for $\gamma = 0$ and $\gamma = \gamma_i$), and in the interval $[\gamma_i, \gamma_{max}]$, we obtain $\dfrac{\partial^2 num_{EE}}{\partial\gamma^2} < 0$ (except for $\gamma = \gamma_i$). As $\lim_{\gamma\to\infty} \dfrac{\partial^2 num_{EE}}{\partial\gamma^2} = 0, 0$ is a vertical asymptote in the interval $[\gamma_i, \gamma_{max}]$ and the strict concavity holds.

With those numerical evidence, we confirm Conditions C5 and C6, at least for the system parameters. This way, all six conditions are demonstrated, and the numerator of the utility functions described in Equations 8.37 and 8.38 are sigmoidal and, according to Rodriguez (2003), those utility functions are strictly quasi-concave regarding the SINR (or power) term.

B. Proof of Theorem 1.1

The proof presented in this section is based on Buzzi and Saturnino (2011). According to it, the existence of the Nash equilibrium can be guaranteed with three conditions:

C1. The strategy set $\{A_k\}$ is a non-empty, convex, and compact subset of some Euclidean space.
C2. The utility function for the kth user ($u_k(p_k, \boldsymbol{p}_{-k})$) is continuous for any $\boldsymbol{p}_{-k}$ and $p_k \in \{A_k\}$.
C3. The utility function for the kth user is quasi-concave regarding p_k (or γ_k).

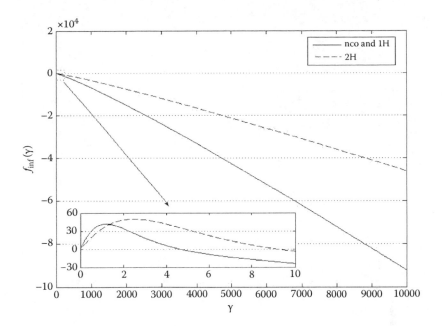

FIGURE 8.12 Plot of $f_{\text{inf}}(\gamma)$. In the detail, SINR interval [0, 10], which includes the inflection points for the considered cases.

Condition C3 is demonstrated in the previous section.

Condition C2 can be proven using the generic definition of $u_k(p_k, \boldsymbol{p}_{-k})$,

$$u_k(p_k, \boldsymbol{p}_{-k}) = \xi_k^{\omega}\left(\gamma_k^{\omega}\right) = \frac{\ell_k \log_2\left(1 + \iota_k \gamma_k^{\omega}\right)\left(1 - e^{-\varsigma\gamma_k^{\omega}}\right)^M}{\varrho_{\text{MT}}\gamma_k^{\omega}\tilde{\mathfrak{T}}_k^{\omega} + p_c}.$$

The first condition is that $\varrho_{\text{MT}}\gamma_k^{\omega}\tilde{\mathfrak{T}}_k^{\omega} + p_c \neq 0$, which is guaranteed since $\varrho_{\text{MT}}\gamma_k^{\omega}\tilde{\mathfrak{T}}_k^{\omega} \geq 0$ and $p_c > 0$. The second condition refers to $1 + \iota_k\gamma_k^{\omega} > 0$, which is also guaranteed since $\iota_k\gamma_k^{\omega} \geq 0$. This way, the continuity of the utility function is demonstrated.

Condition C1 is proven in Buzzi and Saturnino (2011). As the strategy set $\{\mathcal{A}_k\}$ is a segment of the real line, given by $[0, P_{\text{max}}]$, it follows that $\{\mathcal{A}_k\}$ is non-empty, convex, and compact.

This way, the three conditions to the existence of the Nash equilibrium has been demonstrated, and the proposed games achieve at least one Nash equilibrium. As the utility function admits only one maximizer γ_k^* (or p_k^*, given the one-to-one mapping), the achieved equilibrium is also unique (Meshkati et al. 2005).

Now, it is necessary to confirm the possible power levels allocated by the power control algorithm. The kth user power level p_k^* from the power vector $\boldsymbol{p}^*$, obtained after the execution of Algorithm 2, is given by

$$p_k^* = \min\left(\gamma_{k,\text{EE}}^* \tilde{\mathfrak{T}}_k, P_{\text{max}}\right).$$

If the kth user is able to allocate $p_{k,\text{EE}}^* = \gamma_{k,\text{EE}}^* \tilde{\mathfrak{T}}_k$, then $p_k^* = p_{k,\text{EE}}^*$ and ξ_k is maximized, since any other power level implies EE reduction. When $p_{k,\text{EE}}^* > P_{\text{max}}$, the kth user cannot achieve the optimum EE level and then allocates $p_k^* = P_{\text{max}}$. This is caused by the game objective, to maximize ξ_k,

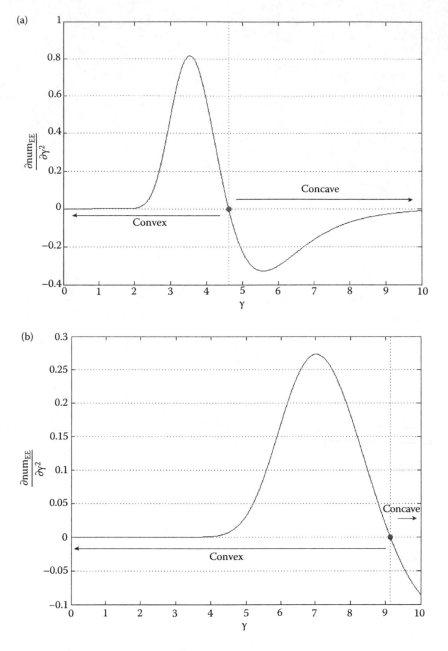

FIGURE 8.13 Plot of the second derivative $\dfrac{\partial^2 \text{num}_{\text{EE}}}{\partial \gamma^2}$ normalized by ℓw for: (a) one-hop and non-cooperative users and (b) two-hop users.

and the quasi-concavity of the utility function ξ_k. Since in the interval $\left[0, p^*_{k,\,\text{EE}}\right]$ ξ_k is strictly increasing, it is obvious that ξ_k is also strictly increasing in the interval $[0, P_{\max}]$ when $p^*_{k,\,\text{EE}} > P_{\max}$. This way, trying to selfishly increase its own utility, the kth user sets $p^*_k = P_{\max}$ and satisfies the game objective.

Thus, we demonstrate the existence of the Nash equilibrium and uniqueness of that equilibrium, as well as the power levels at the equilibrium, and prove Theorem 1.

AUTHOR BIOGRAPHIES

Álvaro Ricieri Castro e Souza received his bachelor's degree (with honors) in 2011 and his master's degree in 2013 from the Londrina State University, Paraná, Brazil. His main research area topics include energy efficiency for multiple access wireless communication systems, cooperative networks, game theory, and characterization of the trade-off between energy and spectral efficiencies.

Taufik Abrão (S-IEEE'12; M-SBrT'98) received his BS, MSc, and PhD degrees, all in electrical engineering, from the Polytechnic School of the University of São Paulo, Brazil, in 1992, 1996, and 2001, respectively. Since March 1997, he has been with the Communications Group, Department of Electrical Engineering, Londrina State University, PR, Brazil, where he is currently an associate professor. He was an academic visitor at CSPC Research Group, University of Southampton, UK (2012), and at TSC of the Polytechnic University of Catalonia, Barcelona, Spain (2007–2008). Dr Abrão has participated in several projects funded by government agencies and industrial companies. He is involved in editorial board activities of several journals and he has served as TCP member in a large number of conferences in the field. His research interests lie in communications and signal processing, including multi-user detection and estimation, MC-CDMA and MIMO systems, cooperative communication and relaying, resource allocation, and heuristic and convex optimization. He has co-authored more than 150 research papers published in specialized/international journals and conferences.

REFERENCES

Alouini, M.-S. and Goldsmith, A. (1999). Area spectral efficiency of cellular mobile radio systems. *IEEE Transactions on Vehicular Technology*, *48* (4), 1047–1066.

Arnold, O., Richter, F., Fettweis, G., and Blume, O. (2010). Power consumption modeling of different base station types in heterogeneous cellular networks. *Future Network and Mobile Summit, 2010*, pp. 1–8.

Badruddin, N. and Negi, R. (2004). Capacity improvement in a CDMA system using relaying. In *WCNC'04— IEEE Wireless Communications and Networking Conference*, *1*, pp. 243–248.

Betz, S. and Poor, H. (2008). Energy efficiency in multi-hop CDMA networks: A game theoretic analysis considering operating costs. In *ICASSP'08—IEEE International Conference on Acoustics, Speech and Signal Processing*, pp. 2781–2784.

Boyd, S. and Vandenberghe, L. (2004). *Convex Optimization*. Cambridge, UK: Cambridge University Press.

Buzzi, S. and Poor, H. (2008). Joint receiver and transmitter optimization for energy-efficient CDMA communications. *IEEE Journal on Selected Areas in Communications*, *26* (3), 459–472.

Buzzi, S. and Saturnino, D. (2011). A game-theoretic approach to energy-efficient power control and receiver design in cognitive CDMA wireless networks. *IEEE Journal of Selected Topics in Signal Processing*, *5* (1), 137–150.

Chen, S., Zhang, X., Liu, F., and Yang, D. (2010). Outage performance of dual-hop relay network with co-channel interference. In *VTC'10—IEEE Vehicular Technology Conference*, pp. 1–5. 10.1109/VETECS.2010.5494039.

Cui, S., Goldsmith, A., and Bahai, A. (2005). Energy-constrained modulation optimization. *IEEE Transactions on Wireless Communications*, *4* (5), 2349–2360.

Ding, H., Ge, J., Benevides da Costa, D., and Guo, Y. (2011). Outage analysis for multiuser two-way relaying in mixed Rayleigh and Rician fading. *IEEE Communications Letters*, *15* (4), 410–412.

Dohler, M. and Li, Y. (2010). *Cooperative Communications: Hardware, Channel and Phy*. Chichester, UK: John Wiley & Sons, Ltd.

European Telecommunications Standards Institute. (2007). Spatial Channel Model for Multiple Input Multiple Output (MIMO) Simulations. TR 25.996, 7.0.0. 3GPP TR 25.996 v.7. r.7. Jun. 2007. Available at http://www.etsi.org/deliver/etsi_tr/125900_125999/125996/07.00.00_60/tr_125996v070000p.pdf.

Fudenberg, D. and Tirole, J. (1991). *Game Theory*. Cambridge, MA: MIT Press.

Goldsmith, A. (2005). *Wireless Communications*. Cambridge, UK: Cambridge University Press.

Goodman, D.J. and Mandayan, N.B. (2000). Power control for wireless data. *IEEE Personal Communication Magazine*, *7* (4), 48–54.

Gross, T.J., Abrão, T., and Jeszensky, P.J. (2011). Distributed power control algorithm for multiple access systems based on Verhulst model. *International Journal of Electronics and Communications*, *65*, 361–372.

IEEE standard for local and metropolitan area networks part 16: Air interface for broadband wireless access systems amendment 3: Advanced air interface. (Mar. de 2011). IEEE Std 802.16m-2011 (Amendment to IEEE Std 802.16-2009), 1-1112.

Kim, D.I., Choi, W., Seo, H., and Kim, B.-H. (2011). Partial information relaying and relaying in 3GPP LTE. In: *Cooperative Cellular Wireless Networks*. Cambridge UK: Cambridge University Press.

Kyösti, P., Meinilä, J., Hentilä, L., Zhao, X., Jämsä, T., Schneider, C., Narandzic, M., Milojevic, M., Hong, A., Ylitalo, J., Holappa, V.-M., Alatossava, M., Bultitude, R., de Jong, Y., and Rautiainen, T. (2008). WINNER II channel models—Part I channel models. Technical Report IST-4-027756, Information Society Technologies. Available at http://www.cept.org/files/1050/documents/winner2%20-%20final%20 report.pdf.

Laneman, J.N. (2002). Cooperative diversity in wireless networks: Algorithms and architectures. PhD dissertation, Massachusetts Institute of Technology, Cambridge, MA.

Liebl, G., de Moraes, T.M., and Weitkemper, P. (2011). Advanced relay technical proposals. Technical Report, ARTIST4G—Advanced Radio Interface Technologies for 4G Systems.

Loa, K., Wu, C.-C., Sheu, S.-T., Yuan, Y., Chion, M., Huo, D., and Xu, L. (2010). IMT-advanced relay standards [WiMAX/LTE Update]. *IEEE Communications Magazine*, *48* (8), 40–48.

Meshkati, F., Poor, H., Schwartz, S., and Mandayam, N. (2005). An energy-efficient approach to power control and receiver design in wireless data networks. *IEEE Transactions on Communications*, *53* (11), 1885–1894.

Miao, G., Himayat, N., Li, G., and Talwar, S. (2011). Distributed interference-aware energy-efficient power optimization. *IEEE Transactions on Wireless Communications*, *10* (4), 1323–1333.

Nosratinia, A. and Hedayat, A. (2011). Network architectures and research issues in cooperative cellular wireless networks. In: *Cooperative Cellular Wireless Networks*. Cambridge, UK: Cambridge University Press.

Nourizadeh, H., Nourizadeh, S., and Tafazolli, R. (2006). Performance evaluation of cellular networks with mobile and fixed relay station. In *VTC-2006 Fall—IEEE 64th Vehicular Technology Conference*, pp. 1–5.

Rodriguez, V. (2003). An analytical foundation for resource management in wireless communication. In *Global Telecommunications Conference, 2003. GLOBECOM'03. IEEE*, *2*, pp. 898–902.

Sampaio, L.D., Abrão, T., Angélico, B.A., Lima, M.F., Proenca Jr., M.L., and Jeszensky, P. J. (2012). Hybrid heuristic-waterfilling game theory approach in MC-CDMA resource allocation. *Applied Soft Computing*, *12* (7), 1902–1912.

Saraydar, C., Mandayam, N., and Goodman, D. (2002). Efficient power control via pricing in wireless data networks. *IEEE Transactions on Communications*, *50* (2), 291–303.

Schurgers, C., Raghunathan, V., and Srivastava, M. (2001). Modulation scaling for real-time energy aware packet scheduling. In *GLOBECOM'01—IEEE Global Telecommunications Conference*, *6*, pp. 3653–3657.

Senarath, G., Tong, W., Zhu, P., Zhang, H., Steer, D., Yu, D., Naden, M., Nortel, D.K., Hart, M., Vadgama S., Cai, S., Chen, D., Xu, H., Peterson, R., Fu, I., Wong, W.C., Srinivasan, R., Lee, H.H., Johnsson, K., Sydir, J., Ahmadi, S., Hamzeh, B., Timiri, S., Oh, C., Wang, P., Sun, Y., Chindapol, A., Kong, P., Wang, H., Ahn, J.B., Kang, H., Molisch, A.F., Zhang, J., Kuze, T., and Son, J.J. (2007). Multi-hop relay system evaluation methodology (channel model and performance metric). Technical Report, IEEE. Available at http://ieee802.org/16/relay/docs/80216j-06_013r3.pdf.

Shannon, C.E. (1948). A mathematical theory of communication. *Bell System Technical Journal*, *27*, 379–423 and 623–656.

Shen, G., Zhang, K., Wang, D., Liu, J., Leng, X., Wang, W., and Jin, S. (2008). Multi-hop relay operation modes, In: *IEEE 802.16 Broadband Wireless Access Working Group*. Technical Report, Alcatel Shanghai Bell.

Souza, Á.R., Abrão, T., Sampaio, L.H., and Jeszensky, P.J. (2012a). Energy and spectral efficiencies trade-off with filter optimization in multiple access interference-aware networks. In *SBRC'12—XXX Brazilian Symposium on Computer Networks and Distributed Systems*, Ouro Preto, Brazil, pp. 1–4).

Souza, Á.R., Abrão, T., Sampaio, L.H., Jeszensky, P.J., Perez-Romero, J., and Casadevall, F. (2012b). Energy and spectral efficiencies tradeoff with filter optimization in multiple access interference. *Transactions on Emerging Telecommunications Technologies*. Accepted.

Verdú, S. (1984). Optimum multiuser signal detection. PhD dissertation, University of Illinois, Illinois, USA.

Xu, J. and Qiu, L. (2010). Area power consumption in a single cell assisted by relays. In *GreenCom'10—IEEE/ ACM International Conference on Green Computing and Communications*, pp. 460–465.

Yamao, Y., Otsu, T., Fujiwara, A., Murata, H., and Yoshida, S. (2002). Multi-hop radio access cellular concept for fourth-generation mobile communications system. In *The 13th IEEE International Symposium on Personal, Indoor and Mobile Radio Communications, 2002. 1*, pp. 59–63.

Yang, W., Hua Li, L., Lu SUN, W., and Wang, Y. (2010). Energy-efficient relay selection and optimal relay location in cooperative cellular networks with asymmetric traffic. *The Journal of China Universities of Posts and Telecommunications*, *17* (6), 80–88.

Zappone, A., Buzzi, S., and Jorswieck, E. (2011). Energy-efficient power control and receiver design in relay-assisted DS/CDMA wireless networks via game theory. *IEEE Communications Letters*, *15* (7), 701–703.

9 Green-Inspired Hybrid Base Transceiver Station Architecture with Joint FSO/RF Wireless Backhauling and Basic Access Signaling for Next-Generation Metrozones

*It Ee Lee, Zabih Ghassemlooy, Wai Pang Ng,
and Mohammad-Ali Khalighi*

CONTENTS

9.1 INTRODUCTION

The unequivocality of global man-made climate change has attracted substantial governmental and political intervention [1,2] across the world to address the urgency of this phenomenon. Correspondingly, the conservation efforts striving for greener technology options [3,4], energy-efficient solutions [5,6], and carbon emissions savings opportunities [7,8] have been gradually leveraged upon the research communities [9], industrial companies [4,10], and regulatory bodies [3] toward realizing a global low-carbon society. Statistics released by the Global eSustainability Initiative [7] reveal that worldwide carbon emissions will rise from 40 billion tons (Gt) carbon dioxide equivalent (CO_2e) per annum (pa) in 2002 to approximately 53 $GtCO_2e$ by 2020, in which the information and communications technology (ICT) sector has been identified as one of the key areas in mitigating the world's carbon footprint because of its astounding direct contribution of 2% by consuming 3% of the worldwide energy with a growth rate of 6% pa [5,11]. In addition to the expected growth in mature developed markets, the explosive growth in the number of mobile, fixed, and broadband subscribers on a global basis, which is attributable to the emerging ICT demand in developing countries [7], inevitably requires extensive mobile networks supported by greater amount of power-hungry base transceiver stations (BTSs) and mobile switching centers, thus incurring enormous stress upon network operators to suppress the resulting infrastructure carbon footprint with a projected growth from 133 million tons (Mt) CO_2 to 299 $MtCO_2$ by 2020 at an annual incremental rate of 5%.

The evolution of wireless mobile communications has witnessed a dynamic shift in technology adoption trends among relevant industry drivers since the early 1980s, which in turn influenced the behavior of end users with regard to the acceptance of new technologies and services, spending pattern, and demand for connectivity [12–14]. A microscopic perspective of these observed trends indicates that mobile service provisioning and user requirements for connectivity have transitioned from fundamental coverage and adequate mobility for making simple voice call toward more sophisticated services demanding ubiquity, all-in-one voice, data and video convergent solutions, and enhanced efficiency of transmission (implying higher capacity at lower cost). As efficient radio technologies such as software-defined radio [15,16], smart antennas [17–19], wireless mesh networking [20,21], and interference cancellation techniques [22,23] approach maturity and are deployed in commercial markets, the penetration of green "criteria" into existing solutions would definitely come into effect over the next two decades to justify the telecommunications industry's accountability and responsibility to the environment and society [4,10,24]. This would ultimately result in the transformation into clean, energy-efficient, and complete wireless communication networks, without compromising the quality of service (QoS) for the mobile users or imposing negative impacts upon the deployment costs for network operators, equipment manufacturers, and content providers.

Greenfield deployment of wireless macrocellular networks typically requires up to tens of thousands of BTSs, in order to provide ubiquitous coverage and seamless communication while coping with the capacity demand for voice and high-speed data and video traffic. This in turn poses immense pressure on network operators to suppress the greenhouse emissions and mushrooming operational expenditure (OPEX) because of the dominant energy requirement at radio base stations [7,11,25,26]. At present, an estimated 24,000 BTSs are deployed in existing third-generation (3G) network throughout the United Kingdom with a total power consumption of 300 GWh/year for providing coverage to a population in excess of 80%, in which these statistics would double up to extend the network for national coverage [11]. The rollout of the fourth generation (4G) cellular wireless standards in the near future is poised to set the peak capacity requirements up to 1 Gbps, in order to provide a comprehensive and secure all-Internet protocol (IP)-based solution to a multitude of mobile broadband facilities, such as ultra-broadband Internet access, voice over IP, online gaming services, and streamed multimedia applications. The incessantly exploding mobile data traffic volumes owing to burgeoning smart devices, applications, and changing user behavior have resulted in severe bandwidth capacity crunch, in which the peak-to-average gap is forecasted to rise to 90 times its current size by 2015 with the adoption of 4G technologies [26].

This detrimental impact has attracted extensive research efforts to look into new radio access network (RAN) architectural paradigm [26,27], which optimizes the existing macrocell site infrastructure and introduces new sub-networking layers at the micro-/picocell level. Correspondingly, the benefits of overlaying smaller cells can be attained, which include (1) capacity enhancement by a factor of 1600 [26] compared to other methods, such as improving spectrum efficiency and increasing cell site transmit power; (2) improved service continuity and ubiquity; (3) even distribution of network load; and (4) minimal cost for network extension compared to deploying traditional macro-BTSs for the same purpose, owing to cheaper micro-/picocell architectures. As Metrozones deployments would typically require 3–6 microcells or 8–14 picocells per macrocell site for 3G/4G service provisioning [26], energy-efficient mechanisms and green approaches are vital for sustainable operation of these Metrozones, in order to address two contradictory phenomena—(1) the non-negligible presence of power-hungry and expensive ground-based BTSs, and the mushrooming of access points (APs) for uniform blanket coverage; and (2) the significant CO_2 emissions contributing to the world's carbon footprint.

The remainder of this chapter is organized as follows: Section 9.2 presents the rationale, network architecture, and contributions of our work with regard to the green approach for next-generation Metrozones. In Section 9.3, we describe the system and channel models pertaining to the hybrid free-space optical and radio frequency (FSO/RF) system. Next, a detailed explanation on the proposed hybrid-BTS (H-BTS) system architecture design is provided in Section 9.4, where we will discuss the basic access signalling (BAS) control protocol and resource prioritization algorithm for enabling the sleep-wake-on-demand (SWoD) mechanism, cooperative inter-cell support, and prioritized switching in the hybrid backhaul links. Numerical results from the feasibility studies, encompassing the BTS daily traffic profile simulation and outage analysis of the hybrid FSO/RF systems, are highlighted and justified in Section 9.5. We present our perspectives and directions for future research in Section 9.6, and conclude this chapter in Section 9.7.

9.2 GREEN METROZONES CONCEPT

9.2.1 RATIONALE

The diffusion of smaller cells into traditional macrocellular networks presents a feasible, energy-efficient, and cost-effective alternative, to shift the access network closer to the mobile user terminals (MUTs) as compared to other methods, such as improving the spectrum efficiency and increasing cell site transmit power [9,26]. Correspondingly, manifold gain in the mobile data capacity can be achieved to curtail the impact of the prevailing bandwidth capacity crunch phenomena. This will inevitably drive the deployment of at least an order-of-magnitude more micro-/picocells for providing a uniformly distributed capacity density across a mobile service area [9]. As a result, the requirement for new data-centric RAN architecture solutions with a higher degree of network flexibility and reconfigurability must be addressed.

On the other hand, the existing macrocellular BTS infrastructures are retained and upgraded to enhance network performance and scalability, while promoting a significant reduction in site costs. These BTSs with integrated routing and mesh networking capabilities can be deployed at the macro tier, to enable rapid, flexible and low-cost connectivity to the core network via the packet-based multi-hop communication, as compared to wired backhaul solutions, such as leased T1/E1 copper lines and optical fiber links [28]. The evolving trends in emerging 4G wireless mobile broadband networks reveals preferences for smaller cells to boost the capacity and migration toward cost-effective packet-based wireless backhauling solutions. This has led to the interdependency between macro- and sub-cells, which has attracted the Metrozones concept to cope with the massive growth in the number of mobile subscribers and high-speed data services, while complementing the operators' escalating deployment and operational costs.

9.2.2 NETWORK ARCHITECTURE

In principle, the network topology of Metrozones is conceptually similar to a two-tier infrastructure/backbone wireless mesh network (WMN) [21,26], which comprises two main hierarchy—the macro-cellular tier and sub-networking layer, as illustrated in Figure 9.1.

The macro tier is populated by the BTS infrastructures with no mobility, less power constraint, sophisticated computational functions, and integrated routing capabilities to perform wireless mesh backhauling. A fraction of BTSs with the gateway or the bridge functions has wired connection to the Internet, representing sources/sinks in the WMN, thereby enabling connectivity to the wired backbone facilities at a much lower cost and with flexible networking options [21,28]. These BTSs automatically establish and maintain the connectivity among themselves, to form interconnected self-configuring, self-healing wireless backhaul links within the WMN. The data traffic is en route to and from the wired Internet entry points via multi-hop communication among mesh nodes using efficient routing protocols, thus promoting link reliability and load balancing in the macro tier.

At the sub-networking layer, metro APs (M-APs) providing coverage to smaller micro-/pico-cells within each macrocell form a cluster, which is connected to the BTS via the wireless Metro Ethernet [29], thereby forming a point-to-multipoint connection. Inter-cell coordination is managed in an autonomous manner based on the self-organizing network [30,31] principles, whereby the M-APs form self-configuring, self-optimizing, and self-healing clusters, in order to maximize the network performance and to deliver enhanced user perceived quality through numerous integrated approaches. These include optimization of the network parameters under interference and overload conditions, mitigation of quality degradation that may arise from inaccuracies of network planning or equipment faults, and rapid and efficient fault identification and compensation.

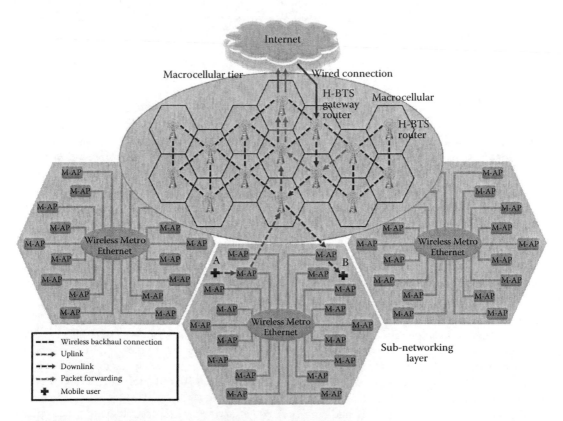

FIGURE 9.1 Network architecture of the green Metrozones concept.

9.2.3 Proposed Green Approach

Gigabit Ethernet (GigE) backhauling solutions in next-generation 4G networks will largely be based on the millimeter-wave (MMW) and licensed E-Band technologies (i.e., 50/60 GHz and 80 GHz bands, respectively), for supporting bandwidth-intensive data operations in the enterprise and urban markets with shorter link requirements of 3–5 km [28,32]. The high operating frequencies of these technologies promote antenna directivity with a very narrow beamwidth (4.7° and 1.2° at 60 and 80 GHz, respectively). This unveils numerous technical advantages in mesh-configured wireless backhaul networks, such as throughput enhancement, interference mitigation, superior security, and a high-frequency reuse rate [32]. Nonetheless, the high susceptibility of MMW radios to rain attenuation presents a greater challenge to network operators in optimizing their backhaul solutions, in order to deliver GigE speeds with a desired carrier-grade availability of 99.999%, not affected by local meteorological conditions [32,33].

FSO communications is a promising broadband wireless access (BWA) candidate in complementing the RF solutions to resolve the existing "last mile" access network problems. This is mainly due to the superior characteristics of such technology option, which include (1) no licensing requirements or tariffs for its utilization; (2) virtually unlimited bandwidth for providing near-optimal capacity and supporting high-speed applications; (3) extensive link range in excess of 5 km; (4) high energy efficiency owing to low power consumption, reduced interference, and fading immunity; (5) high scalability and recon-figurability; and (6) minimal cost and time for deployment [34]. The performance of both FSO and RF links are susceptible to the adverse effects of meteorological and other natural conditions. Therefore, hybrid FSO/RF systems [35–38] present the most prominent alternative to enable these technologies in complementing one another's weaknesses, since fog and rain drastically affect the FSO and RF links, respectively, but only insignificantly vice versa, and rarely occur simultaneously.

In this chapter, we propose a new *H-BTS system architecture* for the green Metrozones, which takes advantage of the symbiotic relationship between the FSO and RF technologies, by integrating these communication links at the macrocellular tier. This corresponds to high data-rate transmission with lower transmit power and less susceptibility to interference, thereby delivering high-capacity, power-efficient wireless backhauling solution under most weather conditions and varying data traffic load. A *radio resource management (RRM) module* encompassing a *resource prioritization mechanism* is designed and introduced into the system hub of the proposed H-BTS architecture. This is to maintain a good control and optimal on-demand resource allocation to both the wireless backhaul and RF access networks and to establish sustainable wireless backhaul link availability via essential switching between the FSO and RF communication links, taking into account various factors such as the fluctuating traffic demand, spectral bandwidth occupancy, network load, QoS, and channel conditions. Furthermore, we consider a *BAS scheme* employing a default low data-rate, low-power radio, which necessitates the discovery, registration, and monitoring of active M-APs, to enable two distinctive features: the *SWoD mechanism* and *cooperative inter-cell support*. The *SWoD mechanism* minimizes the number of operating radio access interfaces (RAIs) and enhances potential energy savings by putting idling/underutilized RAIs and M-APs into sleep mode, particularly in low-traffic scenarios. The *cooperative inter-cell support* off-loads the M-APs located at the macrocell edge to neighboring H-BTSs with more resource availability, thus enabling more even distribution of the network load across a particular topology.

9.3 HYBRID FSO/RF COMMUNICATION SYSTEM

9.3.1 System Description

Figure 9.2 presents the block diagram of a point-to-point hybrid FSO/RF system, which can be conceptually described as a system architecture having a pair of independent, non-ergodic channels with random states [39], in which a source *s* (audio, video, or speech) is transmitted through the parallel

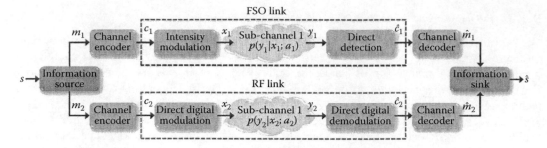

FIGURE 9.2 Block diagram of a point-to-point hybrid FSO/RF system.

fluctuating channels. In the absence of source encoding (i.e., $s = m_1 = m_2$), a channel encoder is required to encode the original information m_1 (respectively, m_2) into a codeword c_1 (respectively, c_2) without knowing the channel state a_1 (respectively, a_2). The encoded signal goes through the modulation process to produce x_1 (respectively, x_2) and then propagates through free space, while a channel decoder is employed at the receiving end along with knowledge of the channel state to produce an estimate of the source $\hat{m}_1$ (respectively, $\hat{m}_2$) from the demodulated codeword $\hat{c}_1$ (respectively, $\hat{c}_2$).

In principle, the channel model of the proposed hybrid FSO/RF system can be described as a combination of two individual line-of-sight (LOS) FSO and RF channels. It is assumed that both the FSO and RF subsystems are single-input–single-output (SISO) systems with relevant parameters as shown in Table 9.1.

TABLE 9.1

Parameters of the FSO and RF Subsystems

FSO System		
Parameter	**Symbol**	**Typical Value**
Laser wavelength	λ_1	1550 nm
Transmitted optical power	P_{FSO}	10 mW
Photodetector responsivity	γ	0.5 A/W
Noise variance	σ_{n1}^2	10^{-14} A^2
Beam divergence angle	ϕ	2 mrad
Receiver diameter	D	20 cm
RF System		
Parameter	**Symbol**	**Typical Value**
Carrier frequency	f_c	60 GHz
Transmitted RF power	P_{RF}	10 mW
Bandwidth	B	250 MHz
Transmit antenna gain	G_{Tx}	44 dBi
Receive antenna gain	G_{Rx}	44 dBi
Attenuation (due to oxygen)	a_{oxy}	15.1 dB/km
Noise power spectral density	N_0	−114 dBm/MHz
Receiver noise figure	N_F	5 dB

Weather-Dependent Parameters of FSO and RF System			
Weather Conditions	**V (km)**	**a_{rain} (dB/km)**	**C_n^2 (m$^{-2/3}$)**
Light fog	0.642	0	2.0×10^{-15}
Moderate rain (12.5 mm/h)	2.80	5.6	5.0×10^{-15}
Heavy rain (25 mm/h)	1.90	10.2	4.0×10^{-15}

9.3.2　FSO CHANNEL MODEL

FSO communications is an emerging low-cost, license-free, and high-bandwidth access technique, which has attracted significant attention for a variety of applications, such as last mile connectivity, optical fiber backup, and enterprise connectivity [40]. Nonetheless, the widespread deployment of such BWA solution is hampered by the combined effects from numerous factors, which include the atmospheric channel, which is highly variable, unpredictable, and vulnerable to different weather conditions, such as scattering, absorption and turbulence [41–43], and the presence of pointing errors [40,44,45].

For the FSO subsystem (as depicted in Figure 9.3), we consider a SISO horizontal FSO communication link, employing intensity modulation with direct detection (IM/DD) and the non-return-to-zero on–off keying (NRZ-OOK) technique. Information-bearing electrical signals are modulated onto the instantaneous intensity of a laser source and then propagate along a horizontal path through a turbulent channel with additive white Gaussian noise (AWGN) in the presence of atmospheric attenuation and pointing errors. At the receiving end, the optical signals are collected by an aperture lens [46] before being focused onto a photodetector, which in turn converts the received optical intensities into a resulting photocurrent.

The received signal y_1 can be described by the conventional channel model [47]:

$$y_1 = h_1 \gamma x_1 + n_1 \tag{9.1}$$

where γ is the photodetector responsivity (in A/W), $x_1 \in \{0, 2P_{\text{FSO}}\}$ is the optical signal intensity, P_{FSO} is the average transmitted optical power, and h_1 is the channel fading coefficient. The noise n_1 arises from various sources, which include the shot noise caused by the signal itself or ambient light, the dark current noise, and the electrical thermal noise, and can be conveniently modeled as a signal-independent AWGN with zero mean and variance σ_{n1}^2 [41].

The channel state h_1 models the optical intensity fluctuations resulting from atmospheric loss and turbulence- and misalignment-induced channel fading, which can be described as a product of three components given by [44]

$$h_1 = h_l h_s h_p \tag{9.2}$$

where h_l, h_s, and h_p denote the attenuation owing to beam extinction arising from both scattering and absorption, scintillation effects, and geometric spread and pointing errors, respectively. The attenuation h_l is a deterministic component that exhibits no randomness in its behavior, thus acting as a fixed scaling factor for a long period (i.e., on the order of hours), which stands in contrast to the bit intervals in ranges of nanoseconds or less [44]. On the other hand, both h_s and h_p are time-variant factors, exhibiting variations in the fading channel on the order of milliseconds, in which their stochastic behavior are described by their respective distributions [47]. Since time scales of

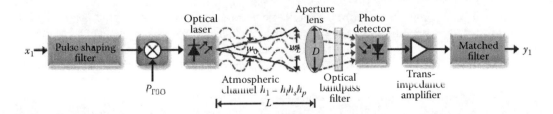

FIGURE 9.3　Block diagram of a single-input–single-output FSO system.

these fading processes are much greater than the bit interval, this block fading (or quasi-static fading) channel is commonly known as the slow-fading channel [48], in which h_1 is assumed constant over a large number of transmitted bits.

For a slow-fading channel with OOK signaling, the received electrical signal-to-noise ratio (SNR) is given by [44]

$$\text{SNR}_{\text{FSO}}(h_1) = \frac{2P_{\text{FSO}}^2 \gamma^2 h_1^2}{\sigma_{n1}^2} \tag{9.3}$$

and is a fluctuating term (i.e., an instantaneous value) owing to the influence of h_1, which is chosen from the random ensemble according to the distribution $f_{h_1}(h_1)$.

9.3.2.1 Atmospheric Loss

Aerosol scattering and absorption caused by rain, snow, and fog result in significant optical power attenuation and beam spreading and link distance reduction, severely impairing the system performance with an increase in the link error probability, thus causing the FSO system to fall short of the desired carrier-grade availability of 99.999% in heavy, visibility-limiting weather [49]. In particular, Kim and Korevaar reported in Ref. [37] that the atmospheric attenuation of laser beam is a random function of the weather, which can vary from 0.2 dB/km in exceptionally clear weather to 350 dB/km in very dense fog.

The atmospheric loss can be modeled by the Beers–Lambert law [43]:

$$h_l = e^{-\sigma L} \tag{9.4}$$

where σ denotes a wavelength- and weather-dependent attenuation coefficient and L resembles the propagation distance. For clear and foggy weather conditions, the attenuation coefficient can be determined from the visibility data through Kim's model [49]:

$$\sigma_{\text{no-precip}} = \frac{3.91}{V}\left(\frac{\lambda_1}{550}\right)^{-q} \tag{9.5}$$

where V is the visibility (in kilometers), λ_1 is the laser wavelength (in nanometers), and q is a parameter related to the particle size distribution and visibility. For rain, an accepted empirical model reported in Ref. [50] is given by

$$\sigma_{\text{rain}} = \frac{2.9}{V}. \tag{9.6}$$

9.3.2.2 Atmospheric Turbulence-Induced Fading

Laser beams propagating through the atmospheric turbulent channel are highly susceptible to the adverse effects of scintillation and beam wander, which are natural phenomenon commonly observed in terrestrial FSO communication systems, owing to refractive index variations along the transmission paths caused by inhomogeneities in both temperature and pressure of the atmosphere [43]. Correspondingly, this produces random fluctuations in both temporal and spatial domains of the received irradiance, known as channel fading, whereby the FSO links may suffer temporary signal degradation or complete system annihilation under the influence of deep signal fades [47,51]. Modeling of the fading effects in an atmospheric turbulent channel using the extended

Huygens–Fresnel principle [43] shows that large transient dips in the optical signals typically last approximately 1–100 ms, which may result in the loss of potentially up to 10^9 consecutive bits at a transmission rate of 10 Gbps [51].

The log-normal distribution is the most widely adopted model for the probability density function (pdf) of the atmospheric turbulence-induced irradiance fluctuations (scintillation) in the weak-to-moderate turbulence regime, which arises because of the random changes in the refractive index of air along the transmission path [43]. The pdf of the irradiance intensity in the turbulent medium is expressed as [41–43]

$$f_{h_s}(h_s) = \frac{1}{h_s \sigma_I(D)\sqrt{2\pi}} \exp\left\{ -\frac{\left[\ln(h_s) + \frac{1}{2}\sigma_I^2(D) \right]^2}{2\sigma_I^2(D)} \right\}. \tag{9.7}$$

Assuming plane wave and aperture averaging, the scintillation index $\sigma_I^2(D)$ can be determined as follows [41]:

$$\sigma_I^2(D) = \exp\left[\frac{0.49\chi^2}{(1+0.18d^2+0.56\chi^{12/5})^{7/6}} + \frac{0.51\chi^2(1+0.69\chi^{12/5})^{-5/6}}{1+0.90d^2+0.62d^2\chi^{12/5}} \right] - 1 \tag{9.8}$$

where $\chi^2 \equiv 0.492(2\pi/\lambda_1)^{7/6} C_n^2 L^{11/6}$ is the Rytov variance for spherical wave, $d \equiv \sqrt{(2\pi D^2)/(4\lambda_1 L)}$, and D is the receiver aperture diameter. The refractive index structure parameter C_n^2 signifies the strength of the atmospheric turbulence and is assumed constant for a horizontal path communication link [41].

The gamma–gamma distribution is a more recent fading model that has evolved from an assumed modulation process, in order to address the large- and small-scale scintillations under moderate-to-strong scenarios. The gamma–gamma pdf is given by [41–43]

$$f_{h_s}(h_s) = \frac{2(\alpha\beta)^{(\alpha+\beta)/2}}{\Gamma(\alpha)\Gamma(\beta)} h_s^{(\alpha+\beta)/2-1} K_{\alpha-\beta}\left(2\sqrt{\alpha\beta h_s} \right) \tag{9.9}$$

where $\Gamma(\cdot)$ denotes the gamma function, and $K_{\alpha-\beta}$ resembles the modified Bessel function of the second kind of order $(\alpha - \beta)$. Parameters α and β represent the effective number of large- and small-scale turbulent eddies, respectively, relating to the scattering effects of the atmospheric turbulent environment, and are valid with the relationship $\sigma_I^2(D) = \frac{1}{\alpha} + \frac{1}{\beta} + \frac{1}{\alpha\beta}$ [43]. For the plane wave model with aperture averaging, both α and β can be calculated in accordance to the following expressions [43]:

$$\alpha = \left[\exp\left(\frac{0.49\chi^2}{(1+0.18d^2+0.56\chi^{12/5})^{7/6}} \right) - 1 \right]^{-1} \tag{9.10}$$

$$\beta = \left[\exp\left(\frac{0.51\chi^2(1+0.69\chi^{12/5})^{-5/6}}{(1+0.9d^2+0.62d^2\chi^{12/5})^{5/6}} \right) - 1 \right]^{-1} \tag{9.11}$$

9.3.2.3 Misalignment Fading

Misalignment-induced fading is another non-negligible effect in FSO systems, as optical terminals typically installed on high-rise buildings are susceptible to building sway, while continuous precise pointing is required to establish link connectivity for successful data transmission with minimum error probability, particularly when narrow beam divergence angle and receiver field of view are employed [40]. Under the influence of wind loads, thermal expansions, and weak earthquakes, building sway causes vibrations in both the transmitter and receiver, in which the stochastic process deviates the optical wave propagation path from the common LOS, thus resulting in decrease of the average received signal [40,52,53]. In addition, pointing errors can arise because of mechanical misalignment, errors in tracking systems, or the presence of mechanical vibrations within the system [45].

A statistical misalignment-induced fading model developed in Ref. [44], based on an earlier work in Ref. [52], provides a tractable pdf for describing the stochastic behavior of the pointing error loss. The model assumes a circular detection aperture of diameter D and a Gaussian spatial intensity profile of beam waist radius w_L on the receiver plane. In addition, both the elevation and horizontal displacement (sway) are considered independent and identically Gaussian distributed with variance σ_{pe}^2. Correspondingly, the pdf of h_p is given by

$$f_{h_p}(h_p) = \frac{\xi^2}{A_0^{\xi^2}} h_p^{\xi^2-1}, \quad 0 \le h_p \le A_0 \tag{9.12}$$

where $\xi = w_{z_{eq}}/2\sigma_{pe}$ is the ratio between the equivalent receiver beam width and the pointing error displacement (jitter) standard deviation at the receiver, $A_0 = [\text{erf}(\upsilon)]^2$, $w_{z_{eq}} = w_L \left[\sqrt{\pi}\text{erf}(\upsilon)/(2\upsilon\exp(-\upsilon^2)) \right]^{1/2}$, and $\upsilon = \sqrt{\pi}D/(2\sqrt{2}w_L)$ [44].

9.3.2.4 Combined Channel Fading Model

Taking into account h_l as a scaling factor and the conditional distribution given by Equation 13 in Ref. [44], the channel state distributions of $h_1 = h_l h_s h_p$ for both weak and strong turbulence regimes are given by Equations 9.13 and 9.14, respectively.

For weak atmospheric turbulence conditions [44],

$$f_h(h) = \frac{\xi^2}{(A_0 h_l)^{\xi^2}} h^{\xi^2-1} \int\limits_{h/A_0 h_l}^{\infty} \frac{1}{h_s^{\xi^2+1}\sigma_I(D)\sqrt{2\pi}} \exp\left\{ -\frac{\left[\ln(h_s) + \frac{1}{2}\sigma_I^2(D) \right]^2}{2\sigma_I^2(D)} \right\} dh_s. \tag{9.13}$$

In the strong turbulence regime [44],

$$f_h(h) = \frac{2\xi^2(\alpha\beta)^{(\alpha+\beta)/2}}{(A_0 h_l)^{\xi^2} \Gamma(\alpha)\Gamma(\beta)} h^{\xi^2-1} \int\limits_{h/A_0 h_l}^{\infty} h_s^{(\alpha+\beta)/2-1-\xi^2} K_{\alpha-\beta}\left(2\sqrt{\alpha\beta h_s} \right) dh_s. \tag{9.14}$$

9.3.3 RF Channel Model

The enormous bandwidth availability of the unlicensed 60 GHz MMW spectrum has attracted a vast variety of wireless applications and services with the potential of high data throughputs, albeit hampered by the adverse atmospheric channel effects [54,55] and transceiver hardware limitations

[56,57]. These include mobile broadband and cellular systems, wireless backhaul networks, fixed wireless access, wireless local area networks, ubiquitous personal communication networks, portable multimedia streaming, and vehicular networks [54,55]. The 60 GHz communications is highly susceptible to atmospheric attenuation, in which the classical Friis free-space path loss formula suggests that this channel effect can potentially result in a 20–40 dB power penalty, as compared to unlicensed communication at operating frequencies below 6 GHz [55]. In addition, this technology option suffers from atmospheric absorption owing to rain drops, water vapor, and oxygen, which depends on the atmospheric conditions such as pressure, temperature, and density, accounting for an additional 7–15.5 dB/km power loss in the received signal [54]. The MMW signal attenuation can be further aggravated by the presence of rain droplets, particularly when the atmosphere becomes saturated, in which different empirical models predict additional atmospheric attenuation between 8 and 18 dB/km for a given rainfall rate of 50 mm/h [54].

A point-to-point RF link operating at an unlicensed carrier frequency of 60 GHz with direct digital modulation using the quadrature amplitude modulation technique [56,57], as illustrated in Figure 9.4, is considered as a complementary link to the FSO channel.

The received signal of the RF channel is given by [36]

$$y_2 = \sqrt{P_{RF}} \sqrt{g_{RF}} h_2 x_2 + n_2 \tag{9.15}$$

where P_{RF} is the RF transmit power, g_{RF} is the average power gain of the RF link, h_2 is the RF fading gain, x_2 is the modulated RF signal, and n_2 is the complex AWGN with variance σ_{n2}^2.

The effective gain of the RF system g_{RF} can be modeled as [58]

$$g_{RF}[dB] = G_{Tx} + G_{Rx} - 20\log_{10}\left(\frac{4\pi L}{\lambda_2}\right) - a_{oxy}L - a_{rain}L \tag{9.16}$$

where G_{Tx} and G_{Rx} represent the transmit and receive antenna gains (in decibels), respectively, λ_2 denotes the wavelength of the RF system, and a_{oxy} and a_{rain} resemble the attenuations owing to oxygen absorption and rain (both in decibels per kilometer), respectively. The RF noise variance is given by $\sigma_{n2}^2[dB] = BN_0 + N_F$, where B is the signal bandwidth, N_0 is the noise power spectral density (in decibel milliwatts per megahertz), and N_F is the receiver noise figure.

The fading gain h_2 can be modeled as Rician distributed, with the pdf given by [59]

$$f_2(h_2) = \frac{2(K+1)h_2}{\Omega} \exp\left(-K - \frac{(K+1)h_2^2}{\Omega}\right) I_0\left(2h_2\sqrt{\frac{K(K+1)}{\Omega}}\right) \tag{9.17}$$

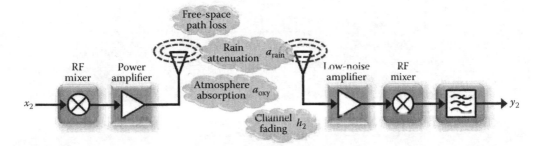

FIGURE 9.4 Block diagram of a point-to-point 60 GHz RF system.

where Ω is the received signal power and the Rician factor $K = A^2/\sigma^2$ is the ratio of the power of the LOS component to the power of the diffuse component. The Rician factor is a useful measure of the communication link quality and is dependent on various factors, such as link distance, antenna height, and the environment, and may also change with time [36,59].

Correspondingly, the SNR of the RF link can be determined from the following relation [36]:

$$\text{SNR}_{\text{RF}}(h_2) = \frac{P_{\text{RF}}g_{\text{RF}}h_2^2}{\sigma_{n2}^2}. \tag{9.18}$$

9.4 HYBRID BTS ARCHITECTURE

9.4.1 Basic Access Signaling Control Protocol

Figure 9.5 illustrates the system hub of the proposed H-BTS system architecture, which comprises the RRM module and BAS scheme, to achieve potential energy savings through resource monitoring, optimal decision making, and priority-based resource allocation. In principle, the BAS scheme utilizes a default low data-rate, low-power radio operating on a dedicated frequency band for exchanging control messages between the H-BTS and M-APs, thus separating the control channel from the high-bandwidth, high-power data transport channel. The proposed out-of-band control signaling scheme promotes enhancement in resource monitoring through the discovery and registration of new M-APs and monitoring of these nodes within the BTS coverage, in which a neighbor list is generated and dynamically updated in accordance to the M-AP activities. The flow diagram in Figure 9.6 provides an overview of the BAS mechanism, in which the main operations are described in the following subsections.

9.4.1.1 M-AP Discovery

An H-BTS ($\mathcal{H}_{(i)}$, for $i = \{1, 2, ...\}$) with resource availability uses the default BAS radio to broadcast control signal (*CTRL_PKT*) on a periodical basis to its M-AP cluster, in which the M-APs are denoted as $\mathcal{M}_{(i,j)}$, for $j = \{1, 2, ...\}$. Upon receiving the control signal, a new M-AP ($\mathcal{M}_{(i,\ell)}$, for $\{\ell \in j\}$) responds by sending a request (*REQ_PKT*) to $\mathcal{H}_{(i)}$ and awaits acknowledgement (*ACK_ID*) to proceed with the next phase.

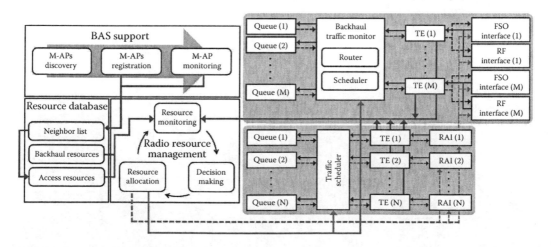

FIGURE 9.5 The proposed H-BTS system architecture.

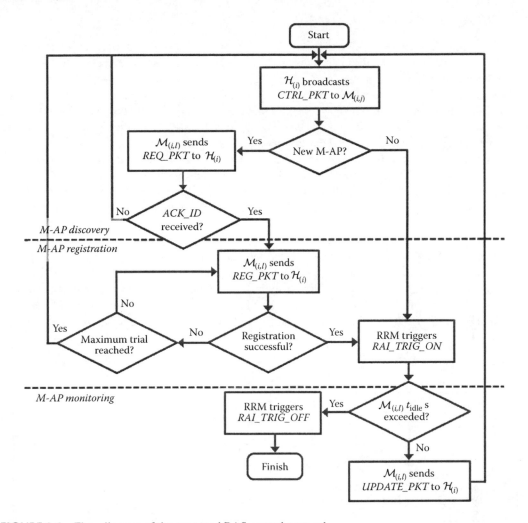

FIGURE 9.6 Flow diagram of the proposed BAS control protocol.

9.4.1.2 M-AP Registration

$\mathcal{M}_{(i,\ell)}$ requires *ACK_ID* to initiate the registration procedure, in which the registration information (*REG_PKT*) comprising M-AP identity (*MAP_ID*), status (*MAP_STAT*), location (*MAP_ADD*), and number MUTs attached (*MAP_USER*) is transmitted to $\mathcal{H}_{(i)}$ and relayed to the resource database for updating the neighbor list. Upon successful completion of M-AP registration, RRM will trigger an RAI (*RAI_TRIG_ON*) to establish a communication link with the newly reported M-AP.

9.4.1.3 M-AP Monitoring

In this phase, the active M-APs will update their information in the resource database by sending update packet (*UPDATE_PKT*) to the H-BTS. If an M-AP remains inactive for an interval exceeding t_{idle} seconds, the node will send a request to the H-BTS to update its status information and disconnect idling RAIs (*RAI_TRIG_OFF*), in which the M-AP is then placed into sleep mode. Correspondingly, underutilized RAIs at H-BTS can be switched off (during low traffic demand) and then reinstated on the basis of the M-AP request via the BAS scheme. This in turn promotes higher energy efficiency by optimizing the operation of RAIs at the sub-networking layer with the proposed SWoD mechanism.

9.4.2 Resource Prioritization Mechanism

The RRM module involves resource/information monitoring, decision making, and priority-based resource allocation, in which these mechanisms are jointly managed and executed by the proposed resource prioritization mechanism as shown in Figure 9.7. In principle, the information/resources observed and managed by the RRM module can be categorized into predetermined and time-varying factors [60]. These are monitored and gathered separately for the macrocellular and sub-networking layers, owing to the vastly varying propagation conditions and different communication purposes with contrasting application and QoS constraints.

Within the sub-networking layer to enable last-mile radio access, the updated neighbor list produced by the BAS scheme (Figure 9.6) and the network load (observed by the traffic estimators at each interface, i.e., TE(1),... TE(N)) present an important time-varying access resource, which influences the decision of the RRM to enforce the SWoD mechanism in one of the following modes:

1. *Sleep mode*: It is enabled under two possible scenarios—(i) $\mathcal{M}_{(i,\ell)}$ inactivity for a time interval exceeding t_{idle} seconds and (ii) the total access traffic demand at $\mathcal{H}_{(i)}$ approaches the lower threshold ($\eta_{(i)} < \eta_{\text{RA, low}}$). In the latter case, $\mathcal{H}_{(i)}$ sends a request to $\mathcal{M}_{(i,\ell)}$ with the lowest radio access traffic, in order to disconnect the communication link (which requires the triggering of *RAI_TRIG_OFF*) and place the targeted $\mathcal{M}_{(i,\ell)}$ into sleep mode. The self-organizing capabilities of the M-AP clusters enable the selected M-AP to aggregate the end

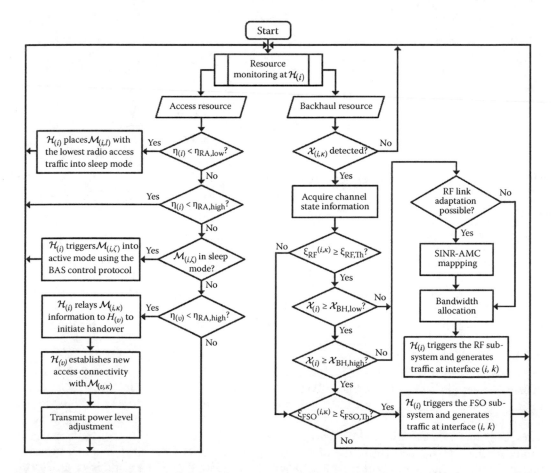

FIGURE 9.7   Flow diagram of the proposed resource prioritization mechanism.

users residing within its coverage area to neighboring micro-/picocells. Upon successful off-loading of the radio access traffic, the M-AP responds with a procedural acknowledgement using the dedicated BAS control protocol and subsequently enters into sleep mode, thereby conserving energy during low traffic demand by minimizing $\mathcal{M}_{(i,\ell)}$ and number of operating RAIs at $\mathcal{H}_{(i)}$.

2. *Active mode*: It is triggered under two possible circumstances—(i) successful registration of newly discovered $\mathcal{M}_{(i,\ell)}$ and (ii) total access traffic demand at $\mathcal{H}_{(i)}$ approaches the upper threshold ($\eta_{(i)} \geq \eta_{\text{RA, high}}$) with sleeping M-APs $\left(\mathcal{M}_{(i,\zeta)}\right)$ within the $\mathcal{M}_{(i,j)}$ cluster. In the second condition, $\mathcal{H}_{(i)}$ triggers $\mathcal{M}_{(i,\zeta)}$ into active mode via the BAS scheme, in which the high-power, high-speed RAIs (of the previously sleeping M-APs) are switched on for distributing the data traffic to the $\mathcal{M}_{(i,j)}$ cluster.

3. *Handover mode*: It is considered when the $\mathcal{M}_{(i,j)}$ cluster is insufficient to cope with the excessive access traffic demand at $\mathcal{H}_{(i)}$, while adjacent $\mathcal{H}_{(v)}$ can utilize its resource availability to enhance load balancing within the cellular network, thus implying the criteria to initiate M-AP handover as follows: (1) $\mathcal{M}_{(i,\zeta)} = \{\varnothing\}$, (2) $\eta_{(i)} \geq \eta_{\text{RA, high}}$, and (3) $\eta_{(v)} < \eta_{\text{RA, high}}$. This cooperative inter-cell support triggers the highly congested $\mathcal{H}_{(i)}$ to relay the information of selected M-AP(s) located at the macrocell edge (resembled by $\mathcal{M}_{(i,\kappa)}$) to $\mathcal{H}_{(v)}$, which then employs the low-power BAS radio to establish new access connectivity with $\mathcal{M}_{(v,\kappa)}$ $\left(\in \mathcal{M}_{(v,j)}\right)$. Hence, the RAIs of $\mathcal{M}_{(v,\kappa)}$ are switched on, and transmit power level adjustments may be required.

At the macro layer, the channel state information $\left(\xi_{\text{RF}}^{(i,k)}, \xi_{\text{FSO}}^{(i,k)}\right)$ and total backhaul traffic demand ($\chi_{(i)}$) are vital time-varying resource metrics monitored by $\mathcal{H}_{(i)}$, which are acquired using a simple feedback path and the traffic estimators (i.e., TE(1),... TE(M)), respectively. Upon detection of a backhaul traffic (for uplink or packet forwarding services) at the interface (i, k), under the normal-to-low data transfer volume condition with $\chi_{(i)} < \chi_{\text{BH, high}}$, the RF subsystem at the interface (i, k) with channel quality exceeding its signal-to-interference-and-noise ratio (SINR) threshold $\left(\xi_{\text{RF}}^{(i,k)} \geq \xi_{\text{RF, Th}}\right)$ is employed as the desired backhaul link, in which $\mathcal{H}_{(i)}$ triggers the high-power backhaul radio and generates data traffic at the interface. The resource prioritization algorithm checks for the possible RF link adaptation and performs SINR-to-adaptive modulation and coding (AMC) mapping for optimizing the link performance. Furthermore, the algorithm enables the complementary FSO subsystem under two scenarios: (1) $\xi_{\text{RF}}^{(i,k)} < \xi_{\text{RF,Th}}$, which may occur because of the rain attenuation, and (2) $\chi_{(i)} \geq \chi_{\text{BH, high}}$ owing to the excessive backhaul load. While adaptive modulation schemes may not be a feasible approach for FSO systems employing the IM/DD method, FSO link adaptation is possible with the error control coding technique, in which the channel encoder adapts the codeword length in accordance to the measured SINR.

9.5 FEASIBILITY STUDIES

We carry out preliminary feasibility studies to examine the time-varying characteristics of the macro-cellular BTSs daily traffic load and to carry out outage analysis for evaluating the performance of the proposed hybrid FSO/RF system under different weather conditions.

9.5.1 DAILY TRAFFIC PROFILE OF THE BTS

The daily traffic pattern of a BTS can be approximated by a modified sinusoidal profile given by [61]

$$\lambda(t) = \frac{1}{2^\psi}\left[1 + \sin\left(\frac{\pi}{12} + \varphi\right)\right]^\psi + \rho(t) \tag{9.19}$$

where $\lambda(t)$ denotes the instantaneous normalized traffic (in unit of Erlangs), $\psi = \{1, 3\}$ determines the abruptness of the traffic profile, φ is a uniform random variable with interval $[0, 2\pi]$, which determines the distribution of the traffic pattern among the BTSs, and $\rho(t)$ is a Poisson-distributed random process that models the random fluctuations of the traffic [61].

Our simulation studies indicate that the above approximation does not model the random fluctuating behavior of the data traffic in a realistic manner, compared to real measurements [62,63]. We suggest that the abruptness in the traffic profile can be generated as a sum of sinusoids, in which the modified expression is given by

$$\Lambda(t) = \sum_{i=1}^{N} \lambda_i(t). \tag{9.20}$$

In our studies, the traffic pattern is adequately modeled as a sum of eight sinusoids (i.e., $N = 8$).

Figure 9.8 shows the daily traffic pattern approximated for four BTSs, in which it is noted that BTSs generally exhibit a lower traffic requirement in the early morning (0:00 h to 4:00 h) and the late evening (20:00 h to 24:00 h) with a normalized traffic < 0.3 Erlangs, compared to peak hours during the day (i.e., 9:00 h to 16:00 h) with data traffic approaching the maximum load. These observations suggest that the proposed SWoD mechanism and cooperative inter-cell support would be of great benefit in promoting potential energy savings during the low traffic period, by placing idling/underutilized RAIs and M-APs into sleep mode. Under heavy traffic scenarios, the proposed features of the H-BTS architecture would enable best decision making for on-demand resource allocation, thereby maintaining even distribution of network load across a particular topology.

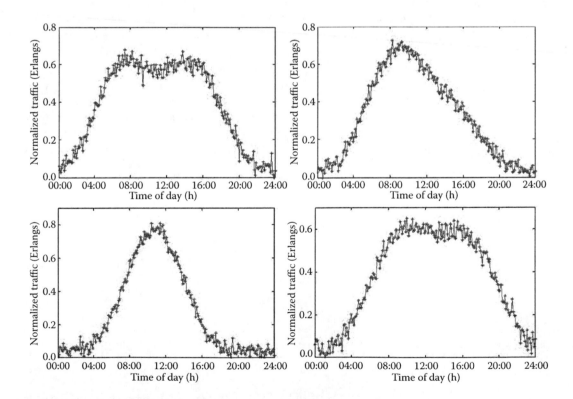

FIGURE 9.8   The resulting daily traffic pattern approximated for four BTSs.

9.5.2 Performance of the Hybrid FSO/RF System

The outage probability P_{out} is an appropriate performance measure for the system under study, which denotes the probability that the instantaneous channel capacity C, for a channel state h, falls below a desired transmission rate R_0, given by the relation [44]

$$P_{out} = \text{Prob}\,(C(\text{SNR}(h)) < R_0). \tag{9.21}$$

In principle, the instantaneous capacity corresponding to a channel state h is given by

$$C(\text{SNR}(h)) = \sum_{x=0}^{1} P_X(x) \int_{-\infty}^{+\infty} f(y|x,h)\log_2\left[\frac{f(y|x,h)}{\sum_{m=0,1} f(y|x=m,h)P_X(m)}\right]dy \tag{9.22}$$

where $P_X(x)$ is the probability of the bit being one ($x = 1$) or zero ($x = 0$), with $P_X(x = 0) = P_X(x = 1) = 0.5$, and the conditional pdfs $f(y|x, h)$ for both the FSO and RF subsystems are given by Equations 9.23 and 9.24, respectively.

For the FSO subsystem:

$$f(y|x, h = h_1) = \begin{cases} \dfrac{1}{\sqrt{2\pi\sigma_{n1}^2}}\exp\left[-\dfrac{y^2}{2\sigma_{n1}^2}\right], & x = 0 \\[3mm] \dfrac{1}{\sqrt{2\pi\sigma_{n1}^2}}\exp\left[-\dfrac{(y - 2P_{\text{FSO}}\gamma h_1)^2}{2\sigma_{n1}^2}\right], & x = 1. \end{cases} \tag{9.23}$$

For the RF subsystem:

$$f(y|x, h = h_2) = \begin{cases} \dfrac{1}{\sqrt{2\pi\sigma_{n2}^2}}\exp\left[-\dfrac{y^2}{2\sigma_{n2}^2}\right], & x = 0 \\[3mm] \dfrac{1}{\sqrt{2\pi\sigma_{n2}^2}}\exp\left[-\dfrac{\left(y - \sqrt{P_{\text{RF}}g_{\text{RF}}}\,h_2\right)^2}{2\sigma_{n2}^2}\right], & x = 1. \end{cases} \tag{9.24}$$

Since the optical slow-fading channel is random and remains unchanged over a long block of bits, the time-varying channel capacity will not be sufficient to support a maximum data rate, when the instantaneous SNR falls below a threshold. Such occurrence, known as system outage, can result in a loss of potentially up to 10^9 consecutive bits at a data rate of 10 Gbps, under deep fade scenarios that may last for ~1–100 ms [51]. Since $C(\cdot)$ is monotonically increasing with the SNR, the P_{out} for the FSO subsystem is the cumulative density function (cdf) of $h = h_1$ evaluated at $h_0 = \sqrt{C^{-1}(R_0)\sigma_{n1}^2/2P_{\text{FSO}}^2\gamma^2}$ and thus can be determined equivalently from the following expression [44]:

$$P_{out} = \int_{0}^{h_0} f_h(h)\,dh. \tag{9.25}$$

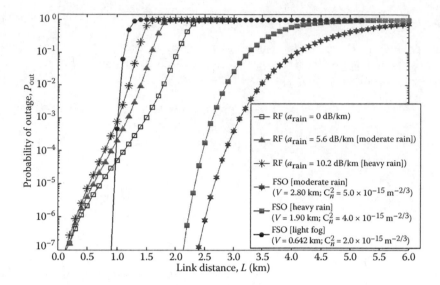

FIGURE 9.9 Outage probability of the FSO and RF links at varying link distance L, for $R_0 = 0.5$ bits/channel use, under different weather conditions.

Similarly, the P_{out} for the RF subsystem can be determined from Equation 9.25 with the cdf of $h = h_2$ evaluated at $h_0 = \sqrt{C^{-1}(R_0)\sigma_{n2}^2/P_{RF}g_{RF}}$, since the fading gain h_2 is modeled as Rician distributed, as defined in Equation 9.17. Figure 9.9 depicts the P_{out} for the FSO and RF links at varying link distance L under different weather conditions, based on the simulation settings as defined in Table 9.1. The weather effects are characterized through numerous parameters including the visibility V, rain attenuation a_{rain}, and turbulence strength C_n^2 [36,38,44]. Our results show that the FSO system is capable of maintaining a link range in excess of 2.0 km at $P_{out} = 10^{-6}$ for $R_0 = 0.5$ bits/channel use, under moderate (12.5 mm/h) and heavy (25.0 mm/h) rain conditions, while the RF link suffers significant link reduction with $L < 0.5$ km. This reveals the vast potential of the FSO link in enabling a very high-speed wireless backhauling under the adverse effects of rain. Under the low-visibility ($V = 0.642$ km) condition, the FSO link suffers severe performance degradation with $L < 1.0$ km at $P_{out} = 10^{-6}$ and experiences system outage probability of 1.0 for $L > 1.5$ km. The RF link can be employed as a complementary alternative to establish backhaul communication, albeit at a relatively lower data rate and link quality. Hence, the inherent advantage of both FSO and RF systems in complementing one another under the effects of rain and fog, respectively, reflects the symbiotic relationship between these technology options.

9.6   FUTURE RESEARCH DIRECTIONS

As the present research work is still at the preliminary stage, more in-depth theoretical analysis and simulation studies will be required to gain a better understanding of the technical characteristics, inherent advantages/limitations, and performance of the hybrid FSO/RF systems, which are primary considerations in performing adaptation and optimization at the link and system level for Metrozones deployment. In addition, the real-time behavior of the Metrozones will be modeled in a more realistic manner, by taking into account varying factors, which include (1) random fluctuations in the daily traffic profile of the BTSs, (2) mobility of the MUTs affecting the user density (hence, the network load) across the topology, and (3) time-series characteristics of the channel state owing to the influence of atmospheric conditions, such as rain, fog, haze, turbulence, and so on. Then, further enhancement to the proposed resource prioritization mechanism and BAS scheme can be made possible. From the ecological perspective, we seek to validate the feasibility of the proposed H-BTS system architecture for the green Metrozones, through the quantification of the total power consumption, potential energy savings, and overall carbon footprint (measured in CO_2e) [8]. Furthermore, capacity and throughput analysis can be carried out to justify the performance improvement achievable with the proposed system.

Nevertheless, the development and deployment of next-generation green Metrozones to enable bandwidth-intensive, ubiquitous, and energy-efficient mobile service provisioning require more elaborative efforts to venture into other research opportunities, in order to seek optimal and holistic approaches for green communications. These include the following:

1. *The introduction of sleep/wake-up modes* in BTSs could promote a substantial amount of energy savings, particularly at relatively low traffic periods, by switching underutilized BTSs into low power state such that they have no traffic to handle, thereby reducing the OPEX of the network and improving the efficiency of network capacity utilization. Studies on practical traffic patterns show that the total traffic generation of different classes in a BTS remains relatively low for a significant portion of day compared to the capacity, and the peak and off-peak times vary with locations [62,64]. These inherent traffic characteristics have been advantageous in implementing the sleep/wake-up technologies in cellular networks. This approach relies upon the design of sophisticated self-sustainable/adaptable networks, with self-organizing, self-adapting, self-managing, and self-optimizing capabilities, and ideally capable of serving mobile subscribers in most situations [6,61]. In Ref. [64], an energy-efficient self-sustainable cellular access network architecture based on the ecological principle of proto-cooperation has been proposed, in which artificial intelligence–enhanced BTSs have been incorporated in the proposed architecture to enable the wake-up technology and intelligent decision making for switching between different power modes (i.e., active, sleep, and off modes) depending on network traffic conditions, in order to promote (and optimize) energy savings in the access networks. In addition, Vereecken et al. [6] addressed the limitations of existing telecommunication networks, which are typically designed to handle peak loads, and highlighted the necessity of adaptable networks to switch off elements under medium and low load conditions, to minimize the power consumption. The authors explained that adaptable networks can be achieved with dynamic topology optimization, dynamic bandwidth allocation, and hybrid hierarchical BTS deployment in the core, fixed line access, and wireless access networks, respectively.

2. *Reduction of the energy consumption of BTSs* can be achieved with the improvement of the BTS hardware design and the inclusion of additional software and system features to strike a balance between energy consumption and performance. In particular, power amplifiers dominate the energy consumption of BTSs by consuming ~50% of the BTSs' total energy requirement, and their energy efficiency depend on the frequency band, modulation, and operating environment [65,66]. The poor power efficiency of power amplifiers is mainly due to the high linearity and maximum output power requirements of these components to maintain the required radio signal quality. At present, power amplifiers based on digital pre-distorted Doherty architectures and gallium-nitride (GaN) transistor technologies present one of the most advanced power amplification technologies in the mobile cellular industry, which have dramatically minimized the energy consumption of state-of-the-art BTSs for BWA, with power efficiency levels in excess of 45% [5,65,66]. Doherty power amplifiers provide enhanced linearization with much convenience by using conventional methods, such as feed-forward and envelope elimination and restoration, whereas GaN structures can operate in the higher temperature and voltage regimes, to provide a higher power output. With the migration of traditional analogue RF amplifiers toward switch-mode power amplifiers, which generate very little power as heat and draw lesser current, further improvements in power efficiency can be attained, with an expected overall component efficiency of approximately 70% [65]. In the recent years, compact BTSs incorporating

distributed architecture have entered the market as a more flexible, cost-effective, and eco-friendly alternative over conventional ground-based macro-BTSs for greenfield deployment and operation in RANs, in order to deliver robust performance at a low cost per bit and to mitigate power loss in feeder cables [25]. The increasingly high degree of integration and sophistication of system-on-chip silicon technology has spurred the emergence of these ultra-compact, ruggedized, single-enclosure BTSs, in which the RF components can now be mounted next to the antennas in the remote radio head units, such that the coaxial cables to the ground are eliminated in favor of optical fiber strands (if wireline backhaul is used). Correspondingly, the throughput, coverage, and functionality comparable to ground-based BTSs can be achieved with the compact BTSs, albeit at a much lower capital and OPEXs and installation cost, and lower power consumption owing to the absence of ground equipments, feeder cable losses, and, particularly, power-hungry cooling units.

3. *Cognitive radio* (also known as smart radio) is defined by Mitola and Maguire [67] as a radio with software-defined intelligent capabilities to take into account every possible parameter measurable by a wireless node or network (cognition), so that the network intelligently adapts/modifies its functionality (reconfigurability) through interaction with the environment in which it operates, such as active negotiation/communication with other spectrum users or passive sensing and decision making within the radio, to meet a certain objective. In the context of green communications, this optimization criterion can be power saving, which can be achieved with the agility and adaptation properties of the cognitive radios, as flexibility and adaptability to different circumstances are crucial for any green communications scheme to attain widespread and effective applicability [65]. While cognitive-enhanced BTSs can improve network performance through environment-aware and self-aware operation capabilities, there are numerous underlying challenges pertaining to cognitive radios that must be addressed, such as hardware complexity, algorithmic problems, and design trade-offs. Gur and Alagoz [68] highlighted two related aspects of cognitive radios from the green networks perspective—(1) the energy management problem within cognitive radios, regarded as a multidimensional optimization problem, which consists of dynamically controlling the system to minimize the average energy consumption under several performance constraint(s), and (2) the energy efficiency optimization capabilities through cognitive radios. These optimization capabilities encompass broad aspects of intelligence support for energy efficiency functionalities, energy savings via duty-cycle optimization and robustness in ad hoc settings, network and physical layer capabilities, cross-layer optimizations, enabler for ubiquitous efficiency optimizations, bandwidth energy trade-offs, and simpler and more efficient evolution paths. It is envisaged that cognitive radios will unlock a multitude of optimal and holistic green approaches, to promote more energy-saving opportunities within the ICT sector and the relevant interconnected systems.

4. *Optimization of network architecture and topology* within existing wireless cellular networks presents an economical viable approach, in order to cope with the demand for high-speed, bandwidth-intensive, ubiquitous, and energy-efficient mobile communication. As cellular networks are becoming more congested with the dense deployment of BTSs, the diffusion of smaller cells (such as micro-, pico-, and femtocells) into conventional macrocellular networks [65] and the strategic placement of repeaters and relays within the networks [5] are particularly advantageous in tackling problems such as dead spots, slow connection speeds, and poor system capacity [14], in order to enable power-efficient and cost-effective network coverage. Careful investigative efforts and extensive development works are essential to seek appropriate solution(s) to address numerous issues related to network planning and deployment, which include (1) intelligent control of transmission power levels to optimize the capacity of the sub-cells within the cellular networks and to promote potential energy savings simultaneously; (2) investigation on the interrelationship/

trade-off between the capacity, network scalability, and power consumption, as have been indicated by Nandiraju et al. [20] that these metrics are dependent upon the network architecture, network topology, traffic pattern, network node density, number of channels, and transmission power level; (3) development of life cycle analytical framework and carbon footprint models to predict the long-term energy consumption and carbon footprint of the next-generation wireless cellular networks [8,62]; and (4) incorporating energy efficiency into commercial telecommunications system modeling and planning software tools [66], in order to enable network operators to strike a balance between optimizing user capacity and minimizing power consumption, particularly for greenfield deployment.

5. *The incorporation of renewable and alternative energy resources* such as sustainable bio-fuels, solar, and wind energy for operating BTSs presents a viable approach to reduce the overall network expenditure and mitigate the carbon emissions, particularly in remote/rural areas where the electricity grids are inaccessible and in developing markets where the grids are poor/unstable and solar or wind resources are abundant [5,8,65,66,69]. To date, network operators have been reluctant in adopting renewable energy-based BTSs, which is mainly due to the fear of little commercial viability and lack of equipment expertise. However, the Global Systems for Mobile communications Association reported that the implementation of green power technology represents a technically feasible and financially attractive solution with a payback period of less than three years at many sites [70].

9.7 CONCLUSIONS

In this chapter, we have proposed a new *H-BTS system architecture* for the green Metrozones. Taking advantage of the symbiotic relationship between the FSO and RF technologies, the *hybrid FSO/RF systems* have been integrated at the macrocellular tier, to enable high-capacity, power-efficient wireless backhauling under most weather conditions and varying data traffic load. In addition, the *RRM module* encompassing the *resource prioritization mechanism* has been introduced into the system hub of the proposed H-BTS architecture. This is to maintain a good control and optimal on-demand resource allocation to both the wireless backhaul and RF access networks, and to establish sustainable wireless backhaul link availability via essential switching between the FSO and RF communication links. Furthermore, the *BAS scheme* employing a default low data-rate, low-power radio has been considered, which necessitates the discovery, registration and monitoring of active M-APs, in order to enable two distinctive features: the *SWoD mechanism* and *cooperative inter-cell support*. The *SWoD mechanism* minimizes the number of operating RAIs and enhances potential energy savings by putting idling/underutilized RAIs and M-APs into sleep mode, particularly in low-traffic scenarios. The *cooperative inter-cell support* off-loads the M-APs located at the macro-cell edge to neighboring H-BTSs with more resource availability, thus enabling more even distribution of the network load across a particular topology. Findings from the present work have indicated that adaptation and optimization at the link and system level are vital for Metrozones deployment, owing to the occurrence of numerous time-varying factors in real networks.

AUTHOR BIOGRAPHIES

It Ee Lee received her BEng (Hons) majoring in electronics and MEngSc degrees from Multimedia University (MMU), Malaysia, in 2004 and 2009, respectively. In 2005, she was awarded the MMU-Agilent scholarship to pursue her MEngSc degree in the research area of visible light communications (VLC). Currently, she is a recipient of the Northumbria University Research Studentship and is working toward a PhD degree on free-space optical (FSO) communication systems in the Optical Communication Research Group at Northumbria University, United Kingdom. Her research interests include channel modeling, performance analysis, and optimization of FSO communication systems, hybrid FSO/RF systems, VLC systems, and optical wireless communications.

Zabih Ghassemlooy received his BSc (Hons) degree in electrical and electronics engineering in 1981 and his MSc and PhD in optical communications from the University of Manchester Institute of Science and Technology, in 1984 and 1987, respectively. From 1987 to 1988, he was a postdoctoral research fellow at the City University, London. In 1988, he joined Sheffield Hallam University as a lecturer, becoming a reader in 1995, and a professor in optical communications in 1997. From 2004 to 2012, he was an associate dean for research in the School of Computing, Engineering, and in 2012, he became associate dean for research and innovation in the Faculty of Engineering and Environment, at Northumbria University at Newcastle, United Kingdom. He also heads the Northumbria Communications Research Laboratories within the faculty. In 2001, he was a recipient of the Tan Chin Tuan Fellowship in Engineering from the Nanyang Technological University in Singapore and was tasked to work on photonic technology. He is the editor-in-chief of *The Mediterranean Journal of Computers and Networks* and *The Mediterranean Journal of Electronics and Communications*. He currently serves on the editorial committees of a number of international journals. He is the founder and the chairman of the IEEE, IET International Symposium on Communication Systems, Network and Digital Signal Processing. His research interests are on photonics switching, optical wireless and wired communications, visible light communications, and mobile communications. He has supervised a large number of PhD students (more than 39) and has published over 440 papers (150 in journals plus 11 book chapters) and presented several keynote and invited talks. He is a co-author of a CRC Press book, *Optical Wireless Communications: System and Channel Modeling with MATLAB®* (2012), and a co-editor of an IET book, *Analogue Optical Fibre Communications*. From 2004 to 2006, he was the IEEE UK/IR Communications chapter secretary, the vice-chairman (2004–2008), the chairman (2008–2011), and chairman of the IET Northumbria Network (October 2011–present).

Wai Pang Ng is a reader at Northumbria University, United Kingdom. He received his BEng (Hons) in communication and electronic engineering and the IEE Prize from Northumbria University in 1997. Then, he pursued his graduate study, in collaboration with BT Labs and completed his PhD in electronic engineering at University of Wales, Swansea, in 2001. Dr. Ng started his career as a senior networking software engineer at Intel Corporation from 2001 to 2004. At Northumbria University, he has published over 70 journals and conferences publications in the area of optical switching, optical signal processing, and radio-over-fiber. He is a chartered engineer of Engineering Council UK and senior member of IEEE. Dr. Ng currently serves as the chair of IEEE UK&RI Communication Chapter. He was also the co-chair of the Signal Processing for Communications Symposium in IEEE International Conference on Communications (ICC) 2009 (Dresden, Germany) and publicity chair of ICC 2015 in London.

Mohammad-Ali Khalighi received his PhD in electrical engineering (telecommunications) from Institut National Polytechnique de Grenoble, France, in 2002. From 2002 to 2005, he has been with GIPSA-lab, Télécom Paris-Tech, and IETR-lab, as a postdoctoral research fellow. He joined Ecole Centrale Marseille and Institut Fresnel in 2005 as an assistant professor. His main research areas of interest include signal processing for optical wireless communication and wireless sensor networks.

REFERENCES

1. United Nations Environment Programme (2011). Towards a green economy: Pathways to sustainable development and poverty eradication. http://www.unep.org/greeneconomy/green economyreport/tabid/29846/default.aspx [Online] (accessed January 17, 2013).
2. United Nations Framework Convention on Climate Change (2012). http://unfccc.int [Online] (accessed January 17, 2013).
3. EARTH Consortium (2012). https://www.ict-earth.eu [Online] (accessed July 30, 2012).
4. Stichting GreenTouch (2012). http://www.greentouch.org [Online] (accessed July 30, 2012).
5. S. Vadgama, Trends in green wireless access, *FUJITSU Scientific and Technical Journal*, 45, 404–408, 2009.

6. W. Vereecken, W. Van Heddeghem, M. Deruyck, B. Puype, B. Lannoo, W. Joseph, D. Colle, L. Martens and P. Demeester, Power consumption in telecommunication networks: Overview and reduction strategies, *IEEE Communications Magazine*, 49, 62–69, 2011.
7. Global e-Sustainability Initiative (2008). SMART 2020: Enabling the low carbon economy in the information age. http://gesi.org/ReportsPublications/Smart2020/tabid/192/Default.aspx [Online] (accessed July 30, 2012).
8. A. Fehske, G. Fettweis, J. Malmodin and G. Biczok, The global footprint of mobile communications: The ecological and economic perspective, *IEEE Communications Magazine*, 49, 55–62, 2011.
9. Mobile VCE (2012). http://www.mobilevce.com [Online] (accessed January 17, 2013).
10. Huawei Technologies (2011). Corporate sustainability report: Enriching life through communication. http://www.huawei.com/en/about-huawei/corporate-citizenship/csr-report/index.htm [Online] (accessed January 17, 2013).
11. C. Forster, I. Dickie, G. Maile, H. Smith and M. Crisp (2009). Understanding the environmental impact of communication systems. http://stakeholders.ofcom.org.uk/market-data-research/other/technology-research/research/sector-studies/environment/ [Online] (accessed January 17, 2013).
12. R. Prasad and F. J. Velez, *WiMAX Networks: Techno-Economic Vision and Challenges*, New York: Springer, 2010.
13. W. Webb, *Wireless Communications: The Future*, Chichester: John Wiley & Sons, 2007.
14. The future of infrastructure: compact base stations, White Paper, In-Stat, 2010.
15. J. Mitola and Z. Zvonar, *Software Radio Technologies: Selected Readings*, Piscataway: Wiley-IEEE Press, 2001.
16. T. Ulversoy, Software defined radio: Challenges and opportunities, *IEEE Communications Surveys and Tutorials*, 12, 531–550, 2010.
17. S. Bellofiore, C. A. Balanis, J. Foutz and A. S. Spanias, Smart-antenna systems for mobile communication networks. Part 1. Overview and antenna design, *IEEE Antennas and Propagation Magazine*, 44, 145–154, 2002.
18. S. Bellofiore, J. Foutz, C. A. Balanis and A. S. Spanias, Smart-antenna system for mobile communication networks. Part 2. Beamforming and network throughput, *IEEE Antennas and Propagation Magazine*, 44, 106–114, 2002.
19. A. Alexiou and M. Haardt, Smart antenna technologies for future wireless systems: Trends and challenges, *IEEE Communications Magazine*, 42, 90–97, 2004.
20. N. Nandiraju, D. Nandiraju, L. Santhanam, H. Bing, J. Wang and D. P. Agrawal, Wireless mesh networks: Current challenges and future directions of web-in-the-sky, *IEEE Wireless Communications*, 14, 79–89, 2007.
21. I. F. Akyildiz, X. Wang and W. Wang, Wireless mesh networks: A survey, *Computer Networks*, 47, 445–487, 2005.
22. W. Chen, K. B. Letaief and Z. Cao, Network interference cancellation, *IEEE Transactions on Wireless Communications*, 8, 5982–5999, 2009.
23. J. Lee, D. Toumpakaris and W. Yu, Interference mitigation via joint detection, *IEEE Journal on Selected Areas in Communications*, 29, 1172–1184, 2011.
24. Nokia Siemens Networks good green business sense, White Paper, Nokia Siemens Networks, 2008.
25. Compact base stations: a new step in the evolution of base station design, White Paper, Senza Fili Consulting, 2010.
26. The emergence of compact base stations in the new RAN architecture paradigm, White Paper, ABI Research, 2010.
27. Metro zone Wi-Fi for cellular data offloading, White Paper, Wavion Wireless Networks, 2010.
28. O. Tipmongkolsilp, S. Zaghloul and A. Jukan, The evolution of cellular backhaul technologies: Current issues and future trends, *IEEE Communications Surveys and Tutorials*, 13, 97–113, 2011.
29. SoC-based mobile broadband evolution, White Paper, DesignArt Networks, 2011.
30. Self organizing network: NEC's proposals for next-generation radio network management, White Paper, NEC Corp., 2009.
31. Self organizing network (SON): Introducing the Nokia Siemens Networks SON suite – an efficient, future-proof platform for SON, White Paper, Nokia Siemens Networks, 2009.
32. WiMAX.com (2012). Backhaul for WiMAX: Top 8 technical considerations. http://www.wimax.com/microwave-backhaul/backhaul-for-wimax-top-8-technical-considerations [Online] (accessed January 18, 2013).
33. D. Jones (2008). 4G: Can't stand the rain. http://www.heavyreading.com/document.asp?doc_id=154434 [Online] (accessed January 18, 2013).

34. A. Mahdy and J. S. Deogun, Wireless optical communications: A survey, In *Proceedings of the IEEE Wireless Communications and Networking Conference (WCNC)*, 2399–2404, 2004.

35. S. Bloom and W. Hartley, The last-mile solution: Hybrid FSO radio, White Paper, AirFiber Inc., 2002.

36. B. He and R. Schober, Bit-interleaved coded modulation for hybrid RF/FSO systems, *IEEE Transactions on Communications*, 57, 3753–3763, 2009.

37. I. I. Kim and E. Korevaar, Availability of free space optics (FSO) and hybrid FSO/RF systems, in *Proceedings of the SPIE, Optical Wireless Communications IV*, 84–95, 2001.

38. W. Zhang, S. Hranilovic and C. Shi, Soft-switching hybrid FSO/RF links using short-length raptor codes: Design and implementation, *IEEE Journal on Selected Areas in Communications*, 27, 1698–1708, 2009.

39. J. N. Laneman, E. Martinian, G. W. Wornell and J. G. Apostolopoulos, Source-channel diversity for parallel channels, *IEEE Transactions on Information Theory*, 51, 3518–3539, 2005.

40. D. Kedar and S. Arnon, Urban optical wireless communication networks: The main challenges and possible solutions, *IEEE Communications Magazine*, 42, S2–S7, 2004.

41. A. K. Majumder and J. C. Ricklin, *Free-Space Laser Communications*, New York: Springer, 2008.

42. Z. Ghassemlooy, W. Popoola and S. Rajbhandari, *Optical Wireless Communications: System and Channel Modelling With MATLAB*, Boca Raton, FL: Taylor & Francis Group, 2012.

43. L. C. Andrews and R. L. Philips, *Laser Beam Propagation Through Random Media*, Washington: SPIE, 2005.

44. A. A. Farid and S. Hranilovic, Outage capacity optimization for free-space optical links with pointing errors, *Journal of Lightwave Technology*, 25, 1702–1710, 2007.

45. D. K. Borah and D. G. Voelz, Pointing error effects on free-space optical communication links in the presence of atmospheric turbulence, *Journal of Lightwave Technology*, 27, 3965–3973, 2009.

46. M. A. Khalighi, N. Schwartz, N. Aitamer and S. Bourennane, Fading reduction by aperture averaging and spatial diversity in optical wireless systems, *IEEE/OSA Journal of Optical Communications and Networking*, 1, 580–593, 2009.

47. X. M. Zhu and J. M. Kahn, Free-space optical communication through atmospheric turbulence channels, *IEEE Transactions on Communications*, 50, 1293–1300, 2002.

48. E. Biglieri, J. Proakis and S. Shamai, Fading channels: Information-theoretic and communications aspects, *IEEE Transactions on Information Theory*, 44, 2619–2692, 1998.

49. I. I. Kim, B. McArthur and E. Korevaar, Comparison of laser beam propagation at 785 nm and 1550 nm in fog and haze for optical wireless communications, In *Proceedings of the SPIE, Optical Wireless Communications III*, 26–37, 2001.

50. D. Atlas, Shorter contribution optical extinction by rainfall, *Journal of Meteorology*, 10, 486–488, 1953.

51. E. J. Lee and V. W. S. Chan, Part 1. Optical communication over the clear turbulent atmospheric channel using diversity, *IEEE Journal on Selected Areas in Communications*, 22, 1896–1906, 2004.

52. S. Arnon, Effects of atmospheric turbulence and building sway on optical wireless-communication systems, *Optics Letters*, 28, 129–131, 2003.

53. S. Arnon, Optimization of urban optical wireless communication systems, *IEEE Transactions on Wireless Communications*, 2, 626–629, 2003.

54. F. Giannetti, M. Luise and R. Reggiannini, Mobile and personal communications in the 60 GHz band: A survey, *Wireless Personal Communications*, 10, 207–243, 1999.

55. R. C. Daniels and R. W. Heath, 60 GHz wireless communications: Emerging requirements and design recommendations, *IEEE Vehicular Technology Magazine*, 2, 41–50, 2007.

56. R. C. Daniels, J. N. Murdock, T. S. Rappaport and R. W. Heath, 60 GHz Wireless: Up close and personal, *IEEE Microwave Magazine*, 11, 44–50, 2010.

57. N. Guo, R. C. Qiu, S. S. Mo and K. Takahashi, 60-GHz millimeter-wave radio: Principle, technology, and new results, *EURASIP Journal on Wireless Communications and Networking*, 68253, 1–8, 2007.

58. J. Schönthier, The 60 GHz channel and its modelling, WP3-Study, BROADWAY IST-2001-32686, version V1.0, May 2003.

59. C. Tepedelenlioglu, A. Abdi and G. B. Giannakis, The Ricean K factor: Estimation and performance analysis, *IEEE Transactions on Wireless Communications*, 2, 799–810, 2003.

60. K. Piamrat, A. Ksentini, J.-M. Bonnin and C. Viho, Radio resource management in emerging heterogeneous wireless networks, *Computer Communications*, 34, 1066–1076, 2011.

61. M. F. Hossain, K. S. Munasinghe and A. Jamalipour, A protocooperation-based sleep-wake architecture for next generation green cellular access networks, In *Proceedings of the 4th International Conference on Signal Processing and Communication Systems (ICSPCS)*, 1–8, 2010.

62. L. M. Correia, D. Zeller, O. Blume, D. Ferling, Y. Jading, I. Gódor, G. Auer and L. Van Der Perre, Challenges and enabling technologies for energy aware mobile radio networks, *IEEE Communications Magazine*, 48, 66–72, 2010.

63. P. E. Heegaard, Evolution of traffic patterns in telecommunication systems, In *Proceedings of the 2nd International Conference on Communications and Networking in China (CHINACOM)*, 28–32, 2007.

64. M. F. Hossain, K. S. Munasinghe and A. Jamalipour, An eco-inspired energy efficient access network architecture for next generation cellular systems, In *Proceedings of the IEEE Wireless Communications and Networking Conference (WCNC)*, 992–997, 2011.

65. Z. Hasan, H. Boostanimehr and V. K. Bhargava, Green cellular networks: A survey, some research issues and challenges, *IEEE Communications Surveys and Tutorials*, 13, 524–540, 2011.

66. A. Amanna, A. He, T. Tsou, X. Chen, D. Datla, T. R. Newman, J. H. Reed and T. Bose, Green communications: A new paradigm for creating cost effective wireless systems. http://filebox.vt.edu/users/ aamanna/web%20page/Green%20Communications-draft%20journal%20 paper.pdf [Online] (accessed January 24, 2013).

67. J. Mitola and G. Q. Maguire, Cognitive radio: Making software radios more personal, *IEEE Personal Communications*, 6, 13–18, 1999.

68. G. Gur and F. Alagoz, Green wireless communications via cognitive dimension: An overview, *IEEE Network*, 25, 50–56, 2011.

69. W. Tuttlebee, S. Fletcher, D. Lister, T. O'Farrell and J. Thompson, Saving the planet—The rationale, realities and research of green radio, *The Journal of the Institute of Telecommunications Professionals*, 4, 2010.

70. Global Systems for Mobile Communications Association (2010). Bi-annual Report November 2010. http://www.gsma.com/mobilefordevelopment/wp-content/uploads/2012/05/GPM_ Bi-Annual_Report_ Nov10.pdf [Online] (accessed January 24, 2013).

10 Green Heterogeneous Small-Cell Networks

Muhammad Zeeshan Shakir, Hina Tabassum,
Khalid A. Qaraqe, Erchin Serpedin,
and Mohamed-Slim Alouini

CONTENTS

10.1 INTRODUCTION

The demand for data rates in upcoming wireless standards is increasing exponentially because of data-hungry wireless devices and bandwidth-intensive applications. Moreover, spectral link efficiency is achieving its theoretical limits; therefore, the only solution to enhance the system performance is to improve and densify the base station (BS) deployment in the required areas. This densification, however, presents its own challenges. Adding a BS in the sparsely deployed area does not have much impact on interference and thus cell splitting gains are easy to achieve, whereas adding a BS in the densely deployed urban area generates severe interference per channel and therefore reduces significantly the cell splitting gains. In addition, the site acquisition cost in a capacity-limited dense urban area may also become prohibitively expensive [1].

To overcome these challenges, heterogeneous small-cell networks (HetSNets) emerge as a striking solution to the challenging demands such as high spectral and energy efficiency, improved cell coverage, and cell edge performance of the future wireless networks. HetSNets combine and support several diverse, low-power, low-cost entities, which are referred to as small cells such as microcells, femtocells, and picocells to complement the existing network and enhance the overall performance of the system [2–4]. The small-cell BSs (SBSs) can be deployed under open or closed access mechanisms to support the macrocell network. Under open access mechanisms, the SBSs are installed by the mobile operators in public areas, for example, airport, public parks, shopping malls, and cafes such that all users in the vicinity are connected to the small-cell access point (SAP). However, under closed access mechanisms, the SBSs are installed in residential places and enterprises for home and office applications such that only a subset of the registered users can access the SAP. HetSNets represent a core wireless network, which is a combination of macrocells and small cells with varying transmit powers and coverage areas in order to augment the network capacity and coverage [5].

10.1.1 Benefit of HetSNets

The idea of bringing the network closer to the subscriber facilitates the subscribers and operators from the following fundamental perspectives:

1. *Macrocell Traffic Off-Loading*: Deploying small cells aims at off-loading the traffic from macrocells [6], improving the indoor coverage and cell-edge user performance by spatial reuse.
2. *Spectral and Energy Efficiency*: Small cells can be deployed with relatively low network overhead and possess high potential for reducing the energy consumption and enhancing the spectral efficiency of the future wireless networks owing to the short distance between the transmitter and the receiver of a typical communication link [7–9].
3. *Reduced CAPEX and OPEX*: Small cells require little or no up-front planning and lease costs, therefore drastically reducing the network operational (OPEX) and capital expenditures (CAPEX) [5,10].

10.1.2 Major Challenges for HetSNets

The implementation of HetSNets implies its own challenges and there are several technical issues that still need to be resolved for their successful operation. Some of the major issues include the following:

1. *Co-tier and Cross-tier Interference*: As the network operators may not control the exact number and locations of the deployed small cells, this unplanned deployment makes the nature of incurred interference more impulsive and unpredictable. In addition, the combination of private and public access policies associated with each small cell and the absence of coordination among them makes it further difficult for the system designers to handle and manage interference through interference coordination techniques. However, the challenging problems of small-cell network access mechanisms are providing secure access to the network resources and mitigation of cross-tier and intra-tier interferences.
2. *Number of Handovers*: Handovers are considered as an essential technique to provide a seamless uniform service where users move in or out of the cell coverage. Handovers are

efficient for balancing the traffic load, shifting cell-edge users from the highly loaded cells to the low loaded ones with a trade-off cost in terms of system overhead. This overhead includes excessive signaling for providing handover between adjacent small cells or between small cells and the macrocell in case of high user mobility. Moreover, this is expected to be more significant in HetSNets owing to the large number of small cells and the different types of backhaul links for each type of cell [2].

3. *Self-Organization*: As small cells are likely to be user-deployed without operator supervision, they should be cognitive and intelligent to configure themselves automatically, perform failure recovery, and optimize their settings according to the network conditions in order to improve coverage and reduce interference [11]. Employing self-organizing features in upcoming heterogeneous networks is a challenging task owing to the various coexisting architectures and a large number of network parameters that need to be optimized.

4. *Backhaul*: The presence of diverse entities in upcoming heterogeneous networks makes the designing of the backhaul network extremely challenging. As an example, the deployment of picocells requires expensive access to utility infrastructure with power supply and wired network backhauling, while femtocells may possess relatively lower backhauling costs and reduced end user quality of service (QoS). A compromising solution can be a mixture of both wireless and wired backhaul technologies [11].

5. *Spectrum Access and Channel Allocation*: The small-cell networks are vulnerable to cross-tier interference. The cross-tier interference can be managed by employing the overlay or underlay strategy. For the overlay strategy, interference can be avoided by employing intelligent resource partitioning schemes. Although such schemes may eliminate the cross-tier interference, they suffer from poor spectrum usage (under- or overusage of spectrum) as a result of static spectrum allocation and poor signal-to-interference ratio of small-cell mobile user caused by high macrocell mobile user activity. As a result, there is considerable interest in the underlay strategy, where the macrocell and small cell share the common spectrum and interference can be mitigated by employing power control (PC) mechanisms [5].

This chapter presents an energy-efficient design of HetSNets that hold the potential to improve the energy savings of the network and yield a low carbon economy. We propose a small-cell deployment in the macrocell where the SBSs are arranged around the edge of the reference macrocell. The deployment is referred to as cell-on-edge (COE) and has been shown to produce significant spectral and energy efficiency gains compared to (i) HetSNets where the small cells are uniformly distributed across the macrocells, that is, uniformly distributed small cells (UDCs), and (ii) macro-only networks (MoNets). The COE deployment improves the energy savings of the HetSNets by ensuring that each of the mobile users in macrocell and small-cell networks are transmitting with adaptive power, thereby saving energy and reducing CO_2 emissions of the mobile communications industry and moving toward greener wireless communications systems.

The following represent the main contributions of this chapter:

- Study the performance analysis of HetSNets with COE configuration in terms of (i) area spectral efficiency (ASE), which is defined as sum of mean achievable rates per unit bandwidth per unit area, and (ii) area green efficiency (AGE), which is defined as the aggregate energy savings per unit area of the heterogeneous network.
- Study the energy economics of the HetSNets in terms of energy savings and corresponding cost savings as a result of the energy-efficient uplink of HetSNets.
- Investigate the reduction in CO_2 emissions of HetSNets and their ecological impact.
- Simulation results, comparisons, and discussions to validate the performance of the HetSNets with COE configuration with respect to the HetSNets with UDC configuration and MoNets included where deemed necessary.

The rest of this chapter is organized as follows. Section 10.2 discusses the significance of energy-efficient designs for wireless networks, while in Section 10.3, methods to improve the energy efficiency of the wireless networks are discussed in detail. Section 10.4 presents some background work on the current trends of designing heterogeneous networks. Section 10.5 presents the proposed HetSNets layout, the bandwidth partition and the channel allocation strategy, and, finally, the energy-aware channel propagation model. Section 10.6 conducts performance analysis in terms of ASE, AGE, ecology, and economy of energy of the HetSNets. This section also discusses the energy savings and the reduction in the CO_2e emissions of the wireless networks. Several simulation studies are presented to compare the performance of HetSNets with COE/UDC configurations with that of MoNets. Section 10.7 presents future directions on designing green networks, their challenges, and issues. Finally, conclusions are presented in Section 10.8.

10.2 SIGNIFICANCE OF ENERGY EFFICIENCY IN WIRELESS NETWORKS

Energy efficiency has been recently recognized as one of the alarming bottlenecks in the telecommunication growth paradigm for two major reasons:

1. Dramatically varying global climate [12]
2. Slowly progressing battery technology [13,14]

Carbon footprint is a key ecological factor that is measured in carbon dioxide equivalent (CO_2e) and defined as the amount of CO_2 emissions calculated according to the global warming potential (GWP-100) indicator as defined by the International Panel on Climate Change [15].* The ever-growing demand for wireless services and ubiquitous network access is increasing the carbon footprint for the mobile communications industry.

The information and communication technology (ICT) sector and the mobile communications industry have been estimated to jointly represent about 2% of global CO_2 emissions and about 1.3% of global CO_2e emissions [16]. Figure 10.1 shows the CO_2 emissions profile for the mobile communications industry. It can be seen clearly that even with the technological advancements in improving the energy efficiency of ICT infrastructure, 6% growth rate is expected every year till 2020 [16]. The fundamental factors contributing to the overall global carbon footprint include production, operation, distribution, and maintenance of the mobile communication networks and services.

In addition, the other essential reasons behind the carbon footprint of mobile communications networks are the number of global mobile customers, the size of the network infrastructure, and the mobile traffic volume (which shows a steep growing pattern over the last few years with the proliferation of smart phones, tablet computers, and other smart devices). Typically, the energy consumption of the wireless BS is far more higher compared to the mobile user terminals. From Ref. [17], more than 50% of the total energy is consumed by the radio access part, where 50%–80% is used for the power amplifier. Radio BS is the most intensive component in terms of energy consumption in 4G mobile communications networks. Currently, there are more than 4 million BSs serving mobile users, each consuming an average of 25 MWh per year [15]. The number of BSs in developing regions is expected to almost double every year especially in India and China. Moreover, energy costs also represent a significant portion of network operators' OPEX. While the BSs connected to the electrical grid may cost approximately $3000 per year to operate, the off-grid BSs in remote areas generally run on diesel power generators and may cost 10 times more [15]. In Ref. [17], it is also pointed out that the energy bill of the mobile communications industry accounts for approximately 18% and 32% of the OPEX in the European and Indian markets, respectively.

---

* CO_2e represents a standard unit to measure the impact of each of the different greenhouse gases in terms of the amount of CO_2 that would create the same amount of warming.

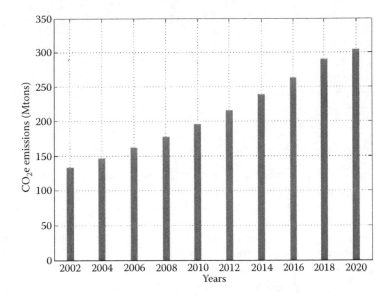

FIGURE 10.1 CO_2 emissions profile of mobile communications industry till 2020.

In addition to the above-mentioned facts about BS power consumption, energy efficiency has recently been deemed important from the mobile user's perspective. Some results of 2010 wireless smart phone customer satisfaction studies from J. D. Power and Associates demonstrate that the iPhone ranked top in all categories except for the battery life (see Ref. [18] and the references therein). According to another more recent report, up to 60% of the mobile users in China complained that the battery consumption is the greatest hurdle while using 3G services [18]. This fact requires a breakthrough battery technology as the battery life of the mobile terminals is the fundamental limitation for power-hungry wireless applications.

10.3   METHODS TO IMPROVE ENERGY EFFICIENCY IN WIRELESS NETWORKS

Heterogeneous networks are considered as an important way to enhance the spectral and energy efficiency of cellular networks by reducing the propagation distance between the source and destination. In this regard, small cellular network deployment strategies are gaining popularity. However, careful network design is still required since deploying too many small cells may in fact saturate and reduce the energy efficiency of the central BS [9]. Numerous alternate methods are available in the literature to enhance the energy efficiency of wireless cellular networks at the design and network planning level. Some of the key technologies are summarized next.

10.3.1   IMPROVED POWER AMPLIFIER TECHNOLOGY

The energy efficiency of the power amplifier, which depends on the frequency band, modulation, and operating environment, plays a vital role in improving the hardware design of a typical BS. Recent improvements in the power amplifier technology not only decrease the power consumption of the hardware installed at BSs but also will make the BSs less dependent on air-conditioning.

10.3.2   POWER-SAVING PROTOCOLS

As the traffic load per hour may vary considerably, the BSs can regularly be under low load conditions, especially during nighttime. Therefore, it is of immense importance to enable inactive operation modes in BSs by designing power-saving protocols, such as sleep modes in which transceivers are switched off whenever there is no need to transmit or receive. LTE standard introduces power-saving protocols such as discontinuous reception (DRX) and discontinuous transmission (DTX) for the mobile terminal. These protocols allow reducing the power consumption of the mobile devices momentarily while remaining connected to the network with reduced throughput. Correia et al. [19] suggest the use of downlink DTX schemes for BSs by enabling micro-sleep modes and deep-sleep modes. They demonstrate that under low load conditions, switching off inactive hardware of BSs can potentially save a lot of power.

10.3.3 COOPERATIVE POWER MANAGEMENT

Traffic fluctuations as a function of time are becoming increasingly serious in upcoming networks especially HetSNets. The cell size adjustment schemes such as cell-breathing schemes where different cells adapt their size depending on the received interference or traffic load conditions are currently under investigation. Several techniques and algorithms to investigate this were explored in Refs. [19,20]. However, in upcoming networks, more collaborative power management schemes are needed, where certain BSs can coordinate with each other for sleeping based on their traffic load and the coverage gaps are filled by the remaining active cells [20–22].

In this regard, a more flexible concept referred to as "cell zooming" was presented in Ref. [23], where BSs can adjust their coverage according to network or traffic conditions, in order to balance the traffic load, while reducing the energy consumption. In case of high loads or increased number of users, it can zoom itself in, whereas the neighboring cells with less number of users can zoom out to cover those users that cannot be served by the highly loaded cell. Cells that are unable to zoom in may even go to sleep to reduce energy consumption, while the neighboring cells can zoom out and help serve the mobile users cooperatively.

10.3.4 RENEWABLE ENERGY RESOURCES

Because of the unavailability of electrical grids, several remote locations of the world constantly rely on diesel-powered generators to run BSs, which not only is expensive but also generates CO_2 emissions. Therefore, using renewable energy resources such as solar and wind energy in place of diesel generators may also be useful in reducing the power consumption of BSs, in particular those at off-grid sites.

10.3.5 COGNITIVE AND COOPERATIVE COMMUNICATIONS

Recently, research on technologies such as cognitive radio and cooperative relaying has received significant attention. While cognitive radio is an intelligent and adaptive wireless communication system that enables us to utilize the radio spectrum in a more efficient manner, cooperative relays can provide a lot of improvement in throughput and coverage for future wireless networks.

Cognitive Radio: The idea of a cognitive radio is to intelligently detect and access the unused spectrum. However, in addition to the efficient spectrum usage, cognitive radios are also foreseen to enhance the efficiency of power consumption in wireless networks. It has been shown in Ref. [24] that up to 50% of power savings can be achieved with the dynamic spectrum management at the operator end. It has been further shown in recent works that structures and techniques based on cognitive radio reduce the energy consumption, while maintaining the required QoS, under various channel conditions [25]. However, because of the complexity of the proposed algorithms/solutions, the implementation issues are still under discussion.

Cooperative Communications: The service provision to extremely remote users via direct transmission is quite expensive in terms of required power. This high-power transmission causes high-power consumption and interference at nearby users and BSs. In this context, relays are well known to help extend the coverage of a BS with high power efficiency and are typically defined as a network element that can be fixed or mobile, much more sophisticated than a repeater, and it has capabilities such as storing and forwarding data, while cooperating in scheduling and routing procedures. In Ref. [26], it is shown that relaying techniques extend the battery life, which is the first step toward energy-efficient networks. Also, in Refs. [27,28], it is illustrated that two-hop communication consumes less energy than direct communication. In particular, there are two possible ways to employ energy-aware relay-assisted networks, as detailed below.

Fixed Relays: It has been demonstrated in Refs. [29,30] that the number of BSs can be increased up to a factor of 1.5 in a unit area while reducing the transmitting power by a factor of 5 and achieving the same signal-to-noise ratio (SNR) level. The underlying message is that the higher density of BSs reduces energy consumption significantly [30]. In this context, the relays emerge as a primary candidate for energy-efficient communication as (i) relays need not be as high as BSs, since they are designed to cover a smaller area with a lower power; (ii) relays can be connected to a BS without a wire; (iii) by deploying relays the cost of installing new BSs can be reduced significantly and present low complexity [29]. It is also illustrated in Ref. [30] that relays provide a flexible way to improve spatial reuse, are less complex than BSs and therefore cheaper to deploy, and reduce the power consumption in the system compared to systems based on direct transmission.

User Cooperation: User cooperation was first introduced in Ref. [31] where it was shown that user cooperation not only increases the data rate but also makes the system more robust. Despite all these advantages, this idea is still unappealing in wireless mobile networks as an increased rate at one user comes with the price of increased energy consumption at the relay node. Therefore, the limited battery life time of mobile users leads to selfish users who do not have incentives to cooperate. Recently, Nokleby and Aazhang [32] addressed the fundamental question whether or not user cooperation is advantageous from the perspective of energy efficiency. The authors developed a game-theoretic approach to give users incentive to act as relays when they are idle, and it is shown that user cooperation has the potential of simultaneously improving both users' bits per energy efficiency under different channel conditions.

10.3.6 PC Mechanisms

Several uplink PC mechanisms such as closed-loop and open-loop PC, slow and fast PC, and fractional PC were discussed in Refs. [33–35]. In general, conventional slow open-loop PC compensates for the long-term channel variations based on shadowing and distance of a mobile user from the serving BS while maintaining the same received target SNR for every user. It can be implemented at each BS by sending slowly updating PC signaling or each mobile may derive its own transmission power according to the path-loss measurement yielded by downlink pilots. In closed-loop PC, mobile users can adapt the fast-fading effects by performing frequent measurements and exchange of control data with its serving BS, making the closed-loop PC less sensitive to error in path-loss estimates.

Nevertheless, it is in favor of both the network operators and society to swiftly address and employ energy efficiency techniques in order to minimize the environmental and financial impact of such a fast-growing and widely adopted technology. This chapter is an effort to study an energy-efficient deployment of small cells in heterogeneous networks with the aim of reducing the energy requirements for the mobile operations.

10.4 HetSNets—INTRODUCTION

This chapter is focused more on quantifying the energy improvements of heterogeneous networks, which is expected to assist network operators in reducing the operational costs as well as environmental effects. In particular, this study will focus on the energy improvements owing to the power savings at the mobile user end. HetSNets, envisioned to enable next-generation wireless networks by providing high data rates, allow off-loading traffic from the macrocell network and provide dedicated capacity to homes and hot spots and thereby facilitate the communication infrastructure. HetSNets consist of infrastructures with multiple radio access technologies (RATs) such as femto-cells, picocells, and so on, each having variable capabilities and functions. However, the distribution of such small cells across the macrocell area is a challenging problem.

Several heterogeneous network deployment strategies are currently under consideration and their performance is commonly calibrated with respect to the achievable profitability, spectral efficiency, and outage probability [7,9,36]. In Ref. [9], the impact of several deployment strategies on the area power consumption of the mobile networks was illustrated. Richter et al. [9] introduced microcells in addition to macrocells to share the traffic load. It was shown that the addition of microcells does not affect the area power consumption of the mobile networks under full traffic load scenarios. Similarly, in Refs. [19,37], the potential improvement in power consumption of the mobile networks was described for a different number of microcells and macrocells. The large-scale deployment of microcells in addition to macrocells may result in significant energy consumption [19,37,38]. The uniform distribution of small cells, which is referred to as UDCs, is considered as one of the tradi-tional approaches to deploy small cells in the current infrastructure [39]. However, the UDC config-uration to cover the macrocell region may not be an efficient choice because of the following facts:

- Energy consumption: The population of the small cells is expected to increase from 100 million to about 500 million mobile users in 2020 [15]. The power consumption of a small cell today is around 10 to 6 W, and it can be assumed that a small cell in 2020 will still con-sume 5 W. Therefore, the 100 million small cells in 2020 will then consume 4.4 TWH in 2020, that is, about an extra 5% on top of the energy consumption of the existing BS infra-structure. Therefore, the huge amount of power consumption of many lightly loaded small cells may reduce the impact of small cells on the spectral efficiency of future networks.
- Interference: Interference management in HetSNets is of paramount importance. In MoNets and HetSNets with UDC configuration, there exist mobile users that are located around the edge of the macrocells that are transmitting with the maximum power in order to achieve the desired signal-to-interference plus noise ratio (SINR). Interference as a result of such edge mobile users may cause significant degradation in the system performance of the cel-lular system with aggressive reuse distance.
- Resources usage: HetSNets with UDC configuration allows the SBSs to be deployed in the macrocell center region, which causes an underutilization of the macrocell BS capabili-ties, that is, existing infrastructure. As an example, the mobile users close to the macrocell BSs should be connected through macrocell BS as long as the link maintains the desired SINR. On the contrary, the mobile users located far away from the macrocell BSs, which may be located around the edge of the macrocell and are not covered by SBSs, are required to transmit with the maximum power in order to maintain the desired quality of the link.

Therefore, introducing energy-efficient deployment strategies that guarantee energy savings is a challenging issue that must be considered while designing HetSNets.

10.5 HetSNets: DESIGN AND MANAGEMENT

This section introduces the HetSNets layout, bandwidth partition, channel allocation, and the energy-aware channel propagation model for the COE design of HetSNets under consideration.

10.5.1 GREEN NETWORK LAYOUT

We consider a two-tier energy-aware HetSNet as illustrated in Figure 10.2, where the integration of macrocell and small-cell networks is illustrated. The first tier of the considered heterogeneous network comprises circular macrocells each of radius R_m [m] with a BS B_m deployed at the center

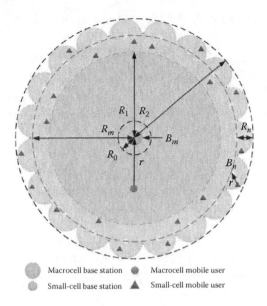

Macrocell base station   Macrocell mobile user

Small-cell base station   ▲ Small-cell mobile user

FIGURE 10.2 Graphical illustration of a two-tier heterogeneous network where a macrocell is surrounded by N small cells around the edge of the macrocell such that the mobile users are transmitting with the adaptive power corresponding to target SINR.

and equipped with an omni-directional antenna. Each macrocell is assumed to have H mobile users uniformly distributed over the region bounded by R_0 and R_m, where R_0 denotes the minimum distance between the macrocell mobile user and its serving BS.

The second tier of the heterogeneous network comprises N circular small cells each of radius R_n [m], with low-power, low-cost, user-deployed SBSs B_n located at the center of each small cell. We consider that the small cells are distributed around the edge of the reference macrocell. The resultant small-cell deployment is referred to as COE configuration [40]. For practical reasons, we calculate the number of small cells per macrocell as follows [40,41]:

$$
N = \begin{cases} \mu \dfrac{(R_2^2 - R_1^2)}{R_n^2} = \mu \dfrac{4R_m}{R_n} & R_m > R_n \\ 0 & R_m \le R_n \end{cases},
$$

where $R_1 = R_m - R_n$, $R_2 = R_m + R_n$ and the factor $0 < \mu \le 1$, referred to as the cell population factor (CPF), controls the number of small cells per macrocell; that is,

$$
\mu = \begin{cases} 0 & \text{off-load small cells} \\ 1 & \text{maximum number of small cells per macrocell}. \end{cases}
$$

The number of mobile users in each small cell is expressed as $F = (H - L)/N$, where $L - \left(H\left(R_1^2 - R_0^2\right)\right)/R_m^2$. To be precise, in the COE configuration, L out of H mobile users are uniformly distributed over the region bounded by R_0 and R_1, whereas the remaining mobile users, that is, $H - L$, are reserved for N small cells. The bandwidth allocated to a macrocell is reused throughout the macrocell network at a channel reuse distance $D' = R_u (R_m + R_n)$ [m], where R_u represents the network traffic load. The total bandwidth allocated to the small-cell tier is reused in each of the N small cells within a macrocell.

10.5.2 BANDWIDTH PARTITION AND CHANNEL ALLOCATION

We consider the spectrum partition based on the proportion of the number of mobile users in the macrocell and small cells [42]. Let w_t [Hz] be the total bandwidth of the available spectrum per cell; then, the total bandwidth may be divided as

$$w_t = w_m + w_n, \qquad (10.1)$$

where $w_m = w_t(L/H)$ [Hz] and $w_n = w_t(NF/H)$ [Hz] are the amount of the spectrum dedicated to the macrocell and small cells, respectively, based on the proportion of active mobile users. The allocated bandwidth for each mobile user in the macrocell and the small cell can be, respectively, calculated as $w_{l_m} = w_m/N_m$ and $w_{f_n} = w_n/N_n$, where N_m and N_n are the number of active serviced channels available per macrocell and small cell, respectively. Moreover, for simplicity, we consider that each channel is allocated to one mobile user at a time and there will be no mobile user that cannot be serviced by the respective macrocell or SBS such that $N_n = F$ and $N_m = L$.

10.5.3 MOBILE USER DISTRIBUTION IN COE CONFIGURATION

All the mobile users in macrocell and small-cell networks are considered as mutually independent and uniformly distributed in their respective cells. The PDF of the location of a macrocell mobile user located at (r, θ) from the serving macrocell BS is expressed as

$$f_r(r) = \frac{2r}{R_1^2 - R_0^2}, \quad f_\theta(\theta) = \frac{1}{2\pi}, \qquad (10.2)$$

where $R_0 \le r \le R_1$ and $0 \le \theta \le 2\pi$. Similarly, the PDF of the location of a small-cell mobile user, which is located at $(\tilde{r}, \tilde{\theta})$ from the serving SBS, can be expressed by

$$f_{\tilde{r}}(\tilde{r}) = \frac{2\tilde{r}}{R_n^2}, \quad f_{\tilde{\theta}}(\tilde{\theta}) = \frac{1}{2\pi}, \qquad (10.3)$$

where $0 \le \tilde{r} \le R_n$ and $0 \le \tilde{\theta} \le 2\pi$ (see Figure 10.2 for the geometrical representation of distances R_2 and R_1).

10.5.4 ENERGY-AWARE CHANNEL PROPAGATION MODEL

The radio environment of a typical wireless cellular network is described by (i) distance-dependent path loss, (ii) shadowing, and (iii) multi-path fading. In our analysis, we consider a two-slope path-loss model for macrocell and small-cell networks [43]. However, for the sake of simplicity and to avoid complex mathematical analysis, this chapter only presents path loss-based analysis. In the following subsections, we introduce a generalized propagation model for both macrocell and small-cell networks.

The received signal power at BS from the mobile user is given by [40]

$$P^{rx}(r) = \frac{K}{r^\alpha (1 + r/g)^\beta} P^{tx}, \qquad (10.4)$$

where

- P^{rx} [W] denotes the average received signal power at the reference macrocell BS/SBS from the desired mobile user, which is located at a distance r from the same reference BS.
- α and β are the basic and additional path-loss exponents, respectively.
- $g = \dfrac{4h_{BS}h_{MU}}{\lambda_c}$ [m] is the breakpoint of a path-loss curve, which depends on the BS antenna height h_{BS} [m], the antenna height of the mobile user h_{MU} [m], and wavelength of the carrier frequency λ_c.
- K is the path-loss constant that depends on the path-loss factor.
- P^{tx} [W] defines the mobile user transmit power for the physical uplink shared channel (PUSCH) such that each of the mobile users in the macrocell/small-cell network adapts its transmit power according to a slow PC mechanism given by* [40,44]

$$P^{tx} = \min\left(P_{max}, P_0 \frac{r^{\alpha}(1 + r/g)^{\beta}}{K} \right).$$ (10.5)

- P_{max} [W] is the maximum transmit power of each of the mobile users in HetSNet.
- P_0 is the cell-specific parameter, which is used to control the target SINR.

The distance at which mobile users require their maximum power P_{max} to fully compensate path loss while maintaining the desired target SNR P_0 is referred to as the *threshold distance* (R_t). Since the mobile users located within R_t can compensate their path loss while saving some proportion of their power, the region within R_t is referred to as the *green area*. However, the mobile users located beyond R_t may transmit with their maximum power to achieve some throughput gains. R_t can be computed by solving the following equation numerically:

$$R_t^{\alpha}(g + R_t)^{\beta} = \frac{KP_{max}}{P_0} g^{\beta}.$$ (10.6)

For special cases, in which $\beta = \alpha$, R_t can be expressed as follows:

$$R_t = \frac{1}{2}\left(-g \pm \sqrt{g} \sqrt{4\left(\frac{P_{max}K}{P_0} \right)^{1/\alpha} + g} \right).$$ (10.7)

Moreover, in the scenarios where $\beta = m\alpha$ and m is an integer, R_t can be determined using standard mathematical software packages such as MATHEMATICA.

10.6 PERFORMANCE ANALYSIS OF GREEN HetSNets

In this section, the performance metrics are introduced to investigate the impact of Green HetSNets on network capacity, energy savings, energy economics, and ecology. The scope of the performance analysis is graphically illustrated in Figure 10.3, which includes the ecology impact to calibrate the carbon footprint of networks, the economics of networks to determine energy–economics relationship and their growth, and, finally, the capacity offered by such networks to meet the demands of future generations of networks (transformation from 4G to 5G). Simulations provided in this section are based on the parameters shown in Table 10.1.

---

* Here, the path loss can be estimated by the mobile users.

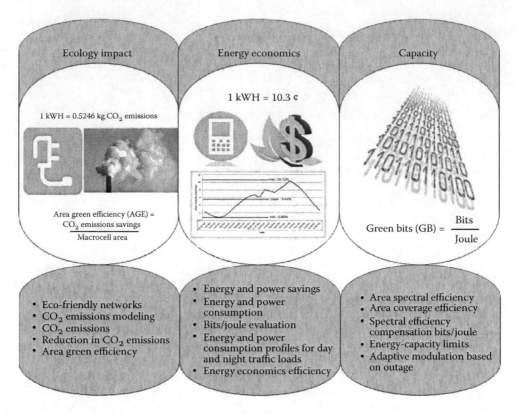

FIGURE 10.3 Performance analysis metrics for Green HetSNets.

TABLE 10.1
Numerical Values of Simulation Parameters

Simulation Parameter	Small Cell	Macrocell
Cell radius (R)	50 m	200–600 m
Path-loss exponent (α)	1.8	2.1
Additional path-loss exponent (β)	1.8	2.1
BS antenna height (h_{rx})	12.5 m	25 m
Mobile antenna height (h_{tx})	2 m	2 m
Reference distance (R_0)	–	100 m
Target SNR (P_0)	0.8 μW	0.8 μW
Breakpoint distance (g)	300 m	600 m
Threshold distance (R_t)	667 m	422 m
System bandwidth (w_t)	20 MHz	
Transmit power (P_{max})	1 W	
Reuse factor (R_u)	2	
Small-cell population factor (CPF)	1	
Path-loss constant (K)	1	

10.6.1 ASE OF HETSNETS

The ASE η of the HetSNets is typically defined as follows [43]:

$$\eta = \frac{C_h}{\pi w_t (D'/2)^2} = \frac{4C_h}{\pi w_t R_u^2 (R_m + R_n)^2} \tag{10.8}$$

where $D' = R_u(R_m + R_n)$, R_u is the frequency reuse factor, and C_h stands for the total achievable Shannon capacity of two-tier HetSNets and it is given by

$$C_h = C_m + C_n = \sum_{l=1}^{L} C_{l_m} + \sum_{n=1}^{N} \sum_{f=1}^{F} C_{f_n}, \tag{10.9}$$

where C_m [bps/Hz] and C_n [bps/Hz] stand for the ergodic capacity of the mth macrocell and N small cells, respectively. C_{l_m} is the Shannon capacity of the lth mobile user in the mth macrocell, whereas C_{f_n} is the capacity of fth mobile user in the nth small cell. More explicitly, C_{l_m} is expressed as [41]

$$C_{l_m} = w_{l_m} \mathbb{E}\left[\log_2\left(1 + \gamma_{l_m}\right) \right], \tag{10.10}$$

where $\mathbb{E}[\cdot]$ represents the expectation operator; γ_{l_m} is the SINR of the lth macrocell mobile user in the mth macrocell and it is expressed as

$$\gamma_{l_m} = \frac{P^{rx}_{l_m, B_m}(r)}{\sum\limits_{i=1, i \neq m}^{M} P^{rx}_{l_i, B_m}(r) + \sigma^2}, \tag{10.11}$$

where σ^2 is the thermal noise power, $P^{rx}_{l_m, B_m}(r)$ is the received power level at the reference macrocell BS B_m from the lth desired mobile user, and $\sum\limits_{i=1, i \neq m}^{M} P^{rx}_{l_i, B_m}(r)$ is the sum of the individual interfering power levels received at the reference macrocell BS B_m from the interfering mobile users $\{l_i\}_{i=1}^{M}$, which are located in each of the (ith) interfering macrocells.

Similarly, for the small-cell networks, C_{f_n} in Equation 10.9 is expressed as

$$C_{f_n} = w_{f_n} \mathbb{E}\left[\log_2\left(1 + \gamma_{f_n}\right) \right], \tag{10.12}$$

where γ_{f_n} is the SINR of the fth mobile user located in the nth small cell and it takes the expression

$$\gamma_{f_n} = \frac{P^{rx}_{f_n, B_n}(\tilde{r})}{\sum\limits_{j=1, j \neq n}^{N} P^{rx}_{f_j, B_n}(\tilde{r}) + \sigma^2}, \tag{10.13}$$

where $P^{\text{rx}}_{f_n,B_n}(\tilde{r})$ is the received power level at the reference SBS B_n from the fth desired mobile user

and $\sum_{j=1,j\neq n}^{N} P^{\text{rx}}_{f_j,B_n}(\tilde{r})$ is the sum of the individual interfering power levels received at the reference

SBS B_n from the interfering mobile users located in the jth interfering small cell.

Figure 10.4 quantifies the ASE of HetSNet as a function of macrocell radius and compares the ASE of the proposed HetSNets with COE configuration with two types of competitive networks, namely, (i) HetSNets with UDC configuration where the small cells are uniformly distributed and also receiving interferences from $N-1$ interfering small cells, and (ii) MoNets where the small cells are not active. Note that SBSs receive interference from $N-1$ small-cell interferers and macrocell BSs receive the interference from $M-1=6$ interferers. The mobile users in each of the macrocell and small-cell networks are transmitting with the adaptive power scheme according to the slow PC mechanism, which is generally defined as Equation 10.5 for both macrocell and small-cell uplink.

It can be seen clearly that the ASE of the HetSNets can be increased significantly when the small cells are active and complementing the macrocell BSs in comparison with the MoNet (compare the green solid curve with circle markers with the red dashed curve with circle markers and the blue dashed/dotted curve with circle markers). It can be observed that the ASE of the HetSNets with UDC configuration and COE configuration outperforms the ASE performance of MoNets by (i) deploying small cells and (ii) reducing the propagation distance between the mobile users and the BSs and thereby decreasing the transmit power requirement. However, the performance of the HetSNets with COE configuration outperforms the ASE performance of the HetSNets with UDC configuration and the improvement in ASE gain is considerable. This is due to the fact that the COE configuration restricts only the cell-edge mobile users to communicate with the small cells, which enhances the overall network ASE compared to the UDC configuration. More precisely, the UDC configuration allows the SBSs to be deployed in the cell center, which causes an underutilization of the macro BS capabilities (existing infrastructure). Thus, the performance degradation caused by the cell-edge mobile users still exists in the UDC configuration. Because of the weaker channel gains of the mobile users in the large macrocells, the degradation of ASE with R_m is obvious.

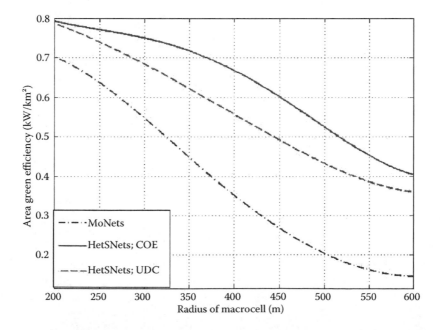

FIGURE 10.4 ASE as a function of the macrocell radius for (i) MoNets, (ii) HetSNet with COE configuration, and (iii) HetSNet with UDCs.

10.6.2   AGE OF HETSNETS

The AGE of a two-tier heterogeneous network is defined as the aggregate power savings per unit area while maintaining a certain desired target SNR at the receiver. Mathematically, the AGE of two-tier heterogeneous network can be expressed as [40]

$$AGE = \frac{\mathcal{P}}{\pi(R_m + R_n)^2},$$

(10.14)

where $\mathcal{P} = \mathcal{P}_m + \mathcal{P}_n$. In order to calibrate the AGE of the energy-aware COE configuration, we guarantee the target SNR of P_0 at both femtocell and macrocell BSs. The total power savings of L macrocell mobile users $\mathcal{P}_m$ can be determined as $\mathcal{P}_m = \sum_{l=1}^{L} \mathcal{P}_{l_m}$, where $\mathcal{P}_{l_m}$ is the power saving of the lth mobile user located in cell m and which takes the expression

$$\mathcal{P}_{l_m} = \begin{cases} 0 & r_{l_m,B_m} > R_t \\ P_{max} - P_0 \dfrac{r_{l_m,B_m}^{\alpha}\left(1 + \dfrac{r_{l_m,B_m}}{g}\right)^{\beta}}{K} & r_{l_m,B_m} < R_t \end{cases},$$

(10.15)

where r_{l_m,B_m} is the distance between the macrocell BS B_m and lth mobile user in lth macrocell.

Similarly, the total power savings of NF femtocell mobile users $\mathcal{P}_n$ can be expressed as $\mathcal{P}_n = \sum_{n=1}^{N} \sum_{f=1}^{F} \mathcal{P}_{f_n}$, where $\mathcal{P}_{f_n}$ is the power saving of the fth mobile user located in cell n and which is given as follows:

$$\mathcal{P}_{f_n} = \begin{cases} 0 & \tilde{r}_{f_n,B_n} > R_t \\ P_{max} - P_0 \dfrac{\tilde{r}_{f_n,B_n}^{\alpha}\left(1 + \dfrac{\tilde{r}_{f_n,B_n}}{g}\right)^{\beta}}{K} & \tilde{r}_{f_n,B_n} < R_t \end{cases}$$

(10.16)

where $\tilde{r}_{f_n,B_n}$ is the distance between the SBS B_n and the fth mobile user in the nth small cell.

Figure 10.5 demonstrates the AGE gains of three types of considered networks: (i) HetSNets with COE configuration, (ii) MoNets, and (iii) HetSNets with UDC configuration. It can be observed that the AGE of the HetSNets with COE configuration is higher in comparison with the AGE of the MoNets and the HetSNets with UDC configuration. The deployment of small cells around the edge of the macrocell in COE configuration mandates a reduction in the number of edge mobile users transmitting with the maximum power, thereby the AGE gains are self-explanatory in COE configuration relative to the other two competitive network configurations. However, this effect is more dominant within the threshold distance $R_t = 422$ m, which is due to the fact that the number of energy-efficient users in both UDC and COE deployments increases with the increase in R_m until R_t. However, the energy efficiency of cell-edge users in COE is higher than the UDC deployment. Nonetheless, this gain of COE increases till $R_t = 422$ m, beyond which all macrocell mobile users start transmitting with P_{max} in UDC and COE configurations and the probability of UDC deployment in non-green area also increases, which is equally beneficial as COE configuration; therefore, the effectiveness of COE slowly disappears beyond R_t.

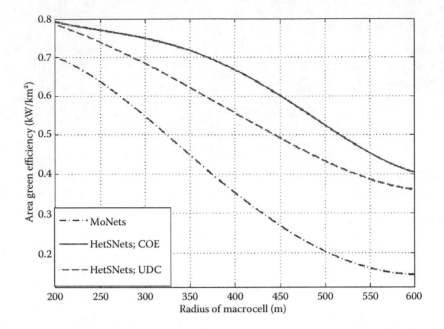

FIGURE 10.5 AGE as a function of number of small cells for (i) MoNets, (ii) HetSNet with COE configuration, and (iii) HetSNet with UDCs.

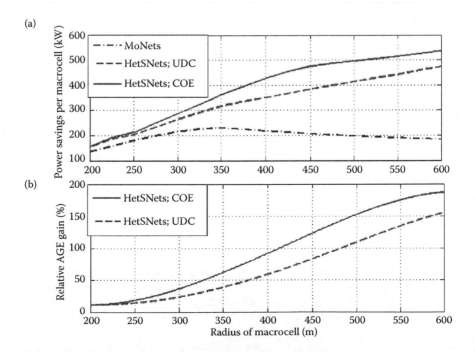

FIGURE 10.6 (a) Total power saved by all users as a function of macrocell radius for (i) MoNets, (ii) HetSNet with COE configuration, and (iii) HetSNet with UDCs. (b) Relative percentage gain in the AGE of (i) HetSNet with COE configuration and (ii) HetSNet with UDCs.

Moreover, the network AGE is found to degrade with the increase of R_m, which is due to the increase in the proportion of mobile users beyond R_t, which are transmitting with their maximum power. Therefore, beyond R_t, the power savings settle down at a specified level as demonstrated in Figure 10.6a. However, a small increase of power savings can still be observed because of the increase in the number of femtocells N at the edge and, in turn, the number of mobile users at the edge of macrocells. The relative AGE gain in percentage with reference to the MoNets configuration is also computed in Figure 10.6b to illustrate the effective increase in AGE.

10.6.3 ENERGY ECONOMICS OF HETSNETS

In this section, we investigate the relationship of energy economics of HetSNets with the energy savings of the wireless networks and the associated cost.

10.6.3.1 Improvement in Energy Savings of HetSNets

The energy savings of the HetSNets can be defined as power savings per unit time per year and may be expressed as

$$\text{Energy savings} = \mathcal{P} \times \phi(t) \times \text{no. of days/year kWH/year} \tag{10.17}$$

where $\phi(t)$ is the number of hours per day a mobile user is active under full load conditions; $\mathcal{P}$ can be calculated using the relationship in Equation 10.14.

Figure 10.7 shows the amount of energy saved by all the mobile users* that are transmitting with adaptive power in the macrocell and the small-cell networks. It can be observed that the HetSNet offers a significant improvement in energy savings in comparison with the MoNets (as illustrated

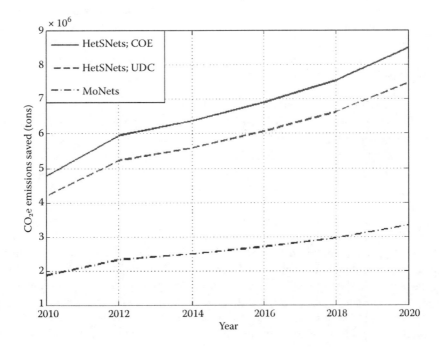

FIGURE 10.7 Energy savings per year corresponding to total mobile subscribers for several wireless network configurations: (i) MoNets, (ii) HetSNet with COE configuration, and (iii) HetSNet with UDCs.

---

* Here, the total number of mobile subscribers is considered as 4.5 billion in 2012 and expected to reach 7.6 billion in 2020 [15].

by the solid curve and the dashed curve with the dashed/dotted curve). However, the performance of HetSNets with COE configuration outperforms the performance of HetSNets with UDC configuration. Under COE configuration, there are no edge mobile users that are transmitting with the maximum power, which is contrary to the UDC configuration where each of the mobile users that are located around the edge of the macrocells are not necessarily transmitting with adaptive power to maintain the desired SINR. This is the reason why the energy savings of the HetSNets with COE configuration increases in comparison with the other two competitive network configurations. As an example, today, the energy savings offered by conventional MoNets even with a PC scheme is 4.4 TWH/year, which is significantly less than the energy savings of the HetSNets with UDC and COE configurations, which offer 9.8 and 11 TWH/year, respectively. Moreover, the improvement in energy savings of the HetSNets COE configuration is expected to reach 12% in comparison with UDC configuration and 60% in comparison with MoNets by 2020.

Reducing energy usage reduces energy costs and may result in a financial cost saving to consumers if the energy savings offset any additional costs of implementing an energy-efficient technology. Therefore, the impact of the energy savings in wireless networks can be calculated in terms of cost of saved energy. The associated cost can be calculated by assuming 1 kWH = 10.3¢:*

$$\text{Cost} = \frac{\text{Energy savings}}{1000 \times 100} \times 10.3 \text{ USD/year.} \qquad (10.18)$$

Today, the cost savings corresponding to the energy savings of MoNets is 0.45 billion USD and it is expected to be 0.6 billion USD in 2020. This can be further increased to 1 billion USD in 2012 and 1.4 billion USD in 2020 for HetSNets with UDC configuration. Finally, the cost corresponding to energy savings of HetSNets with COE configuration may reach 1.1 billion USD in 2012 and 1.6 billion USD in 2020. In short, annual savings of 60% of the cost can be achieved, corresponding to the energy savings of the HetSNets in comparison with MoNets. The economics analysis associated with the energy savings and cost is for the mobile operations only. However, much more energy can be saved by adapting new energy-efficient models for production, distribution, and maintenance of the mobile communications networks and services to reduce the energy consumption and increase the energy savings of the mobile communications industry.

10.6.3.2 Improvement in CO_2e Emissions Savings

In order to determine the ecological impact of energy savings of HetSNets with COE configuration, we calculate the corresponding CO_2e emissions saved in megatons (Mtons). The amount of the CO_2 emissions saved is calculated by assuming 1 kWH = 0.5246 kg CO_2e emissions, which represents the energy used at the point of final consumption. The conversion factors for the conversion of energy consumption to CO_2e emissions for other sources of energy generation can be understood from Ref. [45].

Figure 10.8 shows the uplink CO_2e emissions saved for (i) MoNets, (ii) HetSNets with UDC configuration, and (iii) HetSNets with COE configuration. Here, each of the mobile users is transmitting with the adaptive power to maintain the desired SINR of the link. The CO_2e emissions saved for the systems under consideration are compared with the CO_2e emissions saved for the MoNets with PC, that is, the network where the mobile users are transmitting with adaptive power and small cells are inactive. It can be seen clearly that the CO_2e emissions of the HetSNets reduce significantly in comparison with the MoNets, thereby contributing less amount of CO_2 emissions to the environment and establishing green networks. As an example, the CO_2e emissions saved for the MoNets in 2016 are approximated as being 2.7 Mtons. This can be further increased to 6 Mtons (54% savings) by introducing small cells in HetSNet with UDC configuration. Finally, the significant increase in

---

* This is average cost per kilowatt-hour for commercial type of energy use.

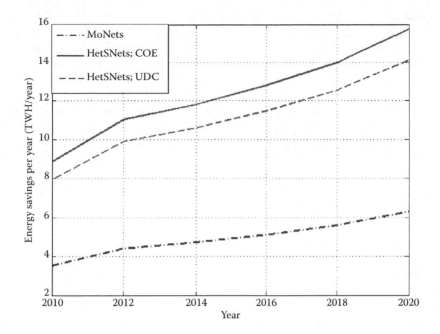

FIGURE 10.8 CO_2e emissions saved per year for total mobile subscribers for several wireless network configurations: (i) MoNets, (ii) HetSNet with COE configuration, and (iii) HetSNet with UDCs.

CO_2e emissions savings for the system can be achieved by introducing small cells around the edge of the macrocells. The proposed HetSNets with COE configuration guarantee to increase the CO_2e emission savings to 6.9 Mtons (62% savings). In short, an annual 57% CO_2e emission can be saved till 2020 by deploying energy-efficient HetSNets such as UDC and COE configurations.

The ICT sector in general and mobile communications industry in particular present considerable potential to decrease the global carbon footprint especially in developing and emerging economies* where future development should not follow the wrong path taken by developed countries. The mobile communication industry must act quickly to demonstrate efforts, enforce policies, and continue to innovate radically in reducing global carbon footprint emissions. Therefore, it is of immense importance for the network designers to critically calibrate the energy efficiency of various deployment strategies along with other performance measures.

10.7 FUTURE RESEARCH FORECAST

The role of telecommunications in the national development process as a means of increasing productivity and efficiency is today more important than ever. Wireless access technologies are becoming progressively less expensive than the various wired alternatives and offer additional advantages of rapid and flexible deployment in both urban and rural areas. One of the most prominent challenges of providing high-rate and reliable transmission in aggressive wireless environments is the scarcity of resources that include but are not limited to power and spectrum. The ever-growing demand for wireless communication services has necessitated the development of systems with high bandwidth and power efficiency. However, wireless networks and in particular mobile networks are increasingly contributing to global energy consumption. Over the last decade, significant work has been done to explore new radio resources and new technologies. Future research shall focus on improving the spectrum and energy efficiency of wireless mobile networks by integrating the innovations of cooperative and cognitive communication systems with the future infrastructure of the

---

* Such as China, Hungary, Indonesia, Poland, and so on. For a complete list and grouping, see Ref. [46].

wireless mobile networks. This highly ambitious and unique goal will provide means for improving end-user data rates, reducing spectrum requirements, and lowering the power consumption in the network by allowing cooperation among the terminals in the network and intelligent utilization of the available spectrum resources. Research in future communication systems shall also aim to develop a unified framework that facilitates the investigation of practical and physical approaches while providing high data rate and uncompromised QoS. A promising approach to resolve this problem is through the deployment of short-range, low-power, and low-cost devices (like BSs in cellular environment) operating in cooperation with the main macrocellular network infrastructure. The resultant wireless network is referred to as heterogeneous wireless network. The heterogeneous network promises energy savings by integrating the femto- and macrocellular networks and thereby reducing CO_2 emissions, OPEX, and CAPEX while enhancing the ASE.

In the last few years, the expansion of communication services to regional, rural, and remote users on an equitable basis compared with their counterparts in metropolitan and large urban centers has been a controversial topic all around the world. According to a report published by the United Nations, more than 3 billion people are currently living in rural areas. In developing countries like China and India, about 70% of the total population live in rural communities, which are spread far away over large geographic areas. For these communities, it is believed that providing communications service is an important step to facilitate development and social equality. Apart from that, rural communication networks are crucial in disaster/emergency response scenarios. However, providing rural communications is often challenging because of the mismatch between costs and demand. Most rural areas have low population density and the demand for services per individual or household can be much lower in rural areas when compared to urban areas. To create a viable business, operators must aim for low-cost solutions. However, the deployment and maintenance of rural communications networks can be costly because of lack of coverage, transportation, difficult terrains, and infrastructure. This is particularly true for wired networks because wires or cables must be laid all the way to the destination. As a result, wireless technologies usually are preferred for rural connectivity. In fact, there are various approaches that consider wireless technologies for rural communications. Nevertheless, the challenges of providing low-cost services to a low-demand market still remain.

In short, to address the challenges of future wireless networks while maintaining profitability and spectrum efficiency, it is essential to consider various paradigm-shifting technologies that mandates to (i) improve the energy efficiency of wireless network and thereby establishing "Greener" networks and (ii) improve the network spectral efficiency while meeting the escalated demand of mobile users. Some of the possible ways to design spectral and energy-efficient systems are as follows:

- Exploiting an energy-efficient architecture in mobile communications networks, which includes but are not limited to minimizing the BS energy consumption, power-saving network protocols, energy-aware cooperation, and relaying among the BSs an energy-efficient design of electronic devices such as power amplifiers.
- Introducing multiple RATs such as femtocells, picocells, and microcells to complement the existing mobile communications networks and establishing a heterogeneous network that aims to decrease the propagation distance between the BS and the mobile users, thereby improving the spectral and energy efficiency of the mobile network.

Heterogeneous wireless networks are envisioned to enhance the operations of next-generation networks by providing high data rates, allowing off-loading traffic from the macrocell, and providing dedicated capacity to homes, enterprises, and urban and rural hot spots. Furthermore, the substantial improvement in network efficiency will yield large cost savings for mobile operators. Although the heterogeneous network is expected to be the most influential solution to meet the challenges imposed for 4G mobile networks of the future generation of wireless networks, the management of cross-tier and intra-tier interferences is critical to achieve such gains and of paramount importance for the successful operation of heterogeneous networks. The capability of the small-cell network to become a self-aware entity in the heterogeneous network can be increased by enabling cognitive radio technology on small-cell BSs. A cognitive enabled small-cell BS can sense spectrum efficiently, manage interference intelligently, and allocate resources effectively by learning and adapting accordingly to the communications environment. This cognitive enabled small-cell technology leads to simpler and easier proliferation of cognitive radio technology into the future generations of cellular network. Future research work is expected to focus on the physical layer for rural and sparsely populated areas, spectrum and energy-efficient designs, smart vehicular communication system design, and intelligent interference management schemes.

The successful development of green wireless networks enables to decrease the energy and spectral demands of future generations of wireless networks. The main aim of the chapter was to promote the green wireless networks that are expected to deploy small cells to support the existing infrastructure of the mobile communication network. The design of such HetNets is referred to as energy-efficient deployment of small cells where intelligent resource allocation, transmission uplink adaptation, and content-aware cooperation will be considered as fundamental design parameters. The energy-efficient deployment of the small cells mandates to improve the energy savings of the networks and yield a low carbon economy. Moreover, the design of the green network can be further improved by classification and mitigation of several types of interferences in HetNets. Significant performance gains in terms of improvement in energy and spectral efficiency for several other small-cell deployments are expected. Such gains will be further exploited to determine the ecology and economy impact of the green wireless networks. The green research framework is expected to empower the current infrastructure of the wireless networks in developing and emerging countries. The green wireless networks will facilitate solving challenging scientific and social technological problems and further catalyze the diversification of the research in this domain.

10.8   CONCLUSIONS

In this chapter, we studied the energy economics of the HetSNets. It has been shown that the significant energy savings can be achieved by (i) deploying small cells around the edge of the macrocells and (ii) employing PC in the uplink where each of the mobile users in the macrocell and the small-cell networks is transmitting with adaptive power. It has been shown that the CO_2e emissions of the proposed HetSNets design are reduced significantly while achieving the ASE and AGE gains of the system. It has been shown that the COE configuration is the smart energy-efficient deployment of small cells across the existing network. However, several alternative types of small-cell deployment strategies should be calibrated in terms of energy savings, associated cost savings, ecological impact, and benefits to the environment. Therefore, the energy consumption and carbon footprint of the cellular communication network should be considered as key variables in designing and planning eco-friendly wireless networks for the future. Therefore, the ICT and in particular mobile communications industry are expected to reduce the global energy consumption and thereby reduce the global carbon footprint and establish green HetSNets.

ACKNOWLEDGMENT

This work was made possible by grants NPRP 4-1293-2-513 and NPRP 09-341-2-128 from QNRF.

AUTHOR BIOGRAPHIES

Muhammad Zeeshan Shakir (M'03) received his BE degree in electrical engineering from NED University of Engineering and Technology, Karachi, Pakistan, in 2002. He obtained his MSc in communications, control and digital signal processing (CCDSP) and PhD in electrical engineering degrees in 2005 and 2010, respectively, from the University of Strathclyde, Glasgow, Scotland. From January 2006 to January 2009, he was a recipient of an industrial scholarship jointly sponsored by the University of Strathclyde, Glasgow, and Picsel Technologies Ltd., Glasgow. In 2009, he joined King Abdullah University of Science and Technology (KAUST), Saudi Arabia, where he was a recipient of a Collaborative Travel Fund award from KAUST Global Collaborative Research. Dr. Shakir holds visiting researcher appointment at the Center for Communication System Research at the University of Surrey, United Kingdom. In July 2012, Dr. Shakir joined Texas A&M University at Qatar where his research interests are in the areas of performance analysis of wireless communication systems, which particularly include heterogeneous network, cooperative communications, information theoretic analysis for multicellular networks, and spectrum sensing for cognitive radios. Dr. Shakir has published more than 25 technical conference and journal papers and contributed three book chapters. He has been serving as secretary to the IEEE DySPAN 1900.7 group since January 2012 where he is involved in the standardization of the project titled "radio interface for white space dynamic spectrum access radio systems supporting fixed and mobile operation."

Hina Tabassum received her bachelor's degree in electronics from the NED University of Engineering and Technology (NEDUET), Karachi, Pakistan, in 2004. During her undergraduate studies she received the Gold Medal from NEDUET and SIEMENS for securing the first position among all engineering universities of Karachi. She then worked as a lecturer in NEDUET for two years. In September 2005, she joined the Pakistan Space and Upper Atmosphere Research Commission, Karachi, Pakistan, and received there the best performance award in 2009. She also completed her master's in communications engineering from NEDUET in 2009. In January 2010, she joined the Computer, Electrical, and Mathematical Sciences & Engineering Division at King Abdullah University of Science and Technology, Thuwal, Makkah Province, Saudi Arabia, where she is currently a PhD candidate. Her research interests include wireless communications with focus on interference modeling, radio resource allocation, and optimization in heterogeneous networks.

Khalid A. Qaraqe (M'97–SM'00) was born in Bethlehem. Dr. Qaraqe received his BS degree in EE from the University of Technology in 1986, with honors. He received his MS degree in EE from the University of Jordan, Jordan, in 1989, and he earned his PhD degree in EE from Texas A&M University, College Station, Texas, in 1997. From 1989 to 2004, Dr. Qaraqe has held a variety positions in many companies and he has over 12 years of experience in the telecommunications industry. Dr. Qaraqe has worked for Qualcomm, Enad Design Systems, Cadence Design Systems/Tality Corporation, STC, SBC, and Ericsson. He has worked on numerous GSM, CDMA, and WCDMA projects and has experience in product development, design, deployments, testing, and integration. Dr. Qaraqe joined the Department of Electrical and Computer Engineering of Texas A&M University at Qatar in July 2004, where he is now a professor. Dr. Qaraqe's research interests include communication theory and its application to design and performance, analysis of cellular systems, and indoor communication systems. He has particular interest in mobile networks, broadband wireless access, cooperative networks, cognitive radio, diversity techniques, and beyond 3G systems.

Erchin Serpedin (SM'04, F'12) received his specialization degree in signal processing and transmission of information from Ecole Supérieure D'Electricité (SUPELEC), Paris, France, in 1992; his MSc degree from the Georgia Institute of Technology, Atlanta, in 1992; and his PhD degree in electrical engineering from the University of Virginia, Charlottesville, in 1999. He is currently a professor in the Department of Electrical and Computer Engineering at Texas A&M University, College Station. He is the author of 2 research monographs, 1 textbook, 90 journal papers, and 150 conference papers, he and serves currently as associate editor for *IEEE Transactions on Information Theory, Physical Communications* (Elsevier), *IEEE Transactions on Communications, Signal Processing* (Elsevier), *EURASIP Journal on Advances in Signal Processing*, and *EURASIP Journal on Bioinformatics and Systems Biology*. He is an IEEE Fellow.

Mohamed-Slim Alouini (S'94, M'98, SM'03, F'09) was born in Tunis, Tunisia. He received his PhD degree in electrical engineering from the California Institute of Technology, Pasadena, California, USA, in 1998. He served as a faculty member in the University of Minnesota, Minneapolis, Minnesota, USA, and then in the Texas A&M University at Qatar, Education City, Doha, Qatar, before joining King Abdullah University of Science and Technology, Thuwal, Makkah

Province, Saudi Arabia, as a professor of electrical engineering in 2009. His current research interests include the modeling, design, and performance analysis of wireless communication systems.

REFERENCES

1. N. Shetty, S. Parekh, and J. Walrand. Economics of femtocells. In *Proc. IEEE Conf. Global Commun., (GLOBECOM'09)*, pages 1–5, Honolulu, HI, USA, Dec. 2009.
2. H. Claussen, L. T. W. Ho, and L. G. Samuel. An overview of the femtocell concept. *Tech. J. Bell Labs*, 13(1):221–264, Sep. 2008.
3. P. Lin, J. Zhang, Y. Chen, and Q. Zhang. Macro–femto heterogeneous network deployment and management: From business models to technical solutions. *IEEE Commun. Mag. Wireless*, 18(3):64–70, Jun. 2011.
4. S. Landstrom, A. Furuskar, K. Johansson, L. Falconetti, and F. Kronestedt. Heterogeneous networks increasing cellular capacity. *J. Ericson Rev.*, 89:4–9, Jan. 2011.
5. V. Chandrasekhar, J. Andrews, and A. Gatherer. Femtocell networks: A survey. *IEEE Mag. Commun.*, 46(9):59–67, Sep. 2008.
6. D. Calin, H. Claussen, and H. Uzunalioglu. On femto deployment architectures and macrocell offloading benefits in joint macro–femto deployments. *IEEE Mag. Commun.*, 48(1):26–32, Jan. 2010.
7. A. J. Fehske, F. Richter, and G. P. Fettweis. Energy efficiency improvements through micro sites in cellular mobile radio networks. In *Proc. IEEE Conf. Global Commun. Workshops, (GLOBECOM'2009)*, pages 1–5, Honolulu, HI, USA, Dec. 2009.
8. F. Cao and Z. Fan. The tradeoff between energy efficiency and system performance of femtocell deployment. In *Proc. Intl. Symp. Wireless Commun. Systems, (ISWCS'2010)*, pages 315–319, York, UK, Sep. 2010.
9. F. Richter, A. J. Fehske, and G. P. Fettweis. Energy efficiency aspects of base station deployment strategies for cellular networks. In *Proc. IEEE 70th Vehicular Technology Conf., (VTC-Fall'2009)*, pages 1–5, Anchorage, AK, USA, Sep. 2009.
10. F. Cao and Z. Fan. The tradeoff between energy efficiency and system performance of femtocell deployment. In *Proc. Intl. Symp. Wireless Commun. Systems, (ISWCS'2010)*, pages 1–5, York, UK, Sep. 2010.
11. D. Lopez-Perez, I. Guvenc, G. De La Roche, M. Kountouris, T. Q. S. Quek, and J. Zhang. Enhanced intercell interference coordination challenges in heterogeneous networks. *IEEE Mag. Wireless Commun.*, 18(3):22–30, 2011.
12. Cool cellular–energy efficient network architectures and transmission methods. http://www.vodafone-chair.com/research/projects cool cellular.html, 2012. Online; accessed Sep. 1, 2012.
13. G. Miao, N. Himayat, Y. Li, and A. Swami. Cross-layer optimization for energy-efficient wireless communications: A survey. *Wiley J. Wireless Commun. Mobile Comput.*, 9(4):529–542, Apr. 2009.
14. K. Lahiri, A. Raghunathan, S. Dey, and D. Panigrahi. Battery-driven system design: A new frontier in low power design. In *Proc. Intl. Conf. on VLSI Design*, pages 261–267, Bangalore, India, Jan. 2002.
15. Energy Aware Radio and Network Technologies (EARTH). http://www.ict-earth.eu, 2012. Online; accessed Sep. 1, 2012.
16. The Climate Group. Smart 2020: Enabling the low carbon economy in the information age. Report for Global e-Sustainability Initiative, GeSI, pages 1–86, New York City, USA, 2008.
17. T. Edler and S. Lundberg. Energy efficiency enhancements in radio access networks. *Ericsson Rev.*, (1):42–51, 2004.
18. D. Feng, C. Jiang, G. Lim, L. Cimini, G. Feng, and G. Li. A survey of energy-efficient wireless communications. To appear in *IEEE Commun. Surv. Tutor.*, 15(1):167–178, 2013.
19. L. M. Correia, D. Zeller, O. Blume, D. Ferling, Y. Jading, I. Gódor, G. Auer, and L. Van der Perre. Challenges and enabling technologies for energy aware mobile radio networks. *IEEE Mag. Commun.*, 48(11):66–72, Nov. 2010.
20. K. Samdanis, D. Kutscher, and M. Brunner. Dynamic energy-aware network re-configuration for cellular urban infrastructures. In *Proc. IEEE Conf. Global Commun., (Globecom'2010)*, pages 1448–1452, Miami, FL, USA, Dec. 2010.
21. K. Samdanis, D. Kutscher, and M. Brunner. Self-organized energy efficient cellular networks. In *Proc. IEEE 21st Intl. Symposium on Personal, Indoor and Mobile Radio Commun., (PIMRC'2010)*, pages 1665–1670, Istanbul, Turkey, Sep. 2010.
22. M. A. Marsan, L. Chiaraviglio, D. Ciullo, and M. Meo. Optimal energy savings in cellular access networks. In *Proc. IEEE Intl. Conf. Commun., (ICC'2009)*, pages 1–5, Dresden, Germany, Jun. 2009.
23. Z. Niu, Y. Wu, J. Gong, and Z. Yang. Cell zooming for cost-efficient green cellular networks. *IEEE Mag. Commun.*, 48(11):74–79, Nov. 2010.

24. O. Holland, V. Friderikos, and A. H. Aghvami. Green spectrum management for mobile operators. In *Proc. IEEE Conf. Global Commun., (Globecom'2010)*, pages 1458–1463, Miami, FL, USA, Dec. 2010.

25. A. He, S. Srikanteswara, K. K. Bae, T. R. Newman, J. H. Reed, W. H. Tranter, M. Sajadieh, and M. Verhelst. System power consumption minimization for multichannel communications using cognitive radio. In *Proc. IEEE Intl. Conf. Microwaves, Commun., Antennas and Electronics Systems, (COMCAS'2009)*, pages 1–5, Tel Aviv, Israel, Nov. 2009.

26. J. N. Laneman and G. W. Wornell. Energy-efficient antenna sharing and relaying for wireless networks. In *Proc. IEEE Wireless Commun. and Networking Conf., (WCNC'2000)*, pages 7–12, 2000.

27. A. Radwan and H. S. Hassanein. Nxg04-3: does multi-hop communication extend the battery life of mobile terminals? In *Proc. IEEE Conf. Global Commun., (Globecom'2006)*, pages 1–5, San Francisco, CA, USA, Dec. 2006.

28. J. Y. Song, H. Lee, and D. H. Cho. Power consumption reduction by multi-hop transmission in cellular networks. In *Proc. IEEE 60th Vehicular Technology Conf., (VTC-Fall'2004)*, pages 3120–3124, Los Angeles, CA, USA, Sep. 2004.

29. R. Pabst, B. H. Walke, D. C. Schultz, P. Herhold, H. Yanikomeroglu, S. Mukherjee, H. Viswanathan, M. Lott, W. Zirwas, M. Dohler, H. Aghvami, D. D. Falconer, and G. P. Fettweis. Relay-based deployment concepts for wireless and mobile broadband radio. *IEEE Mag. Commun.*, 42(9):80–89, Sep. 2004.

30. R. Rost and G. Fettweis. Green communication in cellular network with fixed relay nodes. In *Cooperative Cellular Wireless Communications*. Cambridge University Press, Cambridge, UK, 2011.

31. A. Sendonaris, E. Erkip, and B. Aazhang. User cooperation diversity, Part I: System description. *IEEE Trans. Commun.*, 51(11):1927–1938, 2003.

32. M. Nokleby and B. Aazhang. User cooperation for energy-efficient cellular communications. In *Proc. IEEE Intl. Conf. Commun., (ICC'2010)*, pages 1–5, Cape Town, South Africa, May. 2010.

33. A. M. Rao. Reverse link power control for managing inter-cell interference in orthogonal multiple access systems. In *Proc. IEEE 66th Vehicular Technology Conf., (VTC-Fall'2007)*, pages 1–5, Baltimore, MD, USA, Oct. 2007.

34. A. Simonsson and A. Furuskar. Uplink power control in LTE—Overview and performance: Principles and benefits of utilizing rather than compensating for SINR variations. In *Proc. IEEE 68th Vehicular Technology Conf., (VTC-Fall'2008)*, pages 1–5, Calgary, AB, Canada, Sep. 2008.

35. B. Muhammad and A. Mohammed. Performance evaluation of uplink closed loop power control for LTE system. In *Proc. IEEE 70th Vehicular Technology Conf., (VTC-Fall'2009)*, pages 1–5, Anchorage, AK, USA, Sep. 2009.

36. H. Claussen, L. T. W. Ho, and F. Pivit. Effects of joint macrocell and residential picocell deployment on the network energy efficiency. In *Proc. IEEE 19th Intl. Symp. Personal, Indoor and Mobile Radio Commun., (PIMRC'2008)*, pages 1–6, Cannes, France, Sep. 2008.

37. C. Han, T. Harrold, S. Armour, I. Krikidis, S. Videv, P. M. Grant, H. Haas, J. S. Thompson, I. Ku, C.-X. Wang, T. A. Le, M. R. Nakhai, J. Zhang, and L. Hanzo. Green radio: Radio techniques to enable energy-efficient wireless networks. *IEEE Mag. Commun.*, 49(6):46–54, Jun. 2011.

38. C. Yan, S. Zhang, S. Xu, and G. Y. Li. Fundamental trade-offs on green wireless networks. *IEEE Mag. Commun.*, 49(6):30–37, Jun. 2011.

39. J. Hoydis, M. Kobayashi, and M. Debbah. Green small-cell networks. *IEEE Mag. Veh. Technol.*, 6(1):37–43, Mar. 2011.

40. H. Tabassum, M. Z. Shakir, and M.-S. Alouini. On the area green efficiency (AGE) of heterogeneous networks. In *Proc. Intl. Conf. Global Commun., GLOBECOM'2012*, pages 1–6, Anaheim, CA, USA, Dec. 2012.

41. M. Z. Shakir and M.-S. Alouini. On the area spectral efficiency improvement of heterogeneous network by exploiting the integration of macro–femto cellular networks. In *Proc. IEEE Intl. Conf. Commun., (ICC'2012)*, pages 1–6, Ottawa, ON, Canada, Jun. 2012.

42. V. Chandrasekhar and J. G. Andrews. Spectrum allocation in tiered cellular networks. *IEEE Trans. Commun.*, 57(10):3059–3068, Oct. 2009.

43. M.-S. Alouini and A. Goldsmith. Area spectral efficiency of cellular mobile radio systems. *IEEE Trans. Veh. Technol.*, 48(4):1047–1066, Jul. 1999.

44. 3rd Generation Partnership Project (3GPP)–Technical Specification Group Radio Access Network (TSG RAN) Radio Layer 1. Physical layer aspects for evolved universal terrestrial radio access (UTRA) (Release 9). 3GPP Technical Report TR 25.814 V7.1.0 (2006-09), Oct. 2006.

45. Energy and carbon conversions: fact sheet. http://www.carbontrust.com, 2012. Online; accessed Nov. 1, 2012.

46. International Monetary Fund, World Economic Outlook (WEO) Database Groups and Aggregates Information. http://www.imf.org, Apr. 2012. Online; accessed Sep. 1, 2012.

Section III

Smart Grid

11 Normalized Knowledge Integration to Distribute Intelligent Network Management in Power Utilities

Antonio Martín and Carlos León

CONTENTS

11.1 INTRODUCTION

Electric power systems are one of the most critical and strategic infrastructures of industrial societies and are currently going through a revolutionary change. Electricity networks worldwide are entering a period of change that necessitates improving intelligent methods of control and management. As a result of the electricity evolution, the electricity infrastructure will become more and more interlinked with network infrastructures. However, the same networking capabilities that can provide these benefits have also introduced vulnerabilities that have resulted in these systems having been identified as one of the most vulnerable targets for an operational network. The management of the framework is critically important in the electrical network and smart grid. The backbone of a successful smart grid operation is a reliable, resilient, and secure communication infrastructure. The advanced technology of intelligent agents (IAs) provides a well-researched way of implementing complex distributed, scalable, and open information and communication technology systems in smart grid. Intelligent control systems are an integral part of the critical infrastructures of power utilities. The capabilities of networking these systems provide unprecedented opportunities to improve productivity, reduce impacts on the environment, and help provide energy independence.

It is necessary to develop intelligent techniques that offer more possibilities. To resolve this difficulty we study the integration of advanced artificial intelligence technology into existing network management models. This network provides intelligent linkages between the elements of the grid and participates in the decision-making communication. Application of IAs to perform real-time control functions for the power grid is a way to introduce new information management techniques and information security functions to the power grid. The objective is to draw together all the knowledge management that is used in distribution network automation, to define new ideas, and to introduce solutions now being proposed to facilitate the intelligent control of networks. This chapter presents a technique for the design and implementation of a distributed intelligent system that is designed through the normalization and integration of knowledge management. The goal is the assignment and dispersed intelligent control of network resources, pertaining to hardware as well as software, to help operators manage their networks more effectively and also to promote reliability in network services. We describe an intelligent technique that processes management knowledge collected by IAs and uses it to detect and to resolve network anomalies and faults. Further, this work outlines the development of an intelligent system based on our proposed standard and describes the most important facets, advantages, and drawbacks that were found after prototyping our proposal.

This chapter is organized as follows. In the next section, we describe the evolution of network management and the role of network management functions. It starts with specific applications and work on expert systems in similar domains. We evaluate the approach of using IAs and distributed control in each of these control architectures particularly in the context of strategic energy plans. We make a review of innovative control architectures in electric power systems and we will examine the management network, including the concepts, major approaches, and management models. In the second part, we base this architecture on the requirements identified and suggest that power systems with high penetration of distributed generation may be controlled as a loose aggregation of energy control. In the third part, we formulate flexible control architecture for the future electric power system. We present an analysis of corporate network management requirements and technologies, together with our implementation experience in the development of an integrated management system for a power company network. Finally, we outline the conclusion and future works.

11.2 MANAGEMENT NETWORK OVERVIEW

As the size of communication networks keeps on growing, with more subscribers, faster connections and competing and cooperating technologies, and the divergence of computers, data communications, and telecommunications, the management of the resulting networks becomes more important and time-critical. The purpose of network management is the assignment and control of proper network resources, both hardware and software, to address service performance needs and the network's objectives. Telecommunications and services are in the process of revolutionary yet evolutionary changes owing to the transformation of the regulatory environment that in turn has given rise to rapid improvements in the underlying application, networking, computing, and transmission technologies.

Recent years have seen explosive growth in the areas of power system monitoring using IAs and distributed intelligence. Typical work in related fields include intelligent systems such as the work by Akinyokun and Imianvan [1], who present an intelligent supervisory coordinator for process supervision and fault diagnosis in dynamic physical systems. In Ref. [2], an online intelligent alarm-processing system is developed on the basis of the architecture of the digital substation. Other examples of application of IAs can be seen in the study by Brachman and Levesque [3] who propose a different type of univariate, randomly optimized neural network combined with discrete wavelet transform and fuzzy logic aimed at having a better power quality disturbance classification accuracy. In Ref. [4], some diagnostic requirements and emerging technologies available for insertion into future ship integrated power systems are looked at.

Many researchers have suggested that intelligent sensor network technologies could improve the effectiveness and efficiency of real-time management. The objective of the work by D'Addona and Teti [5] is to develop a novel system that can cover many of the existing challenges in distribution power systems and make distribution grid more intelligent. Doukas et al. [6] introduce a new methodology for settings of transmission system distance protection based on an intelligent analysis of events and their consequences. Goleniewski and Jarrett [7] propose a self-adapting intelligent system used for providing building control and energy-saving services in buildings. This approach [8] develops a novel methodology, which designs a hybrid controller with intelligent algorithms. Hui et al. [9] discuss an intelligent multimodal sensor system developed to enhance the haptic control of robotic manipulations of small three-dimensional objects. In Ref. [10], an expert system is developed by utilizing its fast reasoning mechanism and object-oriented features.

We identify specific requirements and respective capabilities of IAs for each of these control architectures. This work focuses on an intelligent framework and a language for formalizing knowledge management descriptions and combining them with existing Open Systems Interconnection (OSI) management model. We have normalized the knowledge management base necessary to manage the current resources in the telecommunication networks.

11.2.1 Network Management Models

There are two dominant network management models that have been used in the administration and control of most existing networks: Telecommunications Management Network (TMN) and Simple Network Management Protocol (SNMP). Both network management systems operate using client/server architecture. SNMP standards are defined in a series of documents, called request for comments or RFCs proposed by the Internet Engineering Task Force (IETF) and TMN is introduced by the ITU-T (the former CCITT). Of these two, TMN is gaining popularity for large complex networks. In a private network environment, SNMP enjoys near-universal support. In the public environment, however, a more heterogeneous mix of de facto telecommunications industry standards has prevailed, with a move toward TMN support. Moreover, TMN was the first who started, as part of its OSI program, the development of the architecture for network management. The OSI management environment consists of tools and services needed to control and supervise the management networks.

The OSI network management model is a starting point for understanding network management. There are three basic components comprising the elements of the management architecture to support a successful implementation of the OSI Network Management Model (Figure 11.1).

- An information component involved with five major functional areas—fault, configuration, accounting, performance, and security management—in network management that facilitate rapid and consistent progress within each category's individual areas [11].
- A communication component that focuses on how the information is exchanged between the managed systems [12].
- A functional component involved with the various activities performed in support of network management [13].

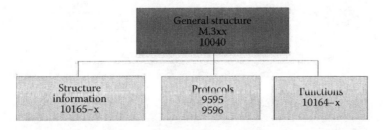

FIGURE 11.1 Overview of OSI network management model.

According to the International Organization for Standardization (ISO), the OSI network management model defines a conceptual model for managing all communication entities within a network. This main concept is the managed object (MO), which is an abstract view of a logical o physical resource to be managed in the network [14]. MOs provide the necessary operations for the administration, monitoring, and control of the telecommunications network. These operations are realized through the use the Common Management Information Protocol (CMIP) [15]. This is a network management protocol built on the OSI communication model. The related Common Management Information Services (CMIS) [16] defines services for accessing information about network objects or devices, controlling them, and receiving status reports from them. For a specific management system, the management process involved will take on one of two possible roles (Figure 11.2):

- A manager or manager role is an element that provides information to users, issues requests to devices in a network, receives responses to the requests, and receives notifications. These notifications are unsolicited information from devices in the network concerning the status of the devices.
- An agent or agent role is a unit that is part of a device in the network that monitors and maintains status about that device. It can act and respond to requests from a manager and can provide unsolicited information (or notifications) to a manager.

Attaining interoperability of network management systems and a common view of managed resources in a managed network environment requires that information models comply with standard models. The management functions currently exchange management information by means of techniques defined in ITU-T X.700. MOs are defined according to the Guidelines for the Definition of Managed Objects (GDMO), which has been established as a means to describe logical or physical resources from a management point of view. GDMO language uses the object orient programming and defines how network objects and their behavior are to be specified, including the syntax and semantics. The guidelines for the definition of MOs, ITU-T Recommendation X.722, allow for a common data structure for MOs in the managed and managing systems. MOs include their names, attributes, and operations that can be performed on attributes, notifications that objects can emit, and behavior descriptions of the objects. GDMO is now widely used to specify interfaces between different components of the TMN, a comprehensive and strategic series of international standards for network management. GDMO property value types are described using the abstract syntax notation one (ASN.1) [17]. ASN.1 describes an abstract syntax for data types and values. Our approach is to explore how the same management information would be modeled within the OSI standard.

The object definitions created using GDMO and related tools form a new management information base (MIB). In addition to being able to pass information back and forth, the manager and the agent need to agree on and understand what information the manager and agent receive in any exchange. This information varies for each type of agent. The collection of this information is

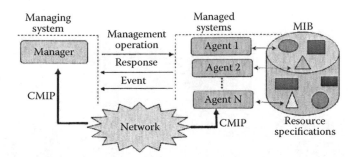

FIGURE 11.2 OSI manager/agent architecture.

referred to as the MIB. A manager normally contains management information describing each type of agent the manager is capable of managing. This information would typically include Internet MIB definitions and ISO definitions for MOs and agents.

After this brief introduction to management elements, we will approach our research in the integration of knowledge management of expert system into MIB in the OSI management model.

11.3   MANAGEMENT KNOWLEDGE DEFINITION

Management information modeling and the denominated MO play a large part in this network management model. MO coordinates all the downstream real-time functions within the distribution network with the intelligent knowledge needed to manage the network in an efficient way. We propose a set of criteria for integrating the knowledge management into the network resource specifications. For this purpose, we modeled network resources as IAs with a consistent language. This language describes objects as remote-controlled devices and communication infrastructure like workstations, LAN servers, switches, and so on. Thus, IAs need a standard structure template on each MO where they need to be hosted.

Some of models are basic information models that define the structures and contents of management information that the management functions act upon. A management information model provides a common characterization of the network resources, enables multiple management functions to interact with each other, and supports different management functions. Management information is exchanged among management systems where management functions are implemented [18]. Basically, GDMO prescribes how a network product manufacturer must describe the resource formally. GDMO uses an object-oriented approach to define the standardized functionality in substation devices. This language provides information models that describe how the information should be organized and how it should be exchanged between different devices.

GDMO is organized in archetypes that are standard formats used in the definition of a particular aspect of the agent. A complete agent definition is a combination of a relationship between a class

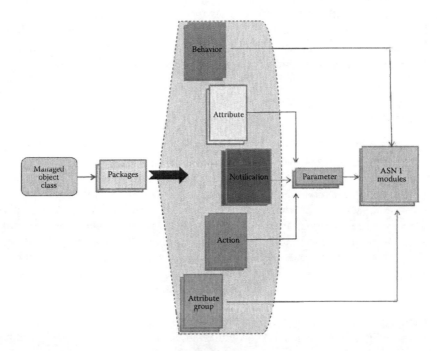

FIGURE 11.3   Relations between GDMO standard templates.

of MOs, package, attribute, group of attributes, action, notification, parameter, connection of name, and behavior (Figure 11.3).

A managed object class (MOC) is the base of the formal definition of an IA. This template is used to define the different kinds of agents that exist in the system. Classes describe what information and services they provide each managed object and GDMO defines format for this information [19]. The set of MOCs and instances under the control of an IA is an MIB, the abstraction of network resources properties and states for management purpose. The attribute values for an instance are accessed by issuing set and get requests to change or retrieve the attribute values, respectively. The definition of an MOC is made uniformly in the standard template, eliminating the confusion that may result when different persons define objects of different forms. This template is used to define the different kinds of objects that exist in the system. This way, we ensure that the classes and the management expert rules defined in system A can be easily interpreted in system B. The MOC structure is shown here:

```
<IA-label> MOC
[DERIVED FROM <IA-label> [,<IA-label>]*;]
[CHARACTERIZED BY
   <IA_properties-label>[,<IA_properties-label>]*;]
[CONDITIONAL PACKAGES
   <IA_properties-label> PRESENT IF condition;
   <IA_properties-label>] PRESENT IF condition]*;]
REGISTERED AS object-identifier;
```

The DERIVED FROM clause specifies the superclass or superclasses from which this MO is derived (inherited). This plays a very important role when determining the relations of inheritance, which makes it possible to reutilize specific characteristics in other classes of MOs. In addition, a great advantage is the reusability of the object classes and therefore of the expert rules that are defined. By using this clause, any attribute, operation, notification, and behavior exposed by MOs, as well as inheritance and containment relationships among MOs and MOCs, can be defined. Packages included in the object class definition are identified by the CHARACTERIZED BY and CONDITIONAL PACKAGES clauses. The CHARACTERIZED BY clause identifies the package or packages that are always present when the MO is included in the system. The CONDITIONAL PACKAGES clause is used to identify those packages that may or may not be included each time the MO of this class is instantiated. Finally, the REGISTERED AS clause identifies the location of the MOC on the OSI registration tree [20].

11.4 IA MODELING

In a heterogeneous and distributed energy context, the management of telecommunication networks and services is becoming increasingly important in operator and service provider environments. This management cannot be performed without the contribution of intelligent management functions, which ensure the most important management operations of provisioning, assurance, and billing. More advanced tools are needed to support this activity. An IA is an autonomous hardware/software system that can react intelligently and flexibly on changing operating conditions and demands from the surrounding processes. IA systems implement distributed decision-making systems in an open, flexible, and extensible way. IA is goal oriented; it carries out tasks and embodies knowledge for the management purpose. IAs can actively and dynamically cooperate for solving problems by using integrated knowledge and intelligence reasoning. Application of IAs to perform soft real-time control functions for the power grid is a way to introduce new information management techniques and information security functions to the power grid. We present a new paradigm to integrate the knowledge management, where knowledge managed and resources specifications are integrated in a same-language definition. For this to occur, IAs are required to have knowledge management of their own local system and at least partial models of the global system.

After describing a number of important aspects that have to be considered when designing a language for behavior descriptions, this section focuses on the syntax and semantics of the language GDMO that is discussed in this chapter. Practical experience with GDMO shows that, from an intelligent point of view, the quality of GDMO specifications is not satisfactory. The MO specification is incomplete to define the management knowledge of a specific resource. As a consequence, a new element is necessary. To solve the current problem, that is, undertaking an intelligent integrated management, we offer an original contribution to include expert rules in the specifications of the network features. To answer these questions, it will be necessary to make changes on the template of the GDMO standard. To formalize the main proposal of the chapter, we analyze the necessary requirement area to undertake the related aspects with the knowledge integration in the MOs. We present an extension of the standard GDMO to accommodate the intelligent management requirements.

Each intelligent object class may be seen as the integration of the following basic components: packages, name bindings, and behavior characteristics. The elements that at the moment form the GDMO standard do not make a reference to the knowledge base of an expert system. Until now, the MOs are not able to use the knowledge that the base of knowledge provides, which collects the management operations and control of a management domain. We observe the need to define new structures for those cases in which it is necessary to express the knowledge. To allow interaction among the IAs for sharing their knowledge, a specific knowledge exchange language is being developed. We propose to extend the GDMO with the following goals: facilitate the normalization and integration of the knowledge base of expert system into resources specifications.

These goals will allow developers to specify the storage location, update the method of intelligent management, and provide a way to specify complex management. Thus, the description of certain aspects of MO knowledge, for example, the definition of expert rules, can be supported. We proposed adding a new property in GDMO standard named "Know." This attribute will define all the aspects about management knowledge in a specific MOC. The set of MOCs and instances under the control of an agent is known as its an MIB, an abstraction of network resource properties and states for the purpose of management. The MIB, which is specified using the structure management information, defines the actual objects to be managed [21].

Thus, the description of certain aspects of MO knowledge, for example, the definition of expert rules, can be supported. To solve this problem, we suggest a refinement of the package template. We propose a new description for the information management definition named GDMO+, wherein we add a new element named KNOW, as shown in Figure 11.4.

Two relationships are essential for the inclusion of knowledge in the component definition of the network: MOC and Package. GDMO includes the basic template MANAGED OBJECT CLASS, which is always implemented, and GDMO also defines an optional template named PACKAGE, which defines a combination of properties for later inclusion in an MO class template. This template allows an IA to have properties that provide normalized knowledge of a management dominion. Management knowledge will be transported between IAs and should be described by using

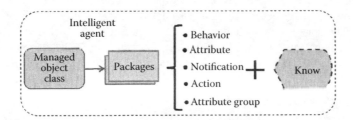

FIGURE 11.4 Template relations in GDMO+ methodology.

the prescriptions of the structure management information language using the CMIP and CMIS. CMIS supports knowledge exchange between network management applications and management agents.

11.4.1 Package Archetype

The PACKAGE template is used to specify the characteristics that represent a consistent set of specifications about a network resource. One purpose of the package is to provide a set of reusable definitions that can be used in several MOC specifications. All the properties that we define in the package will be included later in the Managed Object Class Template, where the package is incorporated. The same package can be referenced by more than one class of MOs. For each MOs class, the following information is defined:

1. Attributes, which are the types of data supported by the class (MO).
2. Operations, which are the actions supported by the class.
3. The behavior of the MO.
4. Notifications, which are the types of unsolicited information an MO can send to a manager.

In the MOC definition, "CHARACTERIZED BY" and "CONDITIONAL PACKAGES" clauses define obligatory packages or conditional packages that are present when the MO is included in the system, and they are used to specialize an IA by adding new characteristics. We will add the KNOW property into the package element. The next definition shows the elements of a package template, in which it is possible to observe the new element KNOW.

```
<IA-properties-label> PACKAGE
  [BEHAVIOUR <behaviour-label> [,<behaviour-label>]*;]
  [ATTRIBUTES
    <attribute-label> propertylist [,<parameter-label>]*
    [,<attribute-label> propertylist [,<parameter-label>]*]*;]
  [ACTIONS <action-label> [<parameter-label>]*
         [<action-label> [<parameter-label>]*]* ;
  [NOTIFICATIONS
    <notification-label> [<parameter-label>]*
    [<notification-label> [<parameter-label>]*]* ;]
  [KNOW <know-label> [,<know-label>]*;]
REGISTERED AS object-identifier;
```

One purpose of the package is to provide a set of reusable definitions that can be used in several IA class specifications. In this way, we implemented a standard agent with knowledge characteristics that can generate a new type of IA, requiring only the management knowledge specification of its principal classes. The package template is used to specify the characteristics that represent a consistent set of specifications about an IA. The package template is a combination of behavior definitions, attributes, attribute groups, operations, notifications, and parameters. Each IA has two types of knowledge: local network management for individual problem solving and community and local network management for coordination of associated node activities. The KNOW attribute will define all the aspects related to management knowledge in a specific intelligent system.

The current template package in GDMO standard is adapted and we add a new feature. In addition to the properties indicated above, we suggest the incorporation of a new property called KNOW and its associated template called KNOW, which contains all the specifications of the knowledge base for the expert system.

11.4.2 KNOWLEDGE MANAGEMENT ARCHETYPE TEMPLATE

There are a number of different knowledge representation techniques for structuring knowledge in an expert system. The three most widely used techniques are expert rules, semantic nets, and frames [22]. For this study, we use expert rules. We represented the knowledge in production rules or simply rules. Rules are expressed as IF–THEN statements that are relatively simple, very powerful, and very natural to represent expert knowledge. A major feature of a rule-based system is its modularity and modifiability, which allow for incremental improvement and fine-tuning of the system with virtually no degradation of performance. Expert rules specify the actions the inference engine should take when the premise or conditions in the rule are true.

In our study case, the template KNOW permits the normalized definition of the specifications of the expert rule to which it is related. This template allows a particular MOC to have properties that provide a normalized knowledge of a management dominion. The structure of the KNOW template is shown here:

```
<IA_know-label> KNOW
  [PRIORITY    <priority> ;]
  [BEHAVIOR <behavior-label> [,<behaviour-label>]*;]
  [IF occurred-event-pattern [,occurred-event-pattern]*]
  [THEN sentence [, sentence]* ;]
REGISTERED AS object-identifier;
```

The first element in a definition is headed. It consists of the name of the management expert rule <know-label> and a key word that indicates the type of template KNOW. After the head, the following elements compose a normalized definition of the management knowledge:

- BEHAVIOR: This construct is used to extend the semantics of previously defined templates. It describes the behavior of the rule. This element is common to the other templates of the GDMO standard.
- PRIORITY: Each production rule in a rule base is a single piece of knowledge that has the capacity to draw a certain conclusion from some evidence. If there are two sources of evidence for some hypothesis, then this value represents the priority of the rule, that is, the order in which competing management actions will be executed by IAs.
- IF: We can add a logical condition that will be applied to the events that have occurred or their parameters. Multiple conditions are joined by logical operators such as AND or OR, and the premise evaluates to true if all or at least one of the conditions evaluates to true for AND and OR, respectively. Rules fire as soon as events are matched with the premises or antecedents of these rules. The premise of a rule examines parameter or slot values, and once the condition evaluates to true, then the action part is executed. Those events should be defined in the notification archetype.
- THEN: An agent's repertoire of tasks represents its capabilities or methods. These are actions and diagnoses that the management platform makes as an answer to network events that have occurred. These actions can consist of setting other parameter or slot values or invoking methods on an instance or a class. Each task can have its procedural "how to do" component represented as expert rules. Those operations should be previously defined in the action template.
- REGISTERED AS is an object identifier. This clause identifies the location of the expert rule on the OSI registration tree. The identifier is compulsory.

The KNOW element allows a treatment similar to the other properties, including the possibility of inheritance of rules between classes.

11.4.3   EXAMPLE OF KNOWLEDGE MANAGEMENT DEFINITION

This section shows a complete example of expert rules integration in the GDMO+ proposed standard. It defines an MOC named *radioTransceiverCTR190* corresponding to a real device in the network of a power utility.

```
radioTransceiver_CTR190 MANAGED OBJECT CLASS
  DERIVED FROM radioTransceptor;
  CHARACTERIZED BY transceiverPackage;
REGISTERED AS {nm-MobjectClass 1};
```

This is a device that both sends and receives radio signals. Its primary purpose is to broadcast the signal. The transmitter and the receiver share common circuitry into a single housing like transponders, transverters, and repeaters. These units typically offer the convenience of having multiple functions like establishing radio channel, controlling signals, monitoring station, monitoring alarm condition, controlling logic to activate operations in response to commands received over said communications network, and so on.

The class *radioTransceiver* includes the compulsory *transceiverPackage*, which contains all the specifications corresponding to the device. We can indicate the two expert rules that have been associated with the defined class by means of the KNOW clause.

```
transceiverPackage PACKAGE
ATTRIBUTES
  reception Power GET,
  sense         GET,
  speedTransmission GET,
  ...;
NOTIFICATIONS
  damageFeeding,
  inferiorLimit,
  repairAction;
KNOW transmissionError, powerError;
REGISTERED AS {nm-package 1};
```

Two typical examples of expert rules used in our GDMO specification are *transmissionError* and *powerError*. These rules are defined by using the KNOW template. The expert rules are used within EXPAriel to capture and detect anomalies or defects in operations produced in the transceiver device and suggest the necessary measures for solving the problem.

```
transmissionError KNOW
  PRIORITY 4;
  BEHAVIOUR transmissionErrorBehaviour;
  IF (?date ?time1 ?local 7_TX_C2 ?remote ALARM)
  (?date ?time2 ?local 7_TX_C2 ?remote ALARM & : (<(ABS(? ?time1 ?time2))
  1.00))
  THEN ("Severity:" PRIORITY),
("Diagnostic: "It damages in the modulate transmission between", ?local,
"and" ?remote),
("Recommendation "Revision transceiver");
REGISTERED AS {nm-rule 1};
powerError KNOW
PRIORITY 3;
BEHAVIOUR powerErrorBehaviour;
IF (?date ? ?local 7_F_ALIM_2 ?remote ALARM)
```

```
   (NOT (?date ? ?local CCA?34_AIS_DE_BB ?remote ALARM))
 THEN ("Severity:" PRIORITY),
 ("Diagnostic:
 It damages in the electric feeding of the station" ?local),
 ("Recommendation: To revise the electric connection", ?local);
REGISTERED AS {nm-rule 2};
```

The first rule, *transmissionError*, is devoted to the detection of errors in the data transmission module of the transceiver CTR190. The second, *powerError*, is in charge of detecting failures in the power supply of the transceiver CTR190. Both rules give recommendations on how to solve the failures.

11.5 A DIAGNOSTIC MODEL FOR A PRIVATE MICROWARE NETWORK CASE

Previous sections performed a requirement analysis for applying IAs in electric power system with intelligent control architecture. Inevitably, the application of these concepts has to be adapted to the real world to account for the business priorities and pressures of a particular utility. In order to validate our approach, we have developed a production system based in IAs that integrates the management knowledge into the resources specification of a network. In order to evaluate the fault management capabilities of the integrated management solution proposed, we have simulated an application that monitors an environment to collect fault event data [23].

We study an example of alarm detection and intelligent resolution of incident concerning a private network. As said before, we have considered a private network. We suppose this network as being heterogeneous and hierarchical. We have used a telecommunications network that belongs to a company in the electrical sector, Sevillana-Endesa's (EA), a Spanish power utility. The intelligent sensor nodes only disseminate data when the network is being monitored and an error occurs in a resource [24]. The use of integrated knowledge in agents can help the system administrator in using the maximum capabilities of the intelligent network management platform without having to use other specification language to customize the application [25].

The dynamic IA platform we have developed is named EXPAriel. The intelligent system development should meet the following requirements: it should be robust, management activity should not interfere with the normal operations of the network, and it should only intervene when necessary [26]. Management applications should be able to perform even when the network is not fully operational, as management is mostly needed in abnormal situations, for example, when connections are broken. We are going to use a SCADA system owing to the management limitations of network communication equipment. The basis of any real-time control is the SCADA system, which acquires data from different sources, preprocesses it, and stores it in a database accessible to different users and applications. SCADA consists of the following subsystems (Figure 11.5):

- Remote terminal units (RTUs) connecting to sensors in the process, converting sensor signals to digital data, and sending digital data to the supervisory system
- Communication infrastructure connecting the supervisory system to the RTUs
- A supervisory (computer) system, gathering (acquiring) data on the process and sending commands (control) to the process, which is our IA

The communication infrastructure takes care of interfacing a set of RTUs to the company network. The Spanish power grid company has a network using wireless technology on the regional high-tension power grid. Part of long-distance traffic in this net is controlled by a wireless intelligent system distributed throughout this private network. Detection mechanisms are implemented in real time in our prototype and have been embedded within the network elements, network protocols, and devices. This architecture covers situations in which sensors deployed in a smart grid place provide

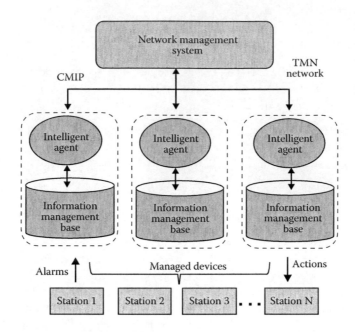

FIGURE 11.5   Elements of the prototype EXPAriel.

information to IAs of the conditions of the surrounding environment. SCADA systems are configured around standard base functions like data acquisition, monitoring and event processing, data
storage archiving and analysis, and so on. RTU is needed to provide an interface between the sensors
and the SCADA network. The RTU is supplied as part of the SCADA system with a communication
structure that uses a standard SCADA protocol over microwave radio or dedicated landlines. This
device adds the transmission of analogy measurements to status and control commands. The RTU
encodes sensor inputs into protocol format and forwards them to the SCADA master. In turn, RTU
receives control commands in protocol format from the master and transmits electrical signals to the
appropriate control relays. RTU continuously monitors all sensors and performs data processing on
information gathered from sensors. The collected data may be processed in different ways, leading to
more or less advanced intelligent RTUs. The fundamental role of an RTU is the acquisition of various types of data from the power process; the accumulation, packaging, and conversion of data in a
form that can be communicated back to the master; the interpretation and outputting of commands
received from the master; and the performance of local filtering, calculation, and processes to allow
specific functions to be performed locally. We are using NetGuardian 832A. It acts as the interface to
the communication system, which receives and sends information to the central control. If we want to
integrate a new RTU in an environment, then we will have to develop a new IA and its corresponding
MIB module. The SCADA level of the telecontrol network has been connected via wireless point-to-
point links from its front end to the remote terminal units (Figure 11.6).

One important element of communication is receiver-specific transformation of information. The
sending agent is aware of the goals and procedures of the receiving agent and will transform the
information appropriately before it is sent. Another element of mediation is notification of other IAs.
IAs can send to the RTUs control commands, time synchronization messages, and device setting and
parameter data. IAs monitor local states and cooperate with one another to ensure an admissible global
state. The communication hierarchical architecture fits into the managers of agents. The supervision
below an RTU includes all network devices and substation and feeder levels like circuit breakers,
reclosers, autosectionalizers, the local automation distributed at these devices, and the communication
infrastructure. This shows how the communication between the expert agents or interfaces, utilizing

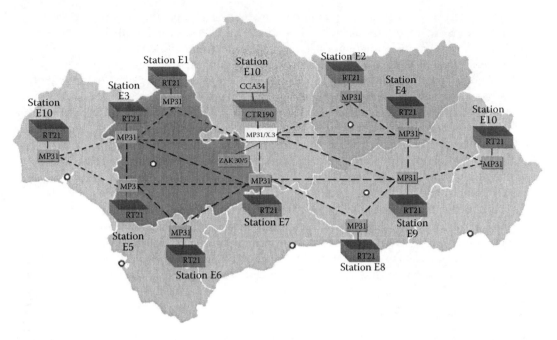

FIGURE 11.6 Power company network map in Andalucía.

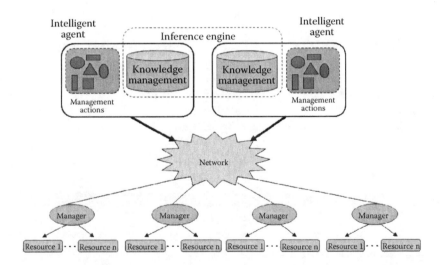

FIGURE 11.7 Architecture system.

RTUs as an inbox for messages received and as a destination for messages sent from each agent in the system, separates the correspondences by the destination agent of each message.

Our system has three major components: an inference engine, a knowledge base, and a user interface (Figure 11.7). Those elements are briefly discussed in the following [27]:

- *The inference engine.* This is the processing unit that solves any given problems by making logical inferences on the given facts and rules stored in the knowledge base. It defines the MOs and the expert rules belonging to the Expert System that manages this network [28]. Our system is implemented in Brightware's ART*Enterprise, an expert system shell.

ART*Enterprise is a set of programming paradigms and tools that are focused on the development of efficient, flexible, and commercially deployable knowledge-based systems. Expert system shells simplify developer interactions by eliminating the developer's concern with operating system requirements. Its use can therefore reduce the design and implementation time of a program considerably. By using an existing general-purpose tool, we were able to build a standard and extensible platform with proven performance and quality. The experience with our prototype is that ART*Enterprise is a useful tool for developing expert systems.

- *The knowledge base*: It is the core of the system; this is a collection of facts and if–then production rules that represent stored knowledge about the problem domain [29]. The knowledge base of our system is a collection of expert rules and facts expressed in the ARTScript programming language ART*Enterprise. The knowledge base contains both static and dynamic information and knowledge about different network resources and common failures. The resultant expert system has about 600 rules and Workstation has been employed to program the expert system. This initial knowledge has been acquired from the experts in the management domain. The knowledge base of our system can be extended by adding new higher-level rules and facts.

11.5.1 HUMAN MACHINE INTERFACE

EXPAriel reports to human operators over a specialized computer called Human–Computer Interface (HCI). IAs sometimes need to communicate with operators. The HCI provides the operator with the best "as operated" view of the network. Isolating and resolving the fault is only a portion of the possible functions of IA, because operation of the network would be improved if, having isolated the fault, as much of the healthy network as possible was re-energized. Basic information describing the operating state of the power network is passed to the communication supervisory system (CSS). This is collected automatically by equipment in various substations and devices, manually input by the operator to reflect the state of any manual operation of no automated devices by field crews, or calculated. In all cases, the information is treated in the same way. This information is categorized as status indications, measured values, and energy values. When a link fails, a number of alarms are generated and passed to the SCADA.

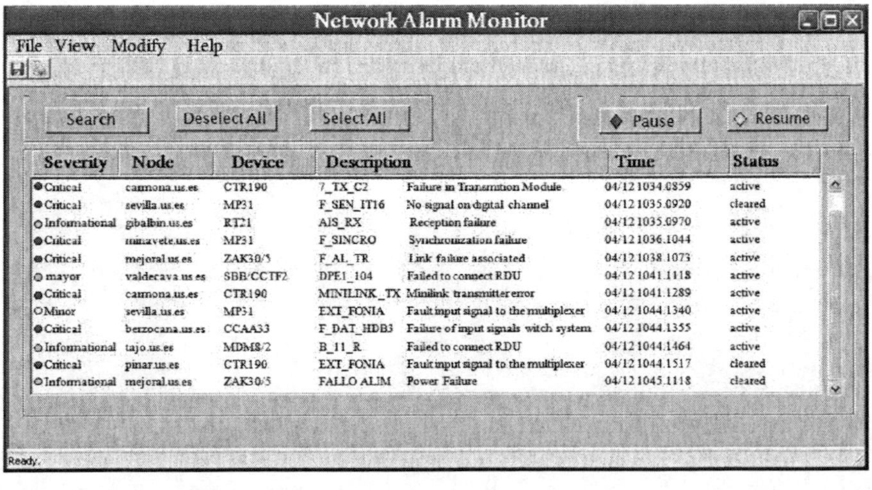

FIGURE 11.8 Network alarm monitor.

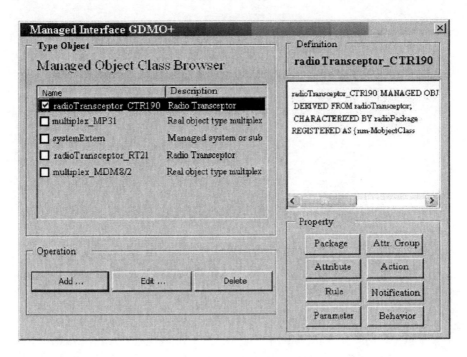

FIGURE 11.9 System object classes dialog box.

```
F1 (31/01 1100.200 sevilla.us.es 7_TX_C2 tajo.us.es ALARM)
F2 (31/01 1103.168 sevilla.us.es 7_TX_C2 tajo.us.es ALARM)
```

Each device provides a time-stamped message on events (starting, tripping, activation, etc.) through the bus. These events are sorted by time and transmitted to the event printers or to the monitoring system (Figure 11.8).

On the other hand, from a developer's perspective, creating an MIB database from a higher-level set of abstractions has become difficult by using traditional database technology. The key is the organization of the distribution management database MIB, access to all supporting IT infrastructure, and applications necessary to populate the model and support the other daily operating tasks. In this context, a human operator can be thought of and can act like another IA. The MIB modeling technique generates an information model comprising many highly interrelated objects. To facilitate the management of the MIB, our HCI provides a variety of tools for the creation, editing, and updating of MIBs and an MIB Browser. The MIB Browser is a tool that allows the administrator to interactively inspect the definitions of management object classes. This includes tools for browsing the inheritance architecture classes generated by both GDMO and ASN.1 compilers. Moreover, our system includes facilities to browse the GDMO classes using a Web browser such as Explorer or Mozilla [30] (Figure 11.9).

The user interface is easy to use and enables mobile management in that remote access only requires access to the Internet and a Web browser. Our user interface also contains a preprocessor for parsing GDMO+ specification files. The user interface component named GDMO Managed Object Definition Tool allows administrators to inspect the definitions of management object classes interactively.

11.6 THE COMMUNICATION SUPERVISORY SYSTEM

The nerve center of any power network is the central control and management function, where the coordination of all operational strategies is carried out. EXPAriel operations use a supervision

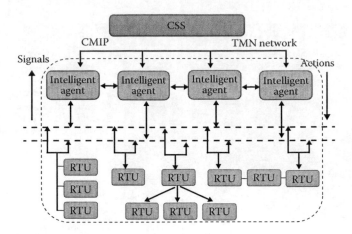

FIGURE 11.10 Communication supervisory system.

system called the communication supervisory system (CSS). This system can monitor, in real time, the network's main parameters, making use of the information supplied by the SCADA, placed on the main company building, and the RTUs are installed at different stations. Collaborative management actions between IAs are achieved through the CSS. Starting from the supplied information, the operator is able to undertake actions through the CSS in order to solve the failures that could appear or to send a technician to repair the stations' equipment. The CSS allows the operator to acquire information, alarms, or digital and analogical parameters of measure, registered on each IA or RTU.

The CSS has the capability of selecting the IA that is best suited for satisfying the client's requirement, without the client being aware of the details about the agent. Further, the IA is able to communicate and negotiate with the other IAs. Collaborative IAs are useful, especially when a task involves several systems on the network. Figure 11.10 shows a hierarchical architecture that represents the whole network, the IA, RTU objects, and the network elements.

Control functions are initiated automatically from software IA and directly affect power system operation. Automatic control is triggered by an event or specific time that invokes the control action. IA automatically initiates a set of sequential management actions to restore the network resource. The control runs completely automatically, without the need for human intervention. The sending agent is aware of the interests of the other IAs and may notify other agents automatically when exception conditions are either detected or may be probable or imminent. IA adds a great deal of intelligence and automation to our system management, making the management network much easier. The concept of grid-based control has been adopted and based upon the capabilities of intelligent systems; for example, modularity, decomposition, local/distributed control, and its implementation are anticipated to be done using sophisticated mechanisms of coordination, cooperation, and competition provided by IA technology. Coordination among IAs involves cooperative management of transactions to ensure consistency and correctness. Even if distributed control and operation are implemented, the results of such action must be communicated to the central coordination point. One has to reason based on the network topology to correlate the alarms and detect the root cause of the failure.

The IA listener is the component responsible for the monitoring of the RTU's message received by the network resources. It is a subsystem of the CSS essentially covering all real-time aspects of the network control process. It receives the RTU's messages and verifies whether the messages should be further redirected to the integrated expert system (real invoked service) or to the network operator. RTUs receive the management actions from the IAs. These operations are the answer to the management's request in the network resources. Many of the collection features and processing of data performed by the RTU are basic IA functions that have been allocated from the IA to the RTU.

11.7   SYSTEM EVALUATION AND TESTING

Validation constitutes an inherent part of the knowledge-based expert system development for EXPAriel and is intrinsically linked to the development cycle. The purpose is to achieve a functionally correct prototype. Validation concerns have the following objectives: to analyze the reliability of the system and to ascertain what the platform knows, does not know, or knows incorrectly. The system developed under GDMO+ specifications is quite different from a traditional expert system, despite the fact that the scenario testing and rules underlying the analysis are similar. As we have already seen in previous sections, EXPAriel provides a lot more than just a detection of the faults, which is all that the traditional expert system is capable of at the moment. In the first place, our system is able to automatically resolve most of the faults produced in the communications network. On the other hand, GDMO+ IAs can decentralize processing and control and improve management efficiency. In order to check the improvements of the GDMO+ IAs with a real application in this section, we have compared EXPAriel with a traditional expert system. For this purpose, EXPAriel has been tested with regard to the following aspects: system validation using test cases, validation by case studies, and validation against human experts.

An important aspect of the design and implementation of an intelligent system is determination of the degree of speed in the answer that the network provides. We will discuss the issue of response time for five agents associated to transceiver resources. Every IA is assigned a particular resource repair task. We test the model by inserting some alarms into the system. We compared our results with those we had obtained with the traditional system. Note that the response time would vary

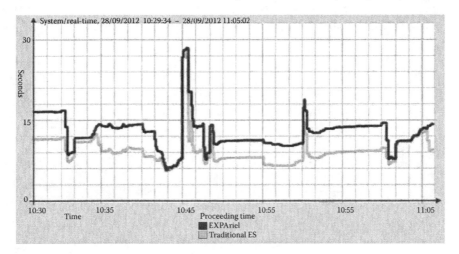

FIGURE 11.11   Performance of EXPAriel and traditional ES.

TABLE 11.1
Comparison of Traditional Expert System and EXPAriel

	Traditional ES		EXPAriel	
	Proceeding Time (s)	Rules per Second	Proceeding Time (s)	Rules per Second
A1	1.525	109.120	1.250	124.0000
A2	6.561	45.773	5.249	55.8202
A3	3.213	69.156	2.975	85.3782
A4	21.758	17.125	17.982	19.2415
A5	0.142	388.983	0.118	432.2034

depending on both the agent and the fault type. Figure 11.11 shows a sample plot of these parameters that was collected as a part of the experiment, which shows that the speed of the EXPAriel system improves the proceeding time and the average of the traditional expert system.

In Table 11.1, we present the average setup time for some measurements. As the table shows, the results for EXPAriel are 15.1% better than the proceeding time and 19.5% better than the executing time rules per second in the traditional expert system.

Another test of significance is the analysis of the number of alarms that have been automatically resolved by EXPAriel and the warnings received by the system operator. As noted in Table 11.2, EXPAriel performs satisfactorily with an approximately 94.6% rate of success in real cases. It is also noted that the performance of EXPAriel may depend considerably on the facts stated. The more information is input, the better the chance of diagnosing the likely causes of the problems in the network.

From these results, we can establish the following conclusions:

- The effectiveness of the filtration process is very high: almost 90% of the whole. This has the advantage of a decreasing percentage in the amount of indications presented to the operator.
- The speed of the system improves, diminishing the number of alarms on which the rest of the rules act.
- The expert system, with over 600 operation rules, has produced excellent results that, after extensive field testing, proved to be capable of filtering 90% of produced alarms with a precision of 95% in locating them (Figure 11.12). As noted above, EXPAriel performs satisfactorily with an approximately 95% rate of success in real cases. The confidence

TABLE 11.2
Performance Management Events and Alarms

Alarms Number	Autonomous Resolution	Autonomous Resolution %	Managed Actions Executed	Operator Warnings
100	99	99	151	1
200	190	95	202	6
400	369	92.25	501	16
600	562	93.66	793	16
800	745	93.12	994	23
1000	946	946.1	1528	49

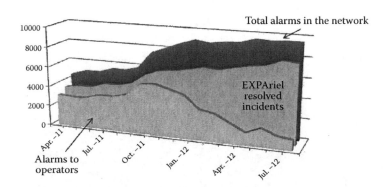

FIGURE 11.12   Filtration process effectiveness.

values provided were also found to be in reasonable relative order. It is also noted that the performance of EXPAriel depends considerably in the facts that happened. The more information is input, the better the chance of diagnosing the likely causes of the problems in the network.

11.8   CONCLUSIONS

Current networks are very complex and demand ever-increasing levels of quality, making it very important to take into account their management. The traditional model of network administration has certain deficiencies that we have tried to overcome by using a model of intelligent integrated management. To improve the techniques of expert management in a communications network, we propose the possibility of integrating and normalizing the expert rules of management within the actual definition of the MOs. In this chapter, we show possibilities to apply and integrate artificial intelligence techniques in network management and supervision by using the proposed standard GDMO. Through the integration of the knowledge within the new extension of the GDMO standard, we can simultaneously define management information and knowledge.

The intelligent control architecture tries to organize the grid in a flexible way that allows dynamic aggregation and de-aggregation of resources at different intelligent control levels. The use of IAs in network supervision can help the administrator in using the maximum capabilities of the network management platform. These IAs not only have to optimally perform local control within the network resource but also must comply with responsibilities toward the main grid. Distributing intelligent power system control and analysis is viewed as one of the fastest growing areas of research and new application development in network management. We have investigated the innovative control architecture in electric power systems, in which we are using IAs.

We conclude by pointing out an important aspect of the obtained integration: the solution not only masks possible faults but also optimizes the management functions and efficiency of the distributed services and their resources by using an artificial intelligent strategy, while ensuring a high degree of functionality in power utilities.

Our work of knowledge integration on the IAs can be viewed as a first step toward automated management by using IAs. Our future work will aim to improve the agent's performance using another method of knowledge representation and reasoning that is different from the rules, that is, via semantic nets, neuronal nets, frameworks, and so on. In addition to fault detection functional area, we would also like to expand the scope of our current work to other functional areas like accounting management, configuration management, performance management, and security management. Finally, it is suggested to further develop the prototype system, adding more modules on the basis of the framework provided by EXPAriel. In this sense, we propose the use of external programs and graphics interface to enhance the functions of the system.

AUTHOR BIOGRAPHIES

Antonio Martín received his DEng degree in computer science engineering from the University of Seville, Seville, Spain, in 2005. He is currently a professor with the Department of Electronic Technology. He is also an associate editor and an editorial board member of different journals. His current research interests include distributed knowledge management, sensors and actuators networks, intelligent agents, and artificial intelligence, and he is also interested in ontology applications, data mining, web services, fuzzy theory, and case base reasoning applications.

Carlos León received his BSc degree in electronic physics in 1991 and his PhD degree in computer science in 1995, both from the University of Seville, Seville, Spain. He has been a full professor of electronic engineering at the University of Seville since 1991 and he is currently CIO of the University of Seville. His research areas include knowledge-based systems and computational intelligence focus on utilities system management and WSN. Dr. León is a member of the IEEE Power Engineering Society and the IASTED Neural Networks technical committee.

REFERENCES

1. Akinyokun, O.C., Imianvan, A.A. Mobile agent system for computer network management. In *Proceedings From the International Conference on Advances in Engineering and Technology*, Pages 796–808. 2006.
2. Bihina Bella, M.A., Eloff, J.H.P., Olivier, M.S. A fraud management system architecture for next-generation networks. *Forensic Science International*, Volume 185, Issues 1–3, 10, Pages 51–58. 2009.
3. Brachman, R.J., Levesque, H.J. *Representation and Reasoning*. San Francisco: Elsevier/Morgan Kaufmann. 2004.
4. Chantaraskul, S., Cuthbert, L. An intelligent-agent approach for congestion management in 3G networks. *Engineering Applications of Artificial Intelligence*, Volume 21, Issue 4, Pages 619–632. 2008.
5. D'Addona, D., Teti R. Intelligent tool management in a multiple supplier network. *CIRP Annals—Manufacturing Technology*, Volume 54, Issue 1, Pages 459–462. 2005.
6. Doukas, H., Patlitzianas, K.D., Iatropoulos, K., Psarras, J. Intelligent building energy management system using rule sets. *Building and Environment*, Volume 42, Issue 10, Pages 3562–3569. 2007.
7. Goleniewski L., Jarrett, K.W. *Telecommunications Essentials, Second Edition: The Complete Global Source*. Boston: Addison Wesley Professional. 2006.
8. Hebrawi, N. GDMO, object modelling and definition for network management. Technology appraisals. 1995.
9. Hui, S.C., Fong, A.C.M., Jha, G. A web-based intelligent fault diagnosis system for customer service support. Original Research Article. *Engineering Applications of Artificial Intelligence*, Volume 14, Issue 4, Pages 537–548. 2001.
10. ISO/IEC DIS 10165-4/ITU-T. Recommendation X.722, Information technology—Open systems interconnection—Structure of management information—Part 4: Guidelines for the Definition of Managed Objects (GDMO), International Organization for Standardization and International Electrotechnical Committee. 1993.
11. ITU-T. Recommendation X.700, Management framework for open systems interconnection (OSI). CCITT Applications. 1992.
12. ITU-T. Rec. M.3010, Principles for a telecommunications management network (TMN). Study Group IV. 1996.
13. ISO/IEC and ITU-T. Information processing systems—Open systems interconnection—Systems management overview. Standard 10040-2, Recommendation X.701. 1998.
14. Hebrawi, N. GDMO, object modelling and definition for network management. Technology appraisals. 1995.
15. ISO/IEC and ITU-T. Information processing systems—Open systems interconnection—Systems management overview. Standard 10040-2, Recommendation X.701. 1998.
16. ITU-T. Recommendation M.3400, TMN management functions. Study Group IV. 1996.
17. Morris, S.B. *Network Management, MIBs and MPLS: Principles, Design and Implementation*. Upper Saddle River, NJ: Pearson Education, Inc. 2003.
18. Huang, C.W. *Introduction to Communication Systems of Network Management*. Taipei, Taiwan: Chinese Taipei Components Certification Board. 2008.
19. Kuo, S.Y., Liao, F.P., and Chen, K.L. *Network Management: Concepts and Practice, A Hands-On Approach*. Taipei, Taiwan: GoTop Book Corporation. 2005.
20. Stallings, W. *SNMP, SNMPv2, and CMIP: The Practical Guide to Network*. Reading, MA: Addison-Wesley. 2000.
21. Clemm, A. *Network Management Fundamentals*. Indianapolis, IN: Pearson Education, Cisco Press. 2006.
22. Brachman, R.J., Levesque, H.J. *Representation and Reasoning*. San Francisco: Elsevier/Morgan Kaufmann. 2004.
23. Ray, P., Parameswaran, N., Lewis, L. Distributed autonomic management: An approach and experiment towards managing service-centric networks. *Journal of Network and Computer Applications*, Volume 33, Issue 6, Advances on Agent-Based Network Management, Pages 653–660. 2010.

24. Baker, D., Nodine, M., Chadha, R., Chiang, C.J. Computing diagnostic explanations of network faults from monitoring data. In *Proceedings of IEEE Military Communication Conference*, CA, USA, pp. 1–7. 2008.

25. Boyapati, P., Shenai, K., Lilly, B.R. Sensors and sensor network. U.S. 1-50400 (D2008-48, 49, 50, 51 and 52). 2008.

26. Akinyokun, O.C., Imianvan, A.A. Mobile agent system for computer network management. In *Proceedings From the International Conference on Advances in Engineering and Technology*, Pages 796–808. 2006.

27. Tacconi, D., Miorandi, D., Carreras, I., Chiti, F., Fantacci, R. Using wireless sensor networks to support intelligent transportation systems, *Ad Hoc Networks*, Volume 8, Issue 5, Vehicular Networks, Pages 462–473. 2010.

28. Waiman, C., Leung, L.C., Tam, P.C.F. An intelligent decision support system for service network planning. *Decision Support Systems*, Volume 39, Issue 3, Pages 415–428. 2005.

29. Power, Y., Bahri, P.A. Integration techniques in intelligent operational management: A review. *Knowledge-Based Systems*, Volume 18, Issues 2–3, Pages 89–97. 2005.

30. Vallejo, D., Albusac, Castro-Schez, J.J., Glez-Morcillo, C., Jimenez, C. A multi-agent architecture for supporting distributed normality-based intelligent surveillance. *Engineering Applications of Artificial Intelligence*, Volume 24, Issue 2, Pages 325–340. 2011.

12 Smart Grid Energy Procurement for Green LTE Cellular Networks

Hakim Ghazzai, Elias Yaacoub,
Mohamed-Slim Alouini, and Adnan Abu-Dayya

CONTENTS

12.1 INTRODUCTION

Smart grid is becoming a new global commercial venture widely seen as a means to upgrade electrical infrastructure to enhance power savings and optimize some green goals of consumers by reducing greenhouse gas emissions and optimally adjusting the power consumed [1]. For this reason, in many different research fields, people opt to introduce smart grid in their works to ensure energy efficiency and minimize costly environmental impacts. In mobile communications, a smart grid can significantly contribute in reducing the power consumption of the network and thus it helps mobile operators cope with global warming. In fact, mobile networks already represent approximately 10% of the total carbon emitted by the information and communication technology (ICT) sector, and this is expected to increase every year because of the extensive growth in the number of subscribers and service usage times [2]. The Fourth Generation Long Terms Evolution (4G LTE) network is the recent standard for wireless communication of high-speed data for mobile phones and data terminals and will become the first truly global mobile phone standard. However, its radio network consumes over 70%–80% of its total power [3]. Therefore, several works were proposed to save energy and make its radio access network (RAN) a green radio network.

Several efforts were proposed to enhance the energy efficiency of the recent LTE network by completely switching off base stations (BSs) during off-peak hours when data traffic is low. Indeed, studies show that an active BS in an idle status (no transmission) consumes more than 50% of the power because of circuit processing, air conditioning, and other factors [4]. Many algorithms have been proposed to reduce the number of active BSs depending on different criteria based on a certain quality of service (QoS) metric (e.g., Refs. [5–7]).

A complementary work is to study these algorithms by introducing a smart grid that contains multiple energy sources (e.g., electricity generated from fossil fuels and/or from renewable energy sources) and powering cellular networks [8]. Recent research focuses on the dynamic operation of cellular BSs that depend on the traffic, real-time pricing provided by the smart grid, and the pollutant level associated with the generation of the electricity [9]. However, the work in Ref. [9] does not consider a particular technology (e.g., LTE) and does not take intercell interference into account.

In this chapter, we present a green optimization problem that improves the energy efficiency in LTE cellular networks not only by applying the BS sleeping strategy but also by introducing the smart grid as a tool for power management. In fact, it aims to optimally procure the required power from different energy retailers in the smart grid in order to

- Minimize the power consumption of the LTE network and reduce its CO_2 emissions
- Maximize the profit of the mobile network operator
- Maintain a certain QoS

This is performed given the nature and the cost of the provided energies in the smart grid as well as the unitary prices of the mobile network operator services. In addition, we take into account both the UpLink (UL) and DownLink (DL) directions, LTE resource allocation, and intercell interference. In our framework, we assume that green energy can be produced from two different renewable sources. In fact, the network operator can procure it from a renewable energy retailer or deploy its own renewable energy equipment by using BSs that can be powered via solar panels or wind turbines. Furthermore, the mobile operator can benefit from these two methods at the same time whenever possible: a certain amount of renewable energy is procured from its installed equipment and another amount is bought from the renewable energy retailer to satisfy the additional power consumption needs of the cellular network. In addition, with smart meters installed at the BS sites, we investigate the scenario when the mobile operator is able to sell the excess of the produced renewable energy to public retailers [1]. In this chapter, we formulate a general optimization problem that encloses all these scenarios, present heuristic solutions for this problem, and compare their performances in practical cases in order to find the optimal configuration for a green LTE network.

Two practical algorithms are introduced. The first one, entitled the iterative algorithm (IA), tries to reduce the number of active BSs by eliminating one BS at each iteration. The second algorithm, the genetic algorithm (GA), is considered as a very strong optimization tool used in several applications for LTE networks such as in resource allocation [10] or energy saving [11] and also for applications associated with smart grid [12]. The proposed GA computes the utility function of the network for multiple BS combinations until reaching an optimal solution that maximizes the utility function.

The rest of this chapter is organized as follows. Section 12.2 reviews some related works and details the contribution of the study. Section 12.3 presents the system model. Section 12.4 describes the green optimization approach. The strategy and the heuristic algorithms to solve the optimization problem are detailed in Section 12.5. In Section 12.6, we present and analyze the simulation results. Future research directions are presented in Section 12.7, and finally, we conclude the chapter in Section 12.8.

12.2 RELATED WORKS AND CONTRIBUTIONS

By improving the energy efficiency of BSs in cellular networks, large savings can be obtained from both environmental and cost points of view [4], since BSs consume more than 80% of the total power of cellular networks. By optimizing their power consumption, the effect of CO_2 emissions is decreased and the power cost is reduced. Many approaches have been investigated to minimize BS power consumption by improving BS energy efficiency or shutting down redundant BSs during low traffic. Renewable energies are also introduced as alternative sources to supply the network. In this section, we review some previous works related to green communications.

12.2.1 RELATED WORKS

Several research works try to ensure power savings in LTE networks by introducing femtocells and relays. In Refs. [13] and [14], using different algorithms, results show that by resizing cells or by introducing relays in the network, the total power consumption of the RAN can be reduced. However, these methods oblige mobile operators to redeploy new femto BSs and relays inside each macrocell of existing networks, which make these solutions very costly.

Abdel Khalek et al. [15] propose an algorithm for joint UL/DL Universal Mobile Telecommunications System (UMTS) radio planning with the objective of minimizing total power consumption in the network. The problem is subdivided into two components that are executed successively: as a first step, the authors investigate the optimal locations of a fixed number of UMTS BSs in the area of interest. The optimal locations of BSs are obtained by solving an optimization problem that aims to minimize the total DL power expenditure and, at the same time, the UL outage that depends on the power capabilities of mobile stations (MSs) under different constraints that maintain an acceptable QoS and satisfy the power budget. As a second step, the authors propose an algorithm to select the minimal cardinality set of BSs with fixed locations. In other words, after deploying BSs in the area of interest, an algorithm to eliminate redundant BSs is applied.

BS sleeping strategy is also investigated in Refs. [5] and [6]. The aim is to switch off lightly loaded BSs to achieve power savings in the LTE network for two different QoS metrics. An example is shown in Figure 12.1. One of the cells is lightly loaded with only one user connected to the BS, whereas the other BS has a higher load with three users connected to it, as shown in Figure 12.1a. If the user in the second cell can be connected to the BS of the first cell without compromising its QoS, then the BS of the second cell can be switched off, which leads to green energy-efficient communications, as shown in Figure 12.1b. In fact, the system model of Ref. [5] is based on an LTE radio network where resource allocation is done through LTE resource blocks (RBs). The algorithm takes into account the UL and the DL directions. The decision criterion for shutting down BSs is based on computing a utility function that maximizes the number of served users in the network. On the other hand, to switch off a BS, El-Beaino et al. [6] calculate the total signal-to-interference plus noise ratio (SINR), which is the sum of SINRs of all active users in the network, and compare it to a

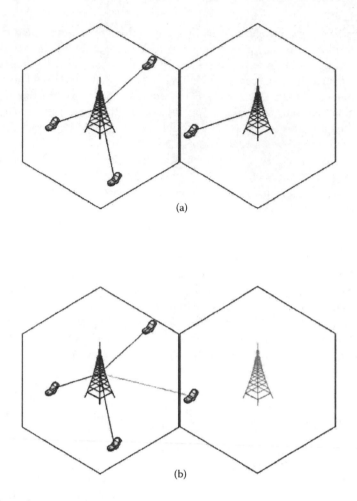

FIGURE 12.1   Green communications example.

fixed SINR threshold: if it is greater than the SINR threshold, the switched-off BS is underutilized and is maintained off. Both algorithms deal with the sleeping strategy to achieve energy savings but they do not consider renewable energy suppliers. The methods in Refs. [5,6], and [15] do not take into account the smart grid aspects of multiple electricity retailers and the use of renewable energy for an environmentally friendly operation of the cellular network.

Xiang et al. [16] apply the BS sleeping strategy to eliminate underutilized BSs. The idea is to find a relationship between the traffic load (user arrival following a Poisson process) and power savings. The solution is obtained by solving an optimization problem that has as an objective to minimize the number of BSs subject to two constraints: maintaining the user connection in the cell and covering the same initial area.

Another approach to ensure energy savings in designing green wireless cellular networks is to investigate energy-efficient communications by taking into account the dynamics of the smart grid that depend on the traffic, real-time price, and the pollutant level associated with the generation of the electricity. Bu et al. [9] introduce the use of coordinated multipoint communication to ensure acceptable QoS in cells whose BSs have been shut down to save power. The active BSs decide from which retailers they procure electricity and how much power is required considering the pollutant level of each retailer and the proposed price. They model the system as a Stackelberg game, which has two levels: a cellular network level plays the role of the follower and a smart grid level plays the role of a leader. They solve the problem by finding the unique Stackelberg equilibrium. The scheme proposed in Ref. [9] can significantly reduce operational expenditure and CO_2 emissions in green wireless cellular networks. However, in Refs. [16] and [9], intercell interference, resource allocation, and fading variations are not taken into account.

In addition, other methods suggest to employ the GA to ensure power savings in LTE networks. For instance, in Ref. [11], the author proposes a method to reduce the power consumption by switching off lightly loaded cubes. In the solution in Ref. [17], which was introduced by Alcatel-Lucent Bell Labs, the shutting-down decision is applied by the GA based on data mining of the network traffic load. This leads to a strong dependency on the traffic load state of adjacent cubes. Besides, some works focus on the application of GA to smart grids (e.g., Ref. [12]), where GA is used to solve non-convex multi-objective optimization problems for ICT applications.

12.2.2 CONTRIBUTIONS

The contributions of this work compared to the existing literature can be summarized as follows:

- The state-of-the-art LTE technology is considered. The DL and UL directions are investigated where orthogonal frequency-division multiple access (OFDMA) and single carrier frequency domain multiple access (SCFDMA) are adopted, respectively. In addition, we consider the existence of intercell interference and we assume a limited power budget for each BS. Furthermore, radio resource allocation is implemented in order to ensure the best use of the available radio resources.
- A novel optimization problem is formulated. The problem aims to maximize the profit of the mobile network operator, minimize the emission of greenhouse gas, and maintain a desired level of QoS at the same time by eliminating lightly loaded BSs and optimizing the power procurement from public and private electricity retailers.
- We implement two heuristic algorithms, IA and GA, with the BS sleeping strategy in order to solve the formulated problem and reach suboptimal solutions, we compare their performances, and we show that even if it requires a higher computational complexity, the GA achieves better results.
- Different scenarios are investigated, depending on the interplay between the cellular network and the smart grid, that is, depending on how the cellular network is procuring the needed power [18]:
  - The mobile operator procures power from a smart grid where different electricity retailers are characterized by different power costs, different sources used to generate power, and hence different pollution levels caused by electricity generation. The procurement of power will depend on the market mechanisms of the smart grid. This scenario is analyzed depending on the mobile operator's attitude toward the environment.
  - The mobile operator possesses its own renewable energy architecture. In this scenario, BSs are equipped with solar panels or wind turbines that provide, fully or partially (depending on the amount that can be generated), the power needed from renewable energy sources. Furthermore, we investigate the scenario where BSs are equipped with smart meters and hence the mobile operator is able to even sell the excess of renewable energy to public electricity retailers.

12.3 SYSTEM MODEL

We consider an LTE network deployed in a uniform geographical area. We divide the area into cells of equal size where a BS is placed. In LTE, the access scheme for the UL is SCFDMA while OFDMA is used in DL where a fixed number of consecutive subcarriers constitutes RBs. A set of RBs are assigned to users according to a given allocation procedure as described in detail in Section 12.4.1.

Briefly, the resource allocation is performed as follows: In the area of interest, we place users following a random distribution and each one communicates with a selected BS. Because of the low power of MS compared to the one of BS, we associate for each user the BS that offers the best available UL channel gain, and from this selected BS, we allocate RBs that provide the best available DL channel gain to that user.

12.3.1 Channel Model

The channel gain for both directions (UL and DL) over subcarrier s between user i and BS j is given by [19]

$$H_{i,s,j,\mathrm{dB}} = (-\kappa - \upsilon \log_{10} d_{i,j}) - \xi_{i,s,j} + 10 \log_{10} F_{i,s,j} \tag{12.1}$$

where the first factor captures propagation loss, with κ being the path-loss constant, $d_{i,j}$ being the distance in kilometers from user i to BS j, and υ being the path-loss exponent. The second factor in Equation 12.1, $\xi_{i,s,j}$, captures log-normal shadowing with zero mean and a standard deviation σ_ξ, whereas the last factor in Equation 12.1, $F_{i,s,j}$, corresponds to Rayleigh fading with a Rayleigh parameter a (usually selected such that $E[a^2] = 1$). In the rest of the chapter, in order to differentiate between UL and DL subcarriers, the notation $H_{k_l,i,j}^{(\mathrm{UL})}$ and $H_{k_l,i,j}^{(\mathrm{DL})}$ will be used, respectively.

12.3.2 Power Consumption Model for BSs

We consider that each BS is equipped with a single omni-directional antenna. The consumed power P_j^{BS} of the jth active BS can be computed as follows [20]:

$$P_j^{\mathrm{BS}} = aP_j^{\mathrm{tx}} + b, \tag{12.2}$$

where P_j^{tx} denotes the radiated power of the jth BS. The coefficient a corresponds to the power consumption that scales with the radiated power owing to amplifier and feeder losses. The term b models an offset of site power that is consumed independently of the average transmit power and is due to signal processing, battery backup, and cooling.

12.3.3 Data Rates in the DownLink

Let $\mathcal{I}_{\mathrm{sub},i}^{(\mathrm{DL})}$ be the set of DL subcarriers allocated to user i, $\mathcal{I}_{\mathrm{RB},i}^{(\mathrm{DL})}$ the set of RBs allocated to user i in the DL, $N_{\mathrm{RB}}^{(\mathrm{DL})}$ the total number of DL RBs, $P_{s,j}^{\mathrm{tx}}$ the total power transmitted by the BS j over subcarrier s, $P_{j,\mathrm{max}}^{\mathrm{tx}}$ the maximum transmission power of BS j, N_{BS} the total number of deployed BSs, N_U the total number of subscribers in the area, and $R_i^{(\mathrm{DL})}$ the achievable DL rate of user i. Then, the OFDMA throughput of user i is given by

$$R_i^{(\mathrm{DL})}\left(P_{j,\mathrm{max}}^{\mathrm{tx}}, \mathcal{I}_{\mathrm{sub},i}^{(\mathrm{DL})}\right) = \sum_{s \in \mathcal{I}_{\mathrm{sub},i}^{(\mathrm{DL})}} B_{\mathrm{sub}}^{(\mathrm{DL})} \cdot \log_2\left(1 + \Gamma_{i,s,j}^{(\mathrm{DL})}\right), \tag{12.3}$$

where $\Gamma_{i,s,j}^{(\mathrm{DL})}$ is the SINR of user i over subcarrier s in cell j and $B_{\mathrm{sub}}^{(\mathrm{DL})}$ is the subcarrier bandwidth. It is expressed as

$$B_{\mathrm{sub}}^{(\mathrm{DL})} = \frac{B^{(\mathrm{DL})}}{N_{\mathrm{sub}}^{(\mathrm{DL})}}, \tag{12.4}$$

with $B^{(DL)}$ being the total usable DL bandwidth and $N_{sub}^{(DL)}$ being the total number of DL subcarriers. In this chapter, we consider equal power transmission over the subcarriers; that is, for all s, we have

$$P_{s,j}^{tx} = \frac{P_{j,max}^{tx}}{N_{sub}^{(DL)}}. \tag{12.5}$$

The DL-SINR of user i over subcarrier s in cell j, $\Gamma_{i,s,j}^{(DL)}$, is given by

$$\Gamma_{i,s,j}^{(DL)} = \frac{P_{s,j}^{tx} H_{i,s,j}^{(DL)}}{I_{s,i}^{(DL)} + \sigma_{s,i}^2}, \tag{12.6}$$

where $H_{i,s,j}^{(DL)}$ is the channel gain of user i over subcarrier s, $\sigma_{s,i}^2$ is the noise power over subcarrier s in the receiver of user i, and $I_{s,i}^{(DL)}$ is the interference on subcarrier s measured at the receiver of user i. The expression of the interference is given by

$$I_{s,i} = \sum_{k=1,k \neq j}^{N_{BS}} \left(\sum_{l=1}^{N_U} \lambda_{l,s,k}^{(DL)} \right) \cdot P_{s,k}^{tx} H_{i,s,k}^{(DL)}, \tag{12.7}$$

where $\lambda_{l,s,k}^{(DL)} = 1$ if DL subcarrier s is allocated to user l in cell k; that is, $s \in \mathcal{I}_{sub,l}^{(DL)}$. Otherwise, $\lambda_{l,s,k} = 0$. In each cell, an LTE RB, and hence the subcarriers constituting that RB, can be allocated to a single user at a given time transmission interval. Hence, in each cell k, we have

$$\sum_{l=1}^{N_U} \lambda_{l,s,k}^{(DL)} \leq 1. \tag{12.8}$$

12.3.4 Data Rates in the Uplink

Let $\mathcal{I}_{sub,i}^{(UL)}$ be the set of UL subcarriers allocated to user i, $\mathcal{I}_{RB,i}^{(UL)}$ the set of UL RBs allocated to user i, $N_{RB}^{(UL)}$ the total number of RBs in the UL, $P_i^{(MS)}$ the total transmit power of user i, and $R_i^{(UL)}$ its achievable rate in the UL. Then, the SCFDMA throughput of user i is given by

$$R_i^{(UL)}\left(P_i^{(MS)}, \mathcal{I}_{sub,i}^{(UL)} \right) = \frac{B^{(UL)} \left| \mathcal{I}_{sub,i}^{(UL)} \right|}{N_{sub}^{(UL)}} \cdot \log_2\left(1 + \Gamma_i^{(UL)}\left(P_i^{(MS)}, \mathcal{I}_{sub,i}^{(UL)} \right) \right), \tag{12.9}$$

where $B^{(UL)}$ is the total UL bandwidth, $\left| \mathcal{I}_{sub,i}^{(UL)} \right|$ is the cardinality of $\mathcal{I}_{sub,i}^{(UL)}$, and $N_{sub}^{(UL)}$ is the number of UL subcarriers. Finally, $\Gamma_i^{(UL)}\left(P_i^{(MS)}, \mathcal{I}_{sub,i}^{(UL)} \right)$ is the SINR of user i after minimum mean squared error frequency domain equalization at the receiver [21]:

$$\Gamma_i^{(UL)}\left(P_i^{(MS)}, \mathcal{I}_{sub,i}^{(UL)} \right) = \left(\frac{1}{\frac{1}{\left| \mathcal{I}_{sub,i}^{(UL)} \right|} \sum_{s \in \mathcal{I}_{sub,i}^{(UL)}} \frac{\Gamma_{i,s,j}^{(UL)}}{\Gamma_{i,s,j}^{(UL)} + 1}} - 1 \right)^{-1}. \tag{12.10}$$

In Equation 12.10, $\Gamma_{i,s,j}^{(UL)}$ is the UL SINR of user i over subcarrier s served by BS j. It is given by

$$\Gamma_{i,s,j}^{(UL)} = \frac{P_{i,s,j}^{(UL)} H_{i,s,j}^{(UL)}}{I_{s,j}^{(UL)} + \sigma_{s,j}^2}, \tag{12.11}$$

where $H_{i,s,j}^{(UL)}$ is the channel gain between user i and BS j over subcarrier s, $\sigma_{s,j}^2$ is the noise power over subcarrier s at BS j, $P_{i,s,j}^{(UL)}$ is the power transmitted by user i over subcarrier s in cell j, and $I_{s,j}^{(UL)}$ is the UL interference on subcarrier s, measured at BS j. The expression of the interference is given by

$$I_{s,j}^{(UL)} = \sum_{k=1, k \neq j}^{N_{BS}} \sum_{l=1}^{N_U} \lambda_{l,s,k}^{(UL)} P_{l,s,k}^{(UL)} H_{l,s,j}^{(UL)}, \tag{12.12}$$

where $\lambda_{l,s,k}^{(UL)} = 1$ if subcarrier s is allocated to user l served by BS k; that is, $s \in \mathcal{I}_{sub,l}^{(UL)}$. Otherwise, $\lambda_{l,s,k}^{(UL)} = 0$.

The LTE standard imposes the constraint that the RBs allocated to a single user should be consecutive with equal power allocation over the RBs [22–24]. Hence, we set

$$P_{i,s,j}^{(UL)} = \frac{P_i^{(MS)}}{\left| \mathcal{I}_{sub,i}^{(UL)} \right|}. \tag{12.13}$$

12.3.5 Mobile Operator Services

In our framework, the network operator offers M different services characterized by their data rate thresholds $R_{m,th}^{(UL)}$ and $R_{m,th}^{(DL)}$ for UL and DL, respectively, and their unitary prices $p^{(m)}$ with $m = 1 \ldots M$. We suppose that each user in the network benefits from one of the M offered services.

12.3.6 Retailers and Pollutant Levels

In our study, we assume that the cellular network is powered by a smart grid where N retailers exist to provide energy with different prices and pollutant levels depending on the nature of the generated energy. The amount of energy $q_j^{(n)}$ procured by the jth BS from retailer n ($n = 1 \ldots N$) depends essentially on the unitary cost proposed by the energy provider ($\pi^{(n)}$) and on its pollutant level, which is represented by a penalty term corresponding to the pollutant emission cost, F, modeled as follows [25]:

$$F\left(q_j^{(n)}\right) = \alpha_n \left(q_j^{(n)}\right)^2 + \beta_n q_j^{(n)}, \tag{12.14}$$

where α_n and β_n are the emission cost coefficients of retailer n. In addition, we suppose that each retailer has a maximum available amount of energy. For instance, the network cannot procure from the renewable energy retailer more than a certain constant $Q_{max}^{(n)}$ $\left(\text{i.e.,} \sum_{j=1}^{N_{BS}} q_j^{(n)} \leq Q_{max}^{(n)} \right)$.

Another way to power the network is to deploy renewable energy equipment on the BS sites, and each BS would be powered by its own installed equipment, for example, solar cells or wind turbine.

The auto-generated amount of energy by BS j is denoted $q_j^{(0)}$. In addition, we suppose that all BSs have a maximum capacity to store the produced energy denoted Q_{max}^{BS} and, at an instant t, the renewable energy available at BS j is denoted $Q_{RE_{max}}^{(j)}$, where $q_j^{(0)} \leq Q_{RE_{max}}^{(j)} \leq Q_{max}^{BS}$ $\forall j = 1 \ldots N_{BS}$.

To summarize, we distinguish two notations of the procured power:

- $q_j^{(n)}$ is the amount of energy provided to BS j and bought from retailer n.
- $q_j^{(0)}$ is the amount of renewable energy generated by the green equipment of BS j, which is free of charge.

On the basis of this system model and these parameters, we formulate an optimization problem where the mobile network operator is able to optimally procure energy from the smart grid to power its BSs and apply the sleeping strategy technique in order to achieve power savings.

12.4 ADMISSION CONTROL AND PROBLEM FORMULATION

Before formulating the optimization problems that will be solved in Section 12.5 using the heuristic algorithms, we define the admission control and resource allocation method.

12.4.1 ADMISSION CONTROL AND RESOURCE ALLOCATION

We assume that the subcarriers constituting a single RB are subjected to the same fading and hence the channel gain on the subcarriers of a single RB is considered to be the same. In addition, the fading is assumed to be independent identically distributed across RBs. In this chapter, we allocate one UL RB and one DL RB for each user.

Hence, when a user i joins the network, it is associated with cell j^* and the UL RB for which the subcarriers $s^{*(UL)}$ satisfy

$$(s^{*(UL)}, j^*) = \arg \max_{(s,j)} \left(1 - \sum_{l=1;\, l \neq i}^{N_U} \lambda_{l,s,j}^{(UL)} \right) H_{i,s,j}^{(UL)}. \tag{12.15}$$

Then, for the DL, it is allocated the RB in cell j^* for which the subcarriers $s^{*(DL)}$ satisfy

$$s^{*(DL)} = \arg \max_{s} \left(1 - \sum_{l=1;\, l \neq i}^{N_U} \lambda_{l,s,j^*}^{(DL)} \right) H_{i,s,j^*}^{(DL)}. \tag{12.16}$$

In Equations 12.15 and 12.16, the first term in the multiplication indicates that the search is on the RBs that are not yet allocated to other users. Then, the rates (Equations 12.3 and 12.9) are computed.

12.4.2 PROBLEM FORMULATION

We assume that N_{BS} BSs are deployed and N_U users are randomly distributed in the area of interest. We denote by N_{out} the number of users that are not able to communicate with network BSs safely ($N_{out} = N_U$). A user i benefiting from the mth service communicates successfully with a BS, if its UL and DL data rates, denoted $R_i^{(UL)}$ and $R_i^{(DL)}$, respectively, are higher than the service data rate thresholds, $R_{m,th}^{(UL)}$ and $R_{m,th}^{(DL)}$, respectively. We associate a binary parameter γ_i, $i = 1 \ldots N_U$ to each user: if the user i is served successfully, then $\gamma_i = 1$ else $\gamma_i = 0$. We can express this assumption as follows:

$$\gamma_i = \begin{cases} 1 \text{ if } R_i^{(UL)} \geq R_{m,th}^{(UL)} \text{ and } R_i^{(DL)} \geq R_{m,th}^{(DL)} \\ 0 \text{ if } R_i^{(UL)} < R_{m,th}^{(UL)} \text{ or } R_i^{(DL)} < R_{m,th}^{(DL)} \end{cases}. \qquad (12.17)$$

In other words, if $\gamma_i = 0$, the ith user is in outage. If we denote $\gamma = \left[\gamma_1 \dots \gamma_{N_U} \right]$, then the number of ones and the number of zeros in γ correspond to the number of served users and the number of users in outage, respectively. Consequently, only the served users pay the equivalent of the used service. On the other hand, the mobile operator can sell the unused amount of renewable energy thanks to smart meters [1]. Hence, the operator network revenue $\mathcal{R}$ is expressed as follows:

$$\mathcal{R} = \sum_{i=1}^{N_U} \gamma_i p_i^{(m)} - \sum_{j=1}^{N_{BS}} \pi^{(0)} \left(Q_{RE_{max}}^{(j)} - q_j^{(0)} \right), \qquad (12.18)$$

where $p_i^{(m)}$ is the cost of the service m used by the ith subscriber and $\pi^{(0)}$ is the unitary cost of the sold renewable energy and is mathematically represented by the negative price $\pi^{(0)} \leq 0$.

In order to include the BS sleeping strategy in the problem formulation, we introduce a binary variable ε_j with $j = 1,\dots, N_{BS}$ to denote the BS state as follows:

$$\varepsilon_j = \begin{cases} 1 \text{ if BS } j \text{ is switched on} \\ 0 \text{ if BS } j \text{ is switched off} \end{cases}. \qquad (12.19)$$

Let $\varepsilon = \left[\varepsilon_1 \dots \varepsilon_{N_{BS}} \right]$. The number of ones and the number of zeros in this vector indicate the number of active and inactive BSs, respectively. Note that each BS can procure energy from different retailers at the same time and it can produce its own energy thanks to the renewable energy equipment installed in its site. Hence, both the total cost of the energy consumption and the CO_2 emission cost function caused by the cellular network depend only on the active BSs and the nature of the procured energy as it is given in the following expressions:

- The total cost of the energy consumption of the network $\mathcal{C}$:

$$\mathcal{C}(\varepsilon, \mathbf{q}) = \sum_{j=1}^{N_{BS}} \sum_{n=1}^{N} \varepsilon_j \pi^{(n)} q_j^{(n)}. \qquad (12.20)$$

- The CO_2 emission penalty function of the network $\mathcal{I}$:

$$\mathcal{I}(\varepsilon, \mathbf{q}) = \sum_{j=1}^{N_{BS}} \sum_{n=1}^{N} \varepsilon_j \left(\alpha_n (q_j^{(n)})^2 + \beta_n q_j^{(n)} \right), \qquad (12.21)$$

where $\mathbf{q} = \left[q_1^{(0)} q_1^{(1)} \dots q_1^{(N)} q_2^{(0)} q_2^{(1)} \dots \dots q_{N_{BS}}^{(N)} \right]^T$ is the vector that contains the procured energy amount by the jth BS from the nth energy source and $\pi^{(n)}$ is the cost of one unit of energy provided by the

nth retailer, where $j = 1,\ldots, N_{BS}$ and $n = 1,\ldots, N$. The function $\mathcal{I}(\boldsymbol{\varepsilon}, \mathbf{q})$ reflects the friendliness to the environment of the mobile network operator and corresponds to the CO_2 emissions caused by its total power consumption. Renewable energy presents a solution for network operators to reduce greenhouse gas emissions. To procure this green energy, the network operator can either buy the required amount from a public renewable energy retailer or procure it by deploying its private renewable energy equipment on its BS sites. If the produced renewable energy is not enough to cover the need of all BSs, the network operator can procure energy from public retailers of the smart grid that have a sufficient amount of energy. The formulated optimization problem accommodates the two previous scenarios at the same time: The mobile operator can consume its own produced energy and buy renewable energy from a public retailer.

We consider that N public retailers are available to provide electricity to the network. Each retailer is assumed to generate electricity from a different source. Hence, we can have one retailer generating electricity from renewable energy sources (having a network of solar panels or wind turbines) and other retailers generating electricity from fossil fuels. Depending on the fossil fuel used, the CO_2 emission caused by electricity generation may differ and the pollution caused to the environment would vary. In addition, the price of electricity generated by renewable energy is expected to be higher than that generated by burning fossil fuels, owing to the higher costs of generating clean green energy. In addition, we consider that $\eta\%$ of the BSs are equipped with renewable energy generators (local solar panels or wind turbines owned by the mobile operator). For instance, if $\eta = 100\%$, all BSs produce renewable energy and this means that the mobile operator produces this energy without any cost (equipment installation costs are not taken into consideration). If $\eta = 50\%$, for example, 50% of the BSs are supplied by the installed equipment while the other 50% of the BSs procure energy from the public retailers. Hence, when the amount of renewable energy is not enough to power all the BSs of the network, the mobile operator has to procure the rest of the energy from the electricity retailers.

Consequently, the mobile operator has to optimally compute the amount of energy to procure from each retailer existing in the smart grid and from the generators installed on BS sites in order to maximize the following utility function:

$$U = (1 - \omega)\mathcal{P}(\gamma, \boldsymbol{\varepsilon}, \mathbf{q}) - \omega\mathcal{I}(\boldsymbol{\varepsilon}, \mathbf{q}), \tag{12.22}$$

where ω is a parameter to be defined, $\mathcal{I}(\boldsymbol{\varepsilon}, \mathbf{q})$ is given in Equation 12.21, and $\mathcal{P}(\gamma, \boldsymbol{\varepsilon}, \mathbf{q})$ is a function that corresponds to the mobile operator's profit. It is given by

$$\mathcal{P}(\gamma, \boldsymbol{\varepsilon}, \mathbf{q}) = \mathcal{R}(\gamma, \mathbf{q}) - \mathcal{C}(\boldsymbol{\varepsilon}, \mathbf{q}). \tag{12.23}$$

It should be noted that the amount of energy produced by BS generators ($q_j^{(0)}$ for $j = 1,\ldots, N_{BS}$) are not included in the cost and CO_2 emission cost function as these green consumed energies are free of charge for the operator and do not pollute the environment.

The objective is now to solve a multi-objective optimization (or Pareto optimization) problem by constructing a single aggregate objective function [26] that corresponds to a weighted linear sum of the objective functions 12.21 and 12.23. These functions are weighted by a parameter ω called the Pareto weight ($0 < \omega < 1$). The elements of the vector $\mathbf{q} = \left[q_1^{(0)} \ldots q_1^{(N)} q_2^{(0)} \ldots \ldots q_{N_{BS}}^{(N)} \right]^T$, the binary variables γ_i, $i = 1,\ldots, N_U$ and ε_j, $j = 1,\ldots, N_{BS}$ are the decision variables of the problem.

When $\omega \to 0$, we are dealing with the utility function given in Equation 12.23. This corresponds to a selfish network operator that aims to maximize its own profit $\mathcal{P}$ regardless of its impact on the environment. When $\omega \to 1$, we deal with the utility function given in Equation 12.21, which corresponds to an environmentally friendly network operator that aims to reduce CO_2 emissions regardless of its own profit. Other values of ω constitute a trade-off between these two extremes.

Hence, the optimization problem is expressed as follows:

$$\underset{\gamma,\varepsilon,\mathbf{q}}{\text{Maximize}} \; U = (1-\omega)\, \mathcal{P}(\gamma,\varepsilon,\mathbf{q}) - \omega \mathcal{I}(\varepsilon,\mathbf{q}), \tag{12.24}$$

$$\text{Subject to: } \sum_{j=1}^{N_{BS}} \varepsilon_j q_j^{(n)} \leq Q_{\max}^{(n)} \; \forall n = 1,\ldots,N, \tag{12.25}$$

$$\sum_{n=0}^{N} q_j^{(n)} = P_j^{BS} \Delta t \, \forall j = 1,\ldots,N_{BS}, \tag{12.26}$$

$$q_j^{(0)} \leq Q_{RE_{\max}}^{(j)} \quad \forall j \in \mathcal{S}_{RE}, \tag{12.27}$$

$$\frac{N_{\text{out}}}{N_U} \leq P_{\text{out}}, \tag{12.28}$$

$$q_j^{(n)} \geq 0 \; \forall j = 1 \ldots N_{BS} \text{ and } \forall n = 0,\ldots,N \tag{12.29}$$

where Δt is the network operation period and $\mathcal{S}_{RE}$ is the set of BSs that are equipped with renewable energy generators. The cardinality of this set depends on the percentage $\eta\%$.

The constraint in Equation 12.25 indicates that the energy consumed by all BSs in the cellular network from energy retailer n cannot exceed the total energy provided by that retailer while Equation 12.26 indicates that the amount of energy drawn by a BS from all retailers and from the renewable energy generated locally should be equal to the power needed for its operation, Equation 12.27 indicates that the energy procured by a BS j from its own power generated locally cannot exceed the amount of energy that can be stored, Equation 12.28 forces the number of users in outage to be less than a tolerated outage probability threshold P_{out}, and Equation 12.29 is a trivial constraint expressing the fact that the energy drawn is a positive amount.

It should be noted that, when a certain retailer n can provide to the mobile network operator enough electricity to power all the BSs in the network, we can set $Q_{\max}^{(n)} = +\infty$ to relax the constraint in Equation 12.25 for that retailer, although in practice the amount of power produced is naturally finite.

12.5 GREEN ALGORITHMS

In this section, we present two algorithms adapted to the BS sleeping strategy (GA and IA). These algorithms aim to find a suboptimal solution to the formulated problem detailed in Section 12.4. In fact, this problem is considered as a combinatorial problem because of the existence of binary variables (γ_i and ε_j) as decision variables that make the optimal and exact solution of this nonlinear optimization problem difficult or even impossible to find [26]. Therefore, we deal with heuristic approaches to find a suboptimal solution to the problem. In fact, several methods can be proposed to solve the optimization problem formulated in Section 12.4. For instance, evolutionary algorithms (e.g., genetic algorithms [27]) or swarm-based optimization algorithms (e.g., particle swarm optimization algorithms [28]), which are directly inspired from the natural evolution and biological systems, are considered as strong optimization tools applied in various applications. In this chapter, we apply one of these algorithms (GA) and another IA that eliminates successively one BS at each time.

12.5.1 Iterative Algorithm for Green Smart Grid Energy Procurement

The basic idea of the algorithm proposed in Ref. [29] is to eliminate redundant BSs without affecting the QoS and by solving at each iteration the formulated problem in Equation 12.24. In fact, at each iteration, the algorithm switches off one BS and verifies whether its absence degrades the QoS. If it is the case, the BS cannot be eliminated. Otherwise, it can be safely switched off.

Consider N_{BS} active BSs deployed in a given area and forming a set S. This means that initially all BSs are switched on and $\varepsilon = [1...1])$. We consider also N_U users placed following a given distribution model to benefit from the operator network services.

As a first step, after allocating the radio resources to all users as described in Section 12.4.1, the algorithm computes the data rates of all users $R_i^{(DL)}$ and $R_i^{(UL)}$ for $i = 1...N_U$ and compares them to the data rate thresholds $R_{m,th}^{(DL)}$ and $R_{m,th}^{(UL)}$ that depend on the type of the service m used by subscriber i. By this way, the algorithm identifies the number of users in outage N_{out} and consequently the entries of the vector γ. Once both vectors ε and γ are known and fixed, the optimization problem formulated in Equation 12.24 becomes a quadratic concave optimization problem that has a unique optimal solution that depends only on one decision variable: the vector q. Next, we initialize the optimal utility function U_0 as the initial maximum utility U_{max}:

$$U_{max} = U_0(\tilde{\mathbf{q}}) = (1 - \omega)\mathcal{P}_0(\tilde{\mathbf{q}}) - \omega\mathcal{I}_0(\tilde{\mathbf{q}}), \tag{12.30}$$

where $\mathcal{P}_0$ and $\mathcal{I}_0$ are the initial profit and CO_2 emission cost function, respectively, that depend on $\tilde{q}_j^{(n)}$, which are the elements of the optimal vector $\tilde{\mathbf{q}}$. Then, at each elimination step where one BS is switched off at a time, we repeat the same procedure as performed in the initialization step but with a reduced number of BSs. For the jth eliminated BS, we compute the corresponding optimal utility function U_j and we compare $\max_j(U_j)$ to the previous utility U_{max} to decide whether eliminating BSs is possible or not. Details of the proposed method are given in Algorithm 1.

Algorithm 1: Iterative Algorithm

- **Step 0:** Compute the utility function $U_{max} = U_0$ when all BSs are switched on (S contains all BSs and $\varepsilon = [1...1]$) and initialize for the current iteration $S^{iter} = S$ and $N_{BS}^{iter} = N_{BS}$.
- **Step 1:**

for $k = 1$ to N_{BS}^{iter} **do**

Eliminate BS k from S^{iter} $\left(\varepsilon^{(k)} = [1...1 \quad \overset{kth\ position}{0} \quad 1...1] \right)$.

Allocate resources (select serving BS and UL and DL RBs) to all users and compute $\gamma^{(k)}$ for the iteration k as shown in Equation 12.17.

if $\dfrac{N_{out}}{N_U} \leq P_{out}$ **then**

Find $\tilde{\mathbf{q}}$ by solving the quadratic optimization problem formulated in Equation 12.24 given $\varepsilon^{(k)}$ and $\gamma^{(k)}$.

Compute the utility function corresponding to the kth iteration: U_k for the optimal value $\tilde{\mathbf{q}}$.

else

BS k cannot be eliminated (we set $U_k = -\infty$).

end if
end for

- **Step 2:** Find the BS k_{op} that, when eliminated, provides the highest utility $\left(U_{k_{op}}^{new} = \max_k U_k \right)$.
- **Step 3:**

if $U_{k_{op}}^{new} \geq U_{max}$ **then**

BS k_{op} is eliminated and **Step 1** is repeated by setting $\mathcal{S}^{iter} = \mathcal{S}^{iter} \setminus \{k_{op}\}$, $N_{BS}^{iter} = N_{BS}^{iter} - 1$ and $U_{max} = U_{k_{op}}^{new}$.

else

No more changes can be made and the final optimal set of active BSs is $\mathcal{S}^{iter}$.

end if

Note that, in this approach, the order of the processed BSs does not influence the achievable result at convergence since the choice of the eliminated BS is obtained after comparing the utilities that are computed in an independent way. In fact, before eliminating one BS, the algorithm computes and compares the utilities of all combinations where, in each combination, one and only one BS is considered off. Then, it switches off the BS leading to the highest utility when it is off. This procedure is repeated to the new set of active BSs. The complexity of Steps 1 and 2 is linear in N_{BS}^{iter}. Taking Step 3 into account, the worst case complexity becomes $N_{BS} + (N_{BS} - 1) + \ldots + 1 = N_{BS}(N_{BS} + 1)/2 \approx \mathcal{O}\left(N_{BS}^2 \right)$. Thus, the algorithm has a quadratic complexity that depends directly on the number of initially deployed BSs. In a complete network with a high BS number, the delay would increase significantly. Hence, to avoid a high computational complexity in this case, the network can be divided into subnetworks with a reduced number of BSs (i.e., several BS groups) such that the algorithm can be applied in each subnetwork independently. Although suboptimal, this approach allows the implementation of the algorithm in parallel in each subnetwork at a tractable complexity in real time.

12.5.2 GENETIC ALGORITHM FOR GREEN SMART GRID ENERGY PROCUREMENT

The heuristic GA proposed in Ref. [30] uses binary strings to represent the solutions. In our case, the binary string corresponds to the vector ε, which refers to a combination of BSs. The idea is to find the optimal binary string ε that maximizes the utility function expressed in Equation 12.22.

Initially, we solve the optimization problem given in Equation 12.24 for ε [1…1] (i.e., all BSs are active) and we compute U_0 as an initial utility value. Then, the GA generates L binary strings of length N_{BS} forming a set called initial population $\mathcal{S}_0$. For each element of $\mathcal{S}_0$, $\varepsilon^{(l)}$, $l = 1…L$ of length N_{BS} and after applying the resource allocation algorithm described in Section 12.4.1 for the lth combination, the algorithm computes the data rates of all users and compares them to the data rate thresholds. By this way, it identifies the users in outage and consequently the value of the vector $\gamma^{(l)}$. Next, we compute the utility function U_l after procuring optimally the power from the available retailers in the smart grid by solving the optimization problem formulated in Equation 12.24. In fact, as $\varepsilon^{(l)}$ and $\gamma^{(l)}$ are known and fixed, the problem becomes a quadratic concave optimization problem that has a unique optimal solution and depends only on one decision variable: the vector $\mathbf{q}^{(l)}$. After computing L utilities U_l associated to each $\varepsilon^{(l)}$, we select the L_b ($L_b < L$) strings having the highest utility on which we apply crossovers and mutations to generate a new population S_1. Many GA models are used in literature. The variation depends on the selection and the reproduction procedure. In our case, L_b "survival" strings are kept to the next generation while $L - L_b$ new strings are produced by crossing over two strings (also called parents) selected randomly from the L_b parents having the highest utilities. The crossing point is chosen randomly between two fixed positions from 1 to N_{BS}. By swapping the obtained fragments, two new strings are produced. After recombination, we can apply the mutation operation with some low probability. The mutation consists of changing randomly a bit value of the generated strings of the new generation. Figure 12.2 describes in detail the generation of a new population in GA after solving our formulated problem. After the process of selection, reproduction, and mutation is complete, the next population can be evaluated. This procedure is repeated until convergence is reached when U_{max} remains constant for a several successive iteration number. At the end, the optimal BS combination is ε_{max}. Details of the proposed method using the GA are given in Algorithm 2.

Algorithm 2: Genetic Algorithm

- **Step 0:** Compute the utility function U_0 for $\varepsilon = \varepsilon^{(0)} = [1...1]$ and set $U_m = U_0$ and $\varepsilon_{max} = \varepsilon^{(0)}$
- **Step 1:** Generate an initial population S composed of L random $\varepsilon^{(l)}$, $l = 1...L$.
- **Step 2:**

  while (**Not** Converged) **and** (Maximum Number of Iterations not reached) **do**
  for $l = 1$ to L **do**

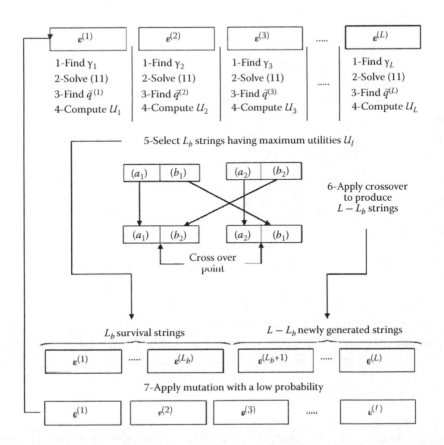

FIGURE 12.2 Flowchart describing one iteration of the GA after solving the optimization problem.

Allocate resources (select serving BS and UL and DL RBs) to all users and compute $\gamma^{(l)}$ and $N_{\text{out}}^{(l)}$ corresponding to the string $\varepsilon^{(l)} \in \mathcal{S}$.

if $\dfrac{N_{\text{out}}^{(l)}}{N_U} \leq P_{\text{out}}$ **then**

Find $\tilde{\mathbf{q}}^{(l)}$ by solving the quadratic optimization problem formulated in Equation 12.24 given $\varepsilon^{(l)}$ and $\gamma^{(l)}$ and compute the corresponding utility U_l.

else

BS k cannot be eliminated (we set $U_l = -\infty$).

end if
end for

Set $U_{\max} = \max_l U_l$ and $\varepsilon_{\max} = \varepsilon_{l_m}$ where l_m indicates the index of the string $\in \mathcal{S}$ that results in the highest utility.

Maintain the best L_b strings $\in \mathcal{S}$ to the next population and from them, generate $L - L_b$ new strings by applying crossovers and mutations to form a new population $\mathcal{S}$.

end while.

Hence, the above approaches aim to

- Minimize the power consumption by switching off redundant BSs.
- Reduce the CO_2 emissions by choosing optimally the amount of energy to be procured from each retailer.
- Maximize the profit of the network operator.
- Maintain a certain QoS.

12.6   RESULTS AND DISCUSSION

In this section, after presenting the simulation model, we analyze the performance of the green IA for the two particular scenarios. Then, we compare their performances for a fixed amount of renewable energy in the network in order to make a fair comparison between these cases. In addition, we investigate the results given by GA and we compare them to the ones provided by IA. Finally, we focus on the impact of the variation of the renewable energy unitary price on the performance of the presented scheme.

12.6.1   SIMULATION MODEL

We consider a 5×5 (km^2) LTE coverage area with uniform user distribution where N_{BS} BSs are placed uniformly according to the cell radius, selected to be 0.5 km. The LTE parameters are obtained from Refs. [24] and [31], and the channel parameters are obtained from Ref. [32]. All BSs and all MSs have the same power model and the same maximal transmit power, respectively. These parameters are detailed in Table 12.1.

In addition, we suppose that the network operator offers $M = 4$ different services. Each one is characterized by its cost (unitary price) $p^{(m)}$, DL and UL data rate thresholds ($R_{m,th}^{(\text{UL})}$ and $R_{m,th}^{(\text{DL})}$, respectively), and the occurrence probability of the service as shown in Table 12.2. The occurrence probability of a given service corresponds to the percentage of users in the network using that service.

TABLE 12.1

Channel and Power Parameters

Parameter	Value
κ (dB)	-128.1
σ_ξ (dB)	8
$(B^{(DL)}, B^{(UL)})$ (MHz)	(10, 10)
P_{BS} (W)	10
a	7.84
υ	3.76
P_{out}	0.02
$\left(N_{RB}^{(DL)}, N_{RB}^{(UL)}\right)$	(50, 50)
$P_{i,max}^{(UL)}$ (W)	0.125
b(W)	71.5

TABLE 12.2

Service Parameters

Services	Service 1	Service 2	Service 3	Service 4
$p^{(m)}$ (MU)	10	5	3	1
$\left(R_{m,th}^{(DL)}, R_{m,th}^{(UL)}\right)$ (kbps)	(1000, 384)	(384, 384)	(256, 56)	(64, 64)
Occurrence probability	0.1	0.1	0.3	0.5

Concerning the energy providers, we assume that $N = 2$ retailers with two different power sources exist in the smart grid and are available to supply the network with power. Each type of energy source n is characterized by its unitary price $\pi^{(n)}$, total available energy $Q_{max}^{(n)}$, and two pollutant coefficients α_n and β_n as shown in Table 12.3. We suppose that the second energy provider has a limited amount of energy $Q_{max}^{(2)}$: for instance, it can correspond to a renewable energy provider producing electricity from wind or solar energy. $Q_{max}^{(2)}$ and $\pi^{(2)}$ are kept as variables to investigate their impacts on the system performance.

If the network operator has its private renewable energy sources, then we assume that the produced power is free of charge and does not pollute the environment and the available power per BS j, $Q_{RE_{max}}^{(j)}$, cannot exceed the storage capacity, which we set to $Q_{max}^{BS} = 100$ W. If the network operator is able to sell the excess of renewable energy, we assume that the unitary price is equal $\pi^{(0)} = -0.05$ (MU). In fact, selling the excess of green energy is mathematically equivalent to "buying" with a negative price [1].

The GA is applied under the following settings: from a population of size $L = 16$, we run the algorithm at most 35 times. From each population, we select $L_b = 0.25\ L$ strings to the next population

TABLE 12.3

Energy Provider Parameters

Retailers	Retailer 1	Retailer 2
$\pi^{(n)}$ (MU)	0.05	$\pi^{(2)}$
$Q_{max}^{(n)}$ (W)	$+\infty$	$Q_{max}^{(2)}$
(α_n, β_n)	(0.02, 0.2)	(0, 0)

while the remaining $0.75L$ strings are obtained by randomly crossing over the L_b strings. The cross-over point is also chosen randomly between the $0.2\,N_{BS}$ and $0.8\,N_{BS}$ positions. The mutation appears rarely with probability of 5%. In the simulations, we focus on two particular scenarios that are derived from the formulated optimization problem expressed in Equation 12.24. Then, we compare their performance for snapshots of users during $\Delta t = 1$ second.

- Scenario I: the network operator can procure renewable energy only from the private local sources deployed on its BS sites. In the smart grid, we assume that only Retailer 1 is available to provide additional power to the mobile network. For this reason, the constraint in Equation 12.25 is not considered. The Pareto weight is fixed to $\omega = 0.5$ because the mobile operator will trivially procure the cheapest power that corresponds to the renewable energy generated by its equipment. In this scenario, we assume also that the mobile operator is able to sell the surplus of renewable energy.
- Scenario II: the network operator can procure energy only from the smart grid retailers where a single renewable energy retailer (Retailer 2) exists. In this case, $q_j^{(0)} = 0\ \forall j = 1,\ldots,N_{BS}$ as there is no private renewable energy generators on BS sites and thus the constraint in Equation 12.27 is not considered. In this scenario, the mobile operator has to allocate optimally the procured power depending on its attitude toward the environment defined by the Pareto weight ω.

12.6.2 Simulation Results of Scenario I

For Scenario I, we focus on the performance of the IA algorithm versus the number of users when all BSs are equipped with renewable energy generators (i.e., $\eta = 100\%$). In Figure 12.3, in order to evaluate the impact of the smart grid and the introduction of the notion of negative price, we plot,

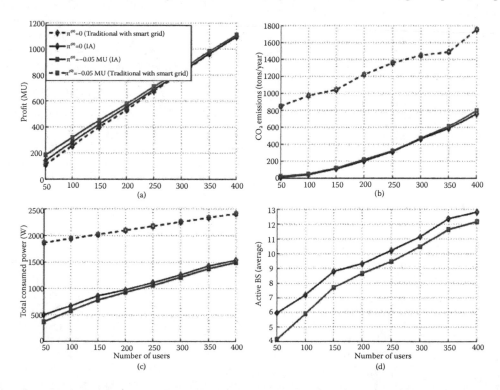

FIGURE 12.3 Performance of the IA scheme with Scenario I versus number of users. (a) Profit. (b) CO_2 emissions. (c) Total consumed power. (d) Number of active BSs.

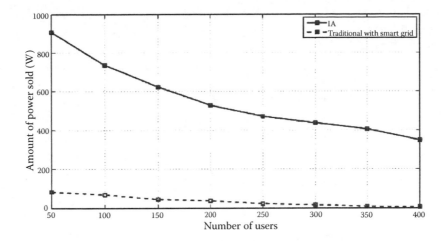

FIGURE 12.4 Amount of renewable power sold by the mobile operator versus the number of users.

in Figure 12.3a through d, the profit, the amount of CO_2 emissions, the total consumed power, and the number of active BSs after applying the BS sleeping strategy, respectively. First, comparing with the traditional scenario when all BSs are kept active (i.e., without sleeping strategy), Figure 12.3b shows that thanks to the BS sleeping strategy and the optimal power allocation, the network operator is able to reduce by more than 80% the emissions of greenhouse gas. On the other hand, we deduce that, with the negative price and the proposed algorithm, the mobile operator is able to get a higher profit by maintaining the same amount of CO_2 emissions. We notice also that, when $\pi^{(0)} = -0.05$ (MU), the network consumes less power as the number of active BSs is lower than the case when the mobile operator does not sell the excess amount of renewable energy. In fact, with negative price, the mobile operator forces to shut down the maximum number of BSs in order to sell the maximum amount of renewable energy. In terms of profit, selling renewable energy is more beneficial than serving subscribers. Of course, the outage probability threshold expressed in the constraint in Equation 12.28 is always satisfied. For instance, for $N_U = 400$ subscribers, the average outage probability, when $\pi^{(0)} = 0$, is equal to 0.71%, while when $\pi^{(0)} = -0.05$ (MU), it is equal to 1.77%, which is very close to P_{out}. Figure 12.4 confirms these results and plots the renewable energy amount sold by the mobile operator versus the number of users in the network. Compared to the traditional scenario, we clearly see that the proposed algorithm with the BS sleeping strategy contributes significantly in the energy efficiency of the network. Indeed, the mobile operator is able to sell more renewable energy as compared to the traditional case. In this way, it helps produce a green energy not only to power its own network but also to supply other infrastructures, and thus it participates to contend with global warming.

12.6.3 SIMULATION RESULTS OF SCENARIO II

In Scenario II, in order to serve 100 users connected to the network, the mobile operator has to procure power from two public retailers available in the smart grid. Each retailer is characterized by its unitary price and pollutant coefficients depending on its energy source. We assume that the unitary cost of an amount of power provided by Retailer 2 is equal to 0.1 (i.e., $\pi^{(2)} = 0.1$). Figure 12.5 plots the profit gained by the company and the CO_2 emissions caused by the consumption of fossil fuels (Retailer 1) for two different maximum amounts of renewable energy $Q_{max}^{(2)} = 100$ W and $Q_{max}^{(2)} = 300$ W (Retailer 2). In addition, we compare the results of the proposed algorithm to those of the traditional approach (without BS sleeping strategy). We show that when the traffic is low, we are able to reduce the CO_2 emissions by more than 90% by switching off redundant BSs and

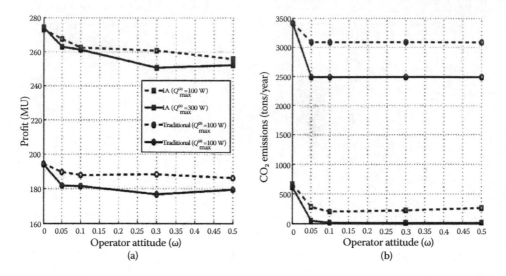

FIGURE 12.5 Performance of the proposed scheme with Scenario II ($N_U = 100$) versus operator attitude. (a) Profit. (b) CO_2 emissions.

consuming the maximum provided renewable energy when $\omega \to 1$ (i.e., when the network operator is environmentally friendly). We notice that the profit when $Q_{max}^{(2)} = 100\,W$ is higher than the one when $Q_{max}^{(2)} = 300\,W$. This is due to the fact that the renewable energy price is the most expensive and the operator is obliged to procure its power from it as $\omega \to 1$. However, when $\omega \to 0$, the mobile company reaches its maximum profit by avoiding the consumption of the most expensive energy (produced by Retailer 2) and this leads to a high CO_2 emission.

Figure 12.6 confirms the precedent results and describes in detail the percentage of the consumed power from the existent retailers. When $Q_{max}^{(2)} = 300\,W$ and $\omega \to 1$, 90% of the consumed power corresponds to the renewable energy, but when $\omega \to 0$, the selfish network operator favors to consume the most pollutant energy as it presents the cheapest cost. Moreover, we note that with the proposed method, renewable energy alone is enough to serve the network mainly for a low number of users.

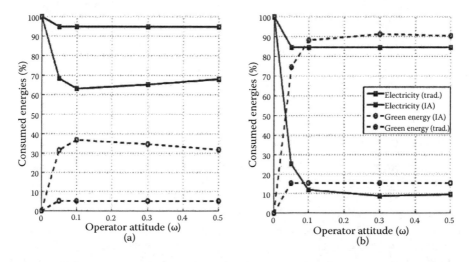

FIGURE 12.6 Consumed energies of Scenario II versus operator attitude ($N_U = 100$). (a) $Q_{max}^2 = 100\,W$. (b) $Q_{max}^2 = 300\,W$.

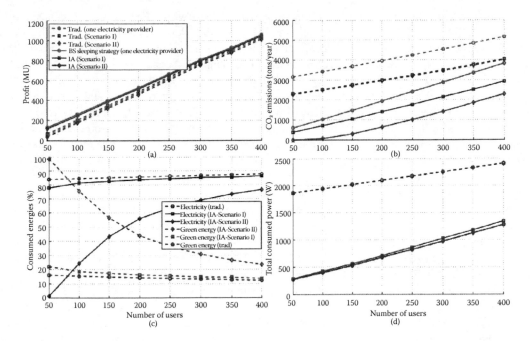

FIGURE 12.7 Comparison of the performance of Scenario I, Scenario II, and a traditional case with a unique electricity provider versus the number of users. (a) Profit. (b) CO_2 emissions. (c) Percentage of procured energies. (d) Total consumed power.

This means that the total energy consumption of the network after shutting down redundant BSs is around 300 W, while in the traditional case, renewable energy (300 W) represents only around 16% of the total required power as shown in Figure 12.6b. However, when $Q_{max}^{(2)} = 100$ W (Figure 12.6a), the consumption of renewable energy does not exceed 40% and 8% of the total consumption for the BS sleeping strategy and for the traditional approach, respectively, as there is no more available power to procure from Retailer 2. Therefore, the network is obliged to consume the pollutant energy to reduce the emissions of greenhouse gas and compensates the lack of renewable energy. When the number of subscribers increases in the network, we notice that although it is slowly getting smaller (going from ≈90% to ≈70% when N_U is increasing from 50 to 300), the gap between the BS sleeping strategy and traditional approach, for this scenario (Scenario II), remains important for the total power consumption and CO_2. However, by having more subscribers in the network, the gap in terms of profit between both approaches is becoming too small. This is explained by the fact that the network operator revenue $\mathcal{R}$ expressed in Equation 12.18 is the same, but the cost of power consumption is increasing because more active BSs are needed to serve the higher subscriber number as shown in Figure 12.7.

12.6.4 COMPARISON BETWEEN SCENARIO I AND SCENARIO II

In this section, we compare the performance of both scenarios in addition to two other cases that do not take into account the existence of multiple retailers. These two cases, denoted in Figure 12.7 by "Trad. (One electricity provider only)" and "BS sleeping strategy (One electricity provider only)," refer to the scenarios when all BSs are kept active and when BS sleeping strategy is applied as described in Ref. [6], respectively. In addition, we assume that the network is powered by only one electricity retailer in order to investigate the impact of the existence of the smart grid and the introduction of renewable energy. In order to make a fair comparison, we assume that the total amount of renewable energy available to the network is equal to 300 W for both scenarios. For the first scenario, this amount is distributed randomly over all BSs while, for the second scenario, this amount is offered by the public renewable energy retailer (Retailer 2). Figure 12.7a through c plots the profit, the amount of CO_2 emissions, and the total consumed power, respectively, versus the number of users (N_U) present in the area of interest. It is clear that Scenario II outperforms Scenario I in terms of energy efficiency and thus in terms of CO_2 emissions. In fact, with Scenario I, the operator is obliged to activate additional BSs to consume more green energy and thus consumes more power. Besides, as the renewable energy per BS is not enough to satisfy all the required power of this BS, the mobile operator is obliged in this case to buy electricity. Therefore, when the total amount of renewable energy is fixed, procuring it from the smart grid is better than producing it locally over all BSs. That is why, in this case, it is advisable to not distribute the whole green power over all BSs but only over a certain number of BSs less than N_{BS}. This can be clearly explained by Figure 12.8. In fact, this figure plots the performance of Scenario I with negative price versus the percentage of BSs equipped with renewable energy generators η for 100 users connected to the network. Three hundred watts is the total amount of renewable energy distributed over η% of BSs. We notice that, in order to maximize the profit (Figure 12.8a) and reduce the amount of CO_2 emissions by 50% compared to the case when η = 100% (Figure 12.8b), the mobile operator has to power only arround 20% of the BSs with renewable energy (i.e., Ren. en.). In fact, at this point and from Figure 12.8c, we deduce that arround 45% of the total consumed power is a green power produced by the generators, which corresponds to the maximum renewable power consumption compared to the other values of η%. In addition, 20% of the active BSs (around 5 or 6 BSs) are enough to serve 100 subscribers, as it is shown in Figure 12.3d. Therefore, we deduce that by applying the BS sleeping strategy algorithm, the network operator keeps only the BSs with green generators and having the highest amount of renewable energy. For this reason, most of the renewable energy is consumed and

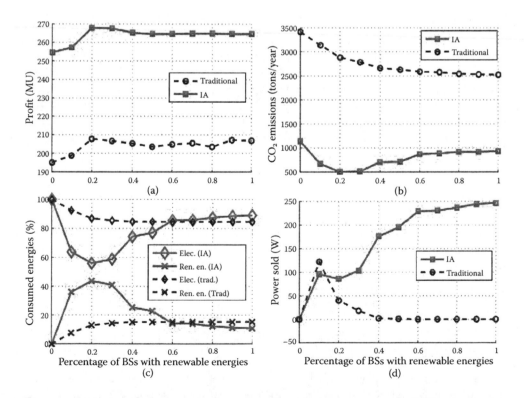

FIGURE 12.8 Performance of Scenario I versus the percentage of BSs with renewable energies η% (N_U = 100). (a) Profit. (b) CO_2 emissions. (c) Total consumed power. (d) Number of active BSs.

this case leads to the lowest surplus of power to be sold as described in Figure 12.8d. For the traditional case, the amount of power sold reaches a peak for $\eta = 10\%$ because the 300 W of green power is concentrated in a low number of BSs, which leads to an important surplus of renewable energy.

12.6.5 COMPARISON BETWEEN IA AND GA

In this section, we study and compare the performance of IA and GA when they are applied to Scenario II versus the operator attitude (ω) in Figure 12.9 and the number of subscribers connected to the network in Figure 12.10. In order to make a fair comparison between the two algorithms (GA and IA), we have run both of them for the same channel realizations and the same user locations and service distributions. In Figure 12.9, we plot the performance of the algorithm versus the Pareto weight. We notice that both algorithms have the same behavior but do not reach the same suboptimal solution. In fact, in all cases, we notice that GA outperforms IA mostly in terms of profit by consuming less power as shown in Figure 12.9d. In addition, we notice that the whole power consumed by GA is a green power mainly for high values of ω while the IA is not able to cover the required power using renewable energy as its total power consumption exceeds $Q_{max}^2 = 300$ W.

From Figure 12.10, we understand that the GA outperforms IA regardless of the number of users. In addition, Figure 12.10d shows that the gap between the total power consumed increases when the number of users increases. In fact, by applying the GA, the mobile operator can save around 35% and more than 60% of the total consumed power compared to IA and the traditional case, respectively, when $N_U = 400$. However, when $N_U = 50$, approximately, there is no difference between the performance of IA and GA.

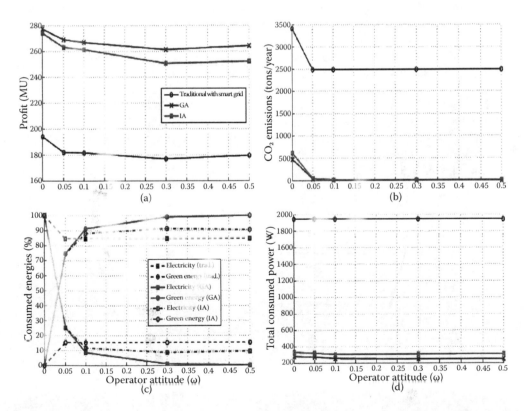

FIGURE 12.9 Performance of the algorithms versus the operator attitude ω ($N_U = 100$ and $Q_{max}^2 = 300$ W). (a) Profit. (b) Amount of CO_2 emissions. (c) Percentage of consumed energies. (d) Total consumed power.

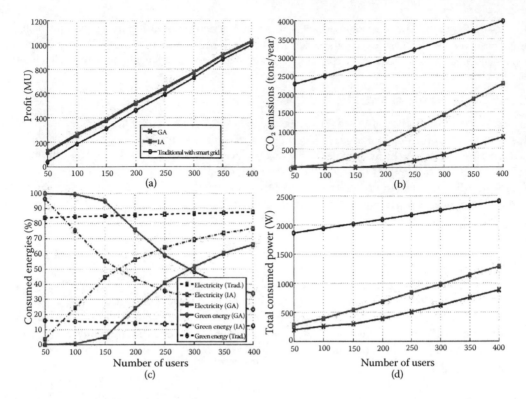

FIGURE 12.10 Performance of the algorithms versus the number of users N_U ($\omega = 0.5$ and $Q^2_{max} = 300$ W). (a) Profit. (b) Amount of CO_2 emissions. (c) Percentage of consumed energies. (d) Total consumed power.

Figure 12.11 shows that GA is able to keep less activated BSs and serve more users at the same time compared to IA and this leads to a less power consumption and a higher profit as it is demonstrated in Figures 12.9 and 12.10. In fact, it is able to find a better combination than IA thanks to the random generation of populations by applying reproduction and mutation.

The randomness of GA provides more chance for the network operator to find a better BS combination (i.e., the best ε vector that gives the highest utility). However, this advantage encounters a high computational complexity that almost depends on the length L. Reducing L not only can significantly increase the fastness of the algorithm but also can lead to worst results as confirmed

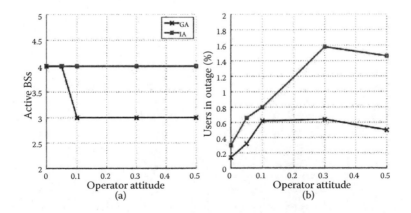

FIGURE 12.11 Strategy of the mobile operator ($Q^{(2)}_{max} = 300$ W and $N_U = 100$). (a) Number of active BSs. (b) Percentage of subscribers in outage.

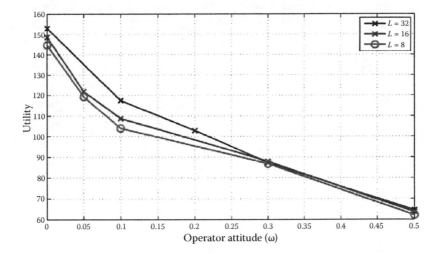

FIGURE 12.12 Comparison of the utilities obtained by GA for a different size of the initial population.

in Figure 12.12 in which we plot the utility function expressed in Equation 12.22 for a different size of population L. The IA is a fast algorithm and reaches always the same suboptimal solution; however, GA is slower and can reach different local maxima of the optimization problem because of its random evolution process.

12.6.6 IMPACT OF THE RENEWABLE ENERGY COST ON THE MOBILE COMPANY BEHAVIOR

In order to investigate the impact of the price of renewable energy on the system performance, we plot in Figure 12.13 the profit and the amount of CO_2 emissions versus $\pi^{(2)}$ that we vary from $0.05 = (\pi^{(1)})$ to higher values. We notice that, when the mobile operator aims only to maximize its profit (i.e., $\omega \to 0$), the profit and the amount of CO_2 emissions are almost constant for all prices strictly higher than $\pi^{(1)}$ and are not affected by the cost variation. This can be explained by the fact that for these values, the mobile operator procures power only from the cheapest retailer (i.e., Retailer 1) as

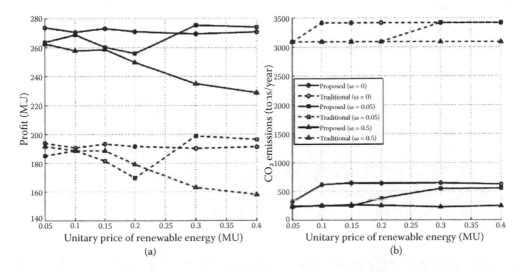

FIGURE 12.13 Performance of the proposed scheme versus the unitary price of renewable energy $\pi^{(2)}$ for $N_U = 100$ and $Q_{max}^{(2)} = 200$ W. (a) Profit. (b) CO_2 emissions.

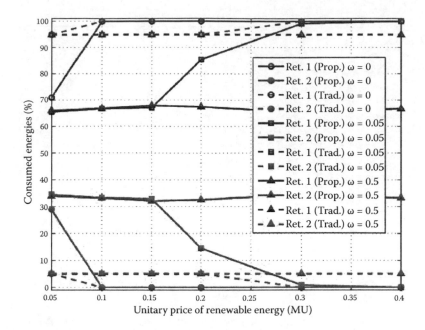

FIGURE 12.14 Percentage of the consumed energies procured from the existing retailers in the smart grid versus the renewable energy unitary price $\pi^{(2)}$ for $N_U = 100$ and $Q_{max}^{(2)} = 200$ W.

demonstrated in Figure 12.14, which plots the percentage of energy procured from existing retailers (Retailers 1 and 2) in the smart grid versus the renewable energy cost. When both retailers offer the same price, the mobile operator procures green power to reduce the greenhouse gas emissions. We understand that, for this case, saving the environment presents a secondary objective for this selfish mobile company. However, when we increase ω, the mobile company tries to procure the maximum renewable energy, which constitutes approximately 35% of the total power if we apply the proposed algorithm. The semi-environmentaly friendly operator ($\omega = 0.05$) can consume renewable energy even if its price becomes 5 times more expensive. This leads to a lower profit but a notable reduction in terms of CO_2 emissions. However, between, 0.15 and 0.3 MU, the amount of green energy procured is becoming less and less until it reaches 0. This is an example of a mobile operator that is doing its best to consume green power but does not accept a big loss in terms of profit. That is why when the renewable power becomes very expensive, it becomes obliged to procure electricity from Retailer 1. The third case corresponds to an environmentally friendly mobile company that procures the whole available renewable power even its profit goes to 0.

Note that there are some fluctuations in the profit curve as it depends also on the number of served users that varies from a channel realization to another and that for all cases, we are maintaining a certain QoS by imposing the constraint expressed in Equation 12.27. In our scenarios, only 2% of the subscribers are allowed to be in outage as mentioned in Table 12.1.

12.7 FUTURE RESEARCH DIRECTIONS

In this section, we describe some interesting future research directions that complement the work presented in this chapter. Future works can include the following:

- In the framework of this chapter, the proposed algorithm is applied for independent snap-shots of users where time variation is not taken into account. As a future work, we can analyze the chronological evolution of the power procurement from the smart grid with the presence of private renewable energy equipment in BS sites where the evolution will significantly depend on the availability of green power during the day and the user traffic behavior. In fact, at each period, the user traffic varies following certain statistics (i.e., some users disconnect, new users connect to the network, and others maintain their communications). The mobile operator has to accommodate the traffic evolution without affecting the required QoS and find an optimal solution to apply the BS sleeping strategy periodically in order to ensure green energy objectives.

- Our framework can be developed by considering heterogenerous networks (i.e., macro-, micro-, and femtocells) that cooperate together in a self-organizing way to manage interference and radio resource management. Relays can also be included in the state of the art since they can significantly help in increasing cell ranges and reducing the power consumption of cellular networks.

12.8 CONCLUSIONS

In this chapter, we investigated heuristic algorithms (IA and GA) that achieve power savings for LTE networks powered by a smart grid where different power retailers exist. The algorithms solve an optimization problem that reduces the network power consumption and hence the CO_2 emissions of the RAN of LTE networks while maximizing the mobile operator profit at the same time. We studied the behavior of the mobile operator toward the environment by eliminating redundant BSs and procuring energy optimally from retailers without affecting the QoS of the network. In addition, we compared this case with another scenario where renewable energy is produced locally by mobile operator equipment and we investigated the impact of negative price. Furthermore, we analyzed the impact of the variation of the renewable energy price of the smart grid on the behavior of the mobile operator.

ACKNOWLEDGMENT

This work was made possible, in part, by NPRP grant # 4-353-2-130 from the Qatar National Research Fund (a member of The Qatar Foundation). The statements made herein are solely the responsibility of the authors.

AUTHOR BIOGRAPHIES

Hakim Ghazzai is currently working toward his PhD degree in electrical engineering at King Abdullah University of Science and Technology in Saudi Arabia. He received his Diplome d'Ingenieur in telecommunication engineering with highest distinction from the Ecole Superieure des Communications de Tunis (SUP'COM), Tunisia, in 2010 and his master's degree in high-rate transmission systems from the same institute in 2011. His general research interests are at the intersection of wireless networks, green communications, and optimization.

Elias Yaacoub received his BE degree in electrical engineering from the Lebanese University in 2002, his ME degree in computer and communications engineering from the American University of Beirut (AUB) in 2005, and his PhD degree in electrical and computer engineering from AUB in 2010. He worked as a research assistant in the AUB from 2004 to 2005 and in the Munich University of Technology in Spring 2005. From 2005 to 2007, he worked as a telecommunications engineer with Dar Al-Handasah, Shair and Partners. Since November 2010, he has been a research scientist at the QUWIC (QU Wireless Innovations Center), rebranded to QMIC (Qatar Mobility Innovations Center) in July 2012. His research interests include wireless communications, antenna theory, sensor networks, energy efficiency in wireless networks, video streaming over wireless networks, and bioinformatics.

Mohamed-Slim Alouini was born in Tunis, Tunisia. He received his Diplome d'Ingenieur from the Ecole Nationale Supérieure des Telecommunications (TELECOM Paris Tech) and his Diplome d'Etudes Approfondies (DEA) in electronics from the Université Pierre et Marie Curie in Paris, both in 1993. He received his MSEE degree from the Georgia Institute of Technology in the United States in 1995 and his doctorate in electrical engineering from the California Institute of Technology in 1998. He also received the Habilitation degree from the Université Pierre et Marie Curie in 2003. Dr. Alouini started his academic career at the Department of Electrical and Computer Engineering of the University of Minnesota in the United States in 1998. In 2005, he joined the Electrical and Computer Engineering program at Texas A&M University at Qatar. He has been a professor of electrical engineering at King Abdullah University of Science and Technology since 2009. His research interests include statistical characterization and modeling of fading channels, performance analysis of diversity combining techniques, MIMO (multiple input–multiple output) and multi-hop/cooperative communications systems, capacity and outage analysis of multiuser wireless systems subject to interference or jamming, cognitive radio systems, and design and performance evaluation of multi-resolution, hierarchical, and adaptive modulation schemes. Dr. Alouini has published several papers on the above subjects, and he is co-author of the textbook *Digital Communication over Fading Channels* published by Wiley Interscience. He is a fellow of the Institute of Electrical and Electronics Engineers (IEEE), a member of the Thomson ISI Web of Knowledge list of Highly Cited Researchers, and a co-recipient of best paper awards in eight IEEE conferences (including ICC, GLOBECOM, VTC, and PIMRC).

Adnan Abu-Dayya received his PhD in electrical engineering from Queens University, Canada, in 1992. He then worked as a manager at Nortel Networks in Canada in the advanced technology group and as a senior consultant at the Communications Research Center in Ottawa, Canada. Before moving to Qatar in March 2007, he worked for 10 years at AT&T Wireless in Seattle, USA, where he served in a number of senior management positions covering product innovations and emerging technologies, systems engineering, and product realization. He was also responsible for developing and licensing the extensive patent portfolio of AT&T Wireless. From April 2007 to December 2008, he was the chairman of the Electrical Engineering Department at Qatar University (QU). He was appointed as the executive director of the QU Wireless Innovations Center (QUWIC) in December 2008 (QUWIC was rebranded to QMIC [Qatar Mobility Innovations Center] in July 2012). Dr. Abu-Dayya has more than 20 years of international experience in the areas of wireless/telecomm R&D, innovations, business development, and services delivery. He has many issued patents and more than 50 publications in the field of wireless communications.

REFERENCES

1. P. Asmus, A. Bae, B. Lockhart, N. Strother, and B. Gohn, Smart grid: Ten trends to watch in 2012 and beyond, Technical Report, Pike Research LLC, Boulder, Colorado, USA, 2012.
2. Smart 2020: Enabling the low carbon economy in the information age, in *The Climate Group, Globe e-Sustainability Initiative (GeSI)*, Brussels, Belgium, 2008.
3. G.P. Fettweis and E. Zimmermann, ICT energy consumption—Trends and challenges, in *Proc. IEEE the 11th International Symposium on Wireless Personal Multimedia Communications (WPMC 2008)*, Oulu, Finland, Sept. 2008.
4. J. Louhi, Energy efficiency of modern cellular base stations, in *Proc. IEEE 29th International Telecommunications Energy Conference (INTELEC 2007)*, pp. 475–476, Oct. 2007.
5. E. Yaacoub, Performance study of the implementation of green communications in LTE networks, in *Proc. IEEE the 19th International Conference on Telecommunications (ICT 2012)*, Jounieh, Lebanon, Apr. 2012.
6. W. El-Beaino, A.M. El-Hajj, and Z. Dawy, A proactive approach for LTE radio network planning with green considerations, in *Proc. IEEE the 19th International Conference on Telecommunications (ICT 2012)*, Jounieh, Lebanon, Apr. 2012.
7. I. Humar, J. Zhang, Z. Wu, and L. Xiang, Energy savings modeling and performance analysis in multi-power-state base station systems, in *Proc. IEEE Green Computing and Communications (GreenCom 2010)/ACM and Int'l Conference on Cyber, Physical and Social Computing (CPSCom 2010)*, pp. 474–478, Hangzhou, China, Dec. 2010.

8. P. Samadi, A. Mohsenian-Rad, R. Schober, V. Wong, and J. Jatskevich, Optimal real-time pricing algorithm based on utility maximization for smart grid, in *Proc. IEEE the 1st International Conference on Smart Grid Communications (SmartGridComm 2010)*, pp. 415–420, Gaithersburg, Maryland, USA, Oct. 2010.

9. S. Bu, F.R. Yu, Y. Cai, and P. Liu, When the smart grid meets energy-efficient communications: Green wireless cellular networks powered by the smart grid, *IEEE Transactions on Wireless Communications*, vol. 11, no. 8, pp. 3014–3024, Aug. 2012.

10. X. Yang, Y. Wang, D. Zhang, and L. Cuthbert, Resource allocation in LTE OFDMA systems using genetic algorithm and semi-smart antennas, in *Proc. IEEE Wireless Communications and Networking Conference (WCNC 2010)*, pp. 1–6, Apr. 2010.

11. K. Lin, Improving energy efficiency of LTE networks by applying genetic algorithm (GA), in *Proc. IEEE the 9th International Conference on Dependable, Autonomic and Secure Computing (DASC 2011)*, pp. 593–597, Dec. 2011.

12. P. Ramaswamy and G. Deconinck, Relevance of voltage control, grid reconfiguration and adaptive protection in smart grids and genetic algorithm as an optimization tool in achieving their control objectives, in *Proc. IEEE International Conference on Networking, Sensing and Control (ICNSC 2011)*, pp. 26 –31, Apr. 2011.

13. D. Chee, M. Suk Kang, H. Lee, and B.C. Jung, A study on the green cellular network with femtocells, in *Proc. IEEE the 3rd International Conference on Ubiquitous and Future Networks (ICUFN 2011)*, pp. 235–240, Dalian, China, Jun. 2011.

14. A. Saleh, O. Bulakci, S. Redana, B. Raaf, and J. Hamalainen, Evaluating the energy efficiency of LTE-advanced relay and picocell deployments, in *Proc. IEEE Wireless Communications and Networking Conference (WCNC 2012)*, pp. 2335–2340, Paris, France, Apr. 2012.

15. A. Abdel Khalek, L. Al-Kanj, Z. Dawy, and G. Turkiyyah, Optimization models and algorithms for joint uplink/downlink UMTS radio network planning with SIR-based power control, *IEEE Transactions on Vehicular Technology*, vol. 60, pp. 1612–1625, May 2011.

16. L. Xiang, F. Pantisano, R. Verdone, X. Ge, and M. Chen, Adaptive traffic load-balancing for green cellular networks, in *Proc. IEEE the 22nd International Symposium on Personal Indoor and Mobile Radio Communications (PIMRC 2011)*, pp. 41–45, Sep. 2011.

17. Alcatel-Lucent, Evolve your wireless broadband network for the new generation of applications and users, (http://www.alcatel-lucent.com/features/light radio/index.html).

18. National Energy Technology Laboratory, Understanding the benefits of smart grid, Technical Report, 2010.

19. A. Goldsmith, *Wireless Communications*. Cambridge University Press, Cambridge, UK, 2005.

20. F. Richter, A. Fehske, and G. Fettweis, Energy efficiency aspects of base station deployment strategies for cellular networks, in *Proc. IEEE the 70th Vehicular Technology Conference (VTC 2009)*, Anchorage, Alaska, USA, Sept. 2009.

21. J. Lim, H.G. Myung, K. Oh, and D.J. Goodman, Channel-dependent scheduling of uplink single carrier FDMA systems, in *IEEE Vehicular Technology Conference (VTC 2006)*, Montreal, Canada, Sept. 2006.

22. H.G. Myung and D.J. Goodman, *Single Carrier FDMA: A New Air Interface for Long Term Evolution*. Wiley, Hoboken, New Jersey, USA, 2008.

23. T. Lunttila, J. Lindholm, K. Pajukoski, E. Tiirola, and A. Toskala, EUTRAN uplink performance, in *International Symposium on Wireless Pervasive Computing (ISWPC 2007)*, Feb. 2007.

24. 3rd Generation Partnership Project (3GPP), 3GPP TS 36.213 3GPP TSG RAN evolved universal terrestrial radio access (E-UTRA) physical channels and modulation, version 8.3.0, Release 8, Technical Report, 3GPP, 2008.

25. K. Senthil and K. Manikandan, Improved tabu search algorithm to economic emission dispatch with transmission line constraint, *International Journal of Computer Science and Communication*, vol. 1, pp. 145–149, Jul.–Dec. 2010.

26. S. Boyd and L. Vandenberghe, *Convex Optimization*. Cambridge University Press, Cambridge, UK, 2004.

27. N. Srinivas and K. Deb, Muiltiobjective optimization using nondominated sorting in genetic algorithms, *Evolutionary Computation*, vol. 2, no. 3, pp. 221–248, 1994.

28. J. Kennedy and R. Eberhart, Particle swarm optimization, in *Proc. IEEE International Conference on Neural Networks (ICNN 1995)*, Perth, Australia, Nov. 1995.

29. H. Ghazzai, E. Yaacoub, M.-S. Alouini, and A. Abu-Dayya, Optimized green operation of LTE networks in the presence of multiple electricity providers, in *Proc. IEEE International Workshop on Emerging Technologies for LTE-Advanced and Beyond-4G in Conjunction With IEEE Global Communications Conference (Globecom 2012)*, Anaheim, CA, USA, Dec. 2012.

30. H. Ghazzai, E. Yaacoub, M.-S. Alouini, and A. Abu-Dayya, A genetic algorithm solution for the operation of green LTE networks with energy and environment considerations, in *Proc. the 19th International Conference on Neural Information Processing (ICONIP 2012)*, Doha, Qatar, Nov. 2012.
31. 3rd Generation Partnership Project (3GPP), 3GPP TS 36.211 3GPP TSG RAN evolved universal terrestrial radio access (E-UTRA) physical channels and modulation, version 11.0.0, Release 11, Technical Report, 3GPP, 2012.
32. 3rd Generation Partnership Project (3GPP), TS 36.213 3GPP TSG RAN physical layer aspects for evolved UTRA, v.11.0.0, release 11, Technical Report, 3GPP, 2012.

13 Applications of Wireless Sensor and Actuator Networks in the Smart Grid
A Double Energy Conservation Challenge

Natalie Matta, Rana Rahim-Amoud,
Leïla Merghem-Boulahia, and Akil Jrad

CONTENTS

13.1 INTRODUCTION

The next-generation electricity network, that is, the smart grid, comes as a response to evolutions in the electricity market. The demand for power is increasing in both developing and developed countries. By 2030, a 50% and a 40% increase in the consumption are expected in the United States and Europe, respectively. It should triple in China and India, and double at the global level [1]. This increase should be accounted for, and quality of service (no interruptions, no variations in voltage, etc.) should be improved and maintained. Furthermore, the availability of fuels can no longer be taken for granted with the ever-increasing demand for energy. Even if these resources are available, and the so-called peak oil is avoided, the long-term impact of carbon emissions caused by the burning of fossil fuels will damage the world's climate [2]. That is why, currently, there is a global consensus to increase energy efficiency, stimulate usage of renewable energy sources, and reduce CO_2 emissions. Thus, we are witnessing the introduction of electric vehicles (EVs) to the consumers market; the proliferation of the use of renewable energy sources such as wind, solar, and tidal; and the establishment of demand side management programs. Nevertheless, the integration of EVs will increase the demand for electricity, the heterogeneous intermittent sustainable energy sources will require efficient management, and two-way communications are needed in order to put in place different energy efficiency programs.

The European Technology Platform defines smart grids as "electricity networks that can intelligently integrate the behaviour and actions of all users connected to it—generators, consumers and those that do both—in order to efficiently deliver sustainable, economic and secure electricity supplies" [3]. In order to achieve these goals, the physical power grid is complemented with a network of communications, making use of the advances in the information and communication technology (ICT) domain. Two-way communication networks for data collection and processing (between smart devices and advanced software) can thus be put in place and connect the different actors in a smart grid.

One major concern that the smart grid aims at dealing with is the reliability of the power network. In fact, power outages are still occurring nowadays and affecting millions of people (e.g., the 2003 US Northeast blackout, the 2009 Brazil and Paraguay blackouts, and the 2012 India blackouts). The objectives are to detect problems at an early stage and to take proper actions as soon as possible in order to reduce both the duration and the impact of a service interruption and avoid large-scale outages. Part of the solution relies in monitoring and controlling the transmission and distribution power networks in order to detect, locate, and restore faults; monitor the state of equipments; and monitor the quality of the power.

In order to put in place the needed monitoring applications, wireless sensor and actuator networks (WSANs) have been considered an attractive technology as they are easily deployed and maintained, have a low cost, and are pervasive. Sensors and actuators can be deployed on power lines, transformers, feeders, circuit breakers, and many other components of the power grid. They monitor the temperature, current, voltage, gas levels, and so on, and relay the collected data to the base station of the network, that is, the sink. The main preoccupation when working with WSANs relies in energy conservation. In fact, having a limited battery life, sensor nodes stop working when their batteries die. In order to prolong the network's lifetime, research has tackled this issue from two perspectives. The first one focused on reducing the sensors' energy consumption, while the second one targeted replenishing the sensors' batteries through energy harvesting.

In summary, a double energy efficiency challenge can be outlined. The first one concerns one of the pillars of the smart grid vision, that is, a more reliable, energy-efficient, and environmentally friendly electricity grid that makes use of EVs, sustainable energy resources, and so on, and limits toxic gas emissions. The second one deals with the energy conservation issue in WSANs. This energy saving also insures a better quality of service in the network by having less packets to process, less traffic congestion, better delivery delays, and so on.

This chapter introduces the smart grid vision and its key concepts in Section 13.2. Section 13.3 points out the role of WSANs' applications in the future electricity grid and expose their energy conservation issue. The needs of management and control of the distribution network are pointed out in Section 13.4, suggesting how the move toward a decentralized approach can be achieved. Finally, some future research directions are presented in Section 13.5.

13.2 THE VISION OF A SMARTER GRID

A power grid is divided into four subsystems, each associated with different voltage levels: generation, transmission, distribution, and consumption. In the smart grid, the physical network is coupled with an information and communication layer making use of advances in the ICT domain to implement advanced features. The smart grid is a conceptual revolution that will affect all aspects of power grid, making it more flexible. This concept comes as a response to changes in the electricity market and should be designed to handle the increased demand while ensuring a better quality of service and security. Its requirements are exposed in the next subsection.

13.2.1 SMART GRID REQUIREMENTS

According to the National Energy Technology Laboratory [4], the transition from the current electricity network to the smart grid should be mainly founded on advances that target the following points:

- *Reliability*: a more reliable grid provides power with the required quality when and where it is needed.
- *Security*: a grid that is resilient to physical and cyber attacks and that protects the privacy of its customers.
- *Efficiency*: a more efficient grid implements cost control, reduces transmission and distribution losses, and optimizes power production.
- *Environment-friendliness*: a more environmentally responsible grid reduces its negative impacts on the environment through the use of sustainable resources for example.
- *Safety*: a safer grid that protects the public, the workers, and the users who depend on it for medical necessities.
- *Economical*: an economic grid offers fair prices and adequate supplies.

To achieve these goals, the smart grid vision is built around key concepts that will be reviewed in the next subsection.

13.2.2 SMART GRID KEY CONCEPTS

On the generation side, distributed, heterogeneous, and intermittent renewable sources need to be managed. They may come to an agreement to cooperate and constitute a Virtual Power Plant (VPP) enabling them to sell their production as an aggregate. Furthermore, the production of electricity from distributed energy resources needs to be monitored and coordinated to insure the equilibrium of the production and the consumption. Instead of having the production follow the demand for electricity, the Demand Side Management (DSM) concept stipulates that consumers adjust their consumption in order to reduce the load of the grid. This can be achieved by deferring appliance use (residential energy management), transferring energy back to the grid, or using energy from storage devices, hence the need to manage distributed storage peripherals. As such, the batteries

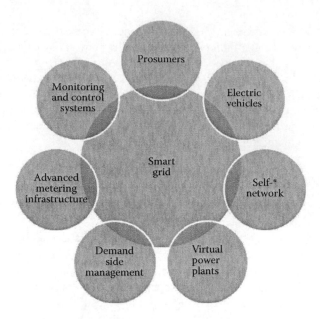

FIGURE 13.1 Key concepts in a smart grid.

of EVs constitute an attractive option for energy storage. Thus, we have witnessed the development of concepts such as vehicle-to-grid (V2G) and vehicle-to-home (V2H) power. This will change the electricity market, as we will have consumers, referred to as *prosumers*, that are also able to sell electricity back to the grid (either from stored energy in their vehicle's battery or from residential sustainable resources such as solar panels). The smart metering infrastructure (Advanced Metering Infrastructure [AMI]) will provide the vital link between consumers and the network. Finally, a more reliable grid needs to be supplied with enough intelligence and strategies to make it autonomous (i.e., *self-**). In order to reduce the frequency and duration of service interruptions, the capacity of self-healing following a fault, and self-adaptation to fluctuations that may affect the voltage quality, should be put in place. That is why monitoring and control of voltage quality and electric components is paramount (Figure 13.1). These concepts are detailed in the following paragraphs.

13.2.2.1 A New Type of Consumer: The Prosumers

The conceptual revolution that lies in a smart grid implicates an important change to the consumption side where consumers will also become producers, known as *prosumers*. Having either a local source of electricity generation (e.g., solar panel), or the capacity of energy storage (e.g., the battery of the user's EV), or both, a prosumer can sell electricity in the new emerging electricity markets [2].

13.2.2.2 Integration of EVs

Currently, environmental changes and the predicted peak oil are contributing to the proliferation of EVs that emit no toxic gas during use.

An EV must be connected to the power grid for its battery to charge. While significantly participating in lowering carbon emissions, EVs constitute an additional load to the grid. In comparison with the daily energy consumption of a typical household, which lies between 20 and 50 kWh, an EV's battery can be recharged with 32 kWh of energy in a few hours [2]. In addition, the high urban density as well as the variability of the demand in time (daily, weekly, seasonal) should be taken into account [5]. This not only creates challenges regarding the management of the battery charging but also makes room for new applications such as the use of the batteries for distributed power storage and the concepts of energy flow back to the grid (V2G and V2H).

13.2.2.3　Management of Distributed Generation and Storage

The mechanism of VPPs is experiencing a growing success that accompanies the liberalization of electricity markets. As defined by Ramchurn et al. [2], VPPs represent the notion of heterogeneous and distributed actors coming together to sell electricity as an aggregate. They allow thousands of renewable generators to act as a single plant. They can be seen as coalitions of distributed energy resources. According to Pike Research consulting, electricity production from this new sector will increase by 65% from 2011 to 2017 to reach 91.7 GW [6]. Nevertheless, a number of challenges arises as to the formation and management of these VPPs. The different actors and their production levels need to be constantly monitored, demand and production balanced, auctions and transactions managed, and so on.

Energy storage aims to support sustainable energy supply. Whether through the use of batteries of vehicles or micro-energy storage devices at homes, energy storage is one way to save energy and reduce the dependence on fossil fuels. However, if micro-storage devices are all charged at the same time using power from the grid, this will lead to a higher demand and, therefore, will require a greater production capacity, leading to additional carbon emissions and possibly to failures in the system [7]. Thus, proper management is required in the context of distributed storage and production.

13.2.2.4　An Autonomous Network

The smart grid should be a *self-** network, capable of self-organization, self-adaptation, and self-healing. Fault recovery and implementation of necessary measures, such as islanding and load shedding to avoid cascading consequences, are examples of the self-healing function of a smart grid.

In order to put in place these functions, monitoring applications must first be integrated into the network. This requires surveillance systems at multiple levels, such as voltage quality monitoring and control of transmission lines and substations, coupled with alarm systems.

13.2.2.5　Monitoring and Control of Electrical Systems

The improvement of the reliability of the electrical network is a key objective for the smart grid, aimed at reducing breakdowns and outages afflicting the power grid today (the latest and largest being the July 2012 India blackout that left hundreds of millions without power).

Surveillance of production sites is necessary in order to closely monitor and balance production levels while ensuring coordination between the elements of distributed generation. This would allow, for example, a generator to be stopped or started according to the levels of production of renewable energy sources.

Real-time monitoring and control of transmission and distribution networks are essential in a smart grid. The network manager must ensure uninterrupted service to its customers with acceptable levels of power quality, while balancing demand and production. For example, the importance of monitoring of transformers is clearly stated in Ref. [8], given that they are one of the most expensive components and play a crucial role in the electrical network as their failures lead to service interruptions.

The smart metering infrastructure ensures the monitoring at the consumers' side. It is introduced in the next paragraph.

13.2.2.6　Smart Metering Infrastructure

The smart metering infrastructure, or AMI, is one of the foundations of a smart grid. This infrastructure is used to collect data from smart meters (which record consumption of electric energy) and transmit it to their dedicated management system, the Meter Data Management System (MDMS). Data collected by the sensors on the status of production, transmission, and distribution are also stored and managed at the MDMS [9]. The AMI provides the crucial link between the power network, consumers, production resources, and energy storage [10]. It establishes bidirectional communication between smart meters, sensors and actuators, and the decision center [11,12].

In this context, aggregators or data concentrators play an important role. These are the entities responsible for the data collection from the AMI and its transport to the MDMS [9].

It is important to note that because the data collected by the meters are private in nature, the AMI compromises the integrity and confidentiality of this information. It is therefore necessary to implement appropriate security techniques to protect users while transmitting their data.

13.2.2.7 Demand Side Management

The DSM concept is a set of measures aimed at improving energy consumption. DSM programs include two main activities [13,14]:

1. Load shifting, also known as Demand Response (DR) programs, which aim at transfering customer loads during periods of peak demand to off-peak periods.
2. Energy efficiency improvement programs that allow customers to use less energy while still receiving the same level of service, through the use of better household appliances for example.

The AMI can be used to put in place DR programs. The smart metering infrastructure can provide the necessary communications to implement the needed functions.

In the next section, we will see how deploying WSANs will help put in place some of smart grid functions that were discussed in this section.

13.3 WSANs FOR THE SMART GRID

Deploying WSANs serves two objectives, namely, collecting information from the environment via the sensor nodes and acting upon this environment via the actuator nodes. In the smart grid, these networks are needed on many levels. This section will expose the main applications of WSANs in the smart grid, while focusing on the transmission and distribution (T&D) subsystem.

13.3.1 THE ROLE OF WSAN APPLICATIONS IN THE SMART GRID

At the different levels of an energy control system, WSAN applications are of great interest. Typically, smart grid WSAN applications can be divided by power grid segment, namely, generation side, T&D side, and consumer side [15–17]. They enable remote monitoring (temperature monitoring, outage detection, power quality disturbances, etc.), as well as DSM, which in turn creates the opportunity for new business processes such as real-time pricing [17].

The opportunities and importance of WSANs in the smart grid were put forward in many research studies such as Refs. [15,16,18–21]. In fact, traditional wired-based monitoring of electrical systems is expensive to install and requires constant maintenance of communication cables, whereas smart grid communications need to be scalable and pervasive. Sensor networks present the advantages of being low-cost, pervasive, flexible, and rapidly deployed. Indeed, the use of WSANs can ensure a wide coverage, even of remote sites, as well as a distributed and decentralized architecture that overcomes the issue of a single point of failure.

According to Erol-Kantarci and Mouftah [18], it is expected that the next-generation power grid migrates from the limited-sensing, centrally controlled operational architecture, to a distributed decision-making and acting system that is supported by wireless sensor communications. Erol-Kantarci and Mouftah [18] expose the advantages, applications, and challenges of deploying wireless multimedia sensor and actor networks in the smart grid. Multimedia sensors can record video, capture audio and still images, and collect ambient scalar data. In T&D facilities, for example, they can provide a larger field view, which improves surveillance. If placed on EVs, they can share with other vehicles' sensors traffic data and images. Actuators have the capacity of operating on devices. A faulty component, such as a wind turbine in a wind farm, can be shut down via an actuator. In a similar way, backup components can be brought online when needed. A major challenge for the sensor network, among others, will be the management of the collected data. Erol-Kantarci and Mouftah [18] suggest that information should be processed at lower levels (when it is sensed for example) to avoid being unnecessarily transmitted to the utility headquarters. That is why local processing and data aggregation (to remove redundancies, compute averages, etc.) will have to be implemented to move toward a distributed decision-making system. The details of such an architecture still need to be defined.

Gungor et al. [19] point out that WSANs have been recognized as a promising technology for the future smart grid, having the ability of enhancing its three subsystems, that is, generation, delivery, and utilization. Some applications include wireless automatic meter reading (WAMR), remote system monitoring, and equipment fault diagnostics. The authors also outline some technical challenges of using WSANs in smart grid applications, such as harsh environmental conditions, reliability and latency requirements, and resource constraints [19].

Next, WSAN applications in T&D networks are presented in more detail.

13.3.2 Monitoring and Control of **T&D** Power Networks

Improving the reliability of the power network is a key objective for the smart grid. Today's electricity system still suffers from power outages and blackouts because of the lack of automated analysis and poor visibility of the utility over the grid [22]. Deploying WSANs will give the utility provider the needed view and control by collecting information from sensors and placing actuators at the different subsystems of the grid.

Real-time monitoring and control of T&D subsystems is much needed in the smart grid. The utility should ensure uninterrupted service to its customers with acceptable power quality levels, while balancing demand and production. In order to fulfill these requirements, the power utility should be able to monitor, control, and manage its T&D subsystems in real time. To this end, WSANs are deployed in T&D networks [20]. The information acquired from these networks can be analyzed to detect and diagnose faults early and to take appropriate actions as soon as possible. Reducing time for detecting disturbances and outages and reducing service interruptions are critical for a power utility since it should maximize its customers' satisfaction. It should also be able to compute quality indices (such as power quality indices, average duration of service interruption, etc.) in order to evaluate the performances of its network [23]. All this can be achieved by analyzing data that is collected via WSANs. However, a WSAN data management architecture for T&D networks is yet missing. It should enable local processing for quick decision making, as well as relay appropriate information for long-term analysis [20].

An example of the use of a WSAN for phase current characteristics measurement on power lines of secondary substations is presented in Ref. [24]. The authors expose how faults can be detected and how short circuits and single phase to earth faults can be located.

In Ref. [25], a decentralized voltage quality monitoring architecture that employs sensor networks is introduced. The advantages of such an architecture over client/server ones are its task distribution, which takes the load off the remote server, and its scalability. The authors consider a scenario where each part of the power grid is monitored by a sensor network and propose a cooperation mechanism between these networks in order to propagate their respective quality indices. However, how the nodes in a single network communicate to evaluate the local index is not treated in Ref. [25]. In addition, the sensors' power consumption issue was not addressed by the authors.

The importance of transformer monitoring is clearly stated in Ref. [8], as transformers are one of the most expensive utility assets and have a crucial role in the power network (their failure could lead to blackouts). Sensors can be used to monitor their temperature and their dissolved gas levels, which can be reported at the substation level. At the transmission level, line conditions can also be monitored. Examples of sensors used include smart voltage sensors as well as voltage, current, continuity, and temperature sensors.

A large-scale wireless sensor network for substation monitoring is considered in Ref. [26]. The paper describes a practical implementation of a network scaled to 122 sensor nodes in a substation in Kentucky, United States. The authors address the issue of the energy consumption of the nodes by applying a level-crossing sampling scheme to reduce the frequency of packet transmissions. This signal sampling mechanism makes the sensor transmit data only when the sensor values change by a predetermined amount. However, they point out that this method is not enough to reduce the energy consumption of the nodes and should be coupled with other power control mechanisms (such as using multiple orthogonal channels on the physical layer, or reducing route updates messages on the routing level).

The use of WSANs in the smart grid is not limited to this. The rest of this section will review some other applications of WSANs in the smart grid.

13.3.3 OVERVIEW OF OTHER APPLICATIONS

In this subsection, we will present an overview of other WSAN applications in the smart grid, such as residential energy management and WAMR on consumer premises, monitoring at the generation facilities, and the use of WSANs in EVs.

13.3.3.1 Energy Management on Consumer Premises

In their proposition for residential energy management in smart grids [27,28], Erol-Kantarci and Mouftah make use of the existing wireless sensor home area network (WSHAN), which would initially be set up for health monitoring and comfort applications, to put in place an appliance coordination scheme in order to reduce peak loads in smart grids. In the proposed ACORD-FI scheme, home appliances communicate with an energy management unit (EMU) by sending their requests via the WSHAN. The EMU is in charge of computing a convenient start time for the requested task, after communicating with the local energy generation unit and the smart meter. Note that the final decision regarding the execution time of the task is left to the consumer.

WAMR is another consumer side application of WSANs in the smart grid where smart meters are embedded with wireless sensors providing two-way communications with the electric utility [19,23,29]. WAMR brings many advantages to the power utility, mainly a reduced operational cost by eliminating the need for human readers, the capability of real-time monitoring of smart meters status (to prevent vandalism, for example) and of energy consumption, and the possibility of implementing a real-time pricing model [19,23]. Furthermore, WSAN-based WAMR is a low-cost, low-power, flexible solution for collecting meter data. An architecture for WiFi-based WSANs for the power meters is proposed in Ref. [29], where each building has a concentrator that acts as the sink for the power meter nodes. At a higher level, building concentrators constitute a WSAN having a central concentrator as the sink at a neighborhood's level. The authors' proposition lacks a lot of details about the specifics of such an architecture, such as the power consumption of the nodes and the management of the data collected from the different energy meters.

13.3.3.2 Monitoring at the Generation Facilities

Whether sustainable (wind, solar, tidal) or traditional (fossil fuels), power generation needs to be monitored because production and demand should always be balanced. On one hand, in case of excess production levels, specially from sustainable sources, storage alternatives might be considered [18]. On the other hand, when demand increases, production should follow.

In addition, monitoring of wind farms and solar parks can be done via a WSAN. For solar panels, temperature, current, and solar radiation data can be used to predict their power output. Information about temperature, pressure, humidity, and wind orientation, which affect the power output of a wind turbine, can be collected [18]. Furthermore, a WSAN can be used to monitor the mechanical condition of the turbines in order to detect any possible failure [30].

13.3.3.3 Electric Vehicles

V2G and V2H communications have emerged with the introduction of EVs in the smart grid. These communications aim to manage the use of an EV's battery as a power storage device and as a

TABLE 13.1
Summary of Main WSAN Applications in the Smart Grid

Generation Side	T&D Side	Consumer Side
Monitoring of production levels	Substation monitoring and control	Residential energy management
Monitoring of wind turbines	Power lines monitoring	Wireless automatic meter reading
Monitoring of solar panels	Fault diagnosis	Vehicle-to-home
	Quality indices computation	Vehicle-to-grid

provider of ancillary services to the grid. In V2G, when the vehicles are idle, their batteries can be used to store excess produced power, which can be recuperated at a later time when needed. With the V2H transmission, electrical load within a home can be shifted by using the owner's EV. Besides enabling the vehicle to be charged from the house through electricity generated at night or by local sustainable resources, V2H also allows the battery to be used as a storage device. The stored power can be used in case of outages or shortages, as well as during peak periods where the demand and the price of electricity are high [31,32].

In order for these applications to be put in place, a vehicle must be able to communicate with a smart home and with the grid, and data must be made available from these three actors. WSANs will be deployed in EVs, in homes, and in T&D power networks, and their data can be used to coordinate, support, and control V2G and V2H power transmissions [33].

Table 13.1 summarizes the main applications of WSANs in the context of the smart grid.

After reviewing the applications of WSANs in the smart grid, the issue of energy conservation in these networks is tackled next.

13.3.4   Energy Conservation in WSANs

In this section, we aim at answering mainly two questions: what is the main source of energy consumption in WSANs, and what are the possible workarounds to this issue?

13.3.4.1   Energy Consumption in WSANs

Energy consumption is a major concern in WSANs. Indeed, the lifetime of the battery is one of the bottlenecks of sensor nodes. Since the sensors are limited by their duration of life owing to their limited supply (a single AAA battery, for example), it is crucial to extend this duration since the sensors are often randomly deployed in a non-accessible way [34]. Often, the lifetime of a node is determined by the lifetime of its energy source [35].

Studies have shown that radio transmissions used to transmit data between nodes are the main source of energy consumption of a sensor node [34–36]. As shown in Ref. [37], the energy required to turn the sensor's antenna on, capture, analyze, and process parameters of the environment is negligible compared to the energy consumption of radio transmissions. Figure 13.2 shows this ratio for a Tmote Sky sensor node [38].

13.3.4.2   Data-Driven Approaches to Energy Conservation

There are several approaches to energy conservation in WSANs. They can be classified into three broad categories [39] (Figure 13.3): the first category is based on duty cycling (e.g., topology control, sleep/wakeup protocols, MAC protocols with low duty cycle), the second one concerns data-driven approaches (e.g., data reduction, energy efficient data acquisition), and the third category consists of mobility-based approaches (e.g., mobile sink or relay).

Because data-driven approaches, and in particular in-network data processing, also allow for information to be processed and evaluated at a node's level, they are interesting in the smart grid

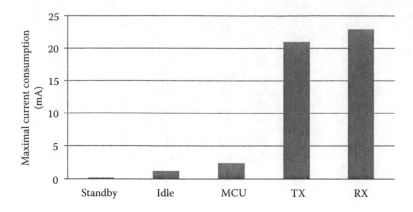

FIGURE 13.2 Maximal energy consumption of a Tmote Sky sensor.

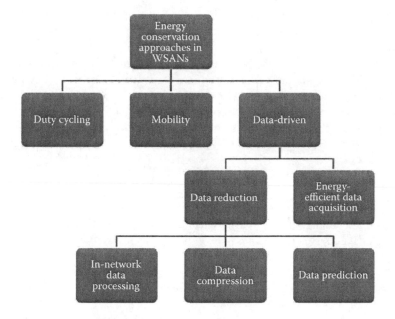

FIGURE 13.3 Approaches to energy conservation in a WSAN.

context. In fact, it is likely that information in the smart grid be processed within the network before being transmitted over long distances to the service provider's control center. This will help avoid wasting unnecessary bandwidth for their transmission and unnecessary storage space and analysis resources. Referring to Figure 13.3, data-driven approaches are mainly classified into data reduction methods and energy-efficient data acquisition methods. The latter is concerned with the reduction of the energy consumed by the sensing component of the sensors, while the former focuses on reducing the quantity of data that are sent to the sink. The second approach is more interesting because the sensing unit is not the main cause of energy consumption of a sensor as we have seen in the previous paragraph. Among these, in-network data processing consists in performing data aggregation at intermediate nodes between the source and the sink, to reduce the amount of data while traversing the network. Its main advantage over data reduction is that it is possible to correlate information from different sources and extract more meaningful data. Finally, data prediction may help reduce the number of communicated messages, but does not allow for information to be processed.

Measurements collected by a sensor node are of great volume. Sending this information in its raw state, without any local treatment, will not only consume energy, but will also require a large amount of bandwidth and cause collisions. Furthermore, because radio transmission and reception are the operations that take the highest percentage of energy (compared to sensing and processing as we have seen earlier), a node's energy consumption can be reduced by making it only send important information to the sink, and only when it is necessary. Similarly, reducing the amount of messages sent by each node entails, on the one hand, a reduction of energy consumption and, on the other hand, less traffic in the network. A reduced traffic rate implies less processing times, shorter delays, and a better quality of service.

A different approach to energy conservation consists in replenishing the sensors' batteries with additional energy via energy harvesting techniques and wireless energy transfer as we will see in the next section.

13.3.4.3 Energy Harvesting and Energy Transfer

Other solutions are proposed in the literature in order to provide extra energy to the sensors. The first one is energy harvesting, which consists in transforming body heat, movement, and vibration to electrical energy [40]. This method is particularly adapted to biomedical sensors that are typically placed on the human body [41]. As for applications related to the electrical network, recent studies have introduced different methods for harvesting energy from electric and magnetic fields [42]. They are summarized in Ref. [43]. The second proposal considers recharging the batteries remotely by the means of wireless energy transmission at short range using evanescent waves.

On the market of wind power, wireless and battery-free temperature sensors, developed by the SENSeOR society, exist. Based on the SAW (Surface Acoustic Waves) technology [44], these sensors are interrogated wirelessly. Their unlimited autonomy makes them suitable for offshore applications where maintenance operations are very difficult and costly. Mobile chargers, equipped with radio frequency-based wireless energy transfer capability, have also been proposed for use to charge multiple sensors in a distribution substation [45].

Research on methods to overcome energy limitations of conventional sensors are of current interest. A combination of low energy consumption coupled with energy harvesting can be applied to extend the operational lifetime of a WSAN.

It is thus clear that WSANs will have their place in the future power network. With different sensing units (sensors, smart meters, etc.) placed in the grid, large quantities of data will be collected. If all these data are sent in their raw state to a centralized location for analysis, then communication bandwidth, storage space, and analysis capacity issues will arise. Response times will increase, and appropriate actions will be delayed. That is why the move toward a more decentralized management of collected information is needed in the smart grid. In particular, this issue will be identified in the context of the distribution network in the next section.

13.4 TOWARD DECENTRALIZED MANAGEMENT AND CONTROL OF THE DISTRIBUTION NETWORK IN A SMART GRID

13.4.1 CONTEXT

Electricity networks, which are becoming increasingly complex and involving multiple actors, can no longer have a fully centralized control architecture. It is therefore necessary to relocate various functions of the network by introducing local intelligence. Indeed, to optimize performance, the manager of the distribution network needs to broaden its range of action and its analysis capacity [5].

Among these functions, we focus on the monitoring and control of the electrical distribution network. Its objective is the prevention of service interruptions and the reduction of their impact and their duration if they occur. According to Gimélec [5], in order to integrate a truly smart network, a distribution network's needs include the following:

- The improvement in the conduct and operation of the network through the collection of data throughout the distribution network (e.g., 1.2 million km of cable for the French distribution network). The network manager must also be able to make instantaneous decisions on the basis of these data.
- The optimization and ease of integration of network assets through the development of fault prediction and failure detection capabilities, both remotely and on the location of the breakdown. These tools are based, among others, on the collection and transmission of data in the network.

To achieve this, measurement and instrumentation equipment such as sensors and actuators can obtain information necessary for the management of energy networks and the protection of their infrastructure. These data can be of varied nature (phase, frequency, voltage, intensity, etc.).

In this context, we are interested in providing local intelligence necessary for the management of the information collected by WSANs deployed in this environment. This management includes the following:

1. The detection of various electrical disturbances (faults, voltage drops, etc.).
2. The reaction to the detected events through appropriate local actions (via actuators or communications with involved entities, e.g., activating a circuit breaker, sending an alarm to the control center, etc.).

Disturbances in power quality can be divided into two categories: variations and events. Variations are measured and evaluated continuously, while events are unforeseen and require action [46]. A distribution network's management system must provide answers in real time and must have the ability to self-repair and to detect and respond to perturbations and unstable conditions with the least impact on consumers [47]. In such a system, sensors detect fluctuations and report possible needed actions (e.g., isolate a certain area). The challenges of this context are discussed in the next subsection.

13.4.2 CHALLENGES

In the context of smart grids, and particularly that of the electrical distribution network, the problem of data deluge is a major issue for network managers as well as for suppliers that has recently been identified by Oracle [48]. Different sensing entities (sensors, meters, etc.) will collect a large amount of data across the grid that requires high bandwidth and big storage in order to be transferred to local managers. This amount can be significantly reduced if data are processed locally to extract information that is useful and important. In addition, it can be converted into actionable intelligence so that the response to any problem is more proactive [48].

In addition, one of the problems that arises is how to introduce the necessary intelligence to put in place the management of data collected by WSANs deployed in a smart grid. In the architecture of a conventional WSAN, data detected by the nodes is sent to the sink without any processing. The sink may or may not be able to analyze these data. In the latter case, it will be relayed to the centralized network management system SCADA for analysis and decision making. This obviously introduces a big delay between the possible detection of an event and the receipt of an order to be executed by an actuator where the event was captured.

Finally, the centralized control architecture will no longer be suitable to the smart grid. Indeed, considering a distribution substation in its ecosystem that includes (1) the integration of renewable energy sources that are intermittent and distributed, (2) the implementation of smart metering infrastructure that is able to collect data related to voltage quality, and (3) EVs that constitute a mobile and variable load on the grid, the distribution network should handle this in a more dynamic and intelligent way that enables local decision making. That is why a decentralized control architecture would be better suited to the distribution network.

These issues can be addressed by moving toward a more decentralized approach for information management, allowing for shorter reaction delays, local decision making, and reduction of transmitted data.

13.4.3   Toward a Decentralized Approach

To address these issues, we have recently proposed an agent-based approach for information management in a WSAN deployed at a substation [49,50].

This approach is based on autonomous agents embodied in the nodes of the network forming a multi-agent system (MAS). MAS is a concept borrowed from the distributed artificial intelligence domain and has a wide variety of applications. In particular, MASs' applications for WSANs have covered various subjects including information processing, routing, and so on (e.g., see Refs. [51–53]).

Through decentralized analysis and agent cooperation, our approach can interpret important data into direct actions and extract correlated information. At the node level, an agent is responsible for evaluating the sensed parameter and determining the priority of that value. On the basis of this evaluation, it will choose the appropriate communication policy, based on which that value will be communicated, or not, to the sink. The reader can refer to Refs. [49,50] for further details.

In summary, we can identify the following points that can be achieved through our proposed approach:

1. In the WSAN itself:
   a. Sensors are able to evaluate the importance of a detected value and act with respect to its priority.
   b. Important messages (carrying important and critical data) are delivered faster to the sink.
   c. When delay requirements allow for it, data from different sensors are aggregated on their way to the sink. The sink can thus learn about the state of several components at once and identify faulty ones.
   d. Nodes send and receive fewer messages and thus consume less energy.
2. Between the sink and the control center:
   a. Communications will require less bandwidth, and less storage space at the control center.
   b. The control center is only informed of important and relevant information, making its decision-making processes run faster to take appropriate actions.

13.5   FUTURE RESEARCH

In recent years, there has been a craze for smart grids considering the number of academic and industrial projects working on the subject, and even "model cities" [54] implementing this type of network. However, smart grids are still considered as a new research area presenting a handful of challenges that interest a variety of scientific communities because of its interdisciplinary nature.

With the evolving technologies in the energy industry and the deployment of sensing devices such as smart meters and sensors, utilities are faced with a growing volume of data that needs to be managed. Data aggregation has proved to be an inevitable technique for information management in smart grids, especially in networks deployed in the electric distribution system such as the AMI.

The main motivation behind using data aggregation is to reduce the high volume of data transmitted over the networks in order to reduce communication overhead, reduce data storage costs, and improve the speed and quality of data analysis to support decision making.

Works on data aggregation in smart grids have been mainly concerned with the AMI. While some have dealt with data aggregation strategies, others have focused on security issues. Indeed, data collection in AMI introduces new challenges related to the privacy and integrity of the collected data [55], which is why many works have been concerned with preserving the confidentiality and integrity of metering data.

Data aggregation for the smart grid is becoming the subject of more and more research. First works in this context mainly adopted a centralized approach to data aggregation. However, as we have seen, a decentralized approach is better adapted to the smart grid context. Although some works are starting to tackle this challenge, there is still a lot of room for improvement and innovative approaches in this area, such as using techniques from the distributed artificial intelligence domain (e.g., MASs).

Security is another research track that is on the rise in smart grids, especially concerning the preservation of the confidentiality and integrity of smart metering data in the AMI. While some works have tackled these issues, advances in practical privacy and integrity preserving algorithms are still expected [55]. Moreover, secure signal processing presents itself as a powerful technological solution that can make the deployment of smart grids more acceptable for end users and promises to be a challenging research area [56].

Regarding renewable energy integration, guidelines adopted by the European Union in 2008 require that renewable energy constitutes 20% of the total energy sources by 2020. The development of renewable energy in itself is still an active research area. Moreover, to improve forecasting and planning tasks, learning and predictive algorithms need to be adapted to make the most of the new information available. This will increase the energy efficiency of the power system (e.g., avoid inefficient energy trading or dispatching too much generation) [57].

Finally, energy storage systems play a key role in enabling the integration of EVs as well as the integration of renewable energy sources and coping with their intermittency. Community energy storage systems, such as the one deployed in Presidio, Texas, are still unusual but are expected to be further deployed over the coming years [58].

In summary, smart grids promise to be a varied and challenging research area where interdisciplinary collaboration can lead to innovative solutions.

13.6 CONCLUSION

The smart grid vision comes as a response to the needs of a more reliable, energy-efficient, and environmentally friendly power network that integrates sustainable energy resources and EVs. In this chapter, we have exposed the different pillar concepts behind the realization of a smart grid. We have also developed the possible applications of WSANs in the smart grid.

A part of a smart grid's reliability improvement challenge can be satisfied by putting in place advanced monitoring and control systems at the T&D levels of the electricity network. These systems should enable the decentralized analysis of data that is followed by appropriately defined actions. These actions are meant to prevent service interruptions and thus reduce the overall impact of equipment failures on the grid. Moreover, the large amounts of data collected by these systems should be reduced before being transmitted to the network manager's centralized control system. This requires less bandwidth and storage and allows for faster processing at the manager's end as only important information needs to be treated.

WSANs are attractive candidates for implementing these systems. However, the nodes' energy consumption can limit the lifetime of these networks. Some of the techniques used to overcome this limitation include reducing the amount of data sent to the sink via in-network processing and replenishing the sensors' batteries via energy harvesting or energy transfer.

As the power grid migrates toward its new smart vision, its control will need to evolve toward a more decentralized architecture. Data collected locally can be processed and aggregated before being reported to the control center in order to reduce the communication load and improve response times.

ACKNOWLEDGMENTS

This work was supported in part through grants from the Troyes University of Technology and the Lebanese University.

AUTHOR BIOGRAPHIES

Natalie Matta received her bachelor of engineering degree in electro-mechanical engineering with a focus on software and networking from the Saint-Joseph University, Lebanon, in 2009. In 2010, she graduated with honors with a master's degree in telecommunication networks jointly from the Saint-Joseph University and the Lebanese University, Lebanon. She is currently pursuing her PhD in computer science in a joint doctoral program between the Troyes University of Technology in France and the Lebanese University in Lebanon. Her current research interests include wireless sensor networks, smart grids and their applications, smart cities, information management, and data aggregation.

Rana Rahim-Amoud received with distinction her computer science and telecommunication engineer degree from the Lebanese University—First Branch, Tripoli, Lebanon, in 2002. She then obtained her master's degree (DEA) in Networking and Telecommunication in 2003 from the USJ University and the Lebanese University, Beirut, Lebanon, and her PhD degree in January 2008 from the University of Technology of Troyes (UTT), France. She was a postdoctoral researcher at the UTT from October 2008 to October 2009. She is currently a researcher and an assistant professor at the Lebanese University. Her research interests include networking, system management, quality of service, MPLS networks, WSN networks, adaptable and autonomous systems, and SmartGrids.

Leïla Merghem-Boulahia received her engineering degree in computer science from the University of Sétif, Algeria, in 1998, and her MS degree in artificial intelligence and her PhD in computer science from the University of Paris 6, France, in 2000 and 2003, respectively. She received the "Habilitation à diriger des recherches" degree in computer science from the University of Compiègne in 2010. She was an associate professor at the University of Technology of Troyes (UTT) in France until 2005. Her main research topics include multi-agent systems, quality of service management, autonomic networks, cognitive and sensor networks, and SmartGrids. Dr. Merghem-Boulahia authored or co-authored 6 book chapters, 13 peer-reviewed international journal articles, and more than 40 peer-reviewed conference papers. She received the best paper award of the IFIP WMNC'2009. She also acted as TPC member of the following conferences and workshops: IEEE ICC, IEEE Globecom, IEEE GIIS, ACM/IEEE ICCVE, IFIP Autonomic Networking, and IFIP NetCon. She has served as a reviewer for internationally well-known journals (*IEEE Communications Letters, IJNM, Communication Networks, European Journal of Industrial Engineering*, and *Annals of Telecommunications*). She is a member of IEEE.

Akil Jrad received his MSc degree in microelectronics from the University Joseph Fourier, Grenoble, France, in 1995, and his PhD degree in electronics from the University of Savoy, France, in 1999. From 1999 to 2001, he was with the Laboratory of Microwaves and Characterization, University of Savoy, France, where he was involved with the conception and realization of NLTLs for the generation and measurement of ultra-fast microwave signals and frequency multipliers. From 2001 to 2008, he joined the Laboratory of Electronics and Applied Physics, Faculty of Science, Lebanese University, Tripoli, Lebanon. Currently, he is a member of the laboratory of the Electronic Systems, Telecommunications and Networks, EDST, Lebanese University, where his current research interests include nonlinear microwave and millimeter-wave circuits analysis and design, networking, system management, and quality of service.

REFERENCES

1. Veolia Environnement. Smart grids. *Scientific Chronicles Magazine*, 18, pages 1–7, Sept. 2010.
2. S. Ramchurn, P. Vytelingum, A. Rogers, and N. Jennings. Putting the "smarts" into the smart grid: A grand challenge for artificial intelligence. *Communications of the ACM*, 55(4):86–97, April 2012.
3. European Technology Platform for the Electricity Networks of the Future. Smart grids. http://www.smart grids.eu, last accessed Jan. 9, 2013.
4. Office of Electricity Delivery. The National Energy Technology Laboratory for the U.S. Department of Energy and Energy Reliability. A vision for the smart grid. Technical report, USA, June 2009.
5. Gimélec. Deuxième livre blanc réseaux électriques intelligents. http://www.gimelec.fr/index.php/pub lications/article/1191-deuxieme-livre-blanc-reseaux-electriques-intelligents.html, March 2011, last accessed Jan. 9, 2013.
6. Pike Research. Revenue from virtual power plants. http://www.pikeresearch.com/newsroom/revenue-from-virtual-power-plants-will-reach-5-3-billion-by-2017, April 2012, last accessed Jan. 9, 2013.
7. P. Vytelingum, T.D. Voice, S.D. Ramchurn, A. Rogers, and N.R. Jennings. Agent-based micro-storage management for the smart grid. In *The Ninth International Conference on Autonomous Agents and Multiagent Systems*, pages 39–46, May 2010.
8. NanoMarkets. Smart grid sensing, monitoring and control systems: Market opportunities 2011 (free sample). http://www.nanomarkets.net/market_reports/report/smart_grid_sensing_monitoring_and_con trol_systems_market_opportunities_2011, March 2011, last accessed Jan. 9, 2013.
9. D. Niyato and Ping Wang. Cooperative transmission for meter data collection in smart grid. *Communications Magazine, IEEE*, 50(4):90–97, April 2012.
10. Office of Electricity Delivery The National Energy Technology Laboratory for the U.S. Department of Energy and Energy Reliability. Advanced metering infrastructure. Technical report, USA, Feb. 2008.
11. A. Bartoli, J. Hernandez-Serrano, M. Soriano, M. Dohler, A. Kountouris, and D. Barthel. Secure lossless aggregation for smart grid M2M networks. In *2010 First IEEE International Conference on Smart Grid Communications*, pages 333–338, Oct. 2010.
12. Z.M. Fadlullah, M.M. Fouda, N. Kato, A. Takeuchi, N. Iwasaki, and Y. Nozaki. Toward intelligent machine-to-machine communications in smart grid. *Communications Magazine, IEEE*, 49(4):60–65, April 2011.
13. McKinsey. The smart grid and the promise of demand-side management. http://www.mckinsey. com/client_service/electric_power_and_natural_gas/latest_thinking/mckinsey_on_smart_grid, 2010, last accessed Jan. 9, 2013.
14. P. Palensky and D. Dietrich. Demand side management: Demand response, intelligent energy systems, and smart loads. *IEEE Transactions on Industrial Informatics*, 7(3):381–388, Aug. 2011.
15. B.E. Bilgin and V.C. Gungor. On the performance of multi-channel wireless sensor networks in smart grid environments. In *2011 Proceedings of 20th International Conference on Computer Communications and Networks (ICCCN)*, pages 1–6, Jul. 31–Aug. 4 2011.
16. M. Erol-Kantarci and H.T. Mouftah. Wireless sensor networks for smart grid applications. In *2011 Saudi International Electronics, Communications and Photonics Conference (SIECPC)*, pages 1–6, April 2011.
17. Organisation for Economic Cooperation and Development (OECD). Smart sensor networks: Technologies and applications for green growth. Technical report, Dec. 2009.
18. M. Erol-Kantarci and H.T. Mouftah. Wireless multimedia sensor and actor networks for the next genera-tion power grid. *Ad Hoc Networks*, 9(4):542–551, 2011.
19. V.C. Gungor, Bin Lu, and G.P. Hancke. Opportunities and challenges of wireless sensor networks in smart grid. *IEEE Transactions on Industrial Electronics*, 57(10):3557–3564, Oct. 2010.
20. S.J. Isaac, G.P. Hancke, H. Madhoo, and A. Khatri. A survey of wireless sensor network applications from a power utility's distribution perspective. In *AFRICON, 2011*, pages 1–5, Sept. 2011.
21. S. Ullo, A. Vaccaro, and G. Velotto. The role of pervasive and cooperative sensor networks in smart grids communication. In *MELECON 2010—2010 15th IEEE Mediterranean Electrotechnical Conference*, pages 443–447, April 2010.
22. Litos Strategic Communication for the U.S. Department of Energy. The smart grid: An introduction. Technical report, Oct. 2008.
23. V.C. Gungor and F.C. Lambert. A survey on communication networks for electric system automation. *Computer Networks*, 50(7):877–897, May 2006.
24. M.M. Nordman and M. Lehtonen. A wireless sensor concept for managing electrical distribution net-works. In *Power Systems Conference and Exposition, 2004. IEEE PES*, pages 1198–1206, vol.2, Oct. 2004.

25. M. di Bisceglie, C. Galdi, A. Vaccaro, and D. Villacci. Cooperative sensor networks for voltage quality monitoring in smart grids. In *PowerTech, 2009 IEEE Bucharest*, 2009.
26. A. Nasipuri, R. Cox, J. Conrad, L. Van der Zel, B. Rodriguez, and R. McKosky. Design considerations for a large-scale wireless sensor network for substation monitoring. In *2010 IEEE 35th Conference on Local Computer Networks (LCN)*, pages 866–873, Oct. 2010.
27. M. Erol-Kantarci and H.T. Mouftah. Using wireless sensor networks for energy-aware homes in smart grids. In *2010 IEEE Symposium on Computers and Communications (ISCC)*, pages 456–458, June 2010.
28. M. Erol-Kantarci and H.T. Mouftah. Wireless sensor networks for cost-efficient residential energy management in the smart grid. *IEEE Transactions on Smart Grid*, 2(2):314–325, June 2011.
29. L. Li, H. Xiaoguang, H. Jian, and H. Ketai. Design of new architecture of AMR system in smart grid. In *2011 6th IEEE Conference on Industrial Electronics and Applications (ICIEA)*, pages 2025–2029, June 2011.
30. I.S. Al-Anbagi, H.T. Mouftah, and M. Erol-Kantarci. Design of a delay-sensitive WSN for wind generation monitoring in the smart grid. In *2011 24th Canadian Conference on Electrical and Computer Engineering (CCECE)*, pages 001370–001373, May 2011.
31. F. Berthold, B. Blunier, D. Bouquain, S. Williamson, and A. Miraoui. PHEV control strategy including vehicle to home (V2H) and home to vehicle (H2V) functionalities. In *Vehicle Power and Propulsion Conference (VPPC), 2011 IEEE*, pages 1–6, Sept. 2011.
32. H. Turker, S. Bacha, D. Chatroux, and A. Hably. Modelling of system components for vehicle-to-grid (V2G) and vehicle-to-home (V2H) applications with plug-in hybrid electric vehicles (PHEVS). In *Innovative Smart Grid Technologies (ISGT), 2012 IEEE PES*, pages 1–8, Jan. 2012.
33. O. Asad, M. Erol-Kantarci, and H.T. Mouftah. Management of PHEV charging from the smart grid using sensor web services. In *2011 24th Canadian Conference on Electrical and Computer Engineering (CCECE)*, pages 001246–001249, May 2011.
34. C. Townsend. Wireless sensor networks: principles and applications. In J.S. Wilson, editor, *Sensor Technology Handbook*, pages 575–589. Newnes, Burlington, 2005.
35. B. Wang, M. Li, H. Beng Lim, D. Ma, and C. Fu. Energy efficient information processing in wireless sensor networks. In S. Chandra Misra, I. Woungang, and S. Misra, editors, *Guide to Wireless Sensor Networks, Computer Communications and Networks*, pages 1–26. Springer, London, 2009.
36. L. Frye and L. Cheng. Topology management for wireless sensor networks. In S. Chandra Misra, I. Woungang, and S. Misra, editors, *Guide to Wireless Sensor Networks, Computer Communications and Networks*, pages 27–45. Springer, London, 2009.
37. E. De Poorter, S. Bouckaert, I. Moerman, and P. Demeester. Non-intrusive aggregation in wireless sensor networks. *Ad Hoc Networks*, 9:324–340, May 2011.
38. Moteiv. TMote Sky datasheet. http://www.eecs.harvard.edu/~konrad/projects/shimmer/references/tmote-sky-datasheet.pdf, last accessed Jan. 15, 2013
39. G. Anastasi, M. Conti, M. Di Francesco, and A. Passarella. Energy conservation in wireless sensor networks: A survey. *Ad Hoc Networks*, 7(3):537–568, 2009.
40. J.A. Paradiso and T. Starner. Energy scavenging for mobile and wireless electronics. *Pervasive Computing, IEEE*, 4(1):18–27, 2005.
41. R. O'Donnell. Prolog to energy harvesting from human and machine motion for wireless electronic devices. In *Proceedings of the IEEE*, 96(9):1455–1456, Sept. 2008.
42. R. Moghe, Yi Yang, F. Lambert, and D. Divan. A scoping study of electric and magnetic field energy harvesting for wireless sensor networks in power system applications. In *Energy Conversion Congress and Exposition, 2009. ECCE 2009. IEEE*, pages 3550–3557, Sept. 2009.
43. A.O. Bicen, O.B. Akan, and V.C. Gungor. Spectrum-aware and cognitive sensor networks for smart grid applications. *Communications Magazine, IEEE*, 50(5):158–165, May 2012.
44. SENSeOR. Technologie saw. http://www.senseor.com/fr/technologie-saw.html, last accessed Jan. 9, 2013.
45. M. Erol-Kantarci and H.T. Mouftah. Suresense: sustainable wireless rechargeable sensor networks for the smart grid. *Wireless Communications, IEEE*, 19(3):30–36, June 2012.
46. M.H.J. Bollen, J. Zhong, F. Zavoda, J. Meyer, A. McEachern, and F.C. Lopez. Power quality aspects of smart grids. In *The International Conference on Renewable Energies and Power Quality*, March 2010.
47. A. Janjic, Z. Stajic, and I. Radovic. Power quality requirements for the smart grid design. *International Journal of Circuits, Systems and Signal Processing*, 5(6):643–651, 2011.
48. Oracle. Big data, bigger opportunities: plans and preparedness for the data deluge. Technical report, July 2012.
49. N. Matta, R. Rahim-Amoud, L. Merghem-Boulahia, and A. Jrad. Enhancing smart grid operation by using a WSAN for substation monitoring and control. *Wireless Days, 2012 IFIP*, 1(6):21–23, Nov. 2012.

50. N. Matta, R. Rahim-Amoud, L. Merghem-Boulahia, and A. Jrad. A wireless sensor network for substation monitoring and control in the smart grid. In *The 2012 IEEE International Conference on Internet of Things*, 203(209):20–23, Nov. 2012

51. A. Rogers, N.R. Jennings, M.A. Osborne, and S.J. Roberts. Information agents for autonomous acquisition of sensor network data. In *2009 AAMAS Workshop on Agent Technologies for Sensor Networks*, pages 9–12, 2009.

52. A. Sardouk, M. Mansouri, L. Merghem-Boulahia, D. Gaiti, and R. Rahim-Amoud. Crisis management using MAS-based wireless sensor networks. *Computer Networks*, New York, USA 57(1):29–45, Jan. 2013.

53. E. Shakshuki, H. Malik, and X. Xing. Agent-based routing for wireless sensor network. In *Proceedings of the Intelligent Computing 3rd International Conference on Advanced Intelligent Computing Theories and Applications*, ICIC'07, pages 68–79. Springer-Verlag, Heidelberg, 2007.

54. Xcel Energy. Smart Grid City. http://smartgridcity.xcelenergy.com/, last accessed Jan. 22, 2013.

55. G. Taban and A.A. Càrdenas. Data aggregation as a method of protecting privacy in smart grid networks. IEEE Smart Grid Newsletter. http://smartgrid.ieee.org/march-2012/525-data-aggregation-as-a-method-of-protecting-privacy-in-smart-grid-networks, March 2012, last accessed Jan. 16, 2013.

56. Z. Erkin, J.R. Troncoso-Pastoriza, R.L. Lagendijk, and F. Pérez-González. An overview of privacy-preserving data aggregation in smart metering systems. *IEEE Signal Processing Magazine*, 30(2):75–86, 2013.

57. IBM. Managing big data for smart grids and smart meters. White Paper, May 2012.

58. B. Roberts. Substation-scale and community energy storage. *IEEE Smart Grid Newsletter*. http://smartgrid.ieee.org/january-2013/739-substation-scale-and-community-energy-storage, January 2013, last accessed Jan. 22, 2013.

14 Smart Grid Networking Protocols and Standards

Binod Vaidya, Dimitrios Makrakis, and Hussein T. Mouftah

CONTENTS

14.1 INTRODUCTION

The energy demand in the globalized world is continuously mounting because of the rapidly increasing population growth. Not only efficient use of energy is a global concern but also controlling environmental impact to offset climate change is exigent. Since global warming and climate change have been a growing worldwide concern, various organizations and governments around the world pledge to develop novel and innovative technologies for the green strategies addressing climate change globally and lowering greenhouse gas emission.

Furthermore, soaring energy prices and widespread fuel demand have intensified the necessity to reassess energy consumption. This challenge has instigated power companies and utilities to initiate new approaches that can address energy needs in the future.

The renewable energy resources including wind, micro hydro, solar, tidal, geothermal, etc., which are also called green energy because they do not release carbon dioxide into the atmosphere, are considered for electric power generation. Recently, enormous efforts have been taken in integrating green sources of energy with the electric power grid and making it cleaner.

In modernizing the electric grid to provide both utilities and consumers with the ability to monitor, control, and predict energy use, Smart grid technology holds great promise. The main objectives of the Smart grid network are to reduce energy consumption to restrain the global environmental consequences, to stabilize energy demand to avoid peaks that put strain on energy production, to curtail energy costs for consumers, to foster customers to generate their own energy and provide them incentive, and to set up energy automation.

Smart grid has introduced computation and communication capabilities into the traditional power grid to make them "smart" and "connected". Smart grid is visualized as a modern, resilient, and reliable electric grid that is cost-effective and that provides a greener environmental frontier. By providing the means to address intelligently and efficiently energy generation and consumption, it will have a vital role in the transition to a low-carbon economy. Energy efficiency, demand response, and direct load control are key components in realizing the value of a Smart grid network deployment, which will allow energy and cost savings with real-time meter data, dynamic pricing, demand response and load control, intelligent appliances, and distributed energy resources.

Many energy providers and utilities all over the world have adopted the concept of Smart energy where smart energy applications are facilitated by two-way communications in the Smart grid network.

The main characteristics of the Smart grid communications networks are as follows: integrated approach with high performance, high reliability, scalability, ubiquity, and security. The communication networks are responsible for gathering and routing data, monitoring all nodes, and acting upon the data received.

In general, the communication network architecture that supports the Smart grid is decomposed into three main networks: home area network (HAN) composed of a variety of smart energy devices in the customer's premise, several access networks such as neighborhood area network (NAN) and field area network (FAN) used to exchange information between the customer and the distribution system, and backhaul network including wide area network (WAN) that connects the utility operations and which is used by the transmission and the bulk generation systems [1–3].

The dissimilarity of network technologies, standards, and protocols used in the Smart grid network is chiefly due to the variety of application requirements that makes it difficult to find an ideal and universal solution. For instance, wireline technologies can be used in either dedicated cable or the power line for Smart grid communications, while there also exist various wireless WAN and LAN technologies. Hence, selecting an appropriate approach for a utility's circumstances is an important decision.

In fact, the Smart grid communications network is similar to the Internet and other broadband networks. For instance, a distribution network of a Smart grid network uses wireless mesh networking that enables two-way intelligent networked communications with smart meters. Wireless mesh networks can be implemented with various wireless technologies including IEEE 802.16, 802.11, 802.15, cellular technologies, or combinations of more than one type.

This chapter is structured as follows. We discuss the Smart grid communications network and Smart grid networking technologies in Section 14.2 and in Section 14.3 respectively. Then, we highlight Smart grid network standards in Section 14.4, while we illustrate challenges and opportunities in Smart grid networks in Section 14.5. And finally, we conclude the chapter in Section 14.6.

14.2   SMART GRID COMMUNICATIONS NETWORK

This section discusses the fundamental building blocks of Smart grid communications network architecture [4]. It defines the major segments, mapping them onto associated power and energy system layers, shown in Figure 14.1.

Typically, Smart grid network architecture is multi-tier, having the following components [4]: utility local area network (LAN), substation LAN; wide area networks (WAN) comprising core network/backbone, regional, or metropolitan area network (MAN) and backhaul; last mile access network; and customer premise.

- *Utility local area network (LAN)*—It is composed of utility operations and enterprise LANs to manage operations, control, and enterprise processes and various Smart grid services including billing and automation. It interconnects to not only the WAN through wireline or wireless communications but also the public Internet to exchange customer data to third-party providers.

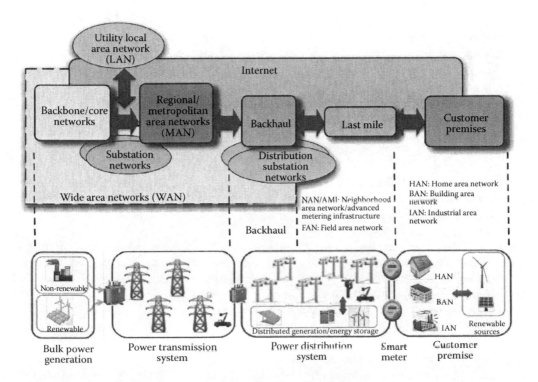

FIGURE 14.1   Smart grid communications network architecture mapping.

- *Substation LAN*—It incorporates a system that controls substation devices and information access to these devices, using any remote connection method using SCADA (supervisory control and data acquisition) protocols.
- *Wide area network (WAN)*—It is composed of the core network/backbone that connects to major service provider backbones or inter-utility backbones, regional and metropolitan area network (MAN), and backhaul network. Backhaul is point-of-presence to the last mile network, which aggregates and transports customers' metering data, substations automation data, and so on, to/from the utility head end to/from the last mile network. The WAN could be either a utility owned or a public service provider network. The WAN also interconnects to the public Internet network, substation LANs, and utility control and enterprise/IT networks. WAN connects the utility operations and is used by transmission and bulk generation systems.
- *Last mile access network*—It allows bi-directional real-time communication between utilities and consumers, which is overlaid on top of the power distribution system. It is usually named as NAN and FAN, depending on the utility network system characteristics, services offered, network topology and demographics, and the vendor technology utilized. The last mile could be an integrated and multipurpose network technology alternative for advanced metering infrastructure (AMI) services (i.e., smart metering, demand response, etc.), distribution automation (intelligent electronic devices [IEDs] in the field), and substation automation. Alternatively, the last mile could be composed of individual network technologies with different purposes and network characteristics for each particular application. On one end, it interfaces with the smart meters—at the customer premise edges, the field IED devices, and sometimes the distribution substation hot spots. On another end, its network access point interfaces with the backhaul network, where the data are collected/aggregated to be transported from the backhaul to the WAN. The last mile may also provide communication to the distributed resources—renewable and non-renewable energy sources—connected to the distribution grid.
- *Customer premise*—It includes the residential or HAN, business/building area network, and industrial area network. HANs are responsible for the communication between the provider–customer interface (i.e., smart meter) and various smart energy devices, (e.g., smart thermostats, load-control devices, in-home displays, smart appliances, plug-in electric vehicle) and the congruous operation of these devices to efficiently manage and monitor the home's electricity supply and demand. These networks are also connected to the ancillary elements outside the customer premises like renewable energy sources and storage devices. It is connected to the public Internet network through the energy service interface (ESI).

In order to achieve the Smart grid vision, different technologies and protocols have been identified that provide abilities to communicate between Smart energy devices and the utility head office in the Smart grid infrastructure [5–9]. The most used protocols and standards for the respective layers in Smart grid network are as follows.

Basically, the core/metro and backhaul networks (i.e., WANs) embrace various wireline and wireless technologies including optical fiber, Metro-Ethernet, cable modem, xDSL, WiMAX IEEE 802.16d/e/m, cellular 3G, 4G-3GPP long-term evolution (LTE), microwave point-to-point (PtP) links, point-to-multipoint (PtMP) links, and so on.

The access network including NAN and FAN may use various communication technologies such as WiFi-IEEE 802.11, WiMAX-IEEE 802.16m, IEEE 802.15.4, 3G, 4G, radio frequency (RF) Mesh, power line communication (PLC), optical fiber, and more. And the home networking may have various protocols and standards such as Ethernet/IEEE 802.3, ZigBee, WiFi, Z-Wave, HomePlug, and so on. Table 14.1 shows various Smart grid networks and their protocols.

TABLE 14.1

Network Protocols and Standards

Network	Protocols/Standards
WAN	IP/MPLS, SONET/SDH/DWDM, GPON/EPON, HFC/RFoG-DOCSIS, Metro-Ethernet, xDSL
	3GPP-3G/1xRTT/EVDO/EDGE/HSDPA, 3GPP-4G/LTE, WiMAX/IEEE 802.16d/e/m, RF Mesh,
	mm-Wave, microwave RF PtP links, satellite broadband
NAN/FAN	FTT-x/FTTN/FTTH, HFC/RFoG-DOCSIS, PLC/BPL
	WLAN/IEEE 802.11g/n, WiMAX/IEEE 802.16d/e/m, ZigBee/IEEE 802.15.4, 3GPP-3G/1xRTT/
	EVDO/EDGE/HSDPA, 3GPP-4G/LTE, RF PtMP links, RF Mesh
HAN	Ethernet/IEEE 802.3, PLC/BPL, HomePlug
	ZigBee/IEEE 802.15.4, WiFi/IEEE 802.11a/b/g/n, Z-Wave, Wireless M-Bus, Wavenis

14.3 SMART GRID NETWORKING TECHNOLOGIES

In this section, diverse communications and networking technologies being developed for the Smart grid will be exemplified. The Smart grid covers a broad range of communications and networking technologies, typically based on wireline and wireless technologies and their relevant protocols and standards.

14.3.1 WIRELINE TECHNOLOGIES

Wireline technologies rely on a direct physical connection to the subscriber's residence and business. Many such technologies including cable modem, xDSL, and PLC have evolved to use an existing form of subscriber connection as the medium for communications. In order words, wireline technologies are used for data transmission over a physical medium such as twisted-pair wire, coaxial cable, fiber optic, and so on. Typically, these technologies can offer higher communication capacity and shorter communication delay. Several wireline technologies can be considered for WAN, NAN/FAN, and HAN [8,9]. Some of them are illustrated as follows.

14.3.1.1 Optical Fiber Communications

Optical fiber communication is a technique of sending information from one place to another by sending light pulses through an optical fiber. The light acts as the carrier wave, which is used in modulation to carry the information signal. The transmission of information involves basic steps, which create an optical signal to carry the information using a transmitter, relaying the signal over the optical fiber, ensuring that the signal does not weaken before it reaches the destination and receiving the data and converting it to electrical signal at the destination. Commonly, fiber optic is used for long-haul connectivity such as WAN, MAN, and so on, as well as last mile access, that is, NAN.

14.3.1.1.1 SONET/SDH/DWDM

Synchronous Optical Networking (SONET) and Synchronous Digital Hierarchy (SDH) are the standard multiplexing protocols used to convey high data rate digital bit streams over optical fiber using light-emitting diodes or lasers. The SONET standard is defined as ANSI standard T1.105, while the SDH standard is formalized as ITU standards G.707, G.783, G.784, and G.803. SONET/SDH enables various ISPs to share the same optical fiber simultaneously without interrupting each other's traffic load. They are physical layer protocols, which offer continuous connections without involving packet mode communication, and are distinguished as time division multiplexing protocols. The basic unit of framing in SONET is STS-1 (synchronous transport signal 1) or OC-1 (optical carrier 1) operating at 51.84 Mbps. Further optical carriers are expressed by OC-x, where x is a multiple of the OC-1. The SDH has a similar structure with STM-n (synchronous transport module-n), where n is a multiple of STM-1 with a rate of 155.520 Mbps.

Dense wavelength division multiplexing (DWDM) is an advanced fiber-optic transmission technique that employs multiple light wavelengths to transmit signals over a single optical fiber. It is a crucial component of optical networks since it maximizes the use of installed fiber cable and allows new services to be quickly and easily provisioned over existing infrastructure.

14.3.1.1.2   GPON/EPON

A passive optical network (PON) is a PtMP fiber to the premises' network architecture, which consists of a central office node (i.e., optical line terminal), one or more user nodes (i.e., optical network units or optical network terminals), and the fibers and splitters between them. Downstream signals are broadcasted to all premises sharing multiple fibers, while upstream signals are combined using a multiple access protocol, usually time division multiple access. Gigabit PON (GPON) is based on the ITU-T G.984 series of recommendations. The standards permit a downstream rate of 2.488 Gbit/s and an upstream rate of 1.244 Gbit/s. G.987 defined 10G-PON with 10 Gbit/s downstream and 2.5 Gbit/s upstream. Ethernet PON (EPON) is based on the IEEE 802.3 standards with symmetric 1 Gbit/s upstream and downstream rates, while 10Gbit/s EPON or 10G-EPON supports 10/1 Gbit/s.

14.3.1.1.3   FTT-x/FTTH/FTTN

Fiber-to-the-x (FTT-x) is any broadband access network using optical fiber for last mile telecommunications. For instance, FTTH (fiber-to-the-home) is one, in which fiber extends till the boundary of the home. Generally, PONs are architecture that delivers triple-play services over FTTH networks directly from the central office. In FTTN (fiber-to-the-neighborhood), fiber is terminated in a cabinet, probably miles away from the customer premises, having ultimate copper connections.

14.3.1.2   Dedicated Copper Communications

Because of low-cost and proven communication technologies, high-tech copper cables are widely used for many applications such as Internet access and Smart grid services.

14.3.1.2.1   IP/MPLS

An IP/MPLS network is a packet-switched network that uses the Internet protocol (TCP/IP) enhanced with the multi-protocol label switching (MPLS) standard. This means that data are directed from one network node to the next based on short path labels rather than long network addresses, avoiding complex lookups in a routing table. The MPLS set of protocols operates above the IP protocols and was introduced to guarantee delivery of traffic, reduce network delay, and guarantee quality of service while still operating in an IP environment.

14.3.1.2.2   xDSL

Digital subscriber line (DSL) provides a connection to the public network (i.e., Internet) through the telephone network. xDSL technologies can be found in various types. Asymmetric DSL has unequal data throughput; that is, the upstream direction is lower than that in the downstream direction. With a symmetric DSL, the downstream and upstream data rates are equal. Very-high-bit-rate DSL (VDSL) is ITU G.993.1 standard that provides data rates up to 52 Mbit/s downstream and 16 Mbit/s upstream over copper wires and up to 85 Mbit/s downstream and upstream on coaxial cable. Similarly, VDSL2 is ITU-T G.993.2 standard that is able to provide data rates exceeding 100 Mbit/s simultaneously in both the upstream and downstream directions. However, the maximum data rate is achieved at a range of approximately 300 m and performance degrades as distance and loop attenuation increase.

14.3.1.2.3   Cable Modem

Cable modem primarily provides broadband Internet access on hybrid fiber-coaxial (HFC) and radio frequency over glass (RFoG) infrastructure. HFC combines optical fiber and coaxial cable while RFoG is deep fiber network design in which the coax portion of the HFC network is replaced by a single-fiber PON. They can offer high data rates: downstream rate ranges from 400 to 100 Mbit/s, while upstream rate ranges from 384 kbit/s to more than 20 Mbit/s.

One of the telecommunications standards is DOCSIS (Data over Cable Service Interface Specification), which allows addition of high-speed data transfer to an existing cable TV (CATV) system. Most cable modems are compliant with one of the DOCSIS versions. The ITU-T has approved various versions of DOCSIS as international standards, for instance, DOCSIS 1.0 as ITU-T J.112 (1998), DOCSIS 1.1 as ITU-T J.112 (2001), DOCSIS 2.0 as ITU-T J.122, and DOCSIS 3.0 as ITU-T J.222.

14.3.1.2.4   Metro-Ethernet

A Metro-Ethernet network is a MAN based on Ethernet standards, which is normally used to connect subscribers to a larger service network or the Internet. The Metro-Ethernet can be used as pure Ethernet, Ethernet over SDH, Ethernet over MPLS, or Ethernet over DWDM. The core in most cases is an existing IP/MPLS backbone but may migrate to newer forms of Ethernet transport in the form of 10 Gbit/s, 40 Gbit/s, or 100 Gbit/s speeds.

14.3.1.2.5   Ethernet/IEEE 802.3

IEEE 802.3 or Ethernet over twisted pair cabling is one of the most common technologies used in LANs that provide Internet access to computers and other devices in a limited area such as a home, school, or office building, usually at relatively high data rates that typically range from 10 to 1000 Mbit/s.

14.3.1.2.6   MoCA

MoCA (multimedia over coax alliance) is a technology that uses existing coaxial cables to connect consumer electronics and home networking devices in homes and allows both data communication and the transfer of audio and video streams. MoCA 2.0 supports two performance modes, basic and enhanced, with 400 and 800 Mbit/s net throughputs at medium access control (MAC), using 700 Mbit/s and 1.4 Gbit/s physical layer (PHY) rates, respectively.

14.3.1.2.7   HomePNA

HomePNA is home networking technology that originally uses balanced pair telephone wire. The original HomePNA 1.0 was developed in the 1990s and then followed by several variations such as HomePNA 2.0—ITU G.9951, G.9952, G.9953; HomePNA 3.0—ITU G.9954; and HomePNA 3.1— ITU G.9954. HomePNA 3.1 added Ethernet over coax operation to overcome limitations of phone jack location. Some advantages of HomePNA 3.1 are no special or new home wiring is required; existing services are not disrupted as HomePNA operates at different frequencies on the same coax or phone wires; some products offer data rates up to 320 Mbit/s.

14.3.1.3   Power Line Communications

PLCs use electrical power cables for data transmission [10–12]. Typically, PLC uses the existing electrical power wiring within a home. Early products were focused on low-speed home networking and targeted home automation that required low bit rate. Typical home control PLC devices are low cost, which enable ubiquitous home automation and device command and control. These devices may be either plugged into standard power outlets or wired permanently. Modern PLC products are offering high-speed data rates. They can interconnect home computers and peripherals, and home entertainment devices having Ethernet port. Power line adapters that are plugged into power outlets establish Ethernet connection using the existing electrical wiring in the home environment. This allows devices to share data and video without any dedicated network cables.

14.3.1.3.1 X10

X10 is a popular technology and open standard for communication among electronic devices used for home automation. It primarily uses power line wiring for signaling and control, where the signals involve brief RF bursts representing digital information. Data are encoded on a 120 kHz carrier, which is transmitted in bursts; one bit is sent at or within a specified proximity to each zero crossing of the 50 or 60 Hz. This method is standardized in IEC 61334 and is also used for electricity meters and SCADA. However, this technology has been limited by its data rate, reliability, and scalability.

14.3.1.3.2 INSTEON

INSTEON is a system for connecting lighting switches and loads without extra wiring and is invented by SmartLabs, Inc. INSTEON has been used in home automation for almost 10 years. This hybrid peer-to-peer mesh network uses proprietary INSTEON power line protocol and INSTEON ISM band RF protocol and is intended chiefly for automated home control [13]. Devices may be single or dual medium. The network is self-configuring, though mapping is required of buttons in controllers to device functions.

The INSTEON power line protocol can coexist with X10 and developers can implement devices that support and interface these protocols together. It achieves sustained raw data, 2880 bps operating the dual medium. The network employs a non-routing topology where all nodes receive and repeat messages. In order to take advantage of common transport protocols such as TCP/IP, INSTEON devices must be fitted with purpose-built serial interfaces such as USB, RS232, or Ethernet and connected with other digital devices, which support bridging to other networks such as a LAN or the Internet.

14.3.1.3.3 KNX

KNX is a mature and open but proprietary home and building control standard owned by an industry alliance, the KNX Association. KNX has gained a number of standards approvals including ISO/IEC 14543-3. KNX power line operates in the 90–125 kHz band and at a bit rate of 1200 bps, while dedicated twisted pair wiring achieves 9600 bps and a maximum cable length of 1000 m. KNX can also be tunneled over an Ethernet LAN or the Internet using the KNXnet/IP specification. Ease of installation is a key feature with KNX providing a single tool, which supports all implementations and all user levels. KNX is mainly found for lighting and HVAC. KNX has been working to introduce solutions for home energy management and Smart grid. KNX RF has the potential to realize a wireless sensor network (WSN), but form factor is restrictive for ubiquitous environmental sensing. KNX support for multiple media is a strong asset, but is let down by a lack of security in KNX RF.

14.3.1.3.4 LonWorks

LonWorks is a proven standard and technology that encompasses all the elements necessary to design, install, monitor, and control a network of diverse devices [14]. This networking platform is built on a protocol created by Echelon Corporation for networking devices over media such as twisted pair, power lines, fiber optics, and RF. It is used for the automation of various functions within buildings such as lighting, HVAC, and intelligent building. Although developers have extended its application field, LonWorks still has issues with flexibility and scalability, which make other applications difficult. The LonWorks network employs a connectionless domain-wide broadcast topology with loop-free learning routers and repeaters. While most of the LonWorks protocol is public and open, layers 3 to 7 of the standards are closed and proprietary. The LonWorks platform is used for many diverse applications that span smart buildings, smart cities, the Smart grid, and other smart controls such as commercial automation and industrial automation.

14.3.1.3.5 UPB

The universal power bus (UPB) protocol uses a home's existing power lines to carry home automation messages. Using power lines instead of wireless signals has its advantages as signals transmitted over power lines can travel further distances. Also, UPB power line signals do not have to account for wall obstructions as wireless signals do. Reliability for UPB transmissions is reported to be greater than 99%. The main advantages of UPB are as follows: it is inexpensive; signals can travel long distances; and it can use a peer-to-peer system or a controller-based system. However, the disadvantages are as follows: it does not have the convenience of wireless systems; the data rate for transferring UPB messages is limited to 480 bits/s; and only a maximum of 250 devices is allowed in a single home. Another disadvantage is that UPB is limited to electrical outlets.

14.3.1.3.6 Broadband over Power Lines

Broadband over power lines (BPL) is a method of PLC that allows relatively high-speed digital data transmission over the existing public electric power distribution wiring and does not need a network overlay as it has direct access to the ubiquitous power utility service coverage areas.

BPL systems are being promoted as a cost-effective way to service a large number of subscribers with broadband. In a BPL system, the data are transmitted over the existing power line as a low-voltage, high-frequency signal, which is coupled to the high-voltage, low-frequency power signal. The frequency transmission band has been chosen to ensure minimum interference with the existing power signal. Typical data rates in current standards are 2–3 Mbps, but vendors have indicated that commercial systems offering up to 200 Mbps could eventually become available. However, there is no clear upgrade path to higher data rates. Most BPL systems at present are limited to a range of 1 km within the low-voltage grid, but some operators are extending this reach into the medium-voltage grid. Experience has shown that BPL requires a high investment cost to upgrade the power transmission network and bypass transformers, to support high-speed and reliable broadband services. In addition, the frequencies used for BPL often interfere with amateur radio transmission. At present, given the cost and the lack of an upgrade path, it seems unlikely that BPL will emerge as a leading broadband technology.

14.3.1.3.7 HomePlug/IEEE 1901

HomePlug is a power line networking standard, which is proposed by the HomePlug Powerline Alliance. HomePlug AV is the most widely used HomePlug specification that has been adopted by the IEEE 1901 group as a baseline technology [15]. HomePlug AV2 technology [16] is the most recent of the HomePlug specifications. More than 45 million HomePlug devices have been deployed worldwide. The Universal Powerline Association, the HD-PLC Alliance, and the ITU-T's G.hn also provide various specifications for power line home networking.

HomePlug Green PHY has been set as the US standard for AMI, demand response, smart appliances, smart homes, and other HAN applications. Furthermore, IEEE P1905 is a working group for a "convergent digital home network." This integrates P1901 communication with other IEEE 802.x standards for arbitrary IP-based communications with the network, devices including vehicles, meter, and the building where charging devices draw power from.

14.3.1.3.8 HomeGrid/ITU G.hn

G.hn is a specification for high-speed communications using the existing wired home networking, which is developed by the ITU-T and the HomeGrid Forum. The G.hn specification defines networking over power lines, phone lines, and coaxial cables with data rates up to 1 Gbit/s. The main benefits of the ITU G.hn standard are lower equipment development costs and lower deployment costs for service providers. The G.hn standard enables devices such as televisions, set-top boxes, residential gateways, and so on, to be AC powered, hence, at least one power line networking interface should be common as it will facilitate integration with home control and demand side management applications for AC-powered appliances.

The ITU-T has developed several G.hn standards such as ITU-T G.9960, ITU-T G.9961, and ITU-T G.9963 (multiple-input/multiple-output [MIMO] technology). The National Institute of Standards and Technology (NIST) has included ITU-T G.hn as one of the standards identified for implementation of the Smart grid infrastructure. By developing dual-mode devices, the G.hn standard can provide an evolution path from other wireline home networking technologies such as MoCA, HomePNA 3.1 over coax and phone wires (ITU G.9954), HomePlug AV, Universal Powerline Association (UPA), and HD-PLC over power line.

14.3.2 Wireless Technologies

Wireless communication refers to transmitting signals and data without cables using electromagnetic waves such as mobile communications, microwave communications, and satellite communications [17,18]. The classification of wireless networks is shown in Table 14.2, and the characteristics of particular wireless link standards are shown in Figure 14.2.

14.3.2.1 2G/3G Cellular Communication

One way to meet the requirements for NAN is through cellular communication. Cellular communication is ubiquitous, is easy to install, and incurs a low maintenance cost. The coverage is excellent because it corresponds to the population concentration and hence ubiquitous. Cellular communication is already established and has 95% coverage extended to consumers and hence no additional efforts for installations are required. Cellular technology is also a price-competitive solution because it leverages the existing carriers and quantity of devices.

Cellular communication could broadly be categorized into two types, namely, GSM (global system for mobile communication) and CDMA (code division multiple access). Each of these platforms has several technology implementations based on their increasing throughput. The variants of GSM are GPRS (general packet radio service), EDGE (enhanced data rates for GSM evolution), and HSDPA/UMTS (high-speed downlink packet access/universal mobile telecommunication system), whereas the variants of CDMA are cdmaOne, CDMA2000/1xRTT, and EV-DO Rev A.

14.3.2.2 Next-Generation (4G) Mobile Communications

The fourth generation (4G) of mobile communications standards is a successor of the third generation (3G) standards. A 4G system provides mobile ultra-broadband Internet access to smart phones and to other mobile devices. Conceivable applications include amended mobile web access, IP telephony, gaming services, high-definition mobile TV, video conferencing, and 3D television.

The 4G refers to IMT-Advanced (International Mobile Telecommunications-Advanced), as defined by ITU-R. An IMT-Advanced cellular system must have peak data rates of up to approximately 100 Mbit/s for high mobility such as mobile access and up to approximately 1 Gbit/s for low mobility such as nomadic/local wireless access, smooth handovers across heterogeneous networks, and the ability to offer high quality of service for next-generation multimedia support. The next-

TABLE 14.2
Classification of Wireless Networks

	WAN	MAN	LAN	PAN
Radio coverage	Dozens of kilometers	<10 km	<100 m	<10 m
Data rate	<250 Mbps	<300 Mbps	<100 Mbps	<256 kbps
Protocols	802.16m, LTE	2G/3G, 802.16, 802.11a/g/n PtP, RF Mesh	802.11a/b/g/n	802.15.4, ZigBee, Z-Wave, Bluetooth, UWB

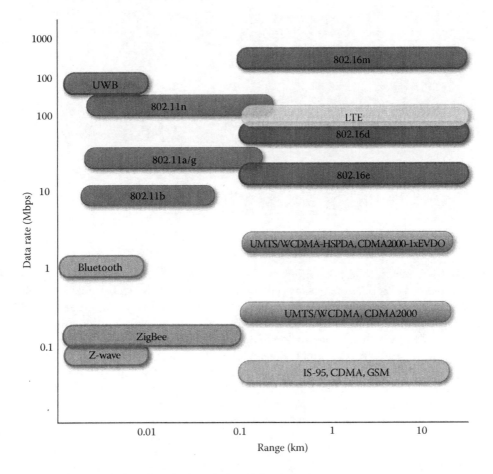

FIGURE 14.2 Characteristics of particular wireless link standards.

generation (4G) mobile networks are finalized by two standards, namely, LTE-Advanced (LTE-A) standardized by the 3GPP and WiMAX 2 (802.16m) standardized by the IEEE.

14.3.2.3 IEEE 802.15.4

IEEE 802.15.4 is a standard that specifies the PHY and MAC for low-rate wireless personal area networks (LR-WPANs). It is the foundation for many LR-WPAN technologies such as the ZigBee, ISA 100.11a, MiWi, and WirelessHART, each of which further extends the standard by developing the upper layers. Furthermore, it can be used with 6LoWPAN and other networks such as Smart Utility Networks/IEEE 802.15.4g.

14.3.2.3.1 ZigBee

ZigBee is a complete protocol stacks standard, using the IEEE 802.15.4 standard for its two lowest layers [19]. ZigBee products are certified by the ZigBee Alliance, which is an industry alliance of more than 230 enterprises. ZigBee incorporates routing and automatic network formation capabilities, allowing the establishment of star and mesh networks to provide communication between devices. ZigBee signals are carried at the 2.4 GHz frequency range and can travel up to 300 feet in an outdoor environment and up to 30 feet in an indoor environment. ZigBee products target the home automation, energy management industrial control, and personal health care markets.

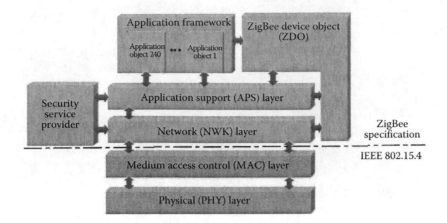

FIGURE 14.3 Overview of ZigBee protocol stack.

The first stack release is ZigBee 2004, while the second stack release is ZigBee 2006. The recent stack release is ZigBee 2007, which contains two stack profiles: stack profile 1 (simply called ZigBee) and stack profile 2 (called ZigBee PRO). ZigBee (stack profile 1) is for home and light commercial use and offers a smaller footprint in RAM and flash, while ZigBee PRO offers more features, such as multi-casting, many-to-one routing, and high security with Symmetric-Key Key Exchange. Both offer full mesh networking and work with all ZigBee application profiles.

The ZigBee protocol stack is composed of four main layers: the application (APL) layer, the network (NWK) layer, the MAC layer, and the PHY layer. The APL and NWK layers are defined by the ZigBee specification, whereas the PHY and MAC layers are defined by the IEEE 802.15.4 standard. Figure 14.3 shows an overview of the ZigBee protocol stack.

The main benefits of ZigBee technology are that it is an open IEEE standard, the signal is carried up to 300 feet in an outdoor environment and up to 30 feet in an indoor environment, it can support over 64,000 devices, it is a reliable protocol, and it includes security functionalities (integrity, encryption, key management). It also has some shortcomings, namely, that ZigBee's use of the 2.4 GHz can cause interference with existing WiFi networks and ZigBee-based home automation devices may not be compatible between different companies.

14.3.2.3.2 ISA 100.11a

ISA 100.11a is a wireless networking technology standard developed by the International Society of Automation (ISA). The ISA 100.11a specification defines higher layer (APP and NWK), while MAC and PHY layers are defined as in IEEE 802.15.4. It is meant for factory and building automation and is confirmed in LAN. The standard ISA 100.11a includes radio link as well as ISA 100.11a over Ethernet and field buses.

14.3.2.3.3 MiWi

MiWi is proprietary wireless protocol, which uses small, low-power digital radios based on the IEEE 802.15.4 standard for WPANs. It is designed for low data transmission rates and short-distance, cost-constrained networks, such as industrial monitoring and control, home and building automation, remote control, lighting control, and automated meter reading.

14.3.2.3.4 WirelessHART

WirelessHART is an approved open standard for WSNs designed primarily for industrial process automation and control systems. The protocol supports operation in the 2.4 GHz ISM band using IEEE 802.15.4 standard radios. Developed as a multi-vendor, interoperable wireless standard, WirelessHART was defined for the requirements of process field device networks. The protocol utilizes a time-synchronized, self-organizing, and self-healing mesh architecture. In April 2010, WirelessHART was approved by the International Electrotechnical Commission (IEC), making it the first wireless international standard as IEC 62591.

14.3.2.3.5 6LoWPAN

The IETF IPv6 over low-power wireless PAN (6LoWPAN) working group has defined the frame format and several mechanisms needed for the transmission of IPv6 packets on top of IEEE 802.15.4 networks. It works on RF bands 800/900 MHz and 2.4 GHz. It is targeted for IP networking applications with low-power radio communication capability that need wireless Internet connectivity at lower data rates for devices with very limited form factor; hence it does not need additional translation gateway for Internet connectivity.

The main benefits of 6LoWPAN are as follows: it entails a massively scalable networking as an end-to-end part of the Internet (i.e., IPv6 addressing); it is applicable to any low-power, low-rate wireless radio; it does not need additional translation gateway for Internet connectivity; it is useful for emerging future networks such as large-scale enterprise automation, Smart grid, M2M networks, and Internet of Things; and it provides end-to-end addressing, security, mobility, traffic multiplexing, reusability, maintainability, and web services.

14.3.2.3.6 IEEE 802.15.4g (SUN)

In an effort to promote open standards for the Smart grid environment and to meet the specific regulations in a global Smart grid deployment environment in a scalable and cost-effective way, the IEEE 802.15.4g Smart Utility Networks (SUN) Task Group is chartered to create a PHY amendment to 802.15.4 to provide a global standard that facilitates very large scale process control applications such as the utility Smart grid network capable of supporting large, geographically diverse networks with minimal infrastructure, with potentially millions of fixed endpoints.

Recently, the IEEE 802.15.4g radio standard adds new PHY support for SUN to IEEE 802.15.4-2011 to support RF Mesh solutions. In addition, the amendment also defines MAC modifications needed to support their implementation. The SUN PHY supports multiple data rates in bands ranging from 450 to 2450 MHz and working in one of these three modes: orthogonal frequency division multiplexing (MR-OFDM), multi-rate and multi-regional offset quadrature phase-shift keying (MR-O-QPSK), and multi-rate and multi-regional frequency shift keying (MR-FSK).

14.3.2.4 Z-Wave

Z-Wave is a wireless standard that uses mesh networking to establish and maintain connection between devices. Mesh networking allows signals to repeat through multiple devices until it gets to the desired home automated device. The result is that Z-Wave networks become more reliable as they grow in size. The distance limit for a Z-Wave signal is 100 feet, but that is in open air. Through walls, the limit is reduced to approximately 50 feet. Z-Wave uses the 800/900 MHz frequency range so it would not interfere with other 802.11 wireless protocols. However, the Z-Wave 400 series is based on 2.4 GHz.

Z-Wave was originally designed for residential applications. All Z-Wave modules are produced by a single manufacturer called Zensys (Sigma Designs). Z-Wave is run by the Z-Wave Alliance, which tests and makes sure Z-Wave devices built by different companies interoperate.

The main benefits of Z-Wave are as follows: it is simpler than ZigBee; wireless and RF signal can travel through walls and at longer distance compared to those signals using the 2.4 GHz band; most Z-Wave devices do not interfere with other wireless signals (especially WiFi, Bluetooth, and ZigBee, which operate at 2.4 GHz); Z-Wave alliance makes sure Z-Wave devices produced by different manufacturers are interoperable; and it is inexpensive. However, it has its share of shortcomings: it is proprietary (i.e., not an open source); there is a wireless distance limit (30 m indoors, 100 m outdoors); signals are not encrypted, which might cause problems for security commands in series 200/300 (Z-Wave series 400 use 128bit AES); and it is limited to 232 devices only.

TABLE 14.3
Various IEEE 802.11 Standards

IEEE Standard	Frequency Band	Speed
802.11	2.4 GHz	1 Mbps, 2 Mbps
802.11a	5 GHz	Up to 54 Mbps
802.11b	2.4 GHz	5.5 Mbps, 11 Mbps
802.11g	2.4 GHz	Up to 54 Mbps
802.11n	2.4 GHz/5 GHz	Up to 300 Mbps
802.11ac	5 GHz	Up to 1 GHz

14.3.2.5   WiFi (IEEE 802.11)

IEEE 802.11, standardized as a wireless LAN technology, also known by the term *WiFi*, is a widely popular wireless technology for home environment. The PHY layer IEEE 802.11 standards existing in functional NICs are the IEEE 802.11b/a/g/n (listed in sequence of completion of standardization, starting with the oldest one). IEEE 802.11i is a security amendment. The most recent protocol amendment is IEEE 802.11n-2009. IEEE 802.11ac is under consideration. Table 14.3 shows the variety of IEEE 802.11 standards.

Among various IEEE 802.11 protocol standards, IEEE 802.11n has some unique features: it can operate at both 2.4 and 5 GHz and it provides use of 20 and 40 MHz channels. It achieves high transmission rates (there are commercial products with a maximum transmission speed of 300 Mbps) by using MIMO technology. Like many other technologies (i.e., PLC, G.hn, LTE, HomePlug, and IEEE 802.11a and g), IEEE 802.11n uses orthogonal frequency division multiplexing (OFDM) as its modulation technique. The main benefits of IEEE 802.11n are as follows: WiFi is a mature technology with a high adoption rate; since many homes and offices already have a WiFi network in place, upgrading to IEEE 802.11n is easy if necessary; data rates of 300 Mbps that exceed the Smart grid requirements; backward compatible with IEEE 802.11a, b, and g; inside range of 70 m, which should cover all but the largest homes; inexpensive chipsets make integration affordable; and MIMO technology helps resilience in the congested ISM bands.

14.3.2.6   Wireless M-Bus

Wireless M-Bus (EN 13757-4: 2005) is a communication protocol that is primarily designed for remote reading of consumption meter (electricity, heat, gas, and water). Wireless M-Bus standard specifies communications between water, gas, heat, and electric meters as well as a meter and an "other" system component, for example, mobile/stationary readout devices, data collectors, and so on. This protocol can be used in industry and build automation. Wireless M-Bus uses a frequency of 868 MHz with a maximum speed from 4.8 to 100 kbps. It is becoming widely accepted in Europe for Smart metering or AMI applications.

14.3.2.7   Wavenis

Wavenis technology is a two-way wireless connectivity platform dedicated for M2M applications and is designed for ultra-low-power energy consumption and long-range transmission of small amounts of data and low traffic communications, specifically for WSN. Wavenis operates in the license-free ISM bands (i.e., 868, 915, and 433 MHz) around the world. Most Wavenis applications communicate at 19.2 kbit/s, though data rates can range from 4.8 to 100 kbit/s. Several Wavenis-based devices have been deployed in many applications such as telemetry, industrial automation, AMI, automatic meter reading, utility meter monitoring, home automation, and active RFID applications.

14.3.2.8 Bluetooth Low Energy

Bluetooth low energy (BLE) is a feature of Bluetooth 4.0 wireless radio technology, aimed for products that require low power consumption and low latency, and applications for wireless devices within a short range (up to 50 m) [20]. BLE technology operates in the same spectrum range (2402–2480 MHz) as classic Bluetooth technology, but it uses a different set of channels. It offers a maximum data transfer rate of 1 Mbps, has a power consumption of approximately 10 mW, and provides security measures; for instance, encryption is done with 128-bit AES block cipher (CCM mode) and authentication: AES CBC-MAC (CCM mode). This facilitates a wide range of applications and smaller form factor devices in the home automation, health care, security, and home entertainment industries. The main advantage of BLE is that it offers high spectral efficiency and low power consumption; however, it is not supported by many devices.

14.3.2.9 UWB

Ultra-wideband (UWB) is a radio technology, which may be used at a very low energy level for short-range, high-bandwidth communications using a large portion of the radio spectrum. UWB radios can use frequencies from 3.1 to 10.6 GHz. UWB signals can co-exist with other short/large-range wireless communications signals because of their nature of being detected as noise to other signals. UWB is used for commercial communications that are capable of delivering very high data rates within short ranges (i.e., 10–20 m). It offers the best solution for bandwidth, cost, power consumption, and physical size requirements for consumer electronic devices, mobile devices, and PC peripheral devices, which provides very high data rate while consuming very little battery power. Most recent applications target sensor data collection and precision locating and tracking applications.

14.3.2.10 WiMAX

WiMAX (Worldwide Interoperability for Microwave Access) is a set of interoperable implementations of the IEEE 802.16 family of wireless network standards, which enables the delivery of last mile wireless broadband access as an alternative to cable and DSL. The original IEEE 802.16 standard, now called "Fixed WiMAX," was published in 2001 and provided 30 to 40 Mbit/s data rates. Mobility support was added in 2005.

WiMAX provides wireless transmission of data in various modes, from a point to multi-point links. It is also called as the last mile connectivity of Broadband Wireless Access (BWA) with a range of around 30 miles and a data transfer rate of up to 280 Mbps with the ability to support data, voice, and video. Its operating range is anywhere from 2 to 66 GHz. It does not require LOS (line of sight). A version of IEEE 802.16, which is IEEE 802.16e, adds mobility features operating in the range of 2–11 GHz license bands. Hence, it allows fixed and mobile non-line of sight (NLOS) applications primarily to enhance OFDMA (orthogonal frequency division multiple access). Figure 14.4 shows overviews of the WiMAX protocol stack.

IEEE 802.16m [21] is an amendment to the released IEEE 802.16-2009 standard. The goal set out in IEEE 802.16m is to develop an advanced air interface to meet the requirements for IMT-Advanced next-generation networks while still supporting the legacy of the IEEE 802.16 OFDMA system. IEEE 802.16m provides data rates of 100 Mbit/s mobile and 1 Gbit/s fixed. WiMAX offers a MAN with a signal radius of approximately 50 km (30 miles), far surpassing the 30 m (100 ft) wireless range of a conventional WiFi LAN.

14.3.2.11 LTE/LTE-A

LTE [22] is a standard for wireless communication of high-speed data for mobile phones and data terminals, which is developed by the 3GPP (3rd Generation Partnership Project). It is based on the GSM/EDGE and UMTS/HSPA network technologies, increasing the capacity and speed using a different radio interface together with core network improvements. Furthermore, with major enhancement,

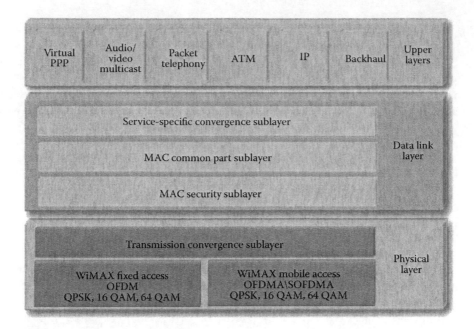

FIGURE 14.4 Overview of WiMAX protocol stack.

the LTE standard is evolved into LTE-A, which was approved into ITU, IMT-Advanced and was finalized by 3GPP in March 2011.

14.3.2.12 Mesh RF

In North America, RF Mesh is one of the dominant technologies for local smart meter communication, with data being aggregated by an upstream concentrator and sent back to a utility via a WAN. Most RF Mesh solutions utilize FSK modulation in the unlicensed ISM frequency bands such as 902–928 MHz within North America and 868–870 MHz for Europe. RF Mesh solution, which is a two-way wireless mesh communication system, is a proven platform for advanced metering, distribution automation, and personal energy management that delivers enhanced connectivity to every application within the utility—all from a single network. Data rate is approximately 100 kbps, a limiting factor to implementing more advanced Smart grid applications. Another common characteristic is while solutions are moving toward IP protocols across each segment of the network, they still predominantly incorporate proprietary technologies, preventing equipment interoperability.

14.3.2.13 mm-Wave

mm-Wave (millimeter wave) is at the high end of microwave frequencies, that is, extremely high frequency, which runs the frequency range from 30 to 300 GHz (wavelength, 1 mm to 1 cm). The mm-Wave is primarily transmitted through waveguide as the waveguide has some advantages: it is completely shielded, it has no trouble transmitting extremely high peak powers, and it has an almost non-existent loss at microwave frequencies. Current applications of mm-Wave include wireless backhaul, inter-satellite link, and some new spectrum for mobile broadband communications.

14.3.2.14 Microwave PtP Links

Microwave links are basically high-capacity PtP wireless transport for backhaul or backbone of telecommunication systems. Licensed microwave link frequencies used for wireless backhaul in a PtP wireless backhaul operate at 6, 11, 18, and 23 GHz bands and at the 80 GHz millimeter wave

E-band, while unlicensed wireless Ethernet bridges, used in PtP wireless bridges, typically operate at 900 MHz, 2.4 GHz, 5.3 GHz, 5.4 GHz, or 5.8 GHz. There is also the 60 GHz millimeter wave band that is used for PtP gigabit wireless bridges.

Microwave links are used as wireless backhaul transport for diverse Smart grid applications such as SCADA, AMI, distribution automation, and demand response. The key advantages of microwave links are as follows: there are a wide range of available capacities, frequencies, and configurations; they incur a lower cost than optical fiber deployment; and they are highly secure.

14.3.2.15 RF PtMP Links

PtMP-based multiple address systems (MAS) consists of a central master station and several remote terminal units (RTUs). The MAS network provides communication between a central host computer and RTUs or other data collection devices. A wide range of outdoor PtMP systems are used, which operate in unlicensed bands (900 MHz, 2.4 GHz, and 5 GHz) and in the licensed 4.9 GHz band that is reserved for public safety applications. PtMP technologies are most widely used in many BWA applications such as wireless Internet service providers—to bring Internet service to homes and businesses in areas where DSL or higher-speed broadband services are not offered by wireline carriers and Smart grid applications—and utilities use them to carry SCADA traffic to monitoring RTU stations.

14.3.2.16 Satellite Broadband

Satellites can be geostationary Earth orbit (GEO), medium Earth orbit (MEO), or low Earth orbit (LEO). Examples for GEO systems are HughesNet and ViaSat, whereas O3b Satellite Constellation is a MEO system and Globalstar and Iridium are LEO systems. Since total round-trip delay (0.75 to 1.25 s) in the GEO system can be relatively long, many applications including remote control of equipment that require a real-time response may be impracticable. In the case of MEO and LEO systems, they offer lower latencies, making the above applications feasible. Satellites can provide fixed, portable, and mobile access and may be the factual choice for remote areas. Data rates range from 2 kbit/s to 1 Gbit/s downstream and from 2 kbit/s to 10 Mbit/s upstream. Satellite communication typically requires a clear line of sight; is adversely affected by moisture, rain, and snow; and may require a fairly large, carefully aimed, directional antenna.

14.4 SMART GRID NETWORKING STANDARDS

In this section, we present Smart grid networking standards such as SCADA standards, OSGP (Open Smart Grid Protocol), and IP (Internet Protocol) Suite for Smart grid.

14.4.1 SCADA STANDARDS

The SCADA is the centralized system of remote control and telemetry used to monitor and control entire sites or complexes of systems spread out over large areas. Most control actions are performed automatically by RTUs or by PLCs (programmable logic controllers). Host control functions are usually restricted to basic overriding or supervisory-level intervention. The modern substation area systems utilize SCADA standards such as Modbus, IEC 61850, and DNP3 (Distributed Network Protocol). These SCADA protocols are designed to exchange information between the master station and the RTU stations. These communication protocols are standardized and recognized by all major SCADA vendors. Many of these protocols now contain extensions to operate over TCP/IP.

14.4.1.1 Modbus

Modbus is a serial communications protocol announced in 1979 for use with its PLCs. Because of its simplicity and robustness, it has since become a de facto standard communications protocol and is now among the most commonly available means of connecting industrial electronic devices. The main motivations for using Modbus in the industrial environment are as follows: it has been developed with industrial specific applications; it is openly published; it is easy to deploy and maintain; and it moves raw bits or words without placing many restrictions on vendors.

Modbus allows for communication between many (approximately 240) devices connected to the same network, for example, a system that measures temperature and humidity and communicates the results to a computer. Modbus is often used to connect a supervisory computer with RTU in SCADA systems.

14.4.1.2   IEC 61850

IEC 61850 is a standard for design of substation automation and a part of the IEC TC57 (International Electrotechnical's Commission Technical Committee 57) reference architecture for electric power systems. The abstract data models defined in IEC 61850 can be mapped to a number of protocols. Current mappings in the standard are to MMS (manufacturing message specification), GOOSE, SMV, and probably web services. These protocols can run over TCP/IP networks or substation LANs using high-speed switched Ethernet to obtain the necessary response times below 4 ms for protective relaying. It is used for protective relaying, substation automation, distribution automation, power quality, distributed energy resources, substation to control center, and other power industry operational functions.

14.4.1.3   Distributed Network Protocol

DNP3 is a set of communications protocols used between components in process automation systems. Their main uses are in SCADA systems for utilities such as electric and water companies. DNP3 was developed for communications between various types of data acquisition and control equipment. DNP3 is one of the protocols that define communications between SCADA master stations (i.e., control centers), RTUs, and IEDs in a SCADA network. It is primarily used for communications between a master station and RTUs or IEDs. The ICCP (Inter-Control Center Communications Protocol), which is a part of IEC 60870-6, is used for inter-master station communications. IEEE has also adopted DNP3 as IEEE Standard 1815-2010 [23], which soon will be upgraded to IEEE 1815-2012.

14.4.2   Open Smart Grid Protocol

The OSGP is a family of specifications published by the European Telecommunications Standards Institute used in conjunction with the ISO/IEC 14908 control networking standard for Smart grid applications [24]. OSGP is optimized to provide reliable and efficient delivery of command and control information for smart meters, direct load control modules, solar panels, gateways, and other Smart grid devices. With over 3.5 million OSGP-based smart meters and devices deployed worldwide, it is one of the most widely used smart meter and Smart grid device networking standards.

14.4.3   IP Suite

The Smart grid begins with a network infrastructure, overlaid atop the existing power grid and enabling data-networked devices to communicate. These devices will deliver a growing range of applications (e.g., smart metering, distribution automation), and they will communicate over a variety of physical media at the utility distribution level. The utilities and their customers benefit from Smart grid network deployments in which devices and media share an open, common, and interoperable language, rather than operating through closed, proprietary, and independent systems. For a diversity of reasons, the Smart grid should embrace IP [25].

The IP suite is expressed as a set of network and transport data message protocols, using IP packets, and a set of routing, IP address mapping and device management protocols. The IP suite is becoming an appealing solution for interoperable end-to-end Smart grid networks; thus, it can be used for various emerging Smart grid applications. This IP suite enables end-to-end Smart grid applications to communicate over a set of interconnected network segments, using various networking technologies.

The IP is still in its early stage in Smart grid communications networks. The main challenge for adopting the IP suite in Smart grid networks is the coexistence and harmonization of the IP suite with existing Smart grid/smart meter-specific protocol stacks, such as the ones developed by ANSI C12.22 and others (IEC 61850, DNP3, etc.), which are specific to the utility industries. Furthermore, HAN standards (ZigBee, Z-Wave, HomePlug, 6LoWPAN) have defined or are in the process of defining the IP layer in their protocol stack. Consequently, the ultimate challenge remains on the choices to fully extend the IP as a unifying protocol layer all the way to the NAN/FAN and AMI networks and its end devices.

14.5 CHALLENGES AND RESEARCH OPPORTUNITIES

In this section, we present challenges and issues in Smart grid networking as well as future research works and opportunities.

As communications and networking technologies are considered as one of the indispensable enabling components of Smart grids, there are numerous challenges that must be focused on to have effusively robust, secure, and effectual Smart grid networks. The emerging networks that enable the Smart grid will be essential to providing comprehensive improvement on existing legacy infrastructure in terms of scalability, reliability, interoperability, and security [26].

14.5.1 SCALABILITY AND RELIABILITY

Smart grid networks will be faced with considerable scalability and reliability issues as they connect hundreds of millions of smart meters.

The Smart grid communications network should offer an indispensable level of reliability to maintain uninterrupted operation for various traffics. A single failure in the network should not affect the entire service. Service interruptions in power grid environments could be power outages, overload conditions, loss of communication links, and so on. Mission-critical Smart grid communications networks must be fault tolerant and self-healing in order to provide exceptional system-level reliability [6,27].

Since a large number of smart meters, data collectors, and other sensing nodes will be encompassed in the Smart grid communications network, Smart grid should be scalable to facilitate the appropriate operation of the electric power grid. Consequently, Smart grid should handle scalability with the integration of advanced web services and reliable protocols with superior functionalities, such as self-configuration and self-healing.

14.5.2 STANDARDS INTEROPERABILITY

A fundamental feature of Smart grids is the interconnection of a possibly large number of different power generating sources, energy distribution networks, and consumers. Thus, interoperability issues in communication capabilities between the networks that connect generation, transmission, and distribution facilities will become crucial.

Interoperability addresses the open architecture of technologies and their software systems to allow their interaction with other systems and technologies. To realize Smart grid capabilities, technology deployments must connect large numbers of smart devices and systems involving hardware and software. Interoperability is defined as the capability of two or more networks, systems, devices, applications, or components to share and readily use information securely and effectively with little or no inconvenience to the user.

The standardization of Smart grid communications network is to make interfaces, messages, and workflows interoperable. Instead of focusing on a particular technology, it is more important to reach an agreement on usage and interpretation of interfaces and messages that can seamlessly bridge different standards, protocols, or technologies. In this regard, the main aim of communication standardization for the Smart grid network is to ensure interoperability between different system components rather than defining these components. Thus, for proper deployments of Smart grid technologies, interoperability is critical to the success of a robust and sustainable grid architecture [28]. Several standards bodies such as the NIST and the IEC are addressing these interoperability issues for the Smart grid [29].

14.5.3 SECURITY

Security is an acute concern in Smart grid networks. Since these networks become more open, the large potential challenge is that of securing these networks properly.

In Smart grid cyber security, even though only a fraction of control systems are connected to the Internet, malware infections such as viruses and worms can penetrate the network. Secure information storage and transportation are particularly vital for the power utilities, especially for billing purposes and grid control. Thus, both control flows and information flows pose a number of challenging privacy and security problems. Significant privacy concerns associated with the uploading of user details from the consumer to the utilities may occur. Furthermore, an intruder can send control signals that could create destabilizing demand patterns and potentially initiate a power outage.

Smart grid security needs to be implemented systematically for monitoring and protection. To mitigate cyber attacks, effectual security mechanisms should be established and standardization efforts regarding the security of the power grid should be made [30]. Application security combined with user or device authentication and role-based network access is essential in shared networking environments.

14.6 CONCLUSIONS

The Smart grid can be visualized as a modern, resilient, and reliable electric grid that is cost-effective and that provides a greener environmental frontier. By providing the means to address intelligently and efficiently energy generation and consumption, it will have a vital role in the transition to a low-carbon economy. Energy efficiency, demand response, and direct load control are key components in realizing the value of a Smart grid network deployment, which will allow energy and cost savings with real-time meter data, dynamic pricing, demand response and load control, intelligent appliances, and distributed energy resources. Many energy providers and utilities all over the world have adopted the concept of smart energy where smart energy applications are facilitated by two-way communications in the Smart grid network. The main characteristics of the Smart grid communications networks are as follows: integrated approach with high performance, high reliability, scalability, ubiquity, and security. The communication networks are responsible for gathering and routing data, monitoring all nodes, and acting upon the data received. In this chapter, we examined the Smart grid communications network and networking technologies. We also emphasized the challenges and issues associated with Smart grid networking as well as research works and opportunities in terms of scalability, reliability, interoperability, and security.

AUTHOR BIOGRAPHIES

Binod Vaidya has been a postdoctoral fellow in the School of Electrical Engineering and Computer Science, University of Ottawa, Canada, since April 2010. Prior to joining the University of Ottawa, he worked as a postdoctoral researcher at Chosun University, South Korea (2007–2008), a research associate at Gwangju Institute of Science and Technology, South Korea (2008–2009), and a researcher at Instituto de Telecomunicações, Portugal (2009–2010). He also worked as a lecturer at the Institute of Engineering, Tribhuvan University, Nepal, for more than 15 years. He has authored or co-authored over 70 papers in international journals, books, conferences, and symposia. He has served as a guest editor for several reputed journals as well as an editorial board member for several international journals.

Dimitrios Makrakis is a professor in the School of Electrical Engineering and Computer Science, University of Ottawa, and the director of the Broadband Wireless and Internetworking Research Laboratory. Prior to joining the University of Ottawa, he was an assistant and later associate professor in the Department of Electrical and Computer Engineering, University of Western Ontario, and the director of the Advanced Communications Engineering Centre (a research facility established in partnership with Bay Networks, presently Nortel, and Bell Canada). Before starting his academic carrier, he worked for the Canadian Government and Canadian industry. He has received the Premiers Research Excellence Award. He has authored or co-authored over 200 papers in international journals, conferences, and books.

Hussein T. Mouftah received his BSc and MSc from Alexandria University, Egypt, in 1969 and 1972 respectively, and his PhD from Laval University, Quebec, Canada, in 1975. He joined the School of Information Technology and Engineering (now School of Electrical Engineering and Computer Science) of the University of Ottawa in 2002 as a Tier 1 Canada Research Chair Professor, where he became a distinguished university professor in 2006. He has been with the ECE Department at Queen's University (1979–2002), where he was prior to his departure a full professor and the department associate head. He has six years of industrial experience mainly at Bell Northern Research of Ottawa (then known as Nortel Networks). He served as editor-in-chief of *IEEE Communications Magazine* (1995–1997) and IEEE ComSoc Director of Magazines (1998–1999), chair of the awards committee (2002–2003), director of education (2006–2007), and member of the Board of Governors (1997–1999 and 2006–2007). He has been a Distinguished Speaker of the IEEE Communications Society (2000–2007). He is the author or co-author of 8 books, 59 book chapters and more than 1200 technical papers, 12 patents, and 140 industrial reports. He is the joint holder of 14 Best Paper and/or Outstanding Paper Awards. He has received numerous prestigious awards, such as the 2007 Royal Society of Canada Thomas W. Eadie Medal, the 2007–2008 University of Ottawa Award for Excellence in Research, the 2008 ORION Leadership Award of Merit, the 2006 IEEE Canada McNaughton Gold Medal, the 2006 EIC Julian Smith Medal, the 2004 IEEE ComSoc Edwin Howard Armstrong Achievement Award, the 2004 George S. Glinski Award for Excellence in Research of the U of O Faculty of Engineering, the 1989 Engineering Medal for Research and Development of the Association of Professional Engineers of Ontario, and the Ontario Distinguished Researcher Award of the Ontario Innovation Trust. Dr. Mouftah is a Fellow of the IEEE (1990), the Canadian Academy of Engineering (2003), the Engineering Institute of Canada (2005), and the Royal Society of Canada RSC Academy of Science (2008).

REFERENCES

1. J. Gao, Y. Xiao, J. Liu, W. Liang, and C.L. Philip Chen, A survey of communication/networking in smart grids, *Future Generation Computer Systems*, vol. 28, pp. 391–404, 2012.
2. W. Wang, Y. Xu, and M. Khanna, A survey on the communication architectures in smart grid, *Computer Networks*, vol. 55, pp. 3604–3629, 2012.
3. T. Sauter and M. Lobashov, End-to-end communication architecture for smart grids, *IEEE Transactions on Industrial Electronics*, vol. 58, pp. 1218–1228, 2011.
4. C. Lima, An architecture for the smart grid, In *IEEE P2030 Smart Grid Communications Architecture SG1 ETSI Workshop*. France: IEEE Standard Association, pp. 1–27, Apr 2011.
5. A. Hardy, F. Bouhafs, and M, Merabti, A survey of communication and sensing for energy management of appliances, *International Journal of Advanced Engineering Sciences and Technologies*, vol. 3, no. 2, pp. 061–077, 2011.
6. V.C. Gungor, D. Sahin, T. Kocak, S. Ergut, C. Buccella, C. Cecati, and G.P. Hancke, Smart grid technologies: Communication technologies and standards, *IEEE Transactions on Industrial Informatics*, vol. 7, no. 4, pp. 529–539, 2011.

7. T. Khalifa, K. Naik, and A. Nayak, A survey of communication protocols for automatic meter reading applications, *IEEE Communications Surveys and Tutorials*, vol. 13, no. 2, pp. 168–182, 2011.

8. V.C. Gungor and F.C. Lambert, A survey on communication networks for electric system automation, *Computer Networks*, vol. 50, pp. 877–897, 2006.

9. C. Xiaoronga, W. Yingb, and N. Yangdanc, The study on the communication network of wide area measurement system in electricity grid, *Physics Procedia*, vol. 25, pp. 1708–1714, 2012.

10. M.D. Darlene, Home automation system using power line communication, Thesis, University of Malaysia Pahang, 2008.

11. S. Galli, A. Scaglione, and Z. Wang, For the grid and through the grid: the role of power line communications in the smart grid, In *Proceedings of the IEEE*, vol. 99, no. 6, pp. 998–1027, Jun 2011.

12. N. Ginot, M.A. Mannah, C. Batard, and M. Machmoum, Application of power line communication for data transmission over PWM network, *IEEE Transactions on Smart Grid*, vol. 1, no. 2, pp. 178–185, 2010.

13. Insteon: the details, http://www.insteon.net/pdf/insteondetails.pdf, Accessed 12 Jul 2012.

14. Introduction to the LonWorks system, http://www.echelon.com/support/documentation/manuals/general/078-0183-01A.pdf, Accessed 12 Jul 2012.

15. HomePlug AV white paper, http://www.homeplug.org/tech/whitepapers/HPAV-White-Paper_050818.pdf, Accessed 15 Jul 2012.

16. HomePlug AV2 Technology white paper, http://www.homeplug.org/tech/whitepapers/HomePlug_AV2_White_Paper_v1.0.pdf, Accessed 15 Jul 2012.

17. V. Aravinthan, B. Karimi, V. Namboodiri, and W. Jewell, Wireless communication for smart grid applications at distribution level—Feasibility and requirements, In *Proceedings of 2011 IEEE Power and Energy Society General Meeting*, pp.1–8, 2011.

18. P.P. Parikh, M.G. Kanabar, and T.S. Sidhu, Opportunities and challenges of wireless communication technologies for smart grid applications, In *Proceedings of 2010 IEEE Power and Energy Society General Meeting*, pp. 1–7, 2010.

19. B.E. Bilgin and V.C. Gungor, Performance evaluations of ZigBee in different smart grid environments, *Computer Networks*, vol. 56, pp. 2196–2205, 2012.

20. M. Honkanen, A. Lappetelainen, and K. Kivekas, Low end extension for Bluetooth, In *Proceedings of IEEE Radio and Wireless Conference 2004*, Sep 2004.

21. IEEE 802.16m-08/003r7, IEEE 802.16m system description document, Jan. 2009, http://ieee802.org/16/tgm/index.html, Accessed 25 Jul 2012.

22. C. Peng, W. Li, Z. Bin, and W. Shihua, Feasibility study of applying LTE to smart grid, In *Proceedings of 2011 IEEE First International Workshop on Smart Grid Modeling and Simulation (SGMS)*, pp. 108–113, 2011.

23. IEEE 1815-2010 Standard for Electric Power Systems Communications—Distributed Network Protocol (DNP3), IEEE, Jul 2010.

24. Open Smart Grid Protocol, http://www.echelon.com/technology/osgp/, Accessed 12 Aug 2012.

25. F. Baker and D. Meyer, Internet protocols for the smart grid, IETF RFC 6272, Internet Engineering Task Force, Internet Society, Jun 2011.

26. Z. Fan, P. Kulkarni, S. Gormus, C. Efthymiou, G. Kalogridis, M. Sooriyabandara, Z. Zhu, S. Lambotharan, and W.H. Chin, Smart grid communications: Overview of research challenges, solutions, and standardization activities, *IEEE Communications Surveys and Tutorials*, vol. 15, no. 1, pp. 21–38, 2013.

27. Y.J. Kim, J. Lee, G. Atkinson, H. Kim, and M. Thottan, SeDAX: A scalable, resilient, and secure platform for smart grid communications, *IEEE Journal on Selected Areas in Communications*, vol. 30, no. 6, pp. 1119–1136, Jul 2012.

28. G.J. FitzPatrick and D.A. Wollman, NIST interoperability framework and action plans, In *The 2010 IEEE Power and Energy Society General Meeting*, pp. 1–4, 2010.

29. P. Chatzimisios, D.G. Stratogiannis, G.I. Tsiropoulos, and G. Stavrou, A survey on smart grid communications: From an architecture overview to standardization activities, In *Handbook of Green Information and Communication Systems* (Eds. M.S. Obaidat, A. Anpalagan, and I. Woungang), Elsevier Inc., Academic Press, 2012.

30. *Guidelines for Smart Grid Cyber Security*, NISTIR 7628, vol. 1–3, NIST, US Dept. of Commerce, Aug 2010.

15 Towards Green Networks Using Optimized Network Device Configurations

Sebastián Andrade-Morelli, Sandra Sendra Compte, Emilio Granell Romero, and Jaime Lloret

CONTENTS

15.1 INTRODUCTION

The popularization of the Internet of things, the increase in mobile devices, the advancement of new technologies, and the implementation of new network services like IP telephony and IPTV [1] are generating a significant increase of network infrastructures such as routers, switches, hubs, and access points (APs). This increase in network infrastructure means an increase in energy consumption in networks [2], which becomes an environmental problem that needs to be addressed, and at the same time an economical problem because the price of energy is significantly increasing. For these economic and environmental reasons, the reduction of energy consumption is a critical issue for companies working in information and telecommunication technologies, especially when they are using data centers [3].

We should note that the network devices work better within a temperature range. The temperature range specified by the manufacturer is usually between 0°C and 40°C [4,5], but this value varies depending on the device and the task that is running. This implies the need to install a good cooling system in those rooms that host these devices, to ensure its proper functioning at peak performance. The more devices hosted in the room, the more cooling power the system must offer, which means more energy consumption in the network [6].

Currently, new energy solutions based on the introduction of IP transmission protocol to all areas of the network are being implemented. The migration of these systems to the next-generation networks saves between 30% and 40% of energy consumption [7]. Furthermore, the incorporation of routing systems and IP switching has improved the energy efficiency of data and voice transmission, reducing the requirements of the network capacity between 60% and 70% [7].

With wireless networks, it is possible to build low-cost telecommunication networks, allowing access to the Internet in areas of difficult access, such as rural areas [8]. A solution to develop networks in places where wireless APs cannot be connected directly to the grid is to place the APs connected to solar panels and batteries [9].

Energy-saving techniques will increase the efficiency of the network and thus will reduce the economic costs and environmental impact. If the network is designed with energy-efficient devices and optimal configurations, less electrical energy must be supplied from the main network distribution center.

In this chapter, we present a study on the power consumption of principal network devices used to implement communication network infrastructures. The conclusions drawn from this chapter determine the optimum parameters to reduce the energy consumption in this kind of networks.

First of all, we will study the energy consumption in routers. We will consider a complex topology to study performance in terms of energy consumption in different routing protocols such as static and dynamic, using different brands and models of routers. We will also do a comparative study of the temperature reaching the router at full capacity, because the heat also consumes energy.

Second, we will study the energy consumption of different brands and models of switches in different situations and settings. Our aim is to determine which devices are most appropriate when we install a communication network. We will consider everything that directly or indirectly involves an increase in the power consumption, such as the number of ports required, the amount of information transmitted, and the possibility of evolution of the network.

Third, we will study the energy consumption in wireless APs. In this case, we will base the study on several situations using different wireless protocols (IEEE 802. 11 a/b/g/n [10]), depending on the brand, the model, and its configuration.

Finally, with the aim of determining a plan of efficient energy distribution in intelligent networks, we will consider a topology where we will make a comparison between energy consumption using devices and configurations that are energetically efficient and using deficient devices in terms of energy consumption.

The rest of the chapter is structured as follows. Section 15.2 presents some related work and the state of the art regarding the estimation of consumption and energy saving in network devices. It is also in this section where several studies about energy consumption are presented, as well as papers describing tools to estimate the energy consumption of the devices and others where authors propose a series of techniques that allow reducing energy consumption in communications networks and devices. For our tests, we have mounted several network topologies, composed of different network devices. The main features of the devices and the different topologies used in our tests are shown in Sections 15.3 and 15.4, respectively. The measurements of our tests are presented in Section 15.5. In Section 15.6, a specific topology is proposed to compare the energy consumption between using energy-efficient devices and using energy-deficient devices. Finally, conclusions and future work are presented in Section 15.7.

15.2 RELATED WORK

The problem of saving energy in communication networks is being widely investigated. There are different kinds of studies and works published previously. This section describes some works that are focused on reducing the energy consumption of communication networks, some of the main projects related to the application and development of techniques for saving energy on the networks, and some studies, surveys, and comparisons of energy-saving techniques that can be applied in sensor and ad hoc networks. Finally, this section presents some tools employed for the estimation of energy consumption in network devices.

15.2.1 Energy-Saving Techniques

In Ref. [11], the authors present the design and evaluation of two types of power management, a mechanism that allows sleeping the network devices during times of inactivity and another that allows adapting the speed of network operation to the offered workload. The first reduces the energy consumed in the absence of packages, and the second reduces the energy consumed when the device is actively processing the packages.

With the development of communication networks, high-speed Internet traffic has grown rapidly. Gao et al. [12] present a new method for reducing power consumption in communication networks, which dynamically modifies the weights of groups and links the traffic on certain nodes in the network, allowing disconnected nodes without workload and, therefore, saving energy.

There are techniques that use artificial intelligence to reduce consumption in an autonomous way. Subrata et al. [13] present a solution based on algorithms to reduce the consumption of energy in network environments using game theory. This solution keeps a certain quality of service while minimizing energy consumption. Another example of intelligent communication network is the work presented in Ref. [6], where the authors provide an algorithm capable of determining possible disconnects that can be made with the aim of reducing energy consumption without losing network connectivity. Similarly, Hermenier et al. [14] propose a strategy of concentration of workload to turn off nodes that are not in use, reducing significantly the energy consumption with minimal impact on performance.

15.2.2 Studies of Energy Consumption

There is also great interest in the reduction of power consumption in wireless networks. Feeney and Nilsson [15] present a study about energy consumption in wireless networks, in particular ad hoc networks. In their study, they perform a series of experiments to determine the consumption of wireless interfaces. And we could determine that the most important factors that influence the consumption of these devices are the relative proportions of broadcast and point-to-point traffic, packet size, and reliance on promiscuous mode operation.

A model for evaluating the energy consumption behavior of a mobile ad hoc network is described in Ref. [16], particularly to examine the energy consumption of two well-known MANET routing protocols. In this work, Feeney determines that energy and bandwidth are substantively different metrics and that resource utilization in MANET routing protocols is not fully addressed by bandwidth-centric analysis.

On the same line, the work of Sendra et al. [17] shows several techniques to reduce energy consumption in wireless networks of sensors, indicating the relationship between energy consumption and the amount of information transmitted by the hardware used.

With respect to consumption in wireless networks, the paper of Andrade-Morelli et al. [18] presents a study on energy consumption in wireless APs in different situations and wireless protocols. In this paper, they have shown that the energy consumption increases when only air–air links are used and when heavy files are transmitted. Also, these increases appear when other features such as MAC filters are used.

Switches and routers are other important network devices. In Ref. [19], Granell et al. present a behavior study of switches in terms of energy consumption. In their study, they did different tests. They measured the consumption using point-to-point communications with a PC emitting Internet Control Messages Protocol (ICMP) in broadcast with and without virtual local area networks (VLANs). With these tests, they could prove that the number of ports of the switches affects its power consumption. There are no significant differences in communications point to point with or without the use of VLANs. However, the consumption increases considerably in broadcast communications.

Routers are important devices in any network and the network devices that consume more energy [20]. For this reason, it is important to make the right choice of models and brands and their configuration. There are routers that consume less than others, and can be equally effective, if not better.

Baliga et al. [21] carried out a study on energy consumption where they proposed a model of optical networks to estimate the energy consumption in networks and the Internet, now and in the future. According to them, 0.4% of the energy consumption in the world is via the Internet, and this percentage will increase gradually in the future.

15.2.3　TOOLS TO MONITOR AND ESTIMATE CONSUMPTION

Not only is the physical deployment of networks a determining factor in energy consumption, configurations and the device software also influence the energy consumption. Barbancho et al. [22] present a smart computer that allows energy saving, using the intelligent routing. These tools are intended to control the consumption of wireless networks.

Network equipment manufacturers offer a series of software tools that allow estimating the energy consumption of the devices such as routers, switches, and wireless APs. The report published by Cisco Systems [23] is a comparison of applications that estimate the energy consumption of the devices. However, the document indicates that the results provided by this software are not reliable (compare their values with those gathered from the actual power consumption). Therefore, we have to treat them as tools of approximation, which can help us to make an initial design of our system, but which can never replace actual measurements on real devices.

With our study, we want to complement the work of the scientific community in terms of reducing energy consumption in communications networks, since we believe that the choice of the devices at the time of the network's design is crucial to achieving and maintaining green communication networks and thus better planning of smart grid networks.

15.3　HARDWARE DESCRIPTION

In this section we will see the technical characteristics of the selected devices to perform our study. The energy consumption was measured with an electronic device called "Kill a Watt." This device is capable of measuring voltage, power, and current with a measurement error of 1%.

First of all, Table 15.1 presents a list with the routers used in our study and their fundamental hardware characteristics. As we can see, the router Allied AR410 is the router that consumes least

TABLE 15.1

Hardware Features for Routers Used in the Study

	Allied AR410	Cisco 1841	Cisco 2620 [24]	Cisco 1700 [25]	3Com Remote Access 531 [26]
Max. data transfer rate (Gbps)	0.1	0.1	0.1	0.1	0.01
Max. power output (W)	17.6	50	75	50	50
Operating temperature (°C)	0–40	0–40	0–40	0–40	0–40
Internal memory (kB)	16	191	32	32	2048
Flash memory (kB)	8192	62,720	8192	8192	2048
Processor (one processor)	Motorola MPC860 50 MHz RISC	Motorola MPC860 50 MHz RISC	Motorola MPC860 50 MHz RISC	Motorola MPC860 50 MHz RISC	Motorola MC68360 25MHz
Routing protocol	RIP and RIP v2, OSPF	RIP and RIP v2, OSPF	RIP and RIP v2, OSPF	RIP and RIP v2, OSPF	—
Data transmission protocol	Eth., Fa. Eth.	Eth., Fa. Eth., serial	Eth., Fa. Eth., Serial	Eth., Fa. Eth., Serial	Eth., Serial, ISDN

Note: Eth. = Ethernet; Fa. Eth. = Fast Ethernet.

energy, a fact demonstrated by our measures. The router Cisco 1841 has more flash memory (62,720 kB), which will influence its energy consumption. And with respect to the internal memory, the router 3Com has the maximum followed by the router Cisco 1841. The other features are similar among all models except for the router 3Com in which it is not possible to use dynamic routing protocols.

Second, the list of the switches used in our study and their main features are shown in Table 15.2. Among these features, we highlight the differences in energy consumption between these devices, where switches Allied AT8124XL [27] and Cisco 3560 [28] present the maximum with 100 W. There are no considerable differences in terms of operating temperature. Regarding flash memory

TABLE 15.2

Hardware Features for Switches Used in the Study

	Allied Telesis		Cisco Catalyst				3Com	
	AT8024 [30]	AT8124XL	2950T-12 [31]	2950T-24	3560-8PC	3560PSE-24	SuperStack II Switch 610 [32]	SuperStack II Hub
Max. power consumption (W)	30	100	30	30	100	100	60	30
Operating temperature (°C)	0–40	0–40	0–45	0–45	0–45	0–45	0–50	0–50
Flash memory (kB)	6144	6144	8192	8192	32,768	32,768	—	—
RAM memory (MB)	16	32	16	16	128	128	—	—
Data transport protocol	Eth., Fa. Eth.	Eth., Fa. Eth.	Eth., Fa. Eth.	Eth., Fa. Eth.	Eth., Fa. Eth.	Eth., Fa. Eth.	Eth., Fa. Eth.	Eth.

Note: Eth. = Ethernet; Fa. Eth. = Fast Ethernet.

TABLE 15.3

Hardware Features for Wireless APs Used in the Study

	Cisco Systems			D-Link	Avaya	Ovislink
	AIR-AP1131 AG-E-K9	Linksys WRT54GL [35]	Linksys WRT320N-EZ	DWL-2000Ap+ [36]	AP-I [37]	WX-1590 [38]
Frequency band (Ghz)	2.4	2.4	2.4/5.0	2.4	2.4	2.4
Operating temperature (°C)	0–40	0–40	0–40	0–55	0–40	0–55
Internal memory (MB)	32	16	32	16	16	16
Flash memory (MB)	16	4	8	4	4	4
Max. data transfer rate (Mbps)	108	54	300	54	11	11
Wireless protocol	IEEE802.11 a/b/g	IEEE802.11 b/g	IEEE802.11 b/g/a/n	IEEE802.1 1b/g	IEEE802.11 b	IEEE802.11 b
Data transmission protocol	Fa. Eth.	Eth., Fa. Eth.	Eth., Fa. Eth.	Eth., Fa. Eth.	Eth.	Eth.

Note: Eth. = Ethernet; Fa. Eth. = Fast Ethernet.

and RAM, Cisco 3560 has the maximum values at 32,758 kb and 128 kb, respectively, a fact that will also influence the energy consumption. All models support the Ethernet and Fast Ethernet Protocol, except for the Hub 3Com [29], which only supports Ethernet links.

Finally, Table 15.3 brings together the list of wireless APs subject to study and its hardware features. All models work in the band 2.4 GHz except for Cisco WRT320N-EZ [33], which is dual band, working on 2.4 GHz and 5.0 GHz. Cisco WRT320N-EZ is the model that supports more wireless protocols (IEEE 802.11a/b/g/n). We observed significant differences in data transfer rates, where the wireless AP Cisco WRT320N-EZ has the highest transfer rate with 300 mbps and wireless APs Avaya and Ovislink have the lowest transfer rate with 11 mbps. The wireless AP Cisco AIR-AP1121AG-EK9 [34] has more flash memory than the rest of the models (16 MB), but it has the same internal memory as the wireless AP Cisco WRT320N-EZ (32 MB).

15.4   TEST BENCH

For our study, we have carried out a series of tests for each of the devices discussed in the previous section. These tests have allowed us to study the energy consumption to identify the models that consume more energy and the influence on the energy consumption of the configuration parameters. In this section, the topologies used for our studies with routers, switches, and wireless APs are presented.

15.4.1   ROUTERS

For the realization of measures on routers, we have used a topology consisting of four routers, where the link between two of them is carried out by a serial connection (the interfaces of these routers

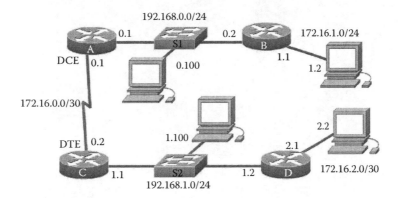

192.168.0.0/24

DCE

172.16.0.0/30

DTE

192.168.1.0/24

172.16.1.0/24

172.16.2.0/30

FIGURE 15.1 Topology used to study the routers.

have been configured as DTE-DCE). In addition, we have used two switches to connect two computers to the network. Moreover, five different subnets are configured with the aim that the routers forwarded the information. Figure 15.1 shows the proposed topology for the measurement of energy consumption in routers. We can see that there are five different subnets separated by routers that make up the network. Also, we highlight the fact that there is a DTE-DCE link, which, as we will see in the next section, implies more consumption.

15.4.2 SWITCHES

To perform actions on the switches, we used a network consisting of eight computers and a switch as shown in Figure 15.2. On this topology, we have measured the switch energy consumption in various situations: in an inactive state, in point-to-point communications between two PCs, in communication within the same VLAN, and when there are several PCs broadcasting packets at the same time.

15.4.3 WIRELESS APS

In the study of energy consumption on wireless APs, we have used two topologies formed by a wireless AP and two PCs connected to it by two types of physical interfaces as shown in Figure 15.3. In both topologies, the wireless AP is responsible for communication between two hosts. In this part

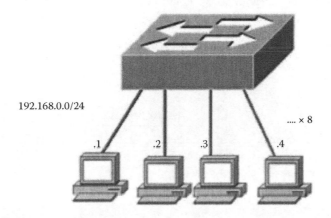

192.168.0.0/24

.... × 8

.1   .2   .3   .4

FIGURE 15.2 Topology used to study the switches.

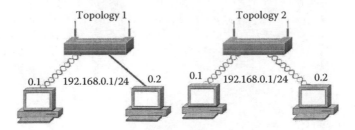

FIGURE 15.3 Topologies used to study the wireless APs.

of the study, we have tried to check the impact on energy consumption, when there is a change of physical environment (air–cable and air–air). In the first case (Topology 1), a mixed network formed by a wireless link and a wired link is used. In the second case (Topology 2), the communication is completely wireless.

15.5   RESULTS OF MEASUREMENTS

In this section, the results of the measurements made on the topologies described in the previous section will be considered. With such tests, we know which models and configurations involve higher energy consumption. We will discuss it with the objective of determining the guidelines to design energetically efficient communications networks.

15.5.1   ROUTERS

In this part of the study aimed at routers, we considered different routing protocols. This will allow us to check whether there are variations in consumption depending on the implemented protocol. Such variations will be reflected not only in the energy consumption generated but also in variations of temperature.

Dynamic routing protocols allow routers to change information in order to modify and update the routing tables dynamically, whereas static routing protocols do not; it is for this reason that static routing protocols are usually used for small networks. In our tests, we have used the two dynamic routing protocols most used [39]: Routing Information Protocol V.2 (RIP V.2) and Open Shortest Path First (OSPF). In each case, we have measured the energy consumption of each network device, when devices are in idle state and while it is executing the routing protocol.

Table 15.4 and Figure 15.4 show the average value of energy consumption in routers. In this case, a combination of RIP protocol with access control lists (ACLs) has been added. An ACL specifies which users or system processes are granted access to objects, as well as what operations are allowed on the system.

TABLE 15.4

Measurement Results for Routers

	Average Power Consumption (W)				
	Cisco 1841	**Cisco 2620**	**Allied AR410**	**Cisco 1700**	**3Com**
Idle mode	16.5	13.85	6.5	9.45	9.05
RIP protocol	17.15	14.38	6.73	9.95	—
OSPF protocol	17.15	14.37	6.65	9.8	—
Static protocol	16.92	14.2	6.6	9.45	11.65
RIP protocol with ACLs	17.30	14.50	—	10.1	—

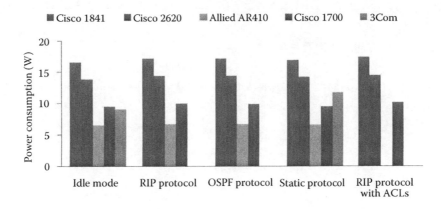

FIGURE 15.4 Power consumption for routers.

15.5.1.1 Idle Mode

In idle mode, the energy consumption depends exclusively on the router model. As shown in Table 15.4, the differences between the different models are important; thus, one of the main aspects to take into account when choosing a router will be the model because it is the router that marks the minimum consumption; from here, the consumption will depend on the configurations and processes running on them. In this case, the router Allied AR410 is the one that consumes least energy.

15.5.1.2 Rip Protocol

This dynamic routing protocol is based on vector distance and determines the routes depending on the number of hops; it is a basic dynamic routing protocol. When this protocol is running on the devices, as we can see in Table 15.4, it causes an increase in the energy consumption between 3% and 5% depending on the model. But the higher difference is shown in Cisco 1841. This consumption is slightly higher, approximately 2% more than that generated by the OSPF protocol.

15.5.1.3 OSPF Protocol

OSPF is a dynamic routing based on Dijkstra's algorithm protocol. In this case, the energy consumption introduced and measured at idle mode is between 2% and 4%. That is an energy saving compared with RIP V2. As in the previous case, Cisco 1841 is the one that consumes most energy.

15.5.1.4 Static Protocol

Static routing protocol is the routing type that consumes less energy, but as mentioned previously, it is only used for small networks. The energy consumption in this case is approximately 0.4 and 0.1 W, which is between 3% and 4% less than that in the dynamic routing protocols. The router 3Com consumes 84% more energy than the rest.

15.5.1.5 RiP Protocol with ACLs

In the case of Cisco routers, we have added the variation of energy consumption when using ACLs, obtaining a slight increase in the consumption of approximately 0.8%.

It is important to mention that the routers that consume more energy independently of the model are those forming the point-to-point link, the connection DTE-DCE. As the router 3Com only supports static routing protocol, we have used the ISDN link to connect A and C.

TABLE 15.5

Temperature Results for Routers

	Temperature (°C)				
	Cisco 1841	**Cisco 2620**	**Allied AR410**	**Cisco 1700**	**3Com**
Idle mode	32.4	30	27	29.3	30
RIP protocol	33.5	32	31	29.75	—
OSPF protocol	33	32.3	32.2	31.25	—
Static protocol	32.8	31.4	29.5	36.20	31

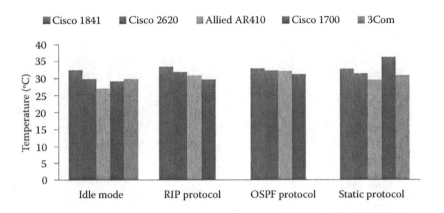

FIGURE 15.5   Temperature in routers.

Last, we highlight, as represented in Figure 15.4, that the router Allied AR410 is making better use of energy in all cases, with a 30% difference with respect to the next router that consumes less (Cisco 1700), which translates into an economic savings of the same amplitude.

Energy consumption is also reflected in the amount of heat that devices dissipate when they are at full performance. Therefore, we did a parallel study on the temperature of the routers to achieve our objective of determining optimal parameters for the design of energy-efficient networks.

The results of the study on temperature are collected in Table 15.5 and are represented in Figure 15.5. We can say that, in general, the temperature in these devices is between 29°C and 33°C, except in the router Cisco 1700 where the maximum temperature reached approximately 37°C.

15.5.2 SWITCHES

As mentioned earlier, another object of study is the energy consumption of switches, subjected to different situations to see possible fluctuations and differences on energy consumption most relevant for our purposes.

As discussed in the previous section, the measurements in these devices were made on a simple topology and in different situations: inactive state, point-to-point communication with and without VLANs, and connecting multiple PCs emitting at the same time.

Table 15.6 and Figure 15.6 show the results of the switches' energy consumption measurements. Next, we present an analysis of the results to find parameters, with the aim of reducing the energy consumption of these devices in the networks.

TABLE 15.6
Measurement Results for Switches

	Power Consumption (W)			
	A	**B**	**C**	**D**
Cisco 2950/24	16.66	17.93	18.03	17.93
Cisco 2950/12	16.03	17.45	17.50	17.53
Cisco 3560/8	15.96	17.67	17.73	17.70
Cisco 3560/24	38.20	39.20	39.30	39.20
Allied AT8024	32.03	34.77	34.80	34.80
Allied AT8124	22.33	24.60	24.67	24.60
3Com SuperStack II	26.27	26.57	26.67	26.63
3Com LinkBuilder FMS II	17.20	18.00	18.10	—

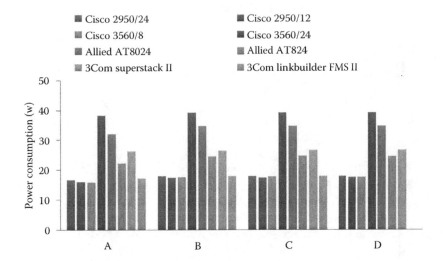

FIGURE 15.6 Power consumption in switches.

15.5.2.1 Devices in Idle Mode

In this case, the energy consumed differs considerably between all devices. Cisco Catalyst 3560PSE-24 consumes more energy than the others; it has a higher processing capacity among the tested devices. We can deduce that switches with a higher number of ports consume more energy. This fact can be seen in the case of Cisco Catalyst 3560PSE where the model with 24 ports consumes 38.2 W while the model with 8 ports consumes 15.96 W. Notably, the energy consumption does not increase proportionally with the number of ports.

15.5.2.2 Point-to-Point Communications

In this second case, we will measure the energy consumed when two computers are connected to the switch and communicating between them. As in the previous case, the switch Cisco Catalyst 3560PSE-24 had a higher energy consumption, while the switch Cisco Catalyst 2950T-12 had a lower energy consumption. Moreover, with respect to the energy consumption in idle mode, we have observed that there is a significant increase in consumption. The higher increase was in the Cisco Catalyst 3560-8PC and Allied AT8124XL, amounting to 10.71% and 10.17%, respectively. The lowest increase was in the Cisco Catalyst 3560PSE-24 and 610 Switch 3Com SuperStack II, amounting to 2.62% and 1.14%, respectively. We have obtained an average increase of 6.79%.

15.5.2.3 Communication of Emission

Another object of measurement was to connect eight PCs to the switch, where one PC sends broadcast messages to the rest continuously. As in the two previous cases, the Cisco Catalyst 3560-8PC had the greatest increase in energy consumption (11.09%), and the 3Com SuperStack II switch 610 had the lowest (1.52%). The average increase in energy consumption for this situation was approximately 7.16%, surpassing 0.37%, the increase in the point-to-point communication; this increase is about the same in all devices.

15.5.2.4 Communication within VLANs

In the latter case, we have measured the energy consumption in the devices when they are using VLANs, which means higher processing in the devices and, therefore, higher consumption. In order to prove the above assumption, we had measured the energy consumption with two PCs on the same VLAN exchanging ICMP echoes and their responses continuously. Table 15.6 shows the values of the measurements obtained. In this case, we have not conducted a study of the model 3Com LinkBuilder FMS II because VLANs cannot be configured in a hub. Figure 15.6 shows the average values of energy consumption. In this case, the energy consumption increase is slightly higher than the point-to-point communication without VLANs.

15.5.3 Wireless APs

Once the wireless APs were configured and the topologies were mounted, we measured the energy consumption in wireless APs. We will determine which wireless AP consumes less energy. For this purpose, we have provided two situations, sending echoes from a PC to another and sending a 2.45 GB file. In each case, we have used all IEEE 802.11 protocols available in each device; we also studied the variation in consumption introducing the MAC filter.

15.5.3.1 Measurements for Cisco Systems Air-AP1131AG-E-K9

The first wireless AP under study was the Cisco-AP1131AG-E-K9. This device supports the protocols IEEE 802.11a, IEEE 802.11b, and IEEE 802.11g and consumes 5.1 W in bootup state and 5.4 W in idle mode. In Tables 15.7 and 15.8, we present the values of energy consumption for topologies 1 and 2, respectively; these values are represented in Figures 15.7 and 15.8. As we can see, the energy

TABLE 15.7

Measurement Results for the Delivery of Echo, Topology 1

	Power Consumption (W)							
	IEEE 802.11a	IEEE 802.11b	IEEE 802.11g	IEEE 802.11n	IEEE 802.11a	IEEE 802.11b	IEEE 802.11g	IEEE 802.11n
	Without MAC Filter				With MAC Filter			
Cisco Systems AIR-AP1131AG-E-K9		10.2	102	—	10.4	10.2	10.3	—
Cisco Linksys WRT54GL	—	6	6	—	—	6.1	6.2	—
Cisco Linksys WRT320N-EZ	5.3	5.3	5.2	5.3	5.3	5.3	5.3	5.3
D-Link DWL-2000Ap+	—	5	5.1	—	—	5.1	5.2	—
Avaya AP-I	—	3.2	—	—	—	3.3	—	—
Ovislink WX-1590	—	5.5	—	—	—	5.4	—	—

TABLE 15.8

Measurement Results for the Delivery of Echo, Topology 2

	Power Consumption (W)							
	IEEE 802.11a	IEEE 802.11b	IEEE 802.11g	IEEE 802.11n	IEEE 802.11a	IEEE 802.11b	IEEE 802.11g	IEEE 802.11n
	Without MAC Filter				With MAC Filter			
Cisco System AIR-AP1131AG-E-K9	10.4	10.3	10.3	—	10.4	10.3	10.4	—
Cisco Linksys WRT54GL	—	6	6.1	—	—	6.1	6.1	—
Cisco Linksys WRT320N-EZ	5.3	5.3	5.3	5.3	5.4	5.3	5.3	5.4
D-Link DWL-2000Ap+	—	5.1	5.1	—	—	5.2	5.2	—
Avaya AP-I	—	3.3	—	—	—	3.4	—	—
Ovislink WX-1590	—	5.5	—	—	—	5.6	—	—

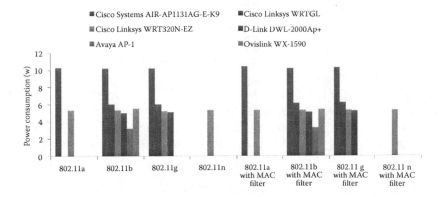

FIGURE 15.7 Measures for the delivery of echo, topology 1.

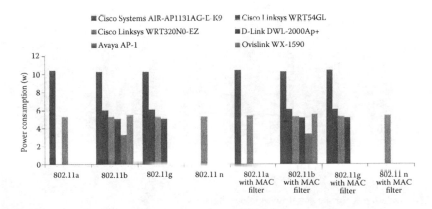

FIGURE 15.8 Measures for the delivery of echo, topology 2.

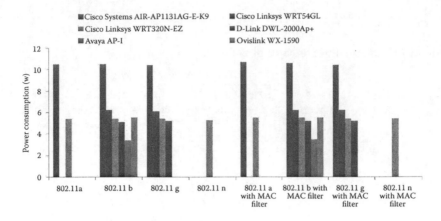

FIGURE 15.9   Measures for file sending, topology 1.

consumed by this device during transmission was always between 10 and 11 W. We observed an increase in energy consumption of 2% in topology 2 in comparison with topology 1. We can also see that protocol 802.11b consumes less energy, sending the file applied to the topology air–cable. As we can see in Figures 15.7 through 15.9, introducing the MAC filter involves a 2% increase in the energy consumption.

15.5.3.2   Measurements for Cisco Linksys WRT54GL

The next wireless AP is the Cisco Linksys WRT54GL. In this case, we can configure protocols IEEE 802.11b and 802.11g and its consumption is 5.2 W during startup and 5.7 W in steady state. Tables 15.7 through 15.10 show the values of consumption of energy for transmission of echoes and for transmission of a 2.45 GB file. As can be seen in Figures 15.7, 15.8, 15.9, and 15.10, the energy consumed by this wireless AP is considerably smaller than that consumed by the previous model, particularly by 40%. Unlike the previous model, in this case, it is the IEEE 802.11g protocol that consumes less energy. As expected, the MAC filter for this model also introduces an increase in power consumption.

15.5.3.3   Measurements for Cisco Linksys WRT320N-EZ

The Cisco Linksys WRT320N-EZ model consumes 4.5 W in bootup state and 4.8 W in idle mode. In Tables 15.7, 15.8, 15.9, and 15.10 and Figures 15.7, 15.8, 15.9, and 15.10, the results obtained for

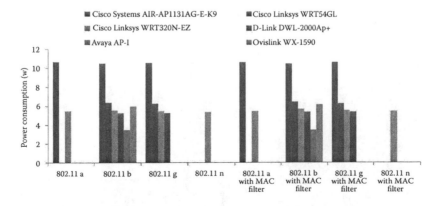

FIGURE 15.10   Measures for file sending, topology 2.

TABLE 15.9
Measurement Results for File Sending, Topology 1

	Power Consumption (W)							
	IEEE 802.11a	IEEE 802.11b	IEEE 802.11g	IEEE 802.11n	IEEE 802.11a	IEEE 802.11b	IEEE 802.11g	IEEE 802.11n
	Without MAC Filter				With MAC Filter			
Cisco Systems AIR-AP1131AG-E-K9	10.5	10.5	10.4	—	10.7	10.6	10.4	—
Cisco Linksys WRT54GL	—	6.2	6.1	—	—	6.2	6.2	—
Cisco Linksys WRT320N-EZ	5.4	5.4	5.4	5.3	5.5	5.5	5.4	5.4
D-Link DWL-2000Ap+	—	5.1	5.2	—	—	5.2	5.2	—
Avaya AP-I	—	3.4	—	—	—	3.5	—	—
Ovislink WX-1590	—	5.5	—	—	—	5.5	—	—

TABLE 15.10
Measurement Results for File Sending, Topology 2

	Power Consumption (W)							
	IEEE 802.11a	IEEE 802.11b	IEEE 802.11g	IEEE 802.11n	IEEE 802.11a	IEEE 802.11b	IEEE 802.11g	IEEE 802.11n
	Without MAC Filter				With MAC Filter			
Cisco Systems AIR-AP1131AG-E-K9	10.7	10.5	10.6	—	10.7	10.5	10.7	—
Cisco Linksys WRT54GL	—	6.4	6.3	—	—	6.5	6.3	—
Cisco Linksys WRT320N-EZ	5.5	5.6	5.5	5.4	5.5	5.7	5.6	5.5
D-Link DWL-2000Ap+	—	5.3	5.3	—	—	5.4	5.4	—
Avaya AP-I	—	3.5	—	—	—	3.5	—	—
Ovislink WX-1590	—	6	—	—	—	6.2	—	—

this model of Cisco for protocols IEEE 802.11a, IEEE 802.11b, IEEE 802.11g, and IEEE 802.11n are displayed and represented. As we can see, this device consumes less energy than what was previously studied. The IEEE 802.11n protocol is the one that presented less power consumption. In relation to the activation of the MAC filter, we observed the same previous effects, and as in the two previous models, the air-to-air topology makes higher power consumption.

15.5.3.4 Measurements for D-Link DWL-2000AP+

The fourth case analyzes the D-Link DWL 2000AP power consumption. For this device, we have studied protocols IEEE 802.11b and IEEE 802.11g. The values obtained for this device are collected and represented in Tables 15.7, 15.8, 15.9, and 15.10 and Figures 15.7, 15.8, 15.9, and 15.10. The power consumption for this device in bootup state and idle mode is 3.4 and 4.8 W, respectively. This model consumes significantly less energy than the models manufactured by Cisco in this study. In particular, it consumes approximately 16% less power than the model of Cisco (Cisco Linksys WRT54GL) while offering the same services. IEEE 802.11b is the protocol that presents the lowest power consumption. In regard to the use of the MAC filter and the differences between the topologies, the conclusions are the same as previous ones.

15.5.3.5 Measurements for Avaya AP-I

This wireless AP has more limited capacities. It only allows the protocol IEEE 802.11b, and as expected, its power consumption is smaller than the other cases, 2.8 W in bootup state and 2 W at steady state. We can see in Figures 15.7, 15.8, 15.9, and 15.10 the considerable difference in power consumption. In this device, the increase on power consumption introduced by the use of the MAC filter is approximately 3%. The conclusions about differences between the air–air link and the air–cable link are the same as previous ones.

15.5.3.6 Measurements for Ovislink WX-1590

The latest model is the wireless AP Ovislink WX-1590 with similar characteristics to the previous one but with limited capabilities; the difference is that this model consumes 40% more power. Its consumption is 4.8 W during boot and 5.4 W in sleep mode, and this device only allows the protocol IEEE 802.11b. For the rest of the parameters studied, we notice the same behavior observed in earlier devices.

15.6 ENERGY-EFFICIENT NETWORK TOPOLOGY

From the studies throughout this chapter, we can determine which network devices from which brand and model and which routing and wireless network protocols are most appropriate for designing an energetically efficient communications network.

To design a network, one must consider not only the energy consumption of the devices but also the needs of our network and the possibility of evolution or enlargement, that is, a potential number of hosts or servers and the need for certain services or capabilities offered by a protocol or another.

In Figure 15.11, we have proposed a complete topology composed of the devices studied. From this topology, we propose models, brands, and protocols that are better adjusted to the needs of the network.

We have selected the router Allied AR410 because it consumes less power and the switch Cisco 2950/24 because the switch must have 18 or more ports. For wireless APs, we have selected the Cisco Linksys WRT320N-EZ, it being the more efficient wireless AP taking into account its capacity and features, as well as the IEEE 802.11n protocol. We have chosen the IEEE 802.11n protocol because it offers more bandwidth than the rest and the consumption difference is minimal. Finally, we recommend using the OSPF dynamic routing protocol because of its reduced power consumption.

In Table 15.11, we can see the network devices with similar characteristics and with the lowest and highest power consumption. Choosing the devices and configuration shown under the column "Less Consumption" from Table 15.11 has enabled us to design an energy-efficient network with an estimated consumption of 79.25 W, against the consumption of 184.8 W that we would have if we had chosen the devices that consume more energy. We have saved 57.12% of energy, that is to say, a 57.12% economic saving.

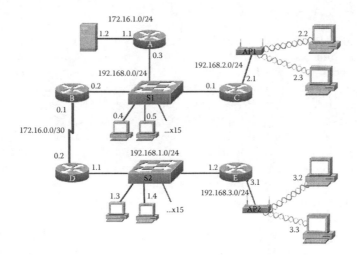

FIGURE 15.11   Proposed topology.

TABLE 15.11
Devices' Consumption in the Proposed Topology

	Less Consumption	More Consumption
Routers	Allied AR410	Cisco 1841
Switches	Cisco 2950/24	Cisco 3560/24
Access points	Cisco Linksys WRT320N-EZ	Cisco AIR-AP1131AG-E-K9
Routing protocol	OSPF	RIP.V2
Wireless protocol	802.11 N	802.11 G
Total power consumption ≈	79.25 W	184.8 W

15.7   CONCLUSION

The design of energy-efficient networks is a hot topic in recent years. This is due to the expansion of the concept of green networks and the interest of companies and researchers in designing networks with efficient energy consumption for economic savings. It is important to know the behavior in terms of energy consumption in the different devices that make up a network.

In this work, we have measured the power consumption in several network devices such as routers, switches, and APs, each in different situations and settings. We find that the most important factor in the design of green networks and the distribution of power in communications networks is the selection of an appropriate model, since power consumption depends directly on the hardware of the device. It is also very important to know the needs of the network when choosing the most efficient protocols and settings.

From the obtained results, we will continue to research and study more complex devices that will help us obtain more parameters to improve the design of energy-efficient networks. In future works, we will extend our study to other devices commonly used in large telecommunications networks and more complex routing protocols such as ISIS and BGP. With these studies, we will determine which models, network protocols, routing protocols, and topologies are most suitable for designing energy-efficient networks to achieve the objective of this field of study, which is designing green networks.

AUTHOR BIOGRAPHIES

Sebastián Andrade-Morelli received his Technical Telecommunication Engineer degree specializing telecommunication systems from the Higher Polytechnic School of Gandia of the Polytechnic University of Valencia. He worked in the city council of Gandia as a programmer and network administrator where he designed a mechanism for remotely monitoring the network using the SNMP protocol. His research focuses on energy saving in data centers and energy distribution between data networks. He is a collaborator of the research group of the Department of Communications at the Polytechnic University of Valencia in Gandia. He has participated in technical activities in several research articles about energy optimization applied to telecommunication networks.

Sandra Sendra Compte received her Technical Engineering degree in telecommunications in 2007. She received her MSc in electronic systems engineering in 2009. Currently, she is working as a researcher in the research line "communications and remote sensing" at the Polytechnic University of Valencia. She has been a CCNA instructor since 2009. She is an IEEE graduate student member. She has several scientific papers published in international conferences and book chapters and several papers in international journals with JCR. She is editor-in-chief of the international journal *WSEAS Transactions on Communications* and associate editor of the international journal *Networks Protocols and Algorithms*. She has been involved in more than 50 program and organization committees of international conferences since 2009 (CENIT 2009, ICAS 2010, INTERNET 2010, IEEE MASS 2011, SCPA 2011, ICDS 2012, Ad-Hoc Now 2012, Aict 2013, Energy 2013, Smartgreen 2013, IEEE EVN-SGA 2013, SCPA 2013, among others).

Emilio Granell Romero completed his Technical Telecommunications Engineering degree at the beginning of 2006 at the Polytechnic University of Valencia (UPV). In 2008, he started his master's in artificial intelligence (imparted by the Department of Information Systems and Computation of UPV). In 2010, he spent four months in a research laboratory in Reykjavik University in Iceland, investigating the use of the Monte Carlo algorithm in tree search for automatic planning. After this experience, he worked as a researcher in the UPV, doing research on reasoning, learning techniques, and applying it into a software player, giving as a result his master's thesis in 2011. During 2012, he completed his grade of Audio, Video and Telecommunications Systems Engineering and participated in several publications for scientific journals and conferences. He is currently finishing his PhD in computer science from the same university and his current research interests are automatic planning, automatic pattern recognition, local area networks, and energy efficiency.

Jaime Lloret Mauri received his MSc in physics in 1997, his MSc in electronic engineering in 2003, and his PhD in telecommunications engineering (DrIng) in 2006. He is currently an associate professor in the Polytechnic University of Valencia. He is the head of the research group "communications and remote sensing" of the Integrated Management Coastal Research Institute and he is the head of the "Active and collaborative techniques and use of technologic resources in the education (EITACURTE)" Innovation Group. He is the director of the University Master "Digital Post Production." He is currently vice-chair of the Internet Technical Committee (IEEE Communications Society and Internet Society). He has been the co-editor of 15 conference proceedings and guest editor of several international books and journals. He is editor-in-chief of the international journal *Networks Protocols and Algorithms*, IARIA Journals board chair (8 journals), and associate editor of several international journals. He is currently the chair of the Working Group for the Standard IEEE 1907.1. He has been the general chair (or co-chair) of 13 international conferences. He is an IEEE senior and IARIA fellow.

REFERENCES

1. C. Bianco, F. Cucchietti, G. Griffa, Energy consumption trends in the next generation access network—A telco perspective. In *The 29th International Telecommunications Energy Conference (INTELEC 2007)*, Sept. 30–Oct. 4, Rome, Italy, 2007.
2. S. Haller, S. Karnouskos, C. Schroth, The internet of things in an enterprise context, *Lecture Notes in Computer Science*, vol. 5468, 2009, pp. 14–28, 2009.

3. Y. Zhang, P. Chowdhury, M. Tornatore, B. Mukherjee, Energy efficiency in telecom optical networks, *IEEE Communications Surveys and Tutorials*, vol. 12, no. 4, pp. 44–458, 2010.
4. Datasheet of Router Allied AR410. Available at Allied Telesis Web site: http://www.alliedtelesis.com/media/fount/datasheet/AR410Series_Datasheet_RevQ.pdf [Last Access: Jan. 9, 2013].
5. Datasheet of Router Cisco 1800. Available at Cisco Web site: http://www.cisco.com/en/US/prod/collateral/routers/ps5853/product_data_sheet0900aecd8016a59b.pdf [Last Access: Jan. 9, 2013].
6. W. Fisher, M. Suchara, J. Rexford, Greening backbone networks: Reducing energy consumption by shutting off cables in bundled links. In *Proceedings of the first ACM SIGCOMM Workshop on Green Networking*, Aug. 30, New Delhi, India, 2010.
7. N. González, L. Moran, J.M. Angioleti, J.A. Varela, Chapter 4. Green telecom networks. In *Green IT*. eKISS n°82. Internal publication of Telefónica, 2009.
8. H. Galperin, Wireless networks and rural development: Opportunities for Latin America, *Information Technologies and International Development*, vol. 2, no. 3, pp. 47–56, 2005.
9. M. Segal, Improving lifetime of wireless sensor networks, *Network Protocols and Algorithms*, vol. 1, no. 2, pp. 48–60, 2009.
10. IEEE Std 802.11 (2007). IEEE Standard for Information Technology—Telecommunications and information exchange between systems—Local and metropolitan area networks—Specific requirements—Part 11: Wireless LAN Medium Access Control (MAC) and Physical Layer (PHY) specifications. pp. 1–1184. Institute of Electrical and Electronics Engineers, Inc. New York, USA, 2007.
11. S. Nedevschi, L. Popa, G. Iannaccone, S. Ratnasamy, D. Wetherall, Reducing network energy consumption via sleeping and rate-adaptation, In *Proceedings of the 5th USENIX Symposium on Networked Systems Design and Implementation (NSDI '08)*, San Francisco, California, USA, April 16–18, 2008, pp. 323–336.
12. S. Gao, J. Zhou, N. Yamanaka, Reducing network power consumption using dynamic link metric method and power off links, IEEE International Student Paper Contest 2009, 5th November 2009.
13. R. Subrata, A.Y. Zomaya, B. Landfeldt, B. Cooperative power-aware scheduling in grid computing environments, *Journal of Parallel and Distributed Computing*, vol. 70, no. 2, pp. 84–91, 2010.
14. F. Hermenier, N. Loriant, J.M. Menaud, Power management in grid computing with Xen, In *International Symposium on Parallel and Distributed Processing and Applications (ISPA 2006)*, Sorrento, Italy, December 4–7, pp. 407–416, 2006.
15. L.M. Feeney, M. Nilsson, Investigating the energy consumption of a wireless network interface in an ad hoc networking environment, In *Proceedings of the IEEE INFOCOM 2001. Twentieth Annual Joint Conference of the IEEE Computer and Communications Societies*, vol. 3, pp. 1548–1557, 2001.
16. L.M. Feeney. An energy consumption model for performance analysis of routing protocols for mobile ad hoc networks, *Journal of Mobile Networks and Applications*, vol. 6, no. 3, pp. 239–249, 2001.
17. S. Sendra, J. Lloret, M. Garcia, J.F. Toledo, Power saving and energy optimization techniques for wireless sensor networks, *Journal of Communications*, Academy Publisher, vol. 6, no. 6, pp. 439–459, 2011.
18. S. Andrade-Morelli, E. Ruiz-Sánchez, E. Granell, J. Lloret, Energy consumption in wireless network access points, In *Proceedings of the Second International Conference on Green Communications and Networking (GreeNETS 2012)*, Gandia (Spain), October 24–26, 2012.
19. E. Granell, S. Andrade-Morelli, E. Ruiz-Sánchez, J. Lloret, Energy consumption study of network access switches to enhance energy distribution, In *Proceedings of the IEEE Workshop on Smart Grid Communications: Design for Performance (SGComm 2012)*, IEEE Globecom 2012, Anaheim (California), December 7, 2012.
20. S. Andrade-Morelli, E. Ruiz-Sánchez, S. Sendra, J. Lloret, Router power consumption analysis: Towards green communications, In *Proceedings of The Second International Conference on Green Communications and Networking (GreeNETS 2012)*, Gandia (Spain), October 24–26, 2012.
21. J. Baliga, R. Ayre, H. Kerry, W.V. Sorin, R.S. Tucker, Energy consumption in optical IP networks, *Journal of Lightwave Technology*, vol. 27, no. 13, pp. 2391–2403, 2009.
22. J. Barbancho, C. León, F.J. Molina, A. Barbancho, Using artificial intelligence in routing schemes for wireless networks, *Computer Communications*, vol. 30, no. 14–15, pp. 2802–2811, 2007.
23. Cisco Report. Server power calculator analysis: Cisco UCS power calculator and IIP power advisor, 2011.
24. Datasheet of Router Cisco 2600. Available at Cisco Web site: http://www.cisco.com/en/US/docs/routers/access/2600/hardware/installation/guide/2600hig.pdf [Last Access: Jan. 9, 2013]
25. Datasheet of Router Cisco 1700. Available at Cisco Web site: http://www.cisco.com/warp/public/cc/pd/rt/1700/prodlit/1760e_ds.pdf [Last Access: Jan. 9, 2013].
26. Datasheet of Router 3Com Remote Access 531. Available at Web site: http://www.mtmnet.com/PDF_FILES/OfficeConnectProductLineSalesBro.pdf [Last Access: Jan. 9, 2013].

27. Datasheet of Switch Allied AT8124XL. Available at Allied Telesis Web site: http://www.alliedtelesis. com/media/datasheets/guides/8118-24-26xl_ig_c.pdf [Last Access: Jan. 9, 2013].
28. Datasheet of Switch Cisco 3560. Available at Cisco Web site: http://www.cisco.com/en/US/prod/ collateral/switches/ps5718/ps5528/product_data_sheet09186a00801f3d7d.pdf [Last Access: Jan. 9, 2013].
29. Datasheet of Hub 3Com SuperStack II Hub. Available at Web site: http://www.mtmnet.com/PDF_ FILES/3C16665A_UserGuide.pdf [Last Access: Jan. 9, 2013].
30. Datasheet of Switch Allied AT8024. Available at Allied Telesis Web site: http://www.alliedtelesis.com/ media/datasheets/8000_family_ds.pdf [Last Access: Jan. 9, 2013].
31. Datasheet of Switch Cisco 2950T. Available at Cisco Web site: http://www.cisco.com/application/pdf/en/ us/guest/products/ps628/c1650/ccmigration_09186a00801cfb71.pdf [Last Access: Jan. 9, 2013].
32. Datasheet of Switch 3Com SuperStack II Switch. Available at Web site: http://www.mtmnet.com/PDF_ FILES/3C16950_SSII1100SalesBro.pdf [Last Access: Jan. 9, 2013].
33. Datasheet of Access Point Cisco Linksys WRT320N-EZ. Available at Cisco Web site: http://homesupport. cisco.com/es-eu/support/routers/WRT320N/download [Last Access: Jan. 9, 2013].
34. Datasheet of Access Point Cisco Systems AIR-AP1131AG-E-K9. Available at Cisco Web site: http://www. cisco.com/en/US/docs/wireless/access_point/1130/installation/guide/1130hig_book.pdf [Last Access: Jan. 9, 2013].
35. Datasheet of Access Point Cisco Systems WRT54GL. Available at Cisco Web site: http://home.cisco. com/es-eu/products/routers/WRT54GL [Last Access: Jan. 9, 2013].
36. Datasheet of Access Point D-Link DWL-2000AP+. Available at D-Link Web site: ftp://ftp.dlink.de/dwl/ dwl-2000applus/documentation/DWL-2000applus_man_en_040401.pdf [Last Access: Jan. 9, 2013].
37. Datasheet of Access Point Avaya AP-I. Available at Avaya Web site: http://downloads.avaya.com/css/P8/ documents/003702825 [Last Access: Jan. 9, 2013].
38. Datasheet of Access Point Ovislink WX-1590. Available at Web site: http://www.ferimex.com/download/ Manual/WX-1590/WX-1590L-en.pdf [Last Access: Jan. 9, 2013].
39. W.R. Stevens, *TCP/IP Illustrated Vol. I: The Protocols*. Pearson Education India, 1994.

16 Greening Data Centers

Amin Ebrahimzadeh and Akbar Ghaffarpour Rahbar

CONTENTS

16.1 INTRODUCTION

Worldwide data centers alone consume 26 GW of electrical power corresponding to about 1.4% of the worldwide energy consumption [1]. This amount of power consumption is growing by approximately 12% each year by increasing Internet traffic. The power consumption in data centers originates from the involved computing, storage, and interconnection equipment; heating, ventilation, and air conditioning (HVAC); uninterruptible power supply (UPS) systems; and lighting facilities.

It is predicted that in the coming years, the large companies operating data centers will have to spend more money on energy than on equipment. It was reported that, in 2009, there existed 44 million servers in the world consuming 0.5% of all electricity and producing 0.2% of all carbon dioxide emissions equal to 80 megatons per year, which is equivalent to the emissions of entire countries like the Netherlands [1]. For example, Microsoft's data center in Quincy consumes 48 MW, approaching enough power for 4000 houses.

On the basis of the Environmental Protection Agency (EPA) report, data centers and servers in the United States consumed about 61 billion kilowatt-hours (kWh) in 2006, which was 1.5% of the total US electricity consumption. This amount of energy is equal to an electricity cost of about $4.5 billion. Energy usage of data centers and servers was doubled in 2006 compared to 2000. This amount of energy consumption by data centers has raised attention in research and manufacturing communities in order to achieve more energy-efficient structures for data centers [2].

There are two general approaches in order to realize the idea of a green data center: using the green elements and making green the process of performance of a data center. From the environmental point of view, the ultimate goal is to minimize the greenhouse gas (GHG) emissions. A basic idea is to apply renewable energy sources together with designing low-power elements while maintaining the same level of performance. In recent years, relocating network equipment to strategic places has become a solution to reduce GHG emission and energy costs. Large companies have relocated their data centers to places close to renewable energy sources or the areas in which the price of energy is low because of supply and demand reasons. Besides, relocating can be applied because of high amount of losses when energy (i.e., electricity) is transmitted in long distances.

Information and communication technology (ICT) companies like Google have transferred a large number of their servers to the banks of the Columbia River in order to take advantage of free cooling [1]. Furthermore, Microsoft relocated its servers to an open air location so that heating can be dissipated more easily to reduce the cost of cooling [3]. In addition to reduction in cooling costs, economic benefits can be an alternate reason to relocate servers for which Amazon is an example [4].

From the engineering point of view, greening a data center means reducing the energy consumption of the process of operating a task while keeping the same level of performance. This chapter mainly aims at this aspect of green data centers. First, green metrics are studied in order to quantify how green a data center is. Then, solutions from the engineering point of view (such as increasing the energy efficiency in network equipment, reducing power consumption in computing resources, increasing server utilization, and reducing the temperature of data centers) are introduced and discussed. In recent research, green data centers have been studied from the energy efficiency point of view [5] in which power-saving techniques using thermal management and smart cooling approaches are mainly reviewed. In order to make a data center green, several efforts have been carried out. In Refs. [6–8], the authors have tried to reduce the loss owing to thermal issues in data centers by applying thermal-aware task scheduling and thermal-aware smart cooling techniques. Increasing server utilization is proposed as an alternate solution [9–11] in which server virtualization and resource consolidation methods are applied. As mentioned, computing processes in servers need a large amount of energy that has to be managed in an efficient way. Reduction of energy consumption in computing resources has been studied in Refs. [12–14]. Besides, several approaches have been studied in Ref. [3] in order to take energy efficiency into account in network equipment. The main goal of this chapter is to discuss the energy efficiency of network equipment of data centers including energy-aware routing algorithms, energy-proportional network design, and energy-proportional computing schemes.

16.2 GREEN METRICS

A data center contains primarily electronic devices called servers that are used for data processing. Furthermore, some equipment is used for storage and communication (i.e., network equipment). This collection of equipment that is gathered in a typical data center is known as information technology (IT) equipment. In addition to this equipment, data centers usually contain power conversion and backup sections to maintain reliable performance and cooling infrastructure to keep the IT equipment in the proper range of temperature and humidity. According to the US EPA report to Congress, data centers use a significant amount of energy to supply three key components: IT equipment, cooling, and power delivery.

Current data centers have been designed for peak load situations, meaning that they consume a high amount of energy even if they are idle. In other words, energy consumption of several components of data centers is not proportional to resource utilization and they consume a constant energy all the time.

In order to determine how much green a data center is, some appropriate performance metrics, which are called green metrics, are required. Green performance metrics make a set of measurements that qualitatively and quantitatively determine the energy efficiency of a data center. Large companies operating data centers such as Google, Microsoft, and Facebook reveal such metrics each year in order to show the efficiency of their servers worldwide.

Generally, a common performance metric is the ratio of useful work to environmental costs [15]. However, useful work cannot be defined considering all existing applications. Besides, environmental costs cannot be defined so easily in an explicit way. It can be defined as the amount of environment-related factors such as temperature, energy or power consumption, humidity, or carbon dioxide emissions.

Green metrics can be classified into two categories: a basic metric that determines the level of efficiency of a data center and an extended metric that is the combination of some basic metrics by which a more comprehensive view can be achieved [15]. Basic and extended metrics are also divided into some categories as shown in Figure 16.1.

Data centers consume some amount of energy in the process of running applications. Generating the required energy increases carbon dioxide emissions. Energy sources of data centers include oil, coal, natural gas, and nuclear. Applying any of these energy sources, a lot of carbon dioxide can be produced and imposed on the environment, resulting in several kinds of ecological damage. The amount of carbon dioxide produced (in kilograms) for generating per unit of energy (kilowatt-hour) can be defined as a performance metric for a typical data center [2].

Humidity is another factor that must be considered when designing a data center infrastructure and must be taken into account carefully during the data center performance. This is because increase in humidity above the predefined threshold of a component manufacturer not only leads to hardware failure but also increases the cooling cost of any data center. In order to measure the amount of humidity in data centers, a parameter known as relative humidity (R.H.) is defined. The R.H. is the ratio of the amount of water that exists in the air at a certain temperature to the maximum humidity that air can hold. The parameter R.H. is expressed as a percentage. Furthermore, another metric called R.H. difference (R.H.D.) is defined as the difference of the return and supply air relative to the data center. A low value for R.H.D. illustrates the fact that low energy is required.

Several thermal metrics are used in order to describe the efficiency of data centers from the temperature point of view such as air flow performance index (AFPI), British thermal unit, cooling system efficiency metrics, and data center temperature. It is proven that electronic equipment have to be within 20°C and 40°C [16]. The difference between the supply air and return air is another important factor that can be considered, for which a low value is favorable.

AFPI includes some metrics that describe the performance of air flow in data centers. The AFPI consists of several metrics such as rack cooling index, return temperature index, supply heat index and return heat index, recirculation index, and capture index [15].

Cooling is one of the major energy-consuming procedures that takes place in a data center. In order to express the efficiency of cooling systems in data centers, several related metrics including

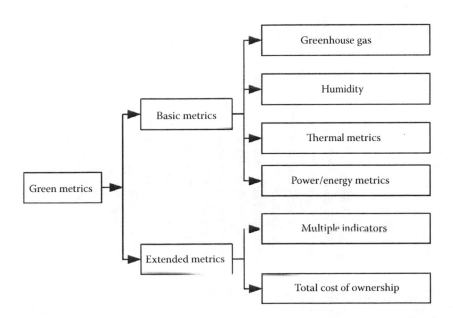

FIGURE 16.1 Classification of green metrics.

cooling system efficiency, air flow efficiency, cooling system sizing, air economizer utilization, and water economizer utilization can be defined [15].

Power and energy metrics are the most common metrics that are used by large companies to show the efficiency of their data centers. Power/energy metrics like data center infrastructure efficiency (DCiE); power usage effectiveness (PUE); HVAC system efficiency; space, watts and performance (SWaP); and data center energy productivity (DCeP) are the most relevant green metrics that are mostly considered [15,17].

The most important energy metric is known as PUE, which is defined as the ratio of the total energy consumption of a data center to the total energy consumption of an IT equipment. The PUE metric measures the total power consumption overhead caused by cooling systems, power delivery, and other supporting facilities. PUE = 1 means that there is no extra power needed to be consumed in supporting facilities. The average value of PUE for typical data centers was reported to be 2, implying that 1 watt of overhead power is consumed in cooling and power delivery parts [2]. The EPA report to Congress suggested that most data centers could reach a PUE of 1.7 by 2011 because of improvements in equipment and by applying liquid cooling systems, whereas some state-of-the-art data centers could reach 1.2. Interestingly, in 2009, Google claimed a PUE value of 1.21 for its six large-scale Google-designed data centers, with one reaching a PUE of 1.15. In addition, Microsoft reported the PUE values for its Chicago data center to be 1.22 per year [1]. According to the reports from the first quarter of 2012, Google claims an average value of 1.11 for the PUE measurements of its data centers, which makes this company more than seven times greener than average. The company's trailing 12-month average PUE for 2011 was 1.14, which is an improvement from 1.16 in 2010. The lowest value of PUE = 1.08 has been recorded in one of Google's data centers. Figure 16.2 shows the value of PUE in Google data centers for the last 4-year interval [18].

Although PUE is a common metric to show the efficiency of data centers, it simply does not imply the level of greenness of a data center, but it indicates how well we can manage our power consumption. Therefore, this metric is not suitable for green marketing. It may be better to define some novel metrics that include the way of producing required energy in data centers as well. This kind of metrics can take into account the amount of carbon dioxide produced per generation of one unit of energy. DCiE is similar to PUE, and it is calculated as the ratio of total energy consumption

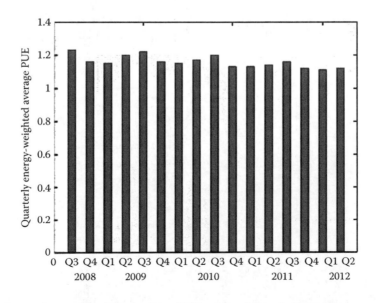

FIGURE 16.2 Quarterly energy-weighted average PUE values of Google data centers.

of an IT equipment to total energy consumption in a data center, which is exactly the inverse of the PUE value [19].

DCeP measures the amount of useful work relative to the amount of energy consumption producing this work [19]. In fact, the DCeP metric indicates the overall work product of a data center per unit of energy consumption.

Multiple indicators include several metrics that explain the greenness of a data center from different points of view. These metrics are divided into two groups, namely, data center indicators and data center sub-level indicators. Finally, total cost of ownership (TCO) represents the cost to the owner to purchase, operate, and maintain a data center. The TCO itself has two subcategories called capital expenses and operational expenses and it can be generally described in terms of dollars per watt for a data center.

16.3 ENERGY CONSUMPTION MODELS IN DATA CENTERS

In order to investigate possible energy-efficient solutions, it is necessary to achieve a viewpoint of how energy/power is consumed in data centers. An analytical model of energy consumption in different components of a data center can help investigate and design a greener infrastructure for a data center. In this section, we aim at modeling the energy consumption in various components of a data center. In Ref. [17], an analytical framework for modeling total data center power has been developed.

In fact, there is a major mismatch between data center utilization and power consumption. It means that most parts of a data center are designed for peak load situation, causing an additional amount of extra power consumption in idle periods. Different parts of a data center such as servers, storage equipment, network equipment, and power delivery systems consume a significant amount of energy even when they are idle and the workload is far below its peak value.

Approximately 60%–70% of power consumption in data centers is related to IT equipment including servers, storage, and network components. Thus, energy savings in IT components could be of high importance in order to achieve an energy-efficient data center structure. In Refs. [20–23], server power has been investigated as the research topic. Many servers consume approximately half of the peak energy while they are idle and energy consumption grows linearly by increasing utilization as illustrated by Equation 16.1. In this equation, $P_{srv.IDLE}$ and $P_{srv.PEAK}$ show the idle and peak power consumptions of a server, respectively. The parameter u_{srv} is the server utilization and P_{srv} denotes the power consumption of the server. This linear approximation is the simplest model of server power consumption. However, more accurate approximations can be achieved via modeling the empirical energy consumption values by a suitable nonlinear function or a proper piecewise linear function.

$$P_{srv} = P_{srv.IDLE} + (P_{srv.PEAK} - P_{srv.IDLE}) \cdot u_{srv} \tag{16.1}$$

Network equipment such as routers and switches account for a major share in energy consumption of data centers. As noted previously, manufacturers tend to ensure that network devices consume energy proportional to their loads. In Figure 16.3, the power consumed by a typical network device is plotted at different loads (i.e., data rates). Ideally, the consumed power must be proportional to the load whose curve is labeled with "Ideal." However, in practice, the real power consumption pattern is similar to the curve marked with "Measured values" in Figure 16.3. In order to achieve a quantitative metric for energy proportionality of a device, a metric named energy proportionality index (EPI) is defined as in Equation 16.2 [24]:

$$EPI = \frac{M - I}{M} \times 100 \tag{16.2}$$

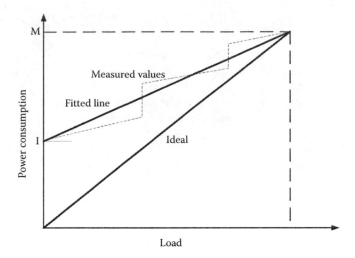

FIGURE 16.3   Ideal and measured power consumption model of network devices.

The value of EPI is expressed as a percentage, implying that when EPI is 100%, then the energy consumption of a device is completely proportional to its load, and it has reached the ideal situation.

Making the energy consumption profile of devices proportional to the workload is referred to as proportional computing. Two methods—dynamic voltage scaling [25] and adaptive link rate (ALR) [26]—have been proposed in the literature and will be discussed in Section 16.4.

There are different energy consumption profiles that a device may exhibit as a function of its utilization level. Energy-agnostic devices (i.e., devices with constant energy consumption when they are active) perform only in two states. They can be either in fully operational mode or in off mode. When a device is in the fully operational mode, its energy consumption is not load dependent and it consumes a fixed amount of power under different utilization levels. Conversely, an energy-aware device is an ideal element whose energy consumption is completely proportional to the workload with zero energy consumption in its idle state. Between these two cases, there are single-step and multistep energy profiles that are piecewise linear approximation of real energy measurements. The energy-agnostic profile is the worst case and a lot of unwanted energy is consumed most of the time, especially in light-load time intervals. The abovementioned energy profiles are shown in Figure 16.4 [3,27].

In designing a data center infrastructure, one may yield some energy profile that is optimized to a certain utilization interval. Thus, the most probable utilization state of a data center must be considered in order to decide which energy profile is desired. Figure 16.5 depicts two separate energy profiles that are optimized for a certain utilization state. It must be determined whether a data center is in low utilization state or not. In state-of-the-art data centers, low utilization state is the most probable; thus, the low utilization-optimized energy profile is desired. On the other hand, if the utilization level of a data center increases by applying server consolidation techniques, then the energy profile that is optimized for high utilization levels may be favorable. As a result, systems that are optimized for the most likely utilization levels must be designed. As an example, Barroso and Hölzle [27] have tried to carry out a study on the distribution of time intervals in terms of CPU utilization levels. The average results of more than 5000 servers show that servers are neither completely idle nor completely fully utilized most of the time; instead, they often work within the range between 10% and 50% of their maximum utilization levels.

A data center requires an infrastructure to deliver reliable and uninterrupted electrical power to IT equipment. The power distribution unit (PDU) is a power conditioning system that is responsible

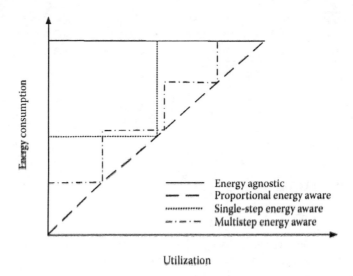

FIGURE 16.4   Energy consumption as a function of utilization.

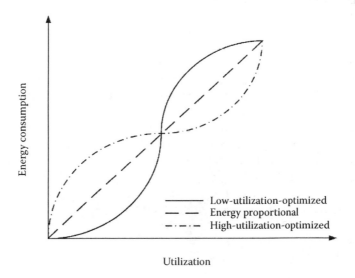

FIGURE 16.5   Energy profiles optimized for certain utilization levels.

for power delivery to servers, converting high voltage electricity to low voltage to be applicable in electronic devices used in servers. At typical servers, there exist some amount of losses in PDUs, and this amount of loss is not negligible. In fact, they incur a constant power loss as well as a power loss proportional to the square of the workload [28]. Power loss in PDU is modeled using Equation 16.3 in which $P_{\text{PDU.Loss}}$ and $P_{\text{PDU.Idle}}$ denote PDU power loss and PDU idle power draw, respectively. The parameter π_{PDU} is the PDU power loss coefficient [28].

$$P_{\text{PDU.Loss}} = P_{\text{PDU.Idle}} + \pi_{\text{PDU}} \left(\sum_{\text{Servers}} P_{\text{srv}} \right)^2 \tag{16.3}$$

UPSs provide temporary power during power system failure and are placed in series between the power supply and PDUs. UPSs impose some power overhead even when operating on utility power. The UPS power loss can be modeled by Equation 16.4, where π_{UPS} denotes the UPS loss coefficient [28]:

$$P_{UPS.Loss} = P_{UPS.Idle} + \pi_{UPS} \sum_{PDUs} P_{PDU} \qquad (16.4)$$

16.4   GREENING DATA CENTERS, SOLUTIONS, AND TECHNIQUES

As mentioned before, there is a major mismatch between a data center utilization level and its power consumption. Many efforts have been carried out in order to increase the energy efficiency of data centers by proposing some techniques. These techniques consider the efficiency of data centers from environmental, economical, and engineering points of view. In this section, we concentrate on the methods that are carried out from the engineering point of view. Energy saving in data centers can be achieved via designing energy-efficient hardware infrastructures, smart power delivery, and smart cooling technologies [5].

Servers, storage devices, and network equipment play a major role in energy consumption in data centers. Thus, making the mentioned elements greener is of high importance. Resource utilization, virtualization, and proportional computing together with energy-aware routing are the proposed techniques in order to increase the energy efficiency of IT equipment.

In resource consolidation [29–31], using resources in a more energy-efficient way considering the current workload has been tried. In this approach, the traditional over-provisioning way of resource allocation is released and some low-traffic routers and devices are switched off and their loads are transferred to other routers. Energy-aware consolidation tries to keep servers well utilized in such a way that idle power costs are effectively diminished. In Ref. [29], the problem of loading servers to a desired utilization level for each resource has been modeled as the modified multidimensional bin packing problem for allocating and migrating workloads to achieve an energy-efficient operation.

Virtualization [32] regroups a set of mechanisms, allowing more than one service to operate on the same piece of hardware, resulting in hardware utilization improvement. Generally, a single device under high traffic load consumes less energy than several low-loaded devices. Virtualization can be applied to several kinds of resources, such as network links, storage devices, and software resources. Sharing servers in data centers is an example of virtualization. Virtualization techniques have been studied comprehensively in a computer architecture perspective [33] and a networking perspective [34].

Two approaches have been proposed to adapt link rates. The first approach, which is usually known as sleeping mode, is to turn off links during idle periods. An alternate scheme is to reduce the link rate during low utilization period, which is referred to as rate switching. In the sleeping mode, only two states of operations are considered: sleep mode and fully operational or active mode [35–37]. However, transition time from the sleep mode to the active mode is an important constraint that must be considered carefully during a system design. A typical value of transition time is reported to be 0.1 ms [36].

In Ref. [35], each node measures packet inter-arrival times and decides on the status of switches and links between switches. If this measured interval is long enough, they execute an energy-efficient procedure in order to take benefit of this long interval without exploiting their equipment in high operational mode. In Ref. [37], the sleep state is deep and arriving packets during sleep mode are dropped. In addition to the choice between two sleep and active modes, one can use a wider range of modes instead of only two on–off schemes. This approach is known as the rate switching method.

16.4.1 ENERGY-PROPORTIONAL APPROACHES

Some techniques based on ALR, as a rate switching technique, have been proposed in Refs. [26,38] in order to reduce the energy consumption of Ethernet networks. However, this approach can be applied in a network of data centers as well. A network interface controller (NIC) consumes a significant amount of energy. Interestingly, various link rates consume different amounts of energy. For example, a typical 1 Gbps Ethernet link consumes about 4 W more than a 100 Mbps link, and a 10 Gbps link consumes from 10 to 20 W [26]. The key idea is to reduce the rate of links whose loads are low.

There are some policies used in the ALR algorithm that determine when to switch a link data rate. On the other hand, there are some mechanisms in ALR that determine how to switch a link data rate. A good policy should maximize the time spent in a low data rate (to achieve more energy saving) while minimizing increased packet delay. As a matter of fact, there is a trade-off between energy savings and packet delay.

In order to implement the ALR in a network, initializing and agreeing upon a link data rate change are necessary. A fast, two-way handshaking is applied using Ethernet medium access control frames. The handshaking process is terminated in 0.1 ms for a 1 Gbps Ethernet. Meanwhile, in order to determine when to switch data rates on a link, several policies have been proposed in ALR such as dual-threshold policy, utilization threshold policy, and time-out threshold policy.

The simplest ALR policy is known as the dual-threshold policy, which is based on nodes' queue length threshold. Two threshold levels (high threshold and low threshold) are determined to prevent trivial oscillations between rates. When the queue length in a node's buffer exceeds the high threshold, then the transmitter sends a request to increase the link data rate to the receiver. Conversely, when the queue length becomes lower than the low threshold, then a request is sent to reduce the line data rate.

In order to prevent oscillation, the utilization threshold policy has been developed. If a link utilization is explicitly monitored and used in decision to transition between data rates, the effect of oscillations on packet delay can be reduced or eliminated. Utilization monitoring is carried out via counting the bytes sent within a specific time interval. However, counting the number of transmitted bytes within a time interval requires additional registers that may increase the complexity of the ALR-capable NICs. Thus, a policy called time-out threshold policy is developed to keep the link rate at a high data rate for a predetermined period. Two timers are used (in each node): one for low data rates and another for high data rates. When it is switched to a higher data rate, transmission stays at a high rate until the timer expires, irrespective of the queue length. After expiring the high-rate timer, the link rate is switched to low rate only if the queue length in the transmitter side is below a certain threshold. Switching from a low rate to a high rate is triggered when queue length increases and exceeds the threshold if the low-rate timer has not expired yet. Comparison of the abovementioned policies can be found in Table 16.1.

The sleeping mode method [22,39–41] takes advantage of putting idle elements to sleep. In fact, server utilization is below 30% of the peak value most of the time. Even when operating on

TABLE 16.1
Comparison among Policies in the ALR Algorithm

Policy	Complexity	Extra Necessary Equipment	Efficiency
Dual threshold	Low	No	Low
Utilization threshold	High	Yes	High
Time-Out threshold	Medium	No	Medium

interactive services (such as transaction processing, file servers, and Web 2.0), it reaches below 10%. Servers consume 60% of their peak energy while they are idle. Although a server has low utilization most of the time, its activity occurs in frequent brief bursts [22]. By applying server consolidation techniques, several services can be time multiplexed on one server, thus increasing average utilization of resources. However, server consolidation by itself does not decrease the difference between peak and average utilization. Data centers still require sufficient capacity for peak demands, which keeps some servers in idle mode.

In Ref. [22], a sleeping mode method that tries to put idle servers to sleep has been proposed. Two key implementation requirements have to be considered via designing the proposed method: fast transition and minimizing power draw in sleep state. A transition speed of 10 ms is sufficient while the proposed scheme yields an overhead of 1 ms, which is negligible.

From the energy saving point of view, there exists a network utilization threshold below which the sleeping mode techniques perform better than the rate switching methods. A comprehensive comparison between sleeping mode and rate switching methods can be found in Ref. [40].

16.4.2 ENERGY-AWARE ROUTING

Reducing energy consumption of data centers can be investigated from a network layer point of view as well. Energy-aware routing aims at aggregating traffic flow over a subset of the network elements while other elements are switched off. The key idea behind energy-aware routing approaches is to switch off the maximum number of network devices within low-traffic periods, while maintaining connectivity and quality of service (QoS).

In Refs. [42,43], the authors have considered a wide area network (WAN) scenario. Given a network topology, a traffic demand, and a power consumption model of each node and link, the possibility of turning off some network elements (nodes and links) is evaluated under connectivity and QoS constraints. The objective is to minimize the total power consumption of a large network in which resources are over-provisioned for fault protection reasons. A network of data centers can be considered in the same situation and the proposed scheme can be applied to such networks. In order to minimize total energy consumption through turning some elements off, a problem is formulated in the form of integer linear programming, which falls in the subclass of capacitated multi-commodity cost flow problem (CMCF) (i.e., the problem in which multiple commodities have to be routed over a graph with capacity constraints). The CMCF problems are known as NP-hard, and therefore, it is impractical to exploit the exact methods to solve a problem consisting of a large number of variables. Thus, heuristic algorithms are proposed in order to solve the abovementioned problem in polynomial time. It is assumed that the power consumption of nodes is greater than the power consumed in links. Thus, the proposed heuristic tends to switch off the largest possible number of nodes rather than links. The proposed heuristics assume that all network elements are initially powered on. Then, they try to switch off network elements iteratively. In each iteration, traffic flowing through the node/link that is going to be switched off is routed via the shortest path for each source and destination pair. Flow conservation and utilization constraint is verified, and if no violation is present, then the selected element can be switched off.

However, a question may arise here: which nodes/links must be selected first? Several policies can be adopted to iterate through the node set such as random, least link, least flow, and opt edge [42]. The node set is sorted initially using one of these policies and then the iterating process through all the nodes begins. The random scheme sorts the nodes in a random order. The least-link policy sorts the nodes on the basis of the number of links that are entering and exiting each node so that the nodes with smaller number of links are examined first. In the least-flow ordering policy, the nodes with the smallest amount of information flowing through are taken into consideration first. In order to describe the opt-edge ordering rule, a conventional topology of a data center network is depicted in Figure 16.6.

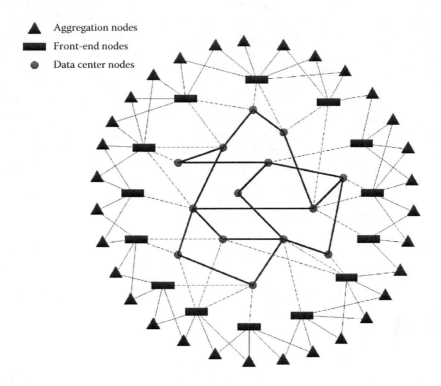

FIGURE 16.6 Generic topology of a data center network.

As shown in Figure 16.6, this topology includes aggregation (client) nodes, front-end (edge) nodes, and data center (core) nodes. User traffic is collected by means of aggregation nodes. For protection purposes, most of these aggregation nodes are connected to two front-end nodes. Front-end nodes are responsible for collecting the traffic from several aggregation nodes in the same area. It must be considered that aggregation nodes are sources and destinations of traffic and thus cannot be switched off. However, one of the two front-end nodes can possibly be powered off while the other is in active mode supporting related aggregation nodes. This is the key idea used in the opt-edge policy. The opt-edge algorithm searches the list of front-end nodes and aggregation nodes to extract a subset of nodes to switch off without violating the abovementioned constraints. This scheme begins from a front-end node and finds all the neighbor front-ends that are still in their active modes by going through aggregation nodes. If all the neighbors of the considered front-end node are powered on, then the current front-end node can be switched off.

In order to turn off links, two sorting policies—random and least flow—can be exploited. The least-flow policy sorts the links in increasing order of flow while the random scheme sorts them in a random order. It means that in the least-flow policy, links with lower amount of flow are considered first. Simulation results show that the opt-edge policy for node sorting together with either one of the random or least-flow link sorting rules yields better results in terms of number of nodes and links that can be switched off, resulting in a significant energy expenditure reduction in the network.

Fisher et al. [44] have reduced the power consumption by turning off cables in a bundled link. It is assumed that each link consists of some cables that can be shut down independently. Given the network topology, bundle size for each link, and traffic demands, an optimization problem is defined in order to shut down the maximum number of cables, while still routing all traffic demands on paths with sufficient bandwidth. Since the ILP formulation of the developed optimization problem is NP-hard, several heuristics have been proposed. These heuristics remove cables in a certain order until no further cables can be removed. The proposed heuristics defer in the order by which the cables are considered as well as in the number of cable combinations that are considered to be shut down. Three different heuristics have been developed: fast greedy heuristic (FGH), exhaustive greedy heuristic (EGH), and bi-level greedy heuristic (BGH). In FGH, a linear programming problem whose objective is to minimize the sum of loads on links is initially solved. Then, the maximum number of cables is removed in such a way that all the problem constraints are satisfied. The algorithm proceeds by selecting the edge with the highest spare capacity and tries to remove one cable from this edge. When the cable is removed, the linear programming problem with the updated link capacities is tried to be solved. If this LP problem has a feasible solution, then the selected cable can be shut down permanently; that is, we can turn it off here and we do not have to check this cable in the next iterations of the algorithm. The procedure is iterated until all edges are checked. The FGH algorithm is simple with low complexity and running time. Exhaustive greedy heuristic is an improved version of FGH. The EGH calculates a penalty value for each candidate cable and the cable with the smallest penalty value is chosen to be switched off. The calculated penalty is defined as the difference of the objective value of the LP after and before cable removal. Furthermore, BGH improves EGH by considering cables for potential removal in pairs. The BGH algorithm is more time consuming than the others since more cable combinations for removal is considered before making each decision. Simulation results show that energy saving is increased while bundle size grows. The performance of the three proposed heuristics is tight in terms of energy savings; thus, it is convenient to prefer the simplest one (i.e., FGH) that has the lowest complexity.

As mentioned before, energy profile is defined as the dependency of energy consumption of a network component to its relative workload (i.e., the percentage of the total capacity of the network component). In Ref. [45], the energy profile aware routing (EPAR) scheme has been proposed, which is based on considering different energy profiles such as linear, cubic, on–off, log10, and log100. In the linear energy profile, energy consumption varies linearly by increasing workload. Switch architecture follows this energy profile. The on–off profile is the simplest profile in which the element is working either in full-power state or in power-off mode. The log10 profile corresponds to the energy consumption behavior of equipment using hierarchical techniques that are based on the idea that data have to be sent as fast as possible so that equipment can return quicker to the low-power idle state. Besides, log100 expresses a middle performance between on–off and log10 models. The cubic energy model represents the behavior of architectures on the basis of the dynamic voltage scaling or dynamic frequency scaling techniques. The key idea of EPAR is to consider the energy consumption of equipment while deciding through which paths traffic should be routed. At the beginning, the energy profile of each element in the network is specified; then, the traffic is routed in such a way that the total energy consumption is minimized while each energy profile is approximated with two constant-slope lines. The developed optimization problem is solved using CPLEX, and results show that by applying the EPAR algorithm and considering cubic energy profile, significant energy savings can be achieved compared to the conventional shortest path routing algorithm.

Garroppo et al. [46] have developed the same energy profile aware formulation for routing. Given a directed graph and traffic matrix together with the energy profile of each node, the objective is to minimize the sum of energy consumption in the network subject to flow conservation, and link and node throughput constraints. In order to solve the related mixed integer linear programming (MILP) problem in polynomial time, a heuristic called Dijkstra-based power-aware routing algorithm (DPRA) is proposed. It is notable that the throughput of each node is defined as the amount of flow entering the node from adjacent nodes plus the amount of flow whose source is that particular node. However, as mentioned before, the MILP problem is solved in Ref. [45] using CPLEX, which is a commercial MILP solver. But the DPRA heuristic yields a near-optimal solution to the related optimization problem. The DPRA partitions each traffic demand in small quantities and tries to accommodate each part iteratively. At each iteration, each link is assigned a cost equal to the amount of increase of the power consumption of the destination node; then, Dijkstra's shortest path algorithm is applied to find the routes with minimum average costs (i.e., energy). Simulation results illustrate that the proposed heuristic can achieve power savings comparable to EPAR.

The same authors of Ref. [46] have developed a novel approach called power-aware routing and network design (PARND) in Ref. [47] by considering jointly both the possibility of switching off links and nodes of the network, together with a power-aware routing strategy. The proposed scheme is a combination of approaches in Refs. [45] and [42]. In this scheme, a multiobjective formulation is developed. The proposed PARND is based on the heuristic in Ref. [42]. The heuristic begins by sorting the nodes by means of a least-flow policy after solving the power-aware routing algorithm (proposed in Ref. [46]) with all elements powered on. Afterward, each node and later all its adjacent links are switched off according to the considered ordering rule and the power-aware routing problem is solved on the restricted network. The procedure is iterated until all nodes are checked. Significant power saving can be achieved by joint utilization of the power-aware routing strategy together with an energy-aware network design.

Servers are the major energy-hungry part of a data center, to which a massive number of them are interconnected (i.e., a network of servers in a data center is called a data center network, or simply DCN hereinafter). Energy efficiency can be studied in an interconnected network of servers in a data center. In addition to efficiency and reliability, fault tolerance is another major factor that is considered in topology design in server networks. The well-known tree architecture in Ref. [48], which is extensively used in DCNs, suffers from the lack of scalability as well as the existence of several points of failure recovery. Thus, a variety of network topologies like Fat-Tree [49] and BCube [50], which aim at increasing connectivity to achieve a 1:1 oversubscription ratio, have been proposed. Note that oversubscription of 1:1 means that all nodes can communicate with other arbitrary nodes at the full bandwidth of their network interface. However, this high capacity is actually over-provisioned since a data center traffic is far below the peak value most of the time.

In Ref. [48], a method to save energy in rich-connected DCNs from the routing point of view has been discussed. The objective is to compute the routing for a given traffic matrix so that as few switches are involved as possible to meet a predetermined performance level. Throughput is the performance metric considered in this study. The problem is formulated in the form of the ILP model and then it is translated to the well-known Knapsack problem [51], which is proven to be NP-hard. In order to achieve a feasible solution for practical network topologies, a heuristic has been proposed. The proposed approach called heuristic routing algorithm initially takes all switches into consideration and computes the routing and corresponding throughput. Then, it gradually tries to eliminate the light-loaded switches from those involved in primary routing. Network elements are being put to sleep and then eliminated. This elimination process is performed on the basis of some elimination order, where elimination order is based on the workload on the switches. Active switches with the lightest traffic can be put to sleep and shut down. The elimination process is terminated when the throughput decreases to a predefined threshold. Efficiency of the proposed scheme is evaluated by simulating the algorithm on BCube and Fat-Tree topologies and a large amount of energy savings is achieved especially when the network is under low-load state.

A recent study aimed at minimizing the energy consumption of the backbone IP over a WDM network [52]. In order to reduce the energy consumption in the IP over a WDM network, light-path bypass in the optical layer is extensively applied because it can reduce the number of required IP router ports. Note that there are two possible implementations of IP over WDM networks, namely, light-path bypass and non-bypass. In light-path non-bypass, data are processed and forwarded in each IP router. Conversely, in the light-path bypass scheme, the IP traffic directly bypasses intermediate nodes without being processed. Based on the light-path bypass concept, an MILP optimization model and two simple but efficient heuristics have been developed in order to minimize energy consumption. The energy-oriented IP over WDM network model, the energy-minimized MILP optimization model, and the proposed heuristics are the novel approaches in this study. Physical layer issues such as power consumption of each component and the layout of optical amplifiers are specifically considered. Given a network topology, demand matrix, the number of available wavelengths, and the capacity of each wavelength, the objective of the developed MILP optimization model is to design an energy-minimized IP over a WDM network, considering energy consumption models

TABLE 16.2

Comparison among Energy-Proportional Approaches

References	Optimization Model	Solution Method	Approach
[52]	MILP	Heuristics and LP relaxation and optimal design	Energy-efficient network design
[45]	MILP	MILP solver	Energy-aware routing
[42–44]	Capacitated multi-commodity flow problem	Heuristics	Turning off network elements
[46]	MILP	Heuristics	Energy-aware routing

of IP router ports, Erbium-doped fiber amplifiers (EDFA), and WDM transponders. The problem aims at finding an optimal virtual topology design, routing of virtual links in the physical layer, and finding the number of required wavelengths, fibers, and EDFAs on each physical link. To solve the developed problem within an acceptable time, two simple heuristics have been proposed, namely, the direct bypass and multi-hop bypass approaches. These heuristics are actually the extension of single-hop and multi-hop grooming strategies [53,54], which are well known in traffic grooming research. Results of simulations show that the energy-minimized model can reduce the energy consumption of IP over a WDM network, ranging from 20% to 45%. Comparisons between different energy-proportional approaches are illustrated in Table 16.2.

16.4.3 Load Distribution

Today, many large organizations operate multiple data centers. Geographical distribution of data centers often provides many opportunities for optimizing energy expenditure and cost via intelligent distribution of computational workload [55,56]. These services place their data centers behind a set of front-end devices. The front-end nodes are responsible for checking each client's request and forwarding it to one of the data centers that can serve it. This procedure is called smart request distribution or load distribution policy. Request distribution policies take advantage of time, zone, variable electricity price, and availability of green energy resources. The optimization-based policy specifies the fraction of the clients' requests that should be sent to each data center in such a way that total energy or total cost or a combination of these factors becomes minimum [56].

However, minimizing energy costs is realized while guaranteeing high performance and availability requirements. These requirements are respected by having the front-end nodes to (1) prevent a data center from overloading and (2) monitor the response time of the data center. The proposed framework's optimization problem searches to find the fraction of requests that should be sent to each data center as depicted in Figure 16.7. The optimization problem is solved using the simulated annealing method [56] or by simple heuristics [56]. Furthermore, in Ref. [55], a greedy cost-aware heuristic policy that is simpler and less computationally intensive than the optimization-based approach is proposed. The key idea is to send the requests to data centers for which the cost of energy is the cheapest, guaranteeing a specific level of service level agreement (SLA). The SLA is expressed as (L, P), which means that at least P percent of requests must be serviced within time L. An MILP optimization problem is updated and solved in a scheduler periodically (e.g., once an hour) as depicted in Figure 16.7. Moreover, at the beginning of each interval, the scheduler needs to estimate and predict the incoming request intensities. In order to predict the load intensities, the auto-regressive moving average (ARMA) model is considered [56].

In short, Le et al. [56] have developed a framework in order to reduce the total energy costs considering only the energy consumption in data centers, while in Ref. [55], the problem is modeled using the energy consumption in network devices in addition to data centers.

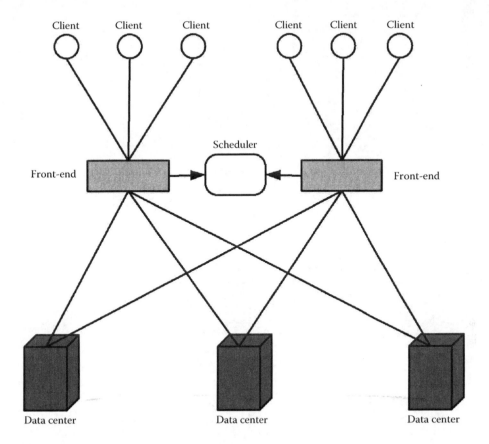

FIGURE 16.7   Load distribution via smart scheduling.

16.4.4   SMART POWER DELIVERY

Power distribution is a major energy-consuming part of data centers. Typical power delivery systems are fed with alternating current (AC) power, which is distributed from a power supply utility to the facility. AC power is stepped down by transformers and then is delivered to uninterruptible power supplies (UPSs), which also take a second feed from a set of diesel generators that will jump into action when utility power fails. UPSs are important infrastructures and are designed for interruptions and other disturbance recovery purposes. Electricity is delivered to PDUs where it is stepped down again and transmitted to servers in order to be fed into electronic devices.

As discussed earlier, significant amount of loss is imposed at different stages of the power delivery system in a data center. Losses due to deficiency of power delivery system are load dependent and vary under different workloads. Thus, designing a smart power delivery system is of high importance. An ideal power delivery system should balance several workloads on PDUs. Since power demands vary in time, it is impractical or extremely hard to balance the PDU loads statically by clever assignment of servers to PDUs. Pelley et al. [57] have explored mechanisms to balance the load through a power delivery infrastructure by dynamically connecting servers to PDUs. The proposed power-routing approaches take advantage of fault tolerance in the power delivery system; that is, each server can draw power from either of two redundant feeds.

The power-routing scheme is based on two main ideas: (1) exploiting shuffle technologies for power distribution to increase connectivity of servers and PDUs, and (2) using a smart scheduling algorithm. Two kinds of topologies exist for power distribution, namely, wrapped topology and shuffled topology, which are illustrated in Figure 16.8. Under the wrapped topology, the reserve

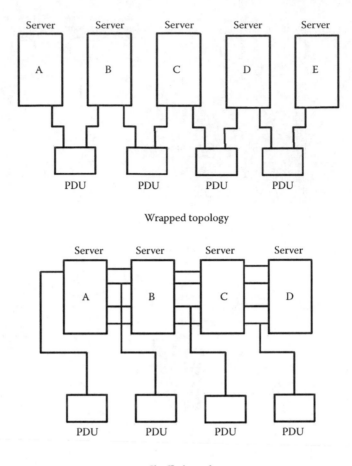

FIGURE 16.8 Wrapped and shuffled topologies in power delivery systems.

capacity for each PDU in case of failure must be 50%. Conversely, in the shuffled topology, the amount of reserve capacity for each PDU is less compared to the wrapped topology. The wrapped topology is an example of high-availability power systems. In the shuffled topology, loads of servers are divided equally among all PDUs. It means that each PDU is responsible for a portion of a server's power requirements. Thus, in case of single PDU failure, the extra load that is added to each PDU (which is called reserve capacity) will be equal to $1/(N-1)$, where N denotes the number of active PDUs. A comparison between the mentioned power delivery systems is shown in Table 16.3.

In addition to applying the shuffled topology, a smart scheduling algorithm is designed to assign a server load across redundant distribution paths while balancing loads over PDUs. If the loads are

TABLE 16.3

Comparison among Power Delivery Systems Topologies

Topology	Connectivity	Reserved Capacity for Failure Recovery
Wrapped	Very rich	50%
Shuffled	Rich	<50%

balanced, then the provisioned capacity of power infrastructure components will decrease, thus resulting in a smaller amount of losses. The proposed scheduling scheme develops a centralized mechanism. When a server demands more power, the scheduler checks if the server's current active power feed has more capacity or not. If there is no free slack, the scheduler tries to create a new allocation schedule for the entire facility. Power feed assignment is an NP-hard problem; thus, it is solved using a heuristic by means of static assignment of primary and secondary feeds to each server. By applying the power-routing scheme, a decrease of 12% in the required PDU capacity can be achieved.

16.4.5 Thermal-Aware Resource Management

Since heat dissipation from data centers increases exponentially, strict cooling policies must be considered. A data center cannot operate without cooling for more than a few minutes before overheating. Moreover, the high temperature of data centers will lead to high hardware failure, thus resulting in an increase of maintenance costs. In fact, every 10°C increase in temperature leads to doubling the system failure rate. Moreover, it is reported that cooling costs can be up to 50% of the total energy cost [58]. Even with more efficient methods (liquid cooling), cooling cost remains a significant portion of the total energy cost. Thus, smart cooling techniques and thermal-aware resource management have become the targets of some studies [58–62].

Detailed research has been carried out in task scheduling using computational fluid dynamics (CFD) models [60,63]. However, the CFD-based techniques are too complex and are not suitable for online scheduling. Other works have proposed simpler models that are appropriate for the online scheme [59,61,64–66]. Wang et al. [62] have developed a heat transfer model of data centers and proposed a less complex method compared to the CFD-based schemes. Data center workloads are modeled as a set of jobs. Each job has some parameters (such as required number of computing nodes, job's arrival time and starting time, required execution time, and the task-temperature profile of the job) by which it is characterized. Note that the task-temperature profile is the temperature increase while the task is executed. Hot jobs are the tasks that increase the temperature of the server significantly when executed. A cold node is a node whose current temperature is below a certain threshold. It is assumed that the increase in temperature relative to each of the jobs is known in advance. The key idea is to assign hot jobs to cold nodes. The hottest jobs are scheduled to the coldest nodes using a simple algorithm by sorting the jobs in decreasing temperature order. Simulation results indicate the fact that the proposed thermal-aware resource management algorithm leads to significant reduction in data center temperature by imposing acceptable reduction in its performance. Note that a data center performance decreases when we try to reduce the temperature because some jobs must wait until they are assigned to some low-temperature computing nodes, so their response time may increase. Therefore, this performance reduction (i.e., increase in response time, which is defined as job execution time and job queuing time) must be kept under an acceptable level. A comparison among various thermal task scheduling algorithms is shown in Table 16.4.

TABLE 16.4

Comparison among Thermal-Aware Task Scheduling Algorithms

References	Problem Formulation	Online/Offline	Complexity
[60,63]	CFD	Offline	High
[64,65]	Sensor based	Online	Acceptable
[61,66]	Genetic algorithm (GA) and quadratic programming (QP)	Online	Acceptable

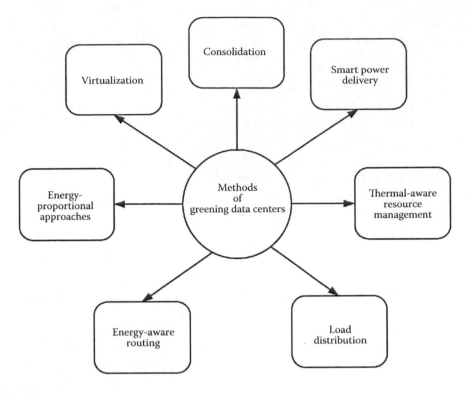

FIGURE 16.9   Classification of data center greening methods.

16.5   SUMMARY AND FUTURE RESEARCH DIRECTIONS

This chapter has investigated and classified the methods proposed for greening data centers as depicted in Figure 16.9. These methods that increase the energy efficiency of data centers include resource consolidation, virtualization, energy-proportional schemes, energy-aware routing, load distribution, smart power delivery systems, and thermal-aware resource management. Energy-proportional schemes are divided into two categories: rate switching and sleeping mode. In addition, some efforts can be carried out in the network layer in order to reduce total energy consumption. These algorithms are called energy/power-aware routing in which the key idea is to shut down idle or low utilized network resources (i.e., links and switches), assuring an acceptable level of QoS. An energy-aware routing problem can be formulated in the form of a linear or non-linear optimization problem that can be solved using evolutionary algorithms or simple heuristics. Load distribution schemes determine how and to where clients' requests must be directed in such a way that total energy and cost becomes minimum. These kinds of methods are known as novel greening solutions. Smart power delivery and thermal-aware resource management are other studied techniques.

For future research, some fast algorithms should be developed in order to accommodate the maximum number of requests using a minimum number of resources in a network. This is because the current proposed algorithms seem to be time consuming and may not be applicable in practical cases. Moreover, the proposed load distribution approaches have considered that the energy consumption of a data center is fully proportional to its load, which is not a realistic assumption. Thus, a more realistic model should be developed and some fast and practical load distribution algorithms must be designed. In addition, fabrication of small on-chip energy-efficient network devices such as routers and optical cross-connects is known as a primary solution for designing and implementing green network infrastructures. Finally, designing spectrally efficient modulation schemes and pulse shapes at the physical layer can lead to utilizing the resources more appropriately.

16.6 CONCLUSION

Data centers are becoming more important in modern infrastructures for high performance computing. Greening data centers requires a broad knowledge of its different components. In this chapter, various green metrics, known as important performance evaluation parameters, have been reviewed. Energy consumption models of different components in data centers have been studied and energy profiles have been classified and the performance of each has been investigated from the energy efficiency point of view.

AUTHOR BIOGRAPHIES

Amin Ebrahimzadeh received his BSc and MSc degrees, both in electrical engineering, from the University of Tabriz, Tabriz, Iran, in 2009 and 2011, respectively. He is currently a PhD student in the Department of Electrical Engineering at Sahand University of Technology, Sahand New Town, Tabriz, Iran. His research interests include optical WDM networks, network optimization, wireless sensor networks, and image transmission.

Akbar Ghaffarpour Rahbar is an associate professor in the Electrical Engineering Department at Sahand University of Technology, Sahand New Town, Tabriz, Iran. Dr. Rahbar received his BSc and MSc degrees in computer hardware and computer architecture, both from Iran University of Science and Technology, Tehran, Iran, in 1992 and 1995, respectively. He received his PhD degree in computer science from the University of Ottawa, Canada, in 2006. He is the director of Computer Networks Research Lab at Sahand University. Dr. Rahbar is a senior member of the IEEE. He is currently on the editorial board of *Wiley Transactions on Emerging Telecommunications Technologies Journal, International Journal of Advances in Optical Communication and Networks*, and *Journal of Convergence Information Technology*. He is editor-in-chief of *Journal of Nonlinear Systems in Electrical Engineering*. His main research interests are optical networks, optical packet switching, scheduling, PON, IPTV, VANET, network modeling, analysis, and performance evaluation, the results of which can be found in over 80 published technical papers (for more information, see http://ee.sut.ac.ir/showcvdetail.aspx?id = 13).

REFERENCES

1. R. H. Katz, Tech titans building boom, *IEEE Spectrum,* vol. 46, pp. 40–54, 2009.
2. U. S. Environmental Protection Agency, EPA report to Congress on server and data center energy efficiency Public Law 109-431, Energy Star Program, August 2007.
3. A. P. Bianzino, C. Chaudet, D. Rossi, and J. L. Rougier, A survey of green networking research, *IEEE Communications Surveys and Tutorials,* vol. 14, pp. 3–20, 2012.
4. A. Qureshi, R. Weber, H. Balakrishnan, J. Guttag, and B. Maggs, Cutting the electric bill for internet-scale systems, *SIGCOMM Computer Communication Review,* vol. 39, pp. 123–134, 2009.
5. Y. Zhang and N. Ansari, Green data centers, in *Handbook of Green Information and Communication Systems*, Elsevier, USA, 2012.
6. C. D. Patel, R. Sharma, C. E. Bash, and A. Beitelmal, Thermal considerations in cooling large scale high compute density data centers, in *The Eighth Intersociety Conference on Thermal and Thermomechanical Phenomena in Electronic Systems, 2002, ITHERM 2002,* pp. 767–776, 2002.
7. T. Qinghui, S. K. S. Gupta, and G. Varsamopoulos, Energy-efficient thermal-aware task scheduling for homogeneous high-performance computing data centers: A cyber-physical approach, *IEEE Transactions on Parallel and Distributed Systems,* vol. 19, pp. 1458–1472, 2008.
8. D. Vanderster, A. Baniasadi, and N. Dimopoulos, Exploiting task temperature profiling in temperature-aware task scheduling for computational clusters in *Advances in Computer Systems Architecture*, vol. 4697, L. Choi, Y. Paek, and S. Cho, Eds., ed: Springer, Heidelberg, pp. 175–185, 2007.
9. G. Jung, K. R. Joshi, M. A. Hiltunen, R. D. Schlichting, and C. Pu, A cost-sensitive adaptation engine for server consolidation of multitier applications, in *Proceedings of the 10th ACM/IFIP/USENIX International Conference on Middleware*, Urbana, Illinois, pp. 1–20, 2009.

10. S. Mehta and A. Neogi, ReCon: A tool to recommend dynamic server consolidation in multi-cluster data centers, in *IEEE Network Operations and Management Symposium, 2008,* NOMS 2008, pp. 363–370, 2008.
11. A. Singh, M. Korupolu, and D. Mohapatra, Server-storage virtualization: Integration and load balancing in data centers, in *International Conference for High Performance Computing, Networking, Storage and Analysis, 2008,* SC 2008, pp. 1–12, 2008.
12. Y. Chen, A. Das, W. Qin, A. Sivasubramaniam, Q. Wang, and N. Gautam, Managing server energy and operational costs in hosting centers, *SIGMETRICS Performance Evaluation Review,* vol. 33, pp. 303–314, 2005.
13. R. Das, J. O. Kephart, C. Lefurgy, G. Tesauro, D. W. Levine, and H. Chan, Autonomic multi-agent management of power and performance in data centers, in *The Proceedings of the 7th International Joint Conference on Autonomous Agents and Multiagent Systems: Industrial Track,* Estoril, Portugal, 2008.
14. J. Nanyan and M. Parashar, Enabling autonomic power-aware management of instrumented data centers, in *IEEE International Symposium on Parallel and Distributed Processing, 2009, IPDPS 2009,* pp. 1–8, 2009.
15. L. Wang and S. Khan, Review of performance metrics for green data centers: A taxonomy study, *The Journal of Supercomputing,* pp. 1–18, 2011.
16. Recommended Data Center Temperature and Humidity. Available: http://www.avtech.com/About/Articles/AVT/NA/All/-/DD-NN-AN-TN/Recommended_Computer_Room_Temperature_Humidity.htm (Accessed June 2012), 2009.
17. D. M. Steven Pelley, Thomas F. Wenisch, and James W. VanGilder, Understanding and abstracting total data center power, in *Workshop on Energy Efficient Design (WEED),* June 2009.
18. http://www.google.com/about/datacenters/inside/efficiency/power-usage.html (Accessed July 2012).
19. The green grids opportunity: Decreasing data center and other IT energy usage patterns, Technical Report, The Green Grid, Feb, 2007.
20. L. A. Barroso and U. Hoelzle, *The Datacenter as a Computer: An Introduction to the Design of Warehouse-Scale Machines*: Morgan and Claypool Publishers, 2009.
21. X. Fan, W.-D. Weber, and L. A. Barroso, Power provisioning for a warehouse-sized computer, *SIGARCH Computer Architecture News,* vol. 35, pp. 13–23, 2007.
22. D. Meisner, B. T. Gold, and T. F. Wenisch, PowerNap: Eliminating server idle power, *SIGPLAN Notices,* vol. 44, pp. 205–216, 2009.
23. S. Rivoire, P. Ranganathan, and C. Kozyrakis, A comparison of high-level full-system power models, in *Proceedings of the 2008 Conference on Power Aware Computing and Systems,* San Diego, California, pp. 3–3, 2008.
24. P. Mahadevan, P. Sharma, S. Banerjee, and P. Ranganathan, A power benchmarking framework for network devices, in *Proceedings of the 8th International IFIP-TC 6 Networking Conference,* Aachen, Germany, pp. 795–808, 2009.
25. M. Weiser, B. Welch, A. Demers, and S. Shenker, Scheduling for reduced CPU energy mobile computing, *Proceedings of the First USENIX Symposium on Operating Systems Design and Implementation (OSDI),* Monterey, California, USA, November 14–17, 1994. USENIX Association 1994, pp. 13–23.
26. C. Gunaratne, K. Christensen, and B. Nordman, Managing energy consumption costs in desktop PCs and LAN switches with proxying, split TCP connections, and scaling of link speed, *International Journal of Network Management,* vol. 15, pp. 297–310, 2005.
27. L. A. Barroso and U. Hölzle, The case for energy-proportional computing, *Computer,* vol. 40, pp. 33–37, 2007.
28. N. Rasmussen, Electrical efficiency modeling for data centers, APC by Schneider Electric, Technical Report #113, 2007.
29. S. Srikantaiah, A. Kansal, and F. Zhao, Energy aware consolidation for cloud computing, in *Proceedings of the 2008 Conference on Power Aware Computing and Systems,* San Diego, California, pp. 10–10, 2008.
30. A. Kansal and F. Zhao, Fine-grained energy profiling for power-aware application design, *SIGMETRICS Performance Evaluation Review,* vol. 36, pp. 26–31, 2008.
31. Q. Zhu, Z. Chen, L. Tan, Y. Zhou, K. Keeton, and J. Wilkes, Hibernator: Helping disk arrays sleep through the winter, *SIGOPS Operating Systems Review,* vol. 39, pp. 177–190, 2005.
32. T. Brey, Impact of virtualization on data center physical infrastructure, Green Grid White Paper, 2011.
33. T. C. C. S. Nanda, A survey on virtualization technologies, Technical Report TR179, Department of Computer Science, SUNY at Stony Brook, 2005.
34. N. M. M. K. Chowdhury and R. Boutaba, A survey of network virtualization, *Computer Networks,* vol. 54, pp. 862–876, 2010.
35. M. Gupta and S. Singh, Greening of the Internet, in *Proceedings of the 2003 Conference on Applications, Technologies, Architectures, and Protocols for Computer Communications,* Karlsruhe, Germany, pp. 19–26, 2003.

36. M. Gupta, S. Grover, and S. Singh, A feasibility study for power management in LAN switches, in *Proceedings of the 12th IEEE International Conference on Network Protocols, 2004, ICNP 2004*, pp. 361–371, 2004.

37. M. Gupta and S. Singh, Using low-power modes for energy conservation in Ethernet LANs, in *The IEEE INFOCOM 2007. 26th IEEE International Conference on Computer Communications*, pp. 2451–2455, 2007.

38. C. Gunaratne, K. Christensen, B. Nordman, and S. Suen, Reducing the energy consumption of ethernet with adaptive link rate (ALR), *IEEE Transactions on Computers,* vol. 57, pp. 448–461, 2008.

39. A. Krioukov, P. Mohan, S. Alspaugh, L. Keys, D. Culler, and R. H. Katz, NapSAC: Design and implementation of a power-proportional web cluster, *Proceedings of the First ACM SIGCOMM Workshop on Green Networking*, New Delhi, India, 2010.

40. S. Nedevschi, L. Popa, G. Iannaccone, S. Ratnasamy, and D. Wetherall, Reducing network energy consumption via sleeping and rate-adaptation, in *Proceedings of the 5th USENIX Symposium on Networked Systems Design and Implementation*, San Francisco, California, pp. 323–336, 2008.

41. Y. Agarwal, S. Savage, and R. Gupta, SleepServer: A software-only approach for reducing the energy consumption of PCs within enterprise environments, in *Proceedings of the 2010 USENIX Conference on USENIX Annual Technical Conference*, Boston, MA, 2010.

42. L. Chiaraviglio, M. Mellia, and F. Neri, Reducing power consumption in backbone networks, in *IEEE International Conference on Communications, 2009, ICC '09*, pp. 1–6, 2009.

43. L. Chiaraviglio, M. Mellia, and F. Neri, Energy-aware networks: Reducing power consumption by switching off network elements in *FEDERICA-Phosphorus Tutorial and Workshop*, Bruges, Belgium, May 18, 2008.

44. W. Fisher, M. Suchara, and J. Rexford, Greening backbone networks: Reducing energy consumption by shutting off cables in bundled links, in *Proceedings of the First ACM SIGCOMM Workshop on Green Networking*, New Delhi, India, pp. 29–34, 2010.

45. J. C. C. Restrepo, C. G. Gruber, and C. M. Machuca, Energy profile aware routing, in *IEEE International Conference on Communications Workshops, 2009,* ICC Workshops 2009, pp. 1–5, 2009.

46. R. G. Garroppo, S. Giordano, G. Nencioni, and M. Pagano, Energy aware routing based on energy characterization of devices: Solutions and analysis, in *2011 IEEE International Conference on Communications Workshops (ICC)*, pp. 1–5, 201.

47. R. G. Garroppo, S. Giordano, G. Nencioni, and M. G. Scutella, Network power management: Models and heuristic approaches, in *Global Telecommunications Conference (GLOBECOM 2011)*, 2011 IEEE, pp. 1–5, 2011.

48. Y. Shang, D. Li, and M. Xu, Energy-aware routing in data center network, in *Proceedings of the First ACM SIGCOMM Workshop on Green Networking*, New Delhi, India, pp. 1–8, 2010.

49. M. Al-Fares, A. Loukissas, and A. Vahdat, A scalable, commodity data center network architecture, *SIGCOMM Computer Communication Review,* vol. 38, pp. 63–74, 2008.

50. C. Guo, G. Lu, D. Li, H. Wu, X. Zhang, Y. Shi, C. Tian, Y. Zhang, and S. Lu, BCube: A high performance, server-centric network architecture for modular data centers, *SIGCOMM Computer Communication Review,* vol. 39, pp. 63–74, 2009.

51. H. Kellerner, U. Pferschy, and D. Pisinger, *Knapsack Problems*: Springer, New York, 2010.

52. S. Gangxiang and R. S. Tucker, Energy-minimized design for IP over WDM networks, *IEEE/OSA Journal of Optical Communications and Networking,* vol. 1, pp. 176–186, 2009.

53. K. Zhu and B. Mukherjee, Traffic groming in an optical WDM mesh network, *IEEE Journal on Selected Areas in Communications,* vol. 20, pp. 122–133, Jan 2002.

54. C. Xin, "Blocking Analysis of Dynamic traffic grooming in mesh WDM optical networks, *IEEE Transcations on Networking,* vol. 15, pp. 721–733, June 2002.

55. H. Tada, M. Imase, and M. Murata, *Evaluation of Effect of Network Energy Consumption in Load Distribution Across Data Centers Broadband Communications, Networks, and Systems*, vol. 66, I. Tomkos, C. J. Bouras, G. Ellinas, P. Demestichas, and P. Sinha, Eds., ed: Springer, Heidelberg, pp. 501–517, 2012.

56. R. B. Kien Le, Margaret Martonosi, and Thu D. Nguyen, Cost-and energy-aware load distribution across data centers, in *Proceedings of HotPower*, 2009.

57. S. Pelley, D. Meisner, P. Zandevakili, T. F. Wenisch, and J. Underwood, Power routing: Dynamic power provisioning in the data center, *SIGPLAN Notices,* vol. 45, pp. 231–242, 2010.

58. W. Lizhe, G. von Laszewski, J. Dayal, and T R. Furlani, Thermal aware workload scheduling with backfilling for green data centers, in *Proceedings of the 2009 IEEE 28th International Performance Computing and Communications Conference (IPCCC)*, pp. 289–296, 2009.

59. J. Moore, J. Chase, P. Ranganathan, and R. Sharma, Making scheduling "cool": Temperature-aware workload placement in data centers, in *Proceedings of the Annual Conference on USENIX Annual Technical Conference*, Anaheim, CA, 2005.

60. C. Jeonghwan, K. Youngjae, A. Sivasubramaniam, J. Srebric, W. Qian, and L. Joonwon, A CFD-based tool for studying temperature in rack-mounted servers, *IEEE Transactions on Computers,* vol. 57, pp. 1129–1142, 2008.

61. Q. Tang, S. K. S. Gupta, and G. Varsamopoulos, Energy-efficient thermal-aware task scheduling for homogeneous high-performance computing data centers: A cyber-physical approach, *IEEE Transactions on Parallel and Distributed Systems,* vol. 19, pp. 1458–1472, 2008.

62. L. Wang, G. V. Laszewski, J. Dayal, X. He, A. J. Younge, and T. R. Furlani, Towards thermal aware workload scheduling in a data center, in *Proceedings of the 2009 10th International Symposium on Pervasive Systems, Algorithms, and Networks*, pp. 116–122, 2009.

63. A. H. Beitelmal and C. D. Patel, Thermo-fluids provisioning of a high performance high density data center, *Distributed and Parallel Databases,* vol. 21, pp. 227–238, 2007.

64. Q. Tang, T. Mukherjee, S. K. S. Gupta, and P. Cayton, Sensor-based fast thermal evaluation model for energy efficient high-performance data centers, in *Fourth International Conference on Intelligent Sensing and Information Processing*, pp. 203–208, 2006.

65. Q. Tang, S. K. S. Gupta, and G. Varsamopoulos, Thermal-aware task scheduling for data centers through minimizing heat recirculation, in *Proceedings of the 2007 IEEE International Conference on Cluster Computing*, pp. 129–138, 2007.

66. T. Mukherjee, Q. Tang, C. Zeisman, S. K. S. Gupta, and P. Cayton, Software architecture for dynamic thermal management for data centers, in *2nd International Conference on Communication Systems Software and Middleware*, Bangalore, 2007.

17 Energy Efficiency Improvement with the Innovative Flexible-Grid Optical Transport Network

Jorge López Vizcaíno, Yabin Ye, Víctor López,
Felipe Jiménez, Raúl Duque, Idelfonso Tafur Monroy,
and Peter M. Krummrich

CONTENTS

17.1   INTRODUCTION

Internet traffic demand has been growing at a rate of 40%–60% per annum for the past few years, forcing operators to upgrade their networks to handle these new capacity requirements. A network upgrade usually implies additional economic efforts resulting from the need for new infrastructure deployments, increased labor costs, and higher electrical power consumption. In this context, telecom operators are burdened with cost optimization as a means to maintain the difficult balance between high-performance services and limited revenues. A reduction of capital expenditures (CAPEX) and operational expenditures (OPEX) is key to achieving this target. Process automation, network data mining, control plane-driven architectures, and many other functionalities are being increasingly deployed for this purpose. Power consumption has recently become a critical budget item because of its growth in volume and cost. Moreover, environmental issues are gaining an ever-increasing relevance as well and receive special attention in the telecom operators' policies.

From the cost perspective, energy optimization advantages are not restricted to lower operational expenses. They will also facilitate higher integration levels on network equipment boards, which will translate into lower costs per traffic capacity and less floor space at the telecom exchanges.

We can then conclude that all the efforts devoted to research and development of energy efficiency mechanisms will be extremely beneficial to ensure the continuous progress of the "Information Society" while preserving the natural environment.

The power optimization problem can be tackled in different ways, such as the following:

- From a *service perspective*, leveraging on the knowledge of service features, usage patterns, geographical distribution, and so on.
- From a *network perspective*, developing architectures and operational procedures that minimize overall power consumption. Benefits are expected at each network segment, ranging from the customer premises, where there are a big number of active elements with moderate power consumption, to the inner network core, where there are a small number of nodes each consuming lots of power. For each segment, optimization can be performed individually at each network layer or in a global fashion, following a multilayer approach (e.g., for IPoWDM [Internet protocol over wavelength division multiplexing] architectures in the network core) where savings are obtained for the overall network segment.
- From an *equipment perspective*, deploying elements with lower power consumption per transmitted bit and with different power modes, adjusted to user activity or traffic demands.

In this chapter, several approaches for energy efficiency improvements in optical transport networks (OTNs) will be described. A new emerging paradigm, known as elastic optical network (EON), is discussed in detail, and its potential for power optimization will be evaluated under different scenarios.

The new architecture is considered advantageous for many other purposes, such as increasing network capacity and spectral efficiency or reducing network expenditures. Moreover, EON helps in the designing of a multilayer architecture, where data communication over the transport infrastructure can be optimally engineered.

The remaining parts of this chapter are organized as follows. Section 17.2 provides background information about the current OTNs and their power consumption. Section 17.3 introduces the new challenges for future transport implementations. Section 17.4 presents some common approaches to improve energy efficiency in OTNs. Section 17.5 describes the operation of the EON solution. Section 17.6 presents the energy efficiency improvements that can be achieved by EON in different network scenarios. Section 17.7 presents some new challenges and future research directions, and Section 17.8 concludes the chapter.

17.2 OTN OVERVIEW

Currently deployed OTNs leverage on frequency multiplexing to exploit the useful bandwidth of the fiber medium, commonly corresponding to the so-called ITU-T (International Telecommunications Union-Telecommunication) C-band. By means of frequency multiplexing (usually called wavelength division multiplexing or WDM), several signals can be simultaneously carried over the same fiber while switched and terminated independently thanks to their different frequency location on the spectral band. The ITU-T is the organization in charge of the standardization of WDM networks, developing relevant recommendations for international vendor interoperability. The G.694.1 recommendation [1] specifies a frequency grid anchored to 193.1 THz, corresponding to a wavelength of 1552.52 nm, which roughly covers 4 THz around this "anchor" frequency as shown in Figure 17.1. Different channel spacing values, ranging from 12.5 to 100 GHz, have been defined by the ITU-T and are commonly referred to as ITU-T grids. Nowadays, most of the deployed DWDM systems operate using either the 50 or the 100 GHz ITU-T grid.

WDM networks were initially built as a set of independent optical routes with some specific locations (interconnection points) hosting nodes from two or more different routes. In these deployments, terminal nodes include optical transponders, mapping the client signals (based on Synchronous Digital Hierarchy [SDH], Gigabit Ethernet [GbE], etc.) into appropriate OTN containers over a specific WDM channel. All the WDM channels are subsequently multiplexed and the composite signal is injected into the optical route. At intermediate nodes, channels can be passed through toward their final destinations, regenerated to improve the signal quality, or added/dropped when a client data link is terminated at that node.

WDM legacy networks lack reconfiguration flexibility and need on-site manual interventions for any service modification (e.g., if a channel needs to be added/dropped at a different location). Adequate procedures for service restoration under failure events have to be designed as well and may include cumbersome tasks, as there are no self-healing mechanisms in the network.

Telecom operators have started deploying intelligent (control plane-driven) optical networks following meshed architectures. End-to-end services over these networks can be easily created and modified thanks to a network control plane, providing a distributed intelligence across the optical mesh. The new capabilities will also provide better exploitation of fiber resources and increased survivability, as service restoration can be done "on the fly" and in a relatively short time.

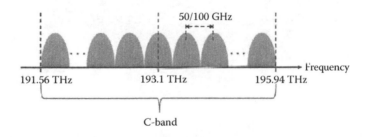

FIGURE 17.1 Optical spectrum in the ITU-T C-band.

17.2.1 Optical Transmission and Switching

Data transmission and switching are the key tasks performed by the transport networks. The extremely wide usable spectrum of the fiber and the possibility of multiplexing a high number of individual channels make it the option of choice as the supporting infrastructure for data networks.

Regarding optical transmission, several mechanisms are being developed to maximize fiber capacity, ranging from the extension of the operating frequencies to the C+L band (wavelength range between 1570 and 1610 nm) to the use of advanced modulation formats or, as under current research, the generation of multicarrier structures with higher spectral efficiency.

Deploying optical transponders with more complex modulation formats seems to be the most straightforward way to increase network capacity, as all other network elements do not need to be necessarily replaced. However, multicarrier structures or "super-channels" will be needed, at some point in time, to cope with higher-speed client signals (like 400 GbE or 1 TbE), which have already been researched by major system vendors.

Concerning optical switching, current transport networks comprise reconfigurable optical add and drop multiplexers (ROADMs) or optical cross-connects (OXCs) that are capable of performing signal switching at the pure optical level. This functionality brings an enormous scalability increase compared to electronic switching mechanisms (e.g., legacy SDH) and very relevant reductions in terms of CAPEX and power consumption.

Operational advantages are also obtained by the use of a distributed intelligence or network control plane, automating the routing and wavelength assignment (RWA) for every end-to-end optical connection (lightpath) as indicated in Figure 17.2.

17.2.1.1 Optical Transmission

Up to recent times, the maximum capacity of a single optical signal was limited to 10 Gbps. Also, in current deployments, every signal is assigned a 50 GHz slot (ITU-T grid) from the overall 4 THz ITU-T C-band, giving a maximum fiber link capacity of around 800 Gbps when the entire available spectrum is in use, that is, when 80 channels of 10 Gbps capacity are being multiplexed on a fiber link.

Straightforward mechanisms are used for optical modulation at 10 Gbps, whereby binary data signals are injected to an optical modulator and a logical "1" causes a laser light to traverse the

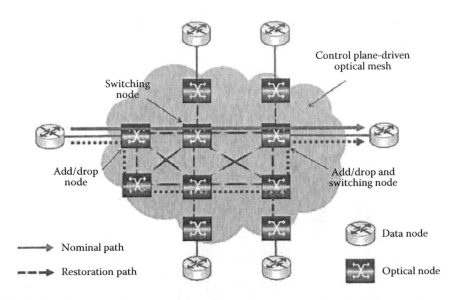

FIGURE 17.2 Routing of an end-to-end optical connection for its nominal and restoration paths.

modulator (ON), while for a logical "0," the laser light does not get out of the modulator block (OFF). This simple modulation technique, known as OOK (ON/OFF Keying), provides good performances for 10 Gbps deployments in an economical way (Figure 17.3a). Optical signal is "directly" detected in this case, based on the photodiode quadratic law, where squared optical field amplitudes (i.e., signal intensity) are converted to electrical levels corresponding to the transmitted signal.

Modulation formats for signal speeds higher than 10 Gbps make use of more than one carrier parameter, in contrast with OOK (or 2-level amplitude shift keying [ASK]) where information resides solely on carrier amplitude. Common approaches modulate the following:

- Carrier phase, also called phase shift keying (PSK) formats. Figure 17.3b and c depict the signal constellation for binary phase shift keying (BPSK) and quadrature phase shift keying (QPSK), respectively.
- Carrier amplitude and phase, also called quadrature amplitude modulation (QAM) formats. For instance, the signal constellation for 8-QAM and 16-QAM is presented in Figure 17.3e and f, respectively.

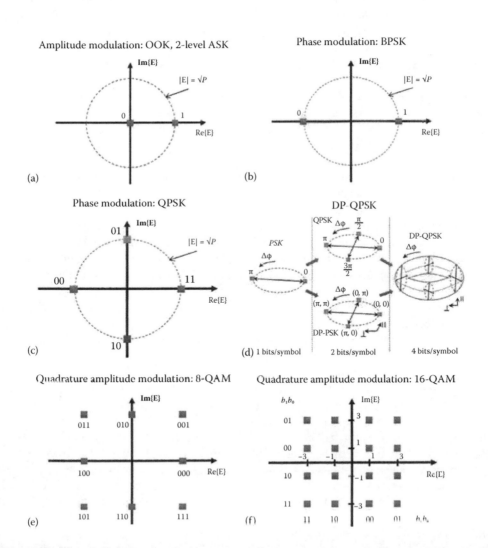

FIGURE 17.3 Signal constellations for different modulation formats: (a) OOK, (b) BPSK, (c) QPSK, (d) DP-QPSK, (e) 8-QAM, (f) 16-QAM.

These new modulation formats are combined, in most recent implementations with polarization multiplexing, where the data stream is divided into two parallel flows and modulates two optical carriers with orthogonal polarization state, as depicted in the signal constellation for dual polarization-QPSK (DP-QPSK) in Figure 17.3d (which is the result of the combination of two signals in quadrature [QPSK] and two signals with orthogonal polarizations [DP-PSK]). This procedure doubles the optical spectral efficiency, as twice as many data bits per hertz can be carried over an optical channel, making it possible to transport 100 Gbps payloads on a single 50 GHz frequency slot. Another key feature of the newest transmission systems lies in the use of coherent detection mechanisms, which allow the receiver to retrieve the full optical field information from the received signal, in contrast with direct detection systems where only the signal power is detected. The availability of a complete signal representation at the receiver side permits easy impairment compensation, which is performed by means of specific digital signal processing (DSP) algorithms.

New compensation capabilities of coherent systems bring some deployment differences with respect to simple OOK ones. Engineering rules have to be followed in both cases in order to stay within the system operating ranges and to prevent the effects of filtering, nonlinearities, and so on, but the main limiting factors will be different. Communications between 10 Gbps OOK transponders are feasible for optical routes comprising many amplified fiber spans; that is, OOK systems have strong tolerance to optical signal-to-noise ratio (OSNR). At the same time, polarization mode dispersion (PMD) is generally not an issue at 10 Gbps speeds, except in very specific scenarios with fiber segments in very bad conditions. However, chromatic dispersion (CD) needs to be periodically compensated in the line.

All fiber impairments become more relevant when the signal (symbol) rate increases. The use of DSP-assisted coherent detection, however, allows the receiver to compensate for any practical values of CD and PMD. This makes OSNR the main limiting factor for the new systems, especially when more complex QAM systems are deployed in order to obtain higher spectral efficiencies and signal speeds.

It can be foreseen that future transmission systems will probably take leverage on the existing blocks for 100 Gbps implementations:

- Complex modulation schemes for the transmission of a bigger number of bps/Hz.
- DSP-assisted coherent detection with increasingly advanced algorithms for impairments compensation: Future use of DSP at the transmitter side will foster the implementation of techniques like electrical pre-filtering (pulse shaping), signal pre-compensation, and so on, achieving extended performances.
- Polarization multiplexing as an easy way to duplicate spectral efficiency and perfectly match coherent detection.
- Advanced forward error correction (FEC) mechanisms to augment the net coding gain and thus deliver a lower bit error rate for a given OSNR value.

These new systems will optimize spectrum usage, carrying more information over the same frequency bands or channels. Compensation algorithms will also simplify design rules and extend deployment scenarios.

Efficiency increase, however, will be restricted to some practical value (e.g., 5 or 6 bps/Hz) where OSNR impact is not high enough to avoid almost any potential deployment. As a consequence, an increase on signal speed above a certain value (e.g., around 200 Gbps) would necessarily imply the use of multiple frequency slots (for the common ITU-T 50 GHz grid) or a single but wider frequency. This introduces the concept of "flexible-grid" networks, and, more generally, the EON paradigm, in which the bit-rate variable transponder (BVT) will play an important role to adjust the transmission capacity to user demands (by software reconfiguration of modulation format and bandwidth), bringing additional advantages like the possibility to reduce the current network grooming level and, as a by-product, the overall power consumption.

17.2.1.2 Optical Switching

Telecom operators are facing increasing traffic demands that foster the extension of their high-capacity optical networks closer to the end users. In the switching part, this extension is enabled by the following advances in new ROADM implementations:

- A higher number of switching degrees, resulting in a denser interconnectivity and larger capacities in terms of tributary signals.
- A simplified node architecture, with fewer elements and optimized cost and power consumption. This is favored by the current trend of deploying coherent technologies for every signal format, as frequency filters can be avoided as a result of the inherent filtering effect of the detection process.
- The colorless, directionless, and contentionless (CDC) features that enable the use of dynamically tunable transponders (colorless); the signal switching to any possible degree (directionless); and the reuse of the same wavelength (not sharing a physical path) by different transponders (contentionless).

Current ROADM capabilities (CDC) are recently being extended to the use of a flexible frequency grid (whereby different spectral widths can be assigned to the optical signals, becoming CDC-Flexible ROADMs). This will provide better usage of spectral resources, as they will be adjusted to individual signal needs.

BVTs and Flexi-grid ROADMs/OXCs are the key parts of the EON paradigm that will be described in detail in Section 17.5.

17.2.2 Energy Consumption in OTNs

Information and communication technology (ICT) is an area with an increasing relevance and a significant impact in the development of our society. As a consequence, the Internet traffic has been growing at significant rates (from approximately 40% to 60% per annum) owing to the increase in the number of users and high-bandwidth applications (e.g., video streaming). This ever-increasing traffic demand resulted in telecom operators and the industry itself to focus their research on enlarging network capacity, which is usually accompanied by higher energy consumption. The ICT sector accounted for 4% of the global energy consumption in 2009, but its contribution is predicted to rise up to 8% in 2020 with the expected traffic growth [2]. Within the ICT sector, the telecommunication networks take a significant part from the total energy consumption. Thus, the Internet (excluding computers, data centers, and home networks) with access rates at around 1–5 Mbps was responsible for 0.4% of the total electricity consumption in a broadband-enabled country in the year 2009 according to the estimations in Refs. [3] and [4], and a higher percentage can be expected in the following years. Among the different parts of telecommunication networks, the core part, the OTN, will play an essential role in coping with the new capacity requirements, and its energy consumption will be significantly affected.

As mentioned before, the implications of a higher power consumption are not only economical, affecting the OPEX and CAPEX, but also ecological, as the carbon footprint of the operators will be increased. In this regard, the ITU estimated that, in 2008, the ICT sector (considering telecommunications, computing, and the Internet, but excluding broadcasting transmitters and receivers) was responsible for 2%–2.5% of the global greenhouse gas (GHG) emissions [5], which have a direct impact on global warming. This percentage is similar to the contribution from the air transport industry, but increasing much faster (the carbon emissions from the ICT sector will be doubled in 2020 [6]). The larger the network becomes, the higher the contribution of telecom networks to both total power consumption and GHG. In this situation, energy efficiency has become one of the key design parameters for network planning and operation, allowing the maintenance of acceptable profit margins while reducing its contribution to global warming.

17.3 FUTURE OTNs

Research around advanced transmission systems will make "beyond 100 Gbps" solutions available in a very short term. First implementations will be compatible with current 50 GHz deployments, offering higher per-channel performance in terms of capacity, reach, and impairments tolerance. Some of the main technological enablers are as follows:

- Advanced transponders with higher spectral efficiency (e.g., DP-16-QAM), including DSP at the transmitter side and more complex algorithms (compared to current DP-QPSK implementations) at the receiver side
- New signal amplification mechanisms, such as hybrid Raman and erbium-doped fiber amplifier (EDFA), to improve OSNR and reduce nonlinear effects
- Special types of fiber that reduce signal attenuation, nonlinear effects, and so on

Network traffic volumes are growing at such incremental rates that any extra capacity provided by the new technologies will soon become insufficient. Overall network capacity could then be augmented by means of a higher grooming level, more frequent regenerations, or, whenever possible, better exploitation of available ROADM degrees. Any of these possibilities come at a cost in equipment expenditures and power consumption, increase operational tasks, and may reflect on lower resilience levels.

The EON approach tries to exploit a different dimension to extend network capacity or optimize infrastructure investments. The key point here is not only to deploy new transmission technologies with higher spectral efficiency but also to make a better use of available bandwidth. Consequently, a relevant increase of "useful" capacity can be obtained with respect to current WDM deployments.

The most relevant benefits brought by EON are the following:

- A *fine adjustment of signal capacity* to the actual traffic demand, allowing the optical network to support more client traffic on every network link. This can also bring relevant power consumption savings with respect to WDM networks, in which the coarse granularity of a wavelength may result in power inefficiency when the demand is much lower than the wavelength capacity, as WDM transponders consume a constant power regardless of the user demand.
- *Grooming levels* can be reduced as unused optical capacity becomes lower. This allows for network simplification and higher reliability, providing significant savings on investments and power consumption.
- *Dynamic adjustments on performance, network resources, and power consumption* are made possible. The BVT capabilities could allow selecting the signal format (modulation, symbol rate, FEC, etc.) best suited to the network conditions, in order to save power (e.g., avoiding regenerators) or spectral resources in Flexi-grid deployments. Also, transmission performance can be fine-tuned through modulation format configuration, amount of FEC overhead, and so on.
- *Network resilience* level will grow as a result of a higher number of feasible routes. Lower blocking ratios will also translate into global capacity augments, which is also beneficial in energy efficiency (e.g., smaller number of fibers can also imply a reduction in the number of some energy-consuming devices that are deployed in the network such as optical amplifiers [OAs]).

- *Future high-speed signals, like 400 GbE or 1 TbE*, could be transported as a single entity thanks to the Flexi-grid functionality. This avoids the need for concatenation of multiple lower-speed individual signals and thus can reduce the switching energy consumption.
- *Logistic advantages and potential economies of scale.* The deployment of BVTs would strongly reduce the network spares inventory, as a single element could be employed for different rates, performance levels, or network conditions. Even though maximum BVT capacity and performance could be much higher than needed for some traffic demands, economies of scale would cause BVT cost to reduce over time and give rise to overall expenditure savings.

The main technological blocks of EONs are already a commercial reality or will become available soon. Wavelength selective switches (WSSs) with Flexi-grid capabilities are being produced by major suppliers, though they are not yet fully supported by system vendors. Likewise, some first implementations of BVT have been announced. They will be compatible with the current ITU-T 50 GHz grid and, thus, could be deployable in the short term. These first BVTs would carry a single client signal over a multicarrier structure although every optical carrier (on a 50 GHz slot) will be individually treated by the network. A comparison of fixed transponders, BVT, and sliceable BVT in terms of necessary spares inventory is shown in Figure 17.4.

It is expected that as soon as Flexi-grid networks start to be deployed, BVT implementations over flexible frequency slots will appear, further improving spectral efficiency as a consequence of

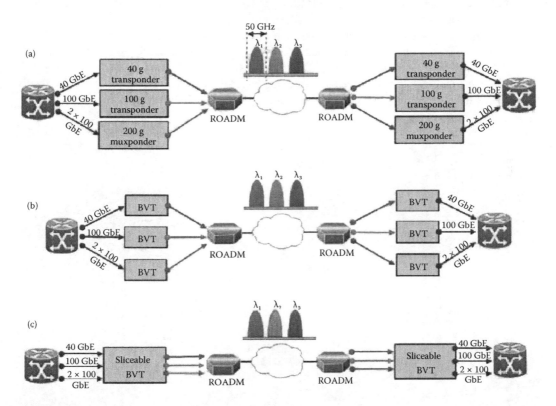

FIGURE 17.4 Comparison of fixed transponders, BVT, and sliceable BVT in terms of necessary spares inventory. (a) Three interfaces with different capacities are connected to 40G, 100G, and 200G transponders/muxponders. Spares are required for all three types of transponders/muxponders. (b) Only 200G BVTs are used for all client interfaces, which brings about a reduction of spares at the expense of the underutilization of the transponder capacity for the first two services. (c) Sliceable BVTs allow generating several elastic optical paths, thus reducing the number of transponders on the network and the amount of necessary spares.

guard bands suppression, as presented in Figure 17.5. In this case, a super-channel structure spanning over a wide frequency band would be switched along the optical network as a single entity.

Although the main EON enablers are available or under advanced research, there are still many open points needing further research. On the one hand, the technology of choice for the future super-channel structures is still under consideration. Major system vendors are currently working on multiple approaches, trying to identify their main benefits and drawbacks. Technical feasibility, economic aspects, and potential for evolution will probably drive the choice of one or a small subset of solutions, helping create a technology ecosystem and fostering generalized network deployments. On the other hand, standardization bodies have started activities on both data and control plane-related issues. Regarding the data plane, the next-generation GbE rate (400 GbE, 1 TbE, etc.) has not been decided yet. The final decision will drive the definition of the new optical transport hierarchy (OTU-5, which stands for optical transport unit 5), while some proprietary frame formats are expected in advance for new optical signals multiplexing several client interfaces, like 4 × 100 GbE over a 400 Gbps optical carrier.

Regarding the control plane, the change of network paradigm will imply relevant modifications on current operation and provisioning procedures:

- The network control plane will have to be extended to account for transponder configurability as well as spectral assignment flexibility.
- Optimization algorithms will be needed to avoid resources fragmentation coming from the use of a flexible width frequency slot, which will create variable spectral gaps between allocated signals, where some new traffic demands may not be accommodated.
- Impairment awareness will also play an important role, as there is a shift from a network based on a few, well-characterized signal formats to an elastic network where a wide range of formats will be available. Figure 17.6 presents a possible application of impairment awareness showing the potential benefits of distance adaptive routing for the network operation (adapting the modulation format according to the path length). Thus, with a BVT, it

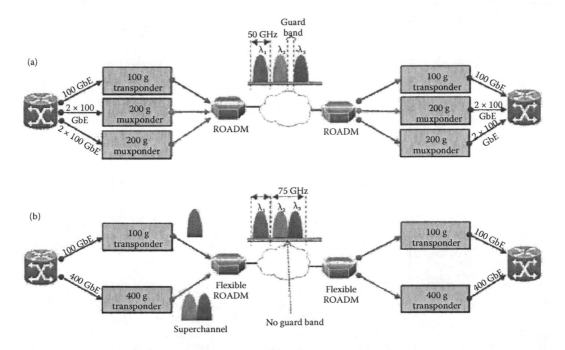

FIGURE 17.5   Comparison of (a) single-carrier channel and (b) super-channel transmission.

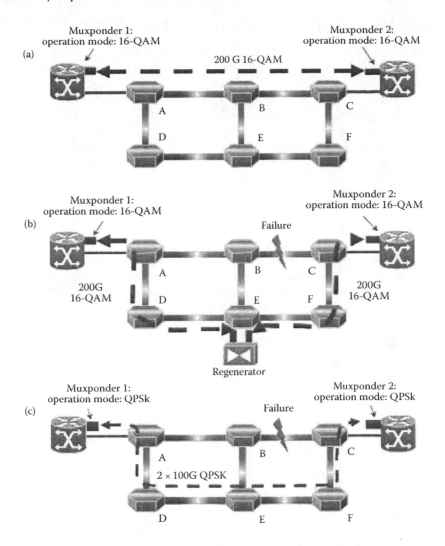

FIGURE 17.6 Impairment-aware, distance-adaptive routing. (a) Initial situation: A spectrally efficient 16-QAM 200 Gbps optical path across route A–B–C is established between muxponders 1 and 2. (b) A failure event between nodes B and C triggers the establishment of a new optical path across route A–D–E–F–C, which requires optical–electrical–optical (OEO) regeneration at node E given that the reach of a 16-QAM muxponder is not long enough for the new route. (c) On an impairment-aware network, the OEO can be avoided on route A–D–E–F–C by using BVTs that allow changing the modulation format as a function of the length of the optical path (in this case, 16-QAM becomes QPSK).

would be possible to select very spectral efficient formats such as 16-QAM (or even higher-order modulation formats) for short distances, whereas more robust modulation formats such as BPSK or QPSK could be used for longer distances, thus reducing the number of necessary regenerations.

All the efforts described above will be worth making owing to the potential advantages brought by the EON, which range from the extension of overall network capacity to the dynamic optimization of resources and lower power consumption. Multiple use cases that leverage on EON technologies can be defined for specific purposes, with power optimization being one of the most straightforward applications of the newly available capabilities. Energy efficiency evaluation results over EON networks will be presented in Section 17.6.4.

17.4 COMMON APPROACHES TO IMPROVE ENERGY EFFICIENCY IN OTNs

Significant research effort has been spent to improve the energy efficiency of OTNs. Several surveys summarize some important approaches to increase energy efficiency in optical networks, such as Refs. [4] and [7] and especially Ref. [8].

At the component or device level, reducing the extra power consumption produced by the equipment inefficiencies and overheads (i.e., ancillary functions that are not essential to the main operation of the network) is a common approach. Improvements can be achieved, among others, by the adoption of new materials for the components, by new architecture designs, by optimizing the software of some devices (e.g., optimizing DSP algorithms at transmitter and receivers), and by augmenting the performance of transmission elements (e.g., through a better efficiency in EDFA pump lasers). Tucker estimates that energy per bit of the network equipment is reduced at a rate of 15% per annum [7]. Nevertheless, this improvement rate is not directly applied to the total energy efficiency of the network, as most of the equipment is not replaced every year and different generations of technologies usually coexist in the network.

A significant improvement in energy efficiency can be achieved by the extensive application of optical bypass to increase the transparency of the network. The benefits come from the reduction in the number of high power-consuming OEO conversions, as the signal can be transported, amplified, and switched directly in the optical domain. This can be achieved thanks to the deployment of OAs and OXCs. In a fully transparent network, the OEO conversion is only necessary for the add/drop functionality in the interface between the data equipment and the optical node (i.e., only at the source and destination nodes). The switching energy required in an OXC is two orders of magnitude lower than an equivalent router switching energy [3], leading to a 25%–45% power consumption reduction [9].

At the network level, many energy-aware routing or adaptive routing solutions have been proposed to improve the energy efficiency of the network by considering the power consumption of the elements in the routing decision and resource allocation phases. Energy savings from 10% to 85% could be obtained [10]. This idea can be complemented by partially deactivating or putting some of the devices into a sleep mode (i.e., low energy state) when the traffic load is below a certain threshold (e.g., at night) and rerouting the traffic along different links [8].

Furthermore, resilience is becoming more and more important for the availability of some services in our society. Dedicated path protection $1 + 1$ ($DP\ 1 + 1$), in which the data are duplicated and transmitted on both paths simultaneously, is the most widely used scheme. Therefore, significant power is consumed in network protection. Some solutions propose the migration from this conventional protection scheme to a more energy-efficient ones by, for instance, sharing backup resources (shared path protection schemes) or putting the backup equipment into sleep mode [11].

Improvements in the network protocols can also help enhance energy efficiency. For instance, computation processing can be reduced in switches and routers by the use of smaller protocol overheads or introducing some extensions into the existing protocols to enable standby or sleep modes (e.g., IEEE Energy Efficient Ethernet [802.3az], where transmission can be set to a low-power idle mode when no packets are being sent).

Energy efficiency can also be improved by using the resources in a more effective manner, for instance, by a finer adjustment of the transmission to the actual traffic demand. In order to get a finer capacity granularity, a straightforward solution would be based on introducing mixed line rate (MLR) operation (e.g., 10, 40, and 100 Gbps), instead of employing a single line rate (SLR) [12]. Moreover, performing traffic grooming (i.e., combining different low-rate traffic demands onto a higher capacity lightpath) together with statistical multiplexing can also be applied to reach a better utilization of the resources, consequently enhancing the energy efficiency of the network [13]. Finally, EON can also offer important advantages to achieve a better utilization of the resources. The application of distance-adaptive working modes together with the Flexi-grid operation can result in better spectral and energy efficiency compared to the current SLR operation mode and fixed grid of current WDM networks. The potential energy efficiency improvements of EON will be shown in detail in Section 17.6.4.

17.5 ELASTIC OPTICAL NETWORKS

17.5.1 INTRODUCTION

As mentioned in Section 17.3, to improve network capacity, new transmission techniques with increased spectral efficiency will be required, such as those based on using a higher modulation order. Furthermore, a channel wider than 50 GHz will become necessary. For instance, the bandwidth of a 400 Gbps channel may use DP 16-QAM, which could require a 75 GHz frequency width, while a 1 Tbps channel may use DP 32-QAM, which could require a width of 150 GHz.

The modification of the grid or channel spacing is the main difference between current WDM and Flexi-grid technologies, in what resource allocation is concerned. As already indicated, the first super-channel implementation will be supported over the current 50 GHz grid, as shown in Figure 17.7. In the case of 1 Tbps, a three-channel bundle would have enough bandwidth to transmit a 1 Tbps channel. However, a 400 Gbps channel would occupy two 50 GHz channels, which is a huge waste of spectrum.

Standardization forums are working on the definition of smaller grids (which will probably be multiples of 6.25 GHz), which yield greater capacity granularity and allow for better spectrum utilization. Figure 17.8 depicts an example of channel allocation using Flexi-grid with 25 GHz spacing. As shown, thanks to its finer spectrum granularity, 75 GHz is available for the allocation of other channels.

Extended flexibility is even more critical if we consider the other main element of EONs, that is, the BVT. This element will have dynamic reconfiguration capabilities, and potential implementations

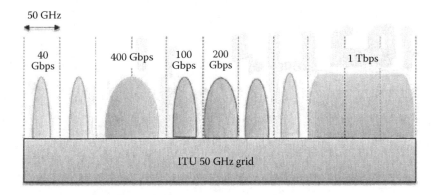

FIGURE 17.7 Example of channel allocation with current fixed grid.

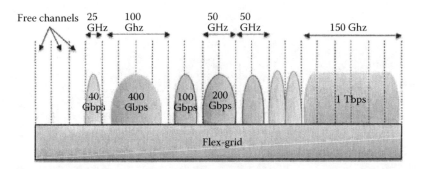

FIGURE 17.8 Example of channel allocation with Flexi-grid.

(like those based on a variable number of modulated carriers) could make use of different spectral widths (depending on lightpath conditions) for the same transmission speed. A good example of BVT based on a multicarrier structure is the OFDM solution being used for the analysis in Section 17.6.

Although EONs bring evident advantages, it is worth mentioning that the new approach will imply changes in different fields like control plane, allocation algorithms, traffic grooming, and survivability mechanisms.

17.5.2 ORTHOGONAL FREQUENCY-DIVISION MULTIPLEXING (OFDM) BASICS

Orthogonal frequency-division multiplexing (OFDM) techniques are well known in telecommunication systems because of their application in wireless and fixed access networks. The introduction of OFDM as the modulation technique in optical networks opens new horizons in transport planning and operation, and the evaluation of their potential benefits as a substitute of traditional WDM is currently a topic under intensive research. The application of OFDM in OTNs has sometimes been referred to in the literature as Spectrum-sLICed Elastic optical path network "SLICE," and it has been experimentally demonstrated in Refs. [14–16].

OFDM is a multicarrier modulation scheme, which uses multiple low-rate subcarriers, each transmitting a portion of the total channel rate. As OFDM uses orthogonal signals in the frequency domain, individual subcarriers overlap, thus reducing the channel spectral occupation. Figure 17.9 shows an example of OFDM operation with four subcarriers, depicting the spectrum savings that can be achieved when subcarriers are overlapped with respect to the fixed-grid allocation. The subcarrier frequencies are chosen in such a manner that they are orthogonal to each other, meaning that cross-talk between the sub-channels at the sampling points is removed. As shown in Figure 17.10, the main concept is based on the fact that the power peak of a subcarrier (on its central frequency) occurs at the same frequency at which its adjacent subcarriers have a zero power value.

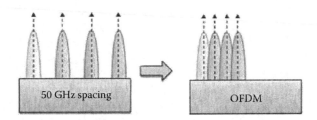

FIGURE 17.9 Spectrum savings when using OFDM.

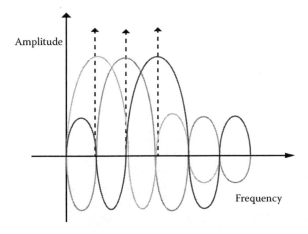

FIGURE 17.10 View of multiple subcarriers in the spectrum domain.

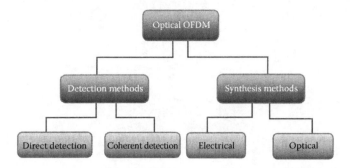

FIGURE 17.11   Classification of O-OFDM schemes.

Optical OFDM (O-OFDM) can be divided into different categories, depending on how the signal is detected and synthesized, as presented in Figure 17.11. A brief description of each method is presented in the following:

- Detection methods
  - *Direct detection.* The optical carrier and the OFDM baseband are transmitted in this technique; thus, a single photodiode, which converts the optical signal into the electrical domain, is used at the receiver to directly detect the signal. This method can be used for low rate signals, but it is more difficult to implement for higher speeds.
  - *Coherent detection.* This method uses a local oscillator, whose continuous wave signal is mixed with the incoming one. The receptor can retrieve signal amplitude and phase information and compensate CD and PMD by means of DSP algorithms. As the incoming signal is mixed with the oscillator signal, the power of the resulting signal is higher than that in a direct method. Moreover, since the oscillator selects a frequency, it intrinsically filters the incoming signal, thus reducing complexity in ROADM architectures.
- Synthesis methods
  - *Electrical.* The OFDM signal is generated in the electrical domain. The bits are parallelized, according to the modulation level and the number of data subcarriers in the inverse fast Fourier transform (iFFT), and mapped into the complex domain. Then, the OFDM frame is constructed (including guard bands, training symbols, and other overheads). After calculating the real and imaginary parts of the iFFT, which are converted into electrical digital signals, each signal traverses a digital-to-analog converter (DAC) module and finally they are used to modulate the transmission laser by means of an I/Q modulator (i.e., it consists of a local oscillator input that is split into in-phase [I] and quadrature [Q] components, which are separated by 90°).
  - *Optical.* In this case, optical subcarriers can be generated by a set of independent lasers or by a frequency comb generator. Each optical subcarrier may contain one (i.e., OOK or PSK) or more bits (i.e., advanced phase and amplitude modulation formats). These subcarriers are then joined fulfilling the orthogonality principle.

17.5.3   PHYSICAL BUILDING BLOCKS IN THE OPERATION OF THE ELASTIC NETWORK

EON architecture allows a more flexible use of the optical spectrum, because spectrum slots can be defined more arbitrarily than in conventional wavelength switched optical networks (WSONs). The slot size will be less than 50 GHz, and most probably 12.5 GHz, so that the ROADM/OXC is capable of adding or dropping almost any signal with a bandwidth multiple of the slot size regardless of its spectral position. A possible architecture of a bandwidth-variable node can be composed

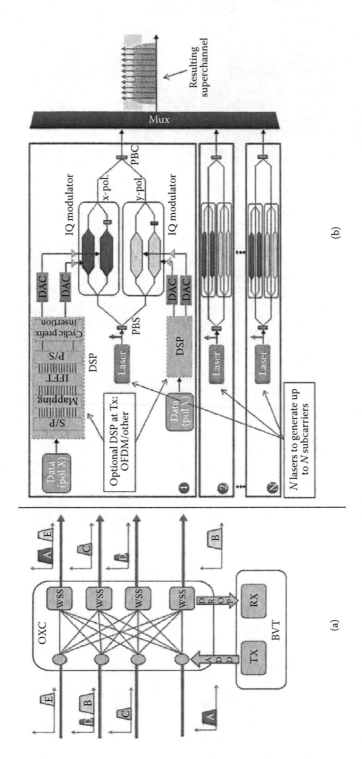

FIGURE 17.12 (a) Possible architecture of a Flexi-grid node with BVT and grid-less WSS. (Adapted from B. Kozicki et al., *Optics Express*, vol. 18, pp. 22105–22118, Oct. 2010.) (b) Possible architecture of a BVT.

of optical splitters in the inputs and flexible-grid WSSs in the outputs, as shown in Figure 17.12a, in which signals with different bandwidths are switched, added, and dropped.

The other key element for the operation of EON is the BVT, which brings additional advantages like the possibility of modifying any of the signal properties by software and adapting the transmission at the user's request. A possible architecture of such a transponder is depicted in Figure 17.12b, where a set of N parallel transmitters are used to generate N subcarriers that are then combined into a super-channel. Each transmitter differs from a conventional WDM one in the presence of a DSP module and two DACs. An interesting advantage of a BVT is the possible adjustment of the transmission rate (TR) to the actual traffic demand, by expanding or contracting the bandwidth of an optical path (i.e., varying the number of subcarriers N) and by modifying the modulation format. Moreover, another benefit of BVTs is the possibility of using a single type of transponder that is able to deliver a wide range of bit rates, thus avoiding the need to purchase different transponder models in the future. This would drive development and production costs down.

In order to serve a traffic demand between two nodes in the elastic network, the BVT creates an appropriately sized signal composed of different subcarriers with the most appropriate modulation format (e.g., the modulation format could be selected according to power consumption, spectral efficiency, or any other parameter). Then, every bandwidth-variable OXC in the route from source to destination sets the appropriately sized spectrum cross-connection creating the end-to-end light-path. In this manner, it is possible to allocate only the required spectrum size for the optical path with the consequent optimization of the spectral resources.

The EON paradigm affects both the switching and transmission elements in current optical networks. However, this does not imply that all legacy network equipment must be replaced from day zero. The deployment of flexible elements can be done gradually in the network. Figure 17.13a shows a WSON where all elements are fixed-grid compliant. On the other hand, Figure 17.13b depicts a network where network elements are migrated gradually. It is divided into two different domains according to the switching operation. In the EON domain, only elastic components are used, while in the WSON domain, only standard fixed-grid modules are used. It is worth mentioning that a Flexi-grid node can operate as a fixed-grid node.

17.5.4 Resource Allocation

Given a set of lightpath demands between a pair of nodes and constraints in the amount of available resources, the routing and resource allocation problem in OTNs is based on, determining the routes and the resources that have to be allocated to serve the demands. Besides the wavelength continuity constraint (i.e., a lightpath usually operates on the same wavelength across the links it traverses) of RWA problems in current fixed-grid WDM networks, EON adds three new dimensions to the problem: (1) elastic bandwidth, (2) spectrum continuity, and (3) presence of different modulation

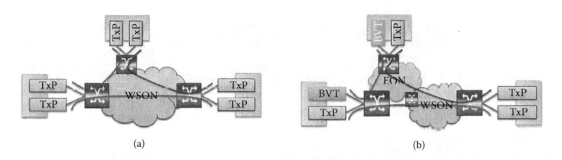

(a)　　　　　　　　　　　　　　　　　(b)

FIGURE 17.13 (a) Current WSON architecture. (b) Partial migration to EON.

formats. Thus, the routing and resource allocation problem in EON is referred to with a different term: Routing, Modulation Level, and Spectrum Allocation (RMLSA).

Since different modulation formats can be selected in the BVT, an RMLSA algorithm must first check which are the possible modulation formats for a service with a given traffic rate, that is, those that can provide the traffic rate requested by the service. The second step is to determine whether, for this modulation format, there are enough common and contiguous available slots along the fiber links in the path to allocate such demand. To solve the spectrum continuity problem, similar algorithms to those used in RWA can be used with appropriate modifications. Some of these possible algorithms can be First-Fit, Worst-Fit, Random, Most-Used, Least-Used, and so on [17]. Once the above-described constraints are satisfied, the last step is to evaluate whether the quality of the signal is enough for the selected path. As previously mentioned, there are different options when choosing the modulation format. Modulation formats with a higher order offer lower spectrum occupation but provide shorter reaches than those with lower order and higher spectrum occupation. Therefore, these algorithms usually define a maximum number of hops, spans, or transmission distance for each modulation format.

Once all demands are allocated, it is possible to modify their bandwidth on the basis of the real user request. However, depending on the technology, not every elastic scenario allows modifying all the transponder parameters. Figure 17.14 shows three different allocation schemes: (1) fixed assignment, (2) semi-elastic assignment, and (3) elastic assignment [18]. The fixed assignment scheme preassigns a bandwidth for each connection and permits changing the TR, but the central frequency and the assigned bandwidth cannot be changed. This method can be considered as the first option to be applied to real systems and would actually reduce the power consumption of the transponders. Semi-elastic allows changing the number of frequency slots for a connection but not the central frequency. Finally, the elastic assignment algorithm is capable of changing both the central frequency and the number of slots assigned to a connection.

The following section presents a set of energy-aware heuristic algorithms for the resource allocation in both current WDM networks and EON and a comparison between these technologies in terms of energy efficiency under different network scenarios.

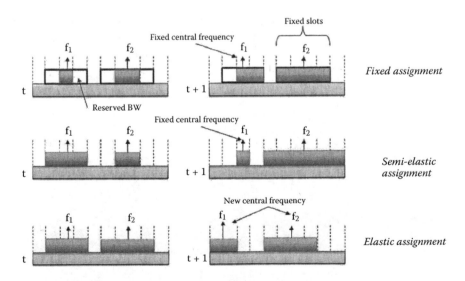

FIGURE 17.14   Examples of allocation schemes in EONs.

17.6 ENERGY EFFICIENCY EVALUATION FOR DIFFERENT SCENARIOS

The following subsections present the methodology (heuristic algorithms), the simulation parameters, and the results in energy efficiency obtained for OTNs in different network scenarios.

17.6.1 NETWORK CONSIDERATIONS

17.6.1.1 Current WDM Networks

In our WDM model, a maximum per-link capacity of 80 wavelengths within the 50 GHz ITU-T grid is assumed. Line rates of 10, 40, and 100 Gbps have been included in the analysis. Two types of operation are considered:

- *SLR*: Transmissions of 10, 40, or 100 Gbps with reaches of 3200, 2200, and 1880 km [19], respectively.
- *MLR*: Possible transmission of the three mentioned line rates (10, 40, and 100 Gbps) in the same fiber. In order to minimize the inter-channel nonlinearities between adjacent signals of different transmission technologies, the C-band has been divided into two independent wavebands, as proposed in Ref. [20], separated by a guard band of 200 GHz (4 channel spacing). The first one is used for 10 Gbps (OOK) transmission, and the second one is used for both 40 and 100 Gbps transmissions, which are based on a "compatible" modulation format (with no intensity variations over time), and thus can be placed on adjacent frequency slots without significantly affecting the signal quality of each other. Consequently, reach values similar to the SLR case can be considered.

17.6.1.2 Elastic Optical Network

The OFDM variation that has been considered for EON in this study assumes transmission on a single polarization with coherent detection where both optical and electrical synthesis are used; that is, different optical orthogonal subcarriers are generated, each of which is composed of several electrical subcarriers like the experience described in Ref. [21]. In our analysis, a frequency slot B of 12.5 GHz has been adopted, resulting in a total of 320 frequency slots in the 4 THz of the C-band. The choice of 12.5 GHz value is justified by "a compromise between operation complexity and flexibility" [22]. In order to maintain the orthogonality condition among the subcarriers, the subcarrier spacing must be equal to the baud rate. Since the subcarriers' spacing is locked in frequency, the TR of a subcarrier ($TR_{subcarrier}$) is determined by the subcarrier bandwidth or frequency slot (B) and the subcarrier modulation order (M), as in Equation 17.1:

$$TR_{subcarrier} = B \cdot \log_2(M). \tag{17.1}$$

Accordingly, the TR of a single subcarrier can be 12.5, 25, 37.5, 50, 62.5, and 75 Gbps for BPSK, QPSK, 8-QAM, 16-QAM, 32-QAM, and 64-QAM respectively. Then, a variable number of contiguous subcarriers are combined for the creation of a super-channel, which can be treated as a single entity in the network. Moreover, a bilateral guard band of one subcarrier per side (25 GHz in total) is considered to separate adjacent channels, so that each channel can be switched by the ROADMs/OXCs. A transmission reach of 4000, 2000, 1000, 500, 250, and 125 km has been assumed for BPSK, QPSK, 8-QAM, 16-QAM, 32-QAM, and 64-QAM, respectively [23].

17.6.2 ENERGY-AWARE ROUTING AND RESOURCE ALLOCATION ALGORITHMS

The routing and allocation of spectral resources may be targeted to optimize some specific parameters or factors in order to get an appropriate use of the limited resources, such as minimizing the traffic load in the links or some other performance indicators. As mentioned in Section 17.5.4, the main difference of the RMLSA with respect to WDM RWA lies in the contiguous spectrum allocation constraint (subcarriers in an optical path must be contiguous), different modulation format possibilities (and thus transmission reaches), and the larger number of subcarriers (320) compared to the wavelengths in WDM (80). Resource optimization problems are *NP*-complete (i.e., it is not possible to find a solution in polynomial time), and heuristic algorithms are usually employed instead. In the present analysis, the targeted optimization is to increase the overall energy efficiency of the network while serving the maximum number of traffic demands.

Heuristic algorithms for static and dynamic operation have been developed to solve the RMLSA for elastic OFDM-based networks and the RWA problem for both SLR and MLR networks in the following cases:

- Static scenario (offline problem) without protection
- Static scenario (offline problem) considering three common protection schemes: dedicated protection 1 + 1 (*DP* 1 + 1), dedicated protection 1:1 (*DP* 1:1), and shared protection (*SP*)
- Dynamic scenario (online problem)

The resource allocation in the static scenario follows a fixed allocation scheme, whereas an elastic allocation is assumed for the dynamic scenario (see Figure 17.14). The following subsections describe the main operation of the energy-aware heuristic algorithms for the different network approaches and scenarios.

17.6.2.1 Energy-Aware Routing and Resource Allocation Algorithms for the Different Network Approaches (EON, WDM SLR, and WDM MLR)

17.6.2.1.1 RMLSA for the Elastic OFDM-Based Network

First, a set of candidate paths (*k* shortest paths) from source to destination nodes is calculated. Then, the allocation is evaluated for the possible combinations of path and modulation format (those allowing for transparent reach), by searching for a common block of contiguous subcarriers in the links of the path following the First-Fit algorithm, that is, assigning the lowest indexes first (i.e., starting from the wavelengths/subcarriers in the lower part of the ITU-T C-band spectrum). The size of this block of subcarriers is determined by the required number of data subcarriers (i.e., traffic demand divided by the capacity of a single subcarrier for the corresponding modulation format) plus the bilateral guard band subcarriers, one at each side of the block. If the allocation is possible in a specific path (i.e., there are free overlapped resources for the subcarriers in all the links of the path), a metric based on the end-to-end lightpath power consumption (MetricPC) is calculated in Equation 17.2 considering the power consumption of the transponders (PC_{TRANS}) and an approximate contribution from the EDFAs and OXCs along the path (PC_{LINKS}). PC_{TRANS} (Equation 17.3) is the product of the number of data subcarriers (NumDataSubc) and the power consumption of a single subcarrier for the corresponding modulation format (PC_{SUBC}) from Table 17.1 in Section 17.6.3.1. PC_{LINKS} is obtained by the product between the proportion of resources that the lightpath would occupy in the links (ratio between the number of subcarriers of the lightpath, NumSubc, and the total number of subcarriers in a fiber, TotalNumSubc) and the addition of the power consumption of the EDFAs (PC_{EDFAs}) and OXCs (PC_{OXCs}) in the path, as in Equation 17.4. Finally, from the feasible combinations of path and modulation format, the most energy-efficient one (lowest MetricPC) can be adopted and resources are assigned correspondingly. Those demands that cannot be allocated in the network are blocked.

$$MetricPC = PC_{TRANS} + PC_{LINKS} \qquad (17.2)$$

$$PC_{TRANS} = NumDataSubc \cdot PC_{SUBC} \qquad (17.3)$$

$$PC_{LINKS} = \frac{NumSubc}{TotalNumSubc}(PC_{EDFAs} + PC_{OXCs}) \qquad (17.4)$$

17.6.2.1.2   RWA for WDM SLR

The presence of an SLR and the possibility of assigning wavelengths that are not contiguous in the optical spectrum result in a simpler resource allocation than that in the RMLSA. Wavelength allocation is evaluated for all the candidate paths following a First-Fit strategy, selecting the least power-consuming path according to MetricPC (Equation 17.2), but considering Equations 17.5 and 17.6 instead of Equation 17.3 and 17.4, respectively. $Num\lambda_s$ is the number of assigned wavelengths (i.e., total traffic demand/wavelength capacity); PC_{WDMT} is the power consumption of the WDM transponder, and $TotalNum\lambda_s$ is the total number of wavelengths considered for the demands allocation in a fiber (80 wavelengths).

$$PC_{TRANS} = Num\lambda_s \cdot PC_{WDMT} \tag{17.5}$$

$$PC_{LINKS} = \frac{Num\lambda_s}{TotalNum\lambda_s}(PC_{EDFAs} + PC_{OXCs}) \tag{17.6}$$

17.6.2.1.3   RWA for WDM MLR

As mentioned in Section 17.6.1.1, the optical spectrum is divided into two wavebands in the MLR approach (i.e., one for 10 Gbps and the other one for both 40 and 100 Gbps connections). This implies some changes in what refers to the resource allocation. The assignment in the first waveband follows the First-Fit algorithm (i.e., assigning the lowest indexes first), whereas in the second waveband, the assignment follows the Last-Fit strategy (i.e., assigning the highest indexes first). These two different allocation strategies would allow for a movement of the 200 GHz guard band to increase the number of wavelengths in a particular band according to traffic conditions, that is, whether more 10 Gbps wavelengths than 40/100 Gbps are required or vice versa.

For each candidate path, all the possible line rate combinations (i.e., according to the traffic demand and transmission reach) are calculated and sorted in ascending order of PC_{TRANS} (Equation 17.7). Then, the evaluation is started from the first possible line rate in the list (the one with the lowest PC_{TRANS}). If wavelengths are available in the links of the path, MetricPC (Equation 17.2) is calculated with PC_{LINKS} (Equation 17.6) and PC_{TRANS} (Equation 17.7), where $10G\lambda_s$, $40G\lambda_s$, and $100G\lambda_s$ correspond to the numbers of WDM transponders of 10, 40, and 100 Gbps, and PCT10G, PCT40G, and PCT100G correspond to the power consumption of WDM transponders of 10, 40 and 100 Gbps, respectively. If the allocation is not successful with the first line rate combination, a movement of the guard band in the links can be considered to increase the number of wavelengths in a band, and the evaluation would be repeated again. If the guard band movement does not result in a feasible allocation, the procedure is repeated for the next line rate combination in the list. After evaluating the allocation in all the candidate paths, the most energy-efficient solution (line rate combination and path) is selected.

$$PC_{TRANS} = (10G\lambda_s \cdot PCT10G + 40G\lambda_s \cdot PCT40G + 100G\lambda_s \cdot PCT100G) \cdot \frac{Num\lambda_s}{TotalNum\lambda_s} \tag{17.7}$$

17.6.2.2   Algorithms for a Static Scenario without Protection

The demands from the traffic matrix are first arranged following the highest demand first ordering strategy, which means that the demands that are supposed to occupy more resources will be served first. Then, the resource allocation is evaluated one by one for all the demands in the list according to the network approach under consideration (elastic OFDM-based, WDM SLR, or WDM MLR networks) as presented in Section 17.6.2.1. Figure 17.15 explains the main flow operation of the heuristic algorithms in a static unprotected scenario.

Static scenario algorithms: Main flow
Sort the connection requests from traffic matrix in descending order according to the traffic demand values
while *list is not empty*
Evaluation of the new lightpath establishment for the demand from the list according to the network approach (see Section 17.6.2.1)
end
Calculate final performance measures

FIGURE 17.15 Main flow of the algorithms in a static scenario.

The energy efficiency measure (Equation 17.8) is obtained by dividing the overall traffic demand successfully served in the network by the total power consumption. TotalTrafficNetwork is the addition of the TR of the demands that were successfully allocated in the network, whereas TotalPowerConsumption is calculated by adding the total power consumption of the transponders, the EDFAs, and the OXCs. In addition to the energy efficiency measure, the service blocking ratio is considered in the evaluation of the energy efficiency, as it determines the amount of resources that are required to operate with a given traffic load. This measure is the ratio between the TR of the demands that were blocked (TRBlockedDemands) and the total traffic demand (TRTotalDemands) as in Equation 17.9.

Another performance measure, energy efficiency per gigahertz (EnergyEffPerGHz) (Equation 17.10) has been used to account for both energy efficiency (Equation 17.8) and spectral efficiency. AvgSpectrumOccupancy is the average spectrum occupancy of the links in the network, and BandwidthCBand is 4 THz.

$$\text{EnergyEfficiency [bits/joule]} = \frac{\text{TotalTrafficNetwork [bits/s]}}{\text{TotalPowerConsumption [W]}} \tag{17.8}$$

$$\text{ServiceBlockingRatio} = \frac{\sum \text{TRBlockedDemands}}{\sum \text{TRTotalDemands}} \tag{17.9}$$

$$\text{EnergyEffPerGHz [bits/joule/GHz]} = \frac{\text{EnergyEfficiency [bits/joule]}}{\text{AvgSpectrumOccupancy} \cdot \text{BandwidthCBand [GHz]}} \tag{17.10}$$

17.6.2.3 Algorithms for a Static Scenario with Protection

The routing and resource allocation for a set of static demands that is resilient to any single link failure, the dominating form of failure in optical networks, has been considered for three common path protection schemes: dedicated path protection 1 + 1 (*DP* 1 + 1), dedicated path protection 1:1 (*DP* 1:1), and shared path protection (*SP*). The traffic demands are sorted in descending order (highest demand first) and then evaluated according to the corresponding protection scheme. In all the algorithms, pre-planned backup routes are calculated offline.

17.6.2.3.1 Dedicated Path Protection (1 + 1 and 1:1)

The allocation is jointly evaluated for the possible combinations of candidate working path (*k* shortest paths) and their corresponding candidate backup paths (*k* link-disjoint paths) according to the considered network approach (see Section 17.6.2.1). Then, for the possible path-pair combinations, a power consumption metric is calculated with the contribution from the working and its backup path pair. Thus, after evaluating all the candidates, it is possible to select the most energy-efficient solution. The spectral resources are reserved for both working and backup paths and pre-cross-connected. In the *DP* 1 + 1 scheme, the transmission is simultaneous in both working and protection paths, which provides the fastest recovery but implies higher energy consumption. On the other hand, in the *DP* 1:1 scheme, even if the resources are reserved along both paths, the transmission occurs only in one of the paths (normally on the working path). If a working and a protection path cannot be provided for a particular traffic demand, then it is blocked.

17.6.2.3.2   Shared Path Protection (SP)

Once the resource allocation is performed for all the traffic demands in their working paths according to the network approach (see Section 17.6.2.1), the remaining spectral resources can be shared by any backup path. In order to evaluate the survivability, the failure of each link of the network is emulated one by one. Then, those lightpaths affected by the link failure are listed and sorted in descending order according to their traffic demand values. After that, the reallocation of those affected traffic demands is evaluated in a modified network graph (i.e., the affected link is pruned from the network graph). If a backup route can be provided for a demand, it is stored in the list of backup routes; otherwise, it is considered as blocked because it cannot be protected against failure of this link.

17.6.2.4   Algorithms for a Dynamic Scenario

In a dynamic scenario, a connection between two nodes can alternate ON and OFF periods. In order to represent dynamic traffic, the demands are assumed to arrive in the network according to a Poisson distribution with a mean arrival rate λ (mean number of connection requests/time unit). The holding time follows an exponential distribution with intensity μ (mean number of finished connections/time unit). In order to analyze the performance at different traffic loads, different values of offered traffic (λ/μ) will be considered.

The routing and resource allocation for the different flows (time-varying traffic demands) is evaluated one by one in order of arrival to the network. The algorithms take into account the time fluctuations that might occur in the TR during the period that the communication is active (i.e., grooming more than one flow with the same source and destination nodes onto the same lightpath) and the release of resources when the communication terminates (i.e., releasing the wavelengths or subcarriers in the path once the holding time of the connection expires). The time-varying demands are referred to as flows, whereas a lightpath refers to the all-optical path that is established between two network nodes provided that enough and common resources are available in all the links of the path. Therefore, a lightpath can transport one or several flows depending on its capacity (i.e., the number of allocated resources, either wavelengths or subcarriers). Two types of events are considered in the dynamic operation of the network:

- *New flow request*: If there is an established lightpath with the same source and destination nodes, the possibility of either grooming the demand using the residual capacity of the lightpath or expanding the lightpath (i.e., additional subcarriers and modulation format increase can be evaluated for EON, whereas additional wavelengths are assigned for the WDM cases) is first checked. Otherwise, if grooming is not possible or there is no established lightpath, the establishment of a new lightpath is evaluated according to the network approach under consideration (Section 17.6.2.1).
- *Flow termination*: Once a flow terminates, if there are no more flows sharing the same lightpath, the allocated resources are released. Otherwise, whether the termination of the flow allows for releasing some of the reserved spectral resources (i.e., wavelengths for WDM or subcarriers for EON) is checked. Furthermore, in EON, whether the modulation

order can be decreased to reduce power consumption could be evaluated. Once the holding time of the connection expires, the energy consumption (Equation 17.11) and the data transmitted (Equation 17.12) in the duration of the flow are calculated.

$$EC_{FLOW} [W \cdot s] = PC_{TRANS} [W] \cdot FlowDurations [s] \tag{17.11}$$

$$Data_{FLOW} [bits] = TR_{FLOW} [bps] \cdot FlowDurations [s] \tag{17.12}$$

The energy efficiency of the network (bits/joule) is defined in Equation 17.13 as the ratio between the total data transmitted (TotalDataTr) and the total energy consumed in the network (TotalEC) during the simulated time. TotalDataTr is obtained by summing the data successfully transmitted in the different flows (Equation 17.12). TotalEC is the addition of the energy consumption of the transponders, the EDFAs, and the OXCs. The contribution from the transponders is obtained by the summation of the energy consumed by each of the flows (Equation 17.11), whereas the ones from the EDFAs and OXCs are obtained by the total energy consumed by these network elements in the total simulated time (assumed to be always ON). Besides, the service blocking ratio is obtained by the ratio between the summation of the TR of the demands that were blocked (TRBlockedDemands) and the addition of the TR of all the flow requests during the simulation (TRTotalDemands). Since the system is not in steady state in the beginning of the simulation, a first set of connection attempts is dropped from the blocking calculation (i.e., the first interval in which the number of blocked connections lower than the average is discarded from the computation).

$$EnEff [bits/joule] = \frac{TotalDataTr [bits]}{TotalEC [W \cdot s]} \tag{17.13}$$

$$ServiceBlockingRatio = \frac{\sum TRBlockedDemands}{\sum TRTotalDemands} \tag{17.14}$$

17.6.3 SIMULATION PARAMETERS

17.6.3.1 Power Consumption Values

Three main energy-consuming devices are considered to calculate the total power consumption of the network (assuming a single-layer design in the transport layer): transponders, OXCs, and EDFAs.

17.6.3.1.1 Transponders

17.6.3.1.1.1 WDM Transponders   Power consumption values of 34, 98, and 351 W [24] have been considered for the transponders with bit rates of 10, 40, and 100 Gbps, respectively. Power figures require an additional 20% overhead for each transponder to take into account the additional contribution to energy consumption in other node elements (e.g., controller cards, supervisory channels, power supply, fans, etc).

17.6.3.1.1.2 BVT Transponder (CO-OFDM Transponder)   A BVT is necessary for the operation of EON. In particular, for our OFDM-based EON, a CO-OFDM (coherent O-OFDM) transponder that is capable of modifying the signal properties (i.e., number of subcarriers and modulation format) by software configuration is considered. Because of the commercial unavailability of CO-OFDM transponders and the novelty of this technology in OTNs, several assumptions have been made to derive some realistic values of power consumption by comparing its architecture with that of a coherent WDM transponder:

- Receiver part: The architecture at the receiver part is similar (both are coherent receivers).
- Transmitter part: The CO-OFDM transponder contains a DSP module and two DACs that are not necessary in the WDM one.

Consequently, the presence of a DSP module at the transmitter part is assumed to be the main distinction between a CO-OFDM transponder and a coherent WDM one, so that the comparison can be mainly based on the difference in DSP complexity. Regarding this matter, a comparison of the number of complex multiplications per second between a DP-QPSK and a DP-OFDM transponder with the same TR (40 Gbps) is presented in Ref. [25], which shows a similar number of complex multiplications per second at both types of transponders regardless of the presence of DSP at both the transmitter and receiver in the OFDM variant (4.2×10^{11} for DP-QPSK and 3.2×10^{11} for DP-OFDM). Since DSP complexity seems to be similar, the power consumption has been assumed to be equivalent for both types of transponders, when operating at the same bit rate. Then, it has also been assumed that the DSP consumption scales linearly with the bit rate. Therefore, adopting the power consumption values for coherent transponders of 250 and 351 W for 40 and 100 Gbps, respectively (125 and 175.5 W for single polarization) [24], the power consumption of a single polarization CO-OFDM transponder can be interpolated as a function of its TR as in Equation 17.15:

$$PC_{OFDM} [W] = 1.683 \cdot TR [Gbps] + 91.333 \qquad (17.15)$$

Based on Equation 17.15, and considering a frequency slot of 12.5 GHz, Table 17.1 presents the power consumption values of a CO-OFDM BVT for the transmission and reception of a single sub-carrier with different modulation formats. An additional 20% of power consumption also needs to be considered as the overhead contribution.

17.6.3.1.2 ROADMs or OXCs

Two variants of OXCs are considered, a bandwidth-fixed and a bandwidth-variable or Flexi-grid OXC/ROADM. The former is capable of switching different channels with the spacing defined by the ITU-T grid, whereas the latter is able to add/drop channels of multiple bandwidths (multiple of a tiny frequency slot). In our model, similar power consumption values have been assumed for both types of OXCs assuming that the main distinction lies in the possibility of operating with different frequency plans and number of channels. The power consumption of an OXC is dependent on the node degree N (number of fibers connected to the nodes) and the add/drop degree α (number of channels added and dropped in the node), including an overhead contribution per node location

TABLE 17.1
Power Consumption of CO-OFDM BVT for Different Modulation Formats

Modulation Format	Subcarrier Capacity (Gbps)	Power Consumption (W)
BPSK	12.5	112.374
QPSK	25	133.416
8-QAM	37.5	154.457
16-QAM	50	175.498
32-QAM	62.5	196.539
64-QAM	75	217.581

(control cards, fans, power supply, etc.) of 150 W [24]. Accordingly, the power consumption of the OXC (PC_{OXC}) can be calculated in Equation 17.16:

$$PC_{OXC} [W] = N \cdot 85 + \alpha \cdot 100 + 150 \tag{17.16}$$

It is important to note that a network with a dynamic operation may require a more complicated control plane and probably higher power consumption. This aspect has not been included in our model yet.

17.6.3.1.3   Optical Amplifiers

An EDFA card consuming 30 W per direction [24] with an overhead contribution per amplifier location of 140 W (including controller cards and fans) has been considered. The total power consumption caused by the OAs will vary depending on the length of the links; thus, a common value for maximum inter-amplifier spacing (80 km) has been taken into account for the computation of power consumption.

17.6.3.2   Network Topologies

Two different topologies have been considered to evaluate the energy efficiency of networks with different dimensions: A long-haul network and an ultra long-haul network. The long-haul network is the Spanish core network model depicted in Figure 17.16a (used by Telefónica I+D for its traffic studies), which has a diameter of approximately 1000 km. This network topology is composed of 30 nodes and 96 bi-directional links. A reference traffic matrix for 2012 (overall traffic of 3.22 Tbps and 96 demands) has been used as a basis for the analysis. Transparent communication has been assumed for this country-sized network (i.e., no signal regeneration at any node).

The ultra-long haul network considered for this study is the GÉANT2, the European research and education network with a continental size (diameter of 7000 km), composed of 34 nodes and 52 bi-directional links as depicted in Figure 17.16b. Its large dimension makes the transparent reach impossible for some connections owing to the accumulation of physical impairments along the path; hence, regenerations are necessary in some network nodes to recover the signal quality. Because of the high energy consumption of regenerators (as they perform OEO conversions), simulations were carried out to determine the optimum placement of a minimum number of regenerators allowing for the realization of all the connections contained in the traffic matrix, at least with a line rate of 10 Gbps (most robust WDM transmission). As a result, regenerators were placed in four network nodes (Switzerland, Denmark, Luxembourg, and Malta). In this manner, a demand between nodes separated by a long distance could be served by several consecutive lightpaths. In such a large

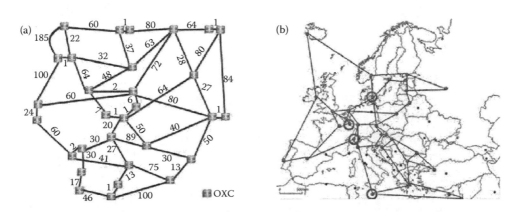

FIGURE 17.16   (a) Telefónica I+D's network model. (b) Pan-European GÉANT2 network model. (From DICONET Project Deliverable: D.2.1: definition of dynamic optical network architectures, Version: 1.00, Sep. 2010.)

network topology, 10 Gbps has been the only TR considered for WDM SLR as using only 40 or 100 Gbps would imply a big number of regenerations. A traffic demand matrix with an overall traffic of 2.394 Tbps has been taken as the traffic reference [26].

17.6.4 Description of Case Studies and Simulation Results

17.6.4.1 Case Study 1—Unprotected Static Scenario (Offline Problem)

In the Telefónica I+D's network model, the initial traffic matrix has been scaled up to a factor of 40 to obtain a total traffic ranging from 3.22 to 128.8 Tbps. At low traffic load conditions, all the network technologies under evaluation (with the only exception of the SLR 100G) show similar EnergyEfficiency, which is justified by the lower energy per bit of the transponders used in SLR 10G and SLR 40G, and the different capacities provided by the EON and MLR, which allow a better adjustment of the TR to the actual requested capacity (i.e., different modulation formats for each OFDM subcarrier and three different line rates in the MLR). In these low-traffic conditions, an SLR 100G is clearly penalized in energy efficiency as the low demand values do not justify the use of such a high capacity for most of the connections. This low utilization of the wavelength capacity has a negative impact on energy efficiency as the transponders consume the same energy as when the wavelengths are completely filled. When the traffic load increases, the fine granularity benefit becomes less relevant so the energy efficiency of SLR 100G is enhanced and becomes closer to that of the elastic network and MLR. The energy efficiency of the elastic network is also improved when traffic increases since it is possible to use higher-order modulation formats that have lower energy consumption per bit (see Table 17.1).

As traffic increases further, the high spectrum occupancy results in the blocking of some demands owing to the unavailability of resources in some links. As mentioned before, the service blocking ratio has to be taken into account in the evaluation of the overall energy efficiency of the network because it actually determines the number of network elements that are necessary to operate at a given traffic load. Deploying more network elements implies not only an increase in cost but also higher energy consumption. This blocking occurs first for the SLR 10G; then for the SLR 40G, MLR, and SLR 100G; and finally for the elastic OFDM-based network. The MLR presents slightly higher blocking than the SLR 100G because of the presence of a guard band (i.e., four wavelengths cannot be used for data transmission).

Figure 17.17 shows the results concerning EnergyEffPerGHz for the different types of network operation, at different traffic load values with no blocking conditions. Thus, it presents an overview of the energy efficiency, spectrum occupancy, and blocking ratio measures. As shown, the elastic network clearly outperforms all the WDM approaches in EnergyEffPerGHz and is able to convey

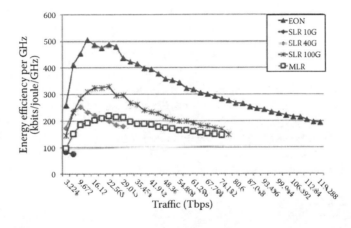

FIGURE 17.17 Energy efficiency per GHz for Telefónica I+D's network model in the static unprotected scenario.

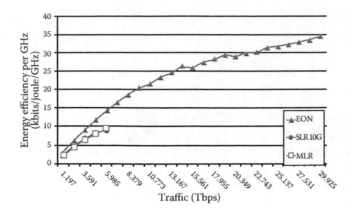

FIGURE 17.18   Energy efficiency per GHz for the GÉANT2 network model in the static unprotected scenario.

higher traffic volumes without blocking (i.e., 119.288 Tbps compared to 74.152 and 77.376 Tbps for the MLR and SLR 100G, respectively).

In the other network model, the GÉANT2, the reference traffic matrix has been scaled by factors from 0.5 to 15 (overall traffic ranging from 1.197 to 35.91 Tbps). The results in EnergyEffPerGHz are presented in Figure 17.18. As shown, the elastic network clearly outperforms the WDM approaches in EnergyEffPerGHz. As in the previous case, only the results without blocking are presented in the figure, making noticeable the significant difference in blocking ratio between the elastic network and the WDM approaches (SLR 10G and MLR). More specifically, in the elastic network, it is possible to transmit 29.925 Tbps in a network with single fiber per link without blocking, whereas the SLR 10G and MLR are able to transmit 5.985 Tbps. By comparing the values in EnergyEffPerGHz from Figure 17.18 with the ones obtained for Telefónica I+D's network model (Figure 17.17), the important role that the network topology and the traffic demand value play in the final energy efficiency of the network can be observed. Thus, in the former network model, the EnergyEffPerGHz ranges from 82 to 505 kbits/joule/GHz, whereas in the latter, these values are much lower, in the range from 3.22 to 34 kbits/joule/GHz. These lower values are justified by the larger dimensions of the GÉANT2, by the consequent deployment of a bigger number of EDFAs and OXCs (main contribution to power consumption), and also by the necessity of using robust and low spectral efficient modulation formats for long distances (higher spectrum occupancy).

In conclusion, considering an unprotected static scenario, the elastic OFDM-based network is shown to be the most energy-efficient solution in both country-sized and continental-sized networks thanks to the fine granularity given by the use of distance adaptive modulation and Flexi-grid operation. This fine granularity results in lower spectrum occupancy and lower blocking ratio.

17.6.4.2   Case Study 2—Protected Static Scenario (Offline Problem)

In this case study, the initial traffic matrix has been scaled up to a factor of 20 to obtain a total traffic ranging from 3.22 to 64.48 Tbps. Figure 17.19 presents the results in EnergyEffPerGHz for the different protection schemes (DP 1 + 1, DP 1:1, and SP) under non-blocking conditions in Telefónica I+D's network. SP and DP *1:1* schemes show higher energy efficiency than DP 1 + 1 because the backup paths only consume energy in case of failure. However, the SP scheme offers lower spectrum occupancy and blocking ratio than the DP 1:1 thanks to the possibility of sharing spectral resources by different backup paths. As can be observed, the results of EnergyEffPerGHz for SP are identical to those of the unprotected scenario in Figure 17.17 because this type of protection does not imply additional power consumption or reservation of spectral resources. However, blocking is higher with SP because when traffic increases (and so is the spectrum occupancy), it may not be

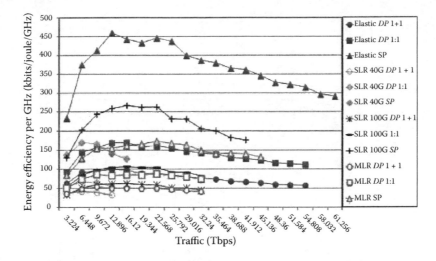

FIGURE 17.19 Energy efficiency per GHz for different protection schemes in Telefónica I+D's network model.

possible to protect all the demands even by sharing the backup spectral resources (e.g., in the elastic network, it is possible to transmit up to 119.14 Tbps without protection and just 61.256 Tbps with *SP*). As can be observed, the elastic network shows superior performance in EnergyEffPerGHz than any of the other approaches for all of the protection schemes under evaluation. Furthermore, the lower blocking provided by the elastic network makes it possible to transmit more traffic in a single fiber, which, as mentioned before, will directly affect the energy efficiency of a network.

Table 17.2 shows the maximum traffic that can be transmitted without blocking for each of the network technologies and protection schemes in Telefónica I+D's network model, including also the maximum traffic in an unprotected scenario (case study 1). Further information about this case study can be found in Ref. [28].

Even though *DP* 1:1 and *SP* schemes commonly show better spectral and energy efficiency than *DP* 1 + 1, the latter is still the most widely used because of its higher resilience and shorter recovery time. Actually, it is the least energy efficient as it requires simultaneous signal transmission on both working and backup paths. In Ref. [29], we propose a novel protection scheme that allows for a reduction in energy consumption while, at the same time, the high availability level of *DP* 1 + 1 is maintained. In this scheme, to reduce the energy consumed by backup resources, the hourly traffic fluctuations are exploited by adapting the rate of the backup transponders to the current required bandwidth requirements. The transmission in the working paths remains fully active (operating for

TABLE 17.2
Maximum Traffic without Blocking in Telefónica I+D's Network Model

Network Type	Maximum Traffic without Protection (Tbps)	Maximum Traffic with *DP* (Tbps)	Maximum Traffic with *SP* (Tbps)
EON	119.288	54.808	61.256
SLR 10G	6.448	3.224	3.224
SLR 40G	29.016	12.896	16.12
SLR 100G	77.376	32.24	41.912
MLR	74.152	32.24	45.136

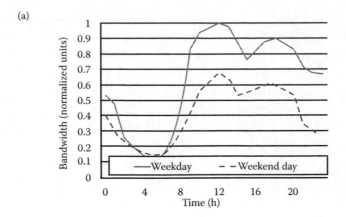

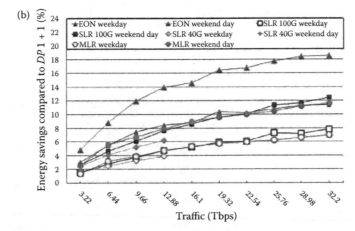

FIGURE 17.20 (a) Traffic variation in Telefónica I+D's network model on weekdays and weekend days. (b) Average energy savings on a weekday and weekend day with respect to the conventional 1 + 1 dedicated protection scheme for different traffic load conditions.

the peak traffic demand value) while the rate of the backup transponders of the protection paths is adjusted to the hourly traffic variation (e.g., it would be possible to partially deactivate some transponders or subcarriers at night when traffic volume is low).

This scheme has been evaluated for the Telefónica I+D's network model under different traffic load conditions (ranging from 3.22 to 32.2 Tbps), considering the traffic variation that occurs in this network during weekdays and weekend days, which is presented in Figure 17.20a.

This protection scheme can bring a considerable reduction in total energy consumption compared to the conventional DP 1 + 1 as shown in the average energy savings presented in Figure 17.20b. In particular, these savings can be especially significant for the elastic network scenario, where up to 11.4% and 18.5% of energy can be saved on a working day and on a weekend day, respectively.

17.6.4.3 Case Study 3—Dynamic Scenario (Online Problem)

In the dynamic scenario, the static traffic matrix is used to specify which network nodes exchange information and the maximum demand value between two nodes. Besides, in order to emulate the dynamic network operation in a more realistic manner, the source and destination nodes for a new connection are randomly selected from the possible source–destination pairs in the traffic matrix with a TR that fluctuates uniformly in an interval from 1% to 100% of the maximum value in the traffic matrix. Simulations were carried out for a total number of connection requests for different values of offered traffic (λ/μ), more specifically from 10 to 610.

For the Telefónica I+D network model, the original traffic matrix is scaled up to a factor of 10 (average flow demand of 167.81 Gbps). Each simulation considered 40,000 connection requests and the system is in steady state after the first 4000 requests. The obtained results on energy efficiency presented in Figure 17.21a show how the elastic network clearly outperforms the other WDM approaches thanks to its better adaptability to dynamic traffic changes. As offered traffic increases, the grooming of several flows into the same lightpath also grows (i.e., the number of simultaneous flows sharing source and destination nodes increases). This fact can be better exploited by the elastic network thanks to its adaptability to traffic variations (e.g., possibility of increasing the modulation order or expanding and contracting the bandwidth). Thus, the difference in energy efficiency between the elastic network and the other networks becomes more noticeable when traffic increases. Moreover, the possibility of increasing the modulation order to enlarge the lightpath capacity results in a better spectral occupancy and a lower blocking for the elastic network, as presented in Figure 17.21b.

For the continental-sized network, the GÉANT2, the original traffic matrix for the static scenario is also scaled up to a factor of 10, resulting in an average flow demand of 16.30 Gbps. Each simulation considered 60,000 connection requests and the system is in steady state after the first 6000 requests. The simulation results are presented in Figure 17.22. As shown, the elastic network offers a superior performance in energy efficiency (Figure 17.22a) and blocking ratio (Figure 17.22b) compared to all the WDM approaches. In particular, SLR 10G presents a considerable high blocking, which makes it unable to handle high traffic loads with the constraint of a single fiber pair per link. Concerning the MLR network, despite presenting a better result in energy efficiency than

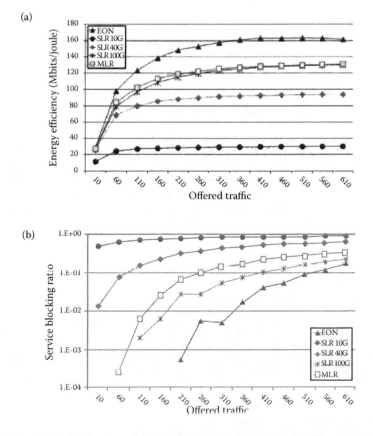

FIGURE 17.21 Simulation results for Telefónica I+D's network model with dynamic traffic conditions: (a) Energy efficiency [Mbits/joule]. (b) Service blocking ratio.

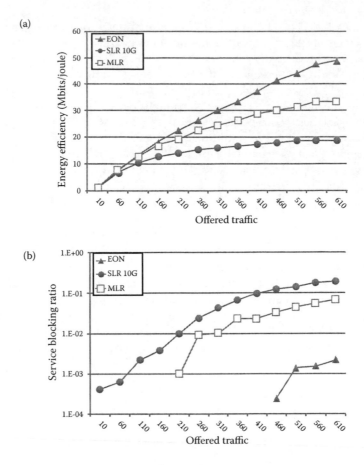

FIGURE 17.22 Simulation results for the GÉANT2 network model with dynamic traffic conditions: (a) Energy efficiency [Mbits/joule]. (b) Service blocking ratio.

the SLR 10G, it is also significantly affected by blocking when the traffic load increases. In this manner, at high traffic load conditions, the elastic network clearly outperforms the other networks in both energy efficiency and blocking ratio. This superior performance is justified by its adaptability to different circumstances and distance adaptive modulation, since a lower modulation order can be selected to extend the transparent reach (reducing the number of regenerations), while a higher modulation order can be used to increase the spectral efficiency, reducing the spectrum occupancy of the network.

17.7 CHALLENGES AND FUTURE RESEARCH

As shown in this chapter, the EON paradigm offers promising potential to improve the energy efficiency of optical networks, but further work is still necessary to make this innovative approach a reality for the operation of future OTNs. The novelty of EON opens new doors for future research at different levels of optical networks, in which energy efficiency must be carefully considered:

- *Node level*: Further work is required to develop optical transponders based on multicarrier structures with more advanced modulation formats, achieving higher spectral efficiency levels and extended resource allocation flexibility. Future implementations should be based on architectures driven by technical feasibility at reduced cost and power consumption figures. The development of power adaptation techniques, such as powering off some transponder elements at low traffic demands, would be beneficial because it increases overall energy efficiency.

Additional power savings will also come from the application of techniques to transponder electronic components (e.g., framers, DSPs, etc.), allowing a dynamic adjustment of consumption to the working load. In such a case, power consumption would be related to the signal symbol rate, and an increase in modulation order (more bits per transmitted symbol) would result in a decrease in energy per bit. Client interfaces should also be adapted to better leverage the flexibility of the new BVT; that is, providing a variable capacity at the node line side will be more advantageous (from the network operation and power consumption point of view) if it can be matched to the client demands.

- *Link level*: An accurate quality of transmission model that considers the channel conditions will be necessary to determine the feasibility of a modulation format for a particular transmission between two nodes. Besides, the possible choice of different modulation formats, together with the selective regeneration functionality at the nodes, will allow for designing the most energy-efficient alternative between fully transparent transmissions with robust modulation formats (lower capacity) and "semi-transparent" transmissions with high-order modulation formats (higher capacity) and regeneration points in some of the intermediate nodes.

- *Network level*: As presented in this chapter, energy-aware routing and resource allocation techniques can help improve the energy efficiency of EON. Further enhancement could be achieved by, for instance, introducing sleep mode capabilities in some links when traffic is low and rerouting that traffic on more loaded links. Regarding protection, introducing a differentiated quality of protection would likely reduce the number of active elements in the network and, thus, the total power consumption.

    In this chapter, we also showed how EON can greatly benefit from the time-varying traffic demands to reduce energy consumption. As a drawback, dynamic reassignment of variable spectrum slots to the traffic demands will create spectral "gaps" across the network (spectrum fragmentation) where new demands could not be accommodated. In this regard, applying some "hitless" (or with negligible impact) spectrum defragmentation techniques can further improve the energy efficiency of EON, as the blocking probability could be reduced, increasing the overall network capacity.

- *Control plane*: The availability of an intelligent control plane is crucial for the operation of current dynamic networks. Further work is required in two areas:

    - *Control plane for EON*. New control plane extensions are required for GMPLS (Generalized Multi-Protocol Label Switching) signaling and routing protocols, as well as Path Computation Element (PCE), taking into account energy information. Energy efficiency has to be taken into account in the energy-aware algorithms that run at the PCE, such as Open Shortest Path First-Traffic Engineering (OSPF-TE). To do so, nodes have to disseminate information about energy consumption, which is not done today with current OSPF implementations.

    - *Multi-layer control plane*. The presence of an advanced multilayer control plane will play an important role in the interaction between the routing and the transmission part. This means that, for the core network, a holistic approach should be followed in order to optimize available resources across the network layers (optical channel, OTN, IP/MPLS, etc.) and minimize power consumption.

17.8   CONCLUSION

As Internet traffic grows, energy efficiency is becoming an increasingly important factor for the planning and design phases of telecommunications networks, as it allows reducing not only the OPEX but also greenhouse gas emissions. In the optical transport, elastic networks technologies have recently been proposed as promising technology candidates to improve the current WDM networks. Elastic networks will play important roles in the new generation of transmission technologies (i.e., possible transmission of super-channels with rates beyond 100 Gbps and bandwidths wider than the 50 GHz ITU-T grid) and in the software-defined networks, that is, possible adjustment of the signal to the user request (e.g., modulation format, bandwidth) and channel conditions (e.g., distance adaptive modulation).

Besides the mentioned advantages, the new technologies will certainly improve the energy efficiency of the network thanks to their adaptability to different traffic conditions and better utilization of resources. An analysis in terms of energy efficiency in different network scenarios and conditions (i.e., different-sized network topologies, static and dynamic traffic operations, unprotected and protected networks, and different traffic loads) showed significant improvements with respect to the conventional WDM networks. Indeed, the variable lightpath capacity granularities and the adaptive modulation formats can bring significant advantages in terms of energy efficiency. The finer granularity allows a better adjustment of the allocated capacity to the actual demand by expanding or contracting the channel bandwidth (or capacity) according to the actual user demand, while the adaptive modulation allows the selection of modulation format according to the demand and distance, minimizing the number of regenerations in the network. The above-mentioned capabilities translate into better spectral efficiency and lower network blocking with respect to WDM networks. This fact is also vital to reduce power consumption, as the deployment of additional network elements would entail not only a higher cost but also an increase in energy consumption. Moreover, EON showed potential benefits for the operation of a future dynamic network thanks to its better adaptability to traffic variations. Finally, the presence of a single type of transponder for different rates, performance levels, or network conditions may bring additional cost advantages (e.g., economies of scale).

ACKNOWLEDGMENT

The research leading to these results has received funding from the European Community's Seventh Framework Programme under grant agreement no. 257740, TREND project; no. 258644, CHRON project; and no. 317999, IDEALIST project.

AUTHOR BIOGRAPHIES

Jorge López Vizcaíno received his bachelor's degree in telecommunications engineering from the Technical University of Madrid in 2008 and his MSc degree in telecommunications from the Technical University of Denmark in 2011. Since 2012, he has been a researcher at the European Research Center of Huawei Technologies in Munich (Germany) and a PhD candidate at the Technical University of Dortmund (Germany). Currently, he is collaborating in the EU project TREND related to energy efficiency. His research interests include network planning, network protocols, energy efficiency, and optical networks.

Yabin Ye received his BE and PhD degrees in electronic engineering from Tsinghua University, Beijing, China, in 1997 and 2002, respectively. From 2002 to 2004, he was a research scientist with the Institute for Infocomm Research, Singapore. From 2004 to 2008, he was a senior researcher with Create-Net, Italy. Currently, he is a senior researcher with Huawei Technologies European Research Center, Munich, Germany. His research interests include optical networking, high-speed transmission technologies, and hybrid optical/wireless access technologies.

Víctor López received his MSc (Hons.) degree in telecommunications engineering from Universidad de Alcalá de Henares, Spain, in 2005 and his PhD (Hons.) degree in computer science and telecommunications engineering from Universidad Autónoma de Madrid (UAM), Madrid, Spain, in 2009. In 2004, he joined Telefónica I+D as a researcher, where he was involved in next-generation networks for metro, core, and access. He was involved with several European Union projects (NOBEL, MUSE, and MUPBED). In 2006, he joined the High-Performance Computing and

Networking Research Group (UAM) as a researcher in the ePhoton/One+ Network of Excellence. He worked as an assistant professor at UAM, where he was involved in optical metro-core projects (BONE, MAINS). In 2011, he joined Telefonica I+D as technology specialist. His research interests include the integration of Internet services over optical networks, mainly sub-wavelength solutions and multilayer architectures.

Felipe Jiménez holds an electronic physics degree UCM (Madrid). He joined Telefónica I+D in 1994 through a PhD student scholarship and became part of the staff in 1995. He started working on the development of Telefónica Spain's nationwide IP network and, since then, has been involved in many internal innovation projects related to access and core architectures as well as in field trials of advanced fiber technologies. He has also been part of TID research groups working on relevant EU projects, like MUSE FP6 or Accordance FP7. Currently, he is collaborating within the EU Trend network of excellence, related to energy efficiency, as well as in FP7 Discus and Idealist projects, devoted to new architectures for optical transmission networks.

Raúl Duque completed his five-year degree in telecommunications engineering at Universidad Carlos III de Madrid (Madrid, Spain) in 2007. In 2006, he joined the New Network Technologies Division at Telefónica I+D where he developed his degree dissertation about a new network resource management mechanism based on the Nominal Route concept. He is currently working in the Core Network Evolution Division at Telefónica I+D, where he is participating in internal research projects for the Telefónica group. His research interests include network performance, optical network planning, resilience techniques, multicast transport algorithms for IPTV, and access control strategies.

Idelfonso Tafur Monroy is professor and head of the metro-access and short-range communications group of DTU Fotonik, the Department of Photonics Engineering at the Technical University of Denmark. His research interests are in hybrid optical fiber-wireless communication systems, cognitive optical networks, coherent digital detection and transmission technologies, digital signal processing receivers for baseband and radio-over-fiber links, optical switching, nanophotonic technologies and systems for integrated metro and access networks, short-range optical links, and communication theory. He is currently involved in the ICT European projects GiGaWaM and EURO-FOS, and he is the technical coordinator of the European project ICT-CHRON on cognitive reconfigurable optical networks.

Peter M. Krummrich received his Dipl.-Ing. and Dr.-Ing. degrees in electrical engineering from TU Braunschweig, Germany, in 1992 and 1995, respectively, where he worked on tunable laser diodes and praseodymium-doped fiber amplifiers. In 1995, he joined Siemens AG where his research interest focused on technologies for ultra high capacity DWDM transmission systems with an emphasis on more robust transmission and enhanced reach such as distributed erbium-doped fiber and Raman amplification, advanced modulation formats, adaptive equalizers, and PMD compensation. Since 2007, he has been working as a university professor at Technische Universitaet Dortmund, heading the chair for high-frequency technology. He is a senior member of IEEE and a member of VDE/ITG.

REFERENCES

1. ITU-T Recommendation G.694.1, Spectral grids for WDM applications: DWDM frequency grid, May 2002.
2. M. Pickavet and R. Tucker, Network solutions to reduce the energy footprint of ICT, in *The European Conference and Exhibition on Optical Communications (ECOC) 2010 Symposium*, Sep. 2010.
3. K. Hinton, J. Baliga, R. Ayre, and R.S. Tucker, The future Internet—An energy consumption perspective, in *OptoElectronics and Communications Conference*, Jul. 2009.
4. J. Baliga, R. Ayre, K. Hinton, W.V. Sorin, and R.S. Tucker, Energy consumption in optical IP networks, *Journal of Lightwave Technology*, vol. 27, no. 13, pp. 2391–2403, Jul. 2009.
5. ITU, ITU and climate change report, Oct. 2008.
6. W. Vereecken, W.V. Heddeghem, B. Puype, D. Colle, M. Pickavet, and P. Demeester, Optical networks: How much power do they consume and how can we optimize this?, In *The European Conference and Exhibition on Optical Communications (ECOC)*, Sept. 2010.

7. R.S. Tucker, Green optical communications—Part II: Energy limitations in networks, *IEEE Journal of Selected Topics in Quantum Electronics*, vol. 27, no. 2, pp. 261–274, Mar/Apr. 2011.
8. Y. Zhang, P. Chowdhury, M. Tornatore, and B. Mukherjee, Energy efficiency in telecom optical networks, *IEEE Communications Surveys and Tutorials*, vol. 12, no. 4, pp. 441–458, Nov. 2010.
9. G. Shen and R.S. Tucker, Energy-minimized design for IP over WDM networks, *Journal of Optical Communications and Networking*, vol. 1, pp. 176–186, Jun. 2009.
10. F. Idzikowski, E. Bonetto, L. Chiaraviglio, A. Cianfrani, A. Coiro, R. Duque, F. Jimenez, E. Le Rouzic, F. Musumeci, W. Van Heddeghem, J. Lopez Vizcaino, Y. Ye, TREND in energy-aware adaptive routing solutions, *Communications Magazine*, in press, 2013.
11. A. Muhammad, P. Monti, I. Cerutti, L. Wosinska, P. Castoldi, and A. Tzanakaki, Energy-efficient WDM network planning with dedicated protection resources in sleep mode, in *IEEE Global Telecommunications Conference (GLOBECOM)*, Miami, FL-USA, pp. 1–5, Dec. 2010.
12. P. Chowdhury, M. Tornatore, and B. Mukherjee, On the energy efficiency of mixed-line-rate networks, *2010 Conference on Optical Fiber Communication (OFC), collocated National Fiber Optic Engineers Conference (NFOEC)*, pp. 1,3, 21–25 March 2010.
13. W.V. Heddeghem, M.D. Groote, W. Vereecken, D. Colle, M. Pickavet, and P. Demeester, Energy-efficiency in telecommunications networks: Link-by-link versus end-to-end grooming, *2010 14th Conference on Optical Network Design and Modeling (ONDM)*, pp. 1,6, Feb. 2010.
14. M. Jinno, H. Takara, B. Kozicki, Y. Tsukishima, T. Yoshimatsu, T. Kobayashi, Y. Miyamoto, K. Yonenaga, A. Takada, O. Ishida, and S. Matsuoka, Demonstration of novel spectrum-efficient elastic optical path network with per-channel variable capacity of 40 Gbps to over 400 Gbps, In *The European Conference and Exhibition on Optical Communications (ECOC) 2008*, Sep. 2008.
15. B. Kozicki, H. Takara, Y. Tsukishima, T. Yoshimatsu, K. Yonenaga, and M. Jinno, Experimental demonstration of spectrum-sliced elastic optical path network (SLICE), *Optics Express*, vol. 18, pp. 22105–22118, Oct. 2010.
16. H. Takara, B. Kozicki, Y. Sone, and M. Jinno, Spectrally-efficient elastic optical path networks, in *OptoeElectronics and Communications Conference (OECC) 2010*, July 2010.
17. B. Mukherjee, *Optical Communication Networks*, McGraw-Hill, New York, 1997.
18. M. Klinkowski, M. Ruiz, L. Velasco, D. Careglio, V. López, and J. Comellas, Elastic spectrum allocation for time-varying traffic in flexgrid optical networks, *IEEE Journal on Selected Areas in Communications*, vol. 30, no. 1, pp. 26–38, Jan. 2013.
19. A. Klekamp, U. Gebhard, and F. Ilchmann, Efficiency of adaptive and mixed-line-rate IP over DWDM networks regarding CAPEX and power consumption, in *Proceedings of the OFC 2012*, Paper OTh3B, March 2012.
20. Y. Tang and W. Shieh, Coherent optical OFDM transmission up to 1 Tbps per channel, *Journal of Lightwave Technology*, vol. 27, no. 16, pp. 3511–3517, Aug. 2009.
21. X. Liu, S. Chandrasekhar, P.J. Winzer, S. Draving, J. Evangelista, N. Hoffman, B. Zhu, and D.W. Peckham, Single coherent detection of a 606-Gb/s CO-OFDM signal with 32-QAM subcarrier modulation using 4×80-Gsamples/s ADCs, in *The 2010 36th European Conference and Exhibition on Optical Communications (ECOC)*, pp. 1–3, 19–23 Sept. 2010.
22. S. Frisken, G. Baxter, D. Abakoumov, H. Zhou, I. Clarke, S. Poole, Flexible and grid-less wavelength selective switch using LCOS technology, *Optical Fiber Communication Conference and Exposition and the National Fiber Optic Engineers Conference (OFC/NFOEC) 2011*, Paper OTuM3, pp. 1,3, 6–10 Mar. 2011.
23. A. Bocoi, M. Schuster, F. Rambach, D.A. Schupke, C.-A. Spinnler, Cost comparison of networks using traditional 10 and 40 Gbps transponders versus OFDM transponders, *Optical Fiber Communication/ National Fiber Optic Engineers Conference (OFC/NFOEC) 2008*, Paper OThB4, pp. 1,3, 24–28 Feb. 2008.
24. C. Dorize, W. Van Heddeghem, F. Smyth, E. Le Rouzic, B. Arzur, GreenTouch draft report on baseline power consumption, GreenTouch Consortium (www.greentouch.org), Version 1.8, Nov. 2011.
25. S.J. Savory, Digital signal processing options in long haul transmission, *Optical Fiber Communication (OFC) 2008*, Paper OTuO3, pp. 1,3, 24–28 Feb. 2008.
26. DICONET Project Deliverable: D.2.1: definition of dynamic optical network architectures, Version: 1.00, Sep. 2010.
27. J. López Vizcaíno, Y. Ye, and I. Tafur Monroy, Energy efficiency analysis for flexible-grid OFDM-based optical networks, *Computer Networks*, vol. 56, pp. 2400–2419, July 2012.
28. J. López, Y. Ye, V. López, F. Jiménez, R. Duque, and P. Krummrich, On the energy efficiency of survivable optical transport networks with flexible-grid, in *The European Conference and Exhibition on Optical Communications (ECOC) 2012*, Sept. 2012.
29. J. López, Y. Ye, V. López, F. Jiménez, R. Duque, P. Krummrich, F. Musumeci, M. Tornatore, and A. Pattavina, Traffic and power-aware protection scheme in elastic optical networks, *International Telecommunications Network Strategy and Planning Symposium (NETWORKS) 2012*, Oct. 2012.

18 Energy Efficiency and Failure Recovery Mechanisms for Communication Networks

Artur Miguel Arsénio and Tiago Silva

CONTENTS

18.1 INTRODUCTION

The Internet has seen a tremendous growth in the past two decades. In fact, the Internet has become essential in our daily lives. Despite its importance, the Internet also has its share in the overall energy consumption of modern society, which according to Ref. [1] is approximately 5% of the total energy consumption of developed countries, and the emission of greenhouse gas (GHG) that originated from the information and communication technology sector will be increased by 130% from 2002 to 2020. Furthermore, the current Internet architecture is not the most suitable for certain applications, such as peer-to-peer traffic and audio and video streaming. These kinds of services or applications require high bandwidth for taking full advantage of their potential. However, nowadays, the current Internet routing system is facing scalability issues, because of "the ever-increasing user population, as well as multiple other factors including multi-homing, traffic engineering, and policy routing" [2].

Hence, energy consumption is a major concern nowadays not only for environmental but also for economic reasons. This makes energy efficiency an important issue that must also be addressed in the design of new network architectures. Research in the green networking area has been proposing solutions that focus on bringing energy awareness to the underlying network infrastructure, which currently lacks effective energy-saving measures. On the other hand, network availability, as well as the capability to recover from failures, is also currently another issue of importance for telecommunication networks' operation. Both issues are not independent, and so there is currently the need for energy-efficient failure recovery mechanisms.

18.1.1 Energy Sustainability

Indeed, one reason for energy consumption being a major concern nowadays is the global warming effect, which leads to major climate changes. A recent report from the European Union estimates a necessary reduction of approximately 15%–30% in the GHG until 2020, in order to keep the temperature increase below 2% [3].

In addition, the energy spent by telecommunications network equipment takes up a huge chunk of the network operations' costs. But building more energy-efficient networks must take into account the impact produced on the network performance. Indeed, many Internet services require high bandwidth, for example, high definition multimedia content. The current network design considers permanent high loads in the network, which is not always the case especially during night hours. This fact makes it possible to induce network elements into an energy-saving mode during the periods of low network load [1].

Energy consumption is mostly caused by two main factors. Primarily, the energy consumption does not vary linearly according to the utilization of network nodes and links, which ideally should be zero in the case of no utilization. On the other hand, the network nodes are always powered on to maintain the network connectivity at all times. By enabling the network elements to enter in an energy-saving mode, it will be possible to greatly reduce the energy consumption when they are idle or underused. This will allow not only a reduction in the emissions of GHG but also a reduction in energy-associated costs.

Therefore, it is important to research and develop efficient energy-saving models that can reduce the energy consumption of the current Internet, without producing a major impact on the performance of the network. There is indeed a growing interest in delivering, efficiently, telecommunication services to operators and, ultimately, to end users. Such solutions should apply both for current IP internet models, as well as for future Internet architecture models (such as the Publish–Subscribe Internetworking Routing Paradigm [PSIRP]), in order to guarantee energy sustainability for the future.

18.1.2 FAILURE RECOVERY ON CONVERGENT, HETEROGENEOUS NETWORKS

Another topic essential for infrastructure networks concerns the usage of redundant mechanisms, which may imply larger energy consumption, in order to compensate and recover from network failures. Indeed, operational expenditure (OPEX) is currently an issue of strategic importance to operators, resulting in energy consumption taking up a significant chunk of the network operations' cost [4]. The costs related to "truck roll" for deploying skilled technicians to the field in case of failure considerably increase the total cost of ownership. Incremental OPEX can directly be related to deployment costs of backup parts and failure probabilities of packages and modules [5]. But the implementation of redundant failure recovery/compensation mechanisms, in order to reduce the cost (as well as quality of service) impact of failure, may by themselves bring additional operational energy costs, besides the obvious infrastructure costs required in order to implement such mechanisms. These issues are increasingly important on scenarios of convergent networks, namely, radio (cellular or WiFi) networks supported by an optical backhaul [6]. Network failures on convergent networks can occur for instance in situations of natural disasters and, to a smaller extent, in cases of optical cable breaks, antenna damage, electronics failure, among other problems that may occur. The work in Ref. [4] proposed several mechanisms for failure recovery precisely for such convergent networks, while this chapter will present a comparative analysis for such failure recovery mechanisms in terms of energy consumption.

Applying energy-saving strategies in wired networks often leads to a performance reduction and even in some cases to a loss of connectivity. The challenge behind a smart energy management system is being able to reduce network energy consumption with a minimal negative impact in its throughput performance. This chapter addresses the work done in this area and proposes a solution that according to experimental results achieves a good trade-off between energy savings and network performance.

The main goals of this work are therefore to evaluate the energy consumption of one current and one future Internet architecture and to develop an energy-saving mechanism that can reduce the energy necessary to properly route all the traffic in a network. Therefore we aim to

- Evaluate the current Internet model.
- Evaluate a completely new architecture, and its applicability in a proposal for a future Internet architecture, namely, PSIRP [7].
- Comparatively evaluate the energy consumption of both systems.
- Design a solution that reduces the overall network energy consumption.
- Evaluate the trade-off between energy savings and network performance in different scenarios.
- Evaluate energy consumption with respect to different failure recovery scenarios on a future Internet architecture proposal, known as FUTON [6].

In order to accomplish this objective, the chapter is organized as follows. Section 18.2 discusses different contributions of several proposals designed to improve the Internet architecture. Some contributions choose to use the "Clean Slate Design" [8], while others try only to improve existing technologies. Some techniques that try to make the Internet architecture more "green" (i.e., to make the Internet more energy efficient) are discussed in this section. Section 18.3 describes the architecture for an integrated solution to reduce the overall network energy consumption, together with a description of the implemented modules. The section also includes an experimental evaluation under different testing scenarios, ending with the presentation and discussion of the evaluation results. Section 18.4 presents a comparative evaluation in terms of energy concerning failure recovery mechanisms for the FUTON architecture. The conclusions are presented in Section 18.5, giving a critical discussion on the main issues addressed by this chapter, together with some directions for further work.

18.2 SUSTAINABLE INTERNET TECHNOLOGY

18.2.1 ENERGY-EFFICIENT FUTURE INTERNET ARCHITECTURES

Despite the tremendous success of the Internet, its current architecture may not be the ideal solution for several challenges, such as security, mobility, manageability, dependability, and scalability [9]. The principle behind the design of the Internet architecture, that network nodes were not able to change their location (i.e., nodes need to have a fixed or static location), no longer applies. But the possibility of dynamic changes (mobility) in network topology is nowadays a reality, since more and more users use mobile devices to access the Internet. Therefore, the future Internet architecture must allow a good integration between mobile devices and fixed devices.

These problems do not have a trivial solution, because it is difficult to address them without increasing the complexity of the architecture. These issues can prevent the achievement of a better performance for some communication technologies, such as fiber optics and radio transmissions [10]. As a consequence of the aforementioned problems, new technologies and approaches need to be developed for the future Internet architecture, allowing good scalability for the upcoming applications and services. Therefore, new solutions and even different paradigms are being researched, such as FUTON, 4WARD, ANA, FARA, NIRA, and PSIRP, which are reviewed here.

There is a growing need for information-centric networking, owing to the increasing usage of overlay networks for information dissemination. In this situation, users will exchange pieces of information among themselves to reduce the load from central servers. Taking this into consideration, the Wired and Wireless World Wide Architecture and Design (4WARD) approach [11] is to make use of virtual networks over multiple physical infrastructures, trying to achieve some sort of separation between the physical and the logical topology of the network and allowing an efficient management of the available network resources [11].

The Autonomic Network Architecture (ANA) makes an important contribution to the future Internet owing to the support of network self-management and self-optimization. Besides this, it provides good flexibility in terms of the utilization of different networking schemes and protocols, also allowing the easy deployment of new ones. It also provides good support for mobility, allowing a better connectivity and performance when moving between different networks, for example, wireless networks [12].

The Publish–Subscribe Internetworking Routing Paradigm (PSIRP) approach uses the publish–subscribe paradigm, whose architecture is based in the information and not in the network nodes. This way, the receivers have full control of the information that they want to consume [7]. Most publish–subscribe architectures are composed of three major components, namely, publishers, subscribers, and routing nodes (brokers). The publishers are responsible for feeding the network with information to be consumed (i.e., publications). The subscribers are the consumers of information by expressing their interest on some published items using subscription messages. The brokers are responsible for forwarding the data between the publishers and the subscribers by matching the interests of the subscribers with the information published. Thus, the brokers or rendezvous points have the responsibility to route, forward, and allow the delivery of data from publishers to subscribers. Using this kind of architecture, the publishers and subscribers do not need to be aware of the existence of each other [7].

An alternative approach consists of the Forwarding directive, Association, and Rendezvous Architecture (FARA). Considering that nowadays IP addresses are used for identifying both networks and communication points, which provides some security but at the cost of mobility, FARA proposes a solution for solving this problem without the creation of a new identifier name space. This way, it is possible to separate entities from their respective location, which offers better support for entity mobility [13].

The New Internet Routing Architecture (NIRA) was designed to allow users the possibility to choose their own domain-level routes. A domain-level route is characterized as the domains that the packet needs to pass until it reaches its destination, differing from router-level route, which is described as the routers that forward the packet to the destination. Also, it avoids the use of a global link-state protocol by configuring link-state messages to be propagated within a provider hierarchy [14].

FUTON [6] addresses convergent network architectures, namely, Radio over Fiber (RoF) architectures, which are aimed toward the development of a hybrid fiber-radio network with the interconnection between multiple remote antenna units (RAUs) and a central unit (CU) on a transparent optical fiber system. FUTON was designed for deployment scenarios on a dense-urban location with ultra-high data rates per user on the order of 1 Gbps for low-mobility users (e.g., pedestrians) and up to 100 Mbps for high-mobility users (e.g., traveling on fast-moving vehicles) [6]. It considers the existence of a large concentration of remote antenna (RA) sites per geographical area, which are inexpensive elements. These RA sites are, for instance, located on the rooftops or lampposts. Each of the RA sites is catered by an entire wavelength, thereby guaranteeing necessary bandwidth. In FUTON, all communications are managed from a central location (CU), where the RoF Manager component is located, which is responsible for monitoring the overall network, triggering failure recovery actions in case of network faults.

The Explicit Control Protocol (XCP) is a window-based protocol, like Transmission Control Protocol (TCP) and Stream Control Transmission Protocol (SCTP), which implements congestion control at the endpoints of a connection, offering high end-to-end throughput. The TCP is commonly used in the current Internet for congestion control, but it is not capable of offering high throughput since it is inversely proportional to the packet drop rate. For this reason, a new congestion control protocol that can provide better performance than TCP in conventional environments, and that can still be efficient, fair, and stable when the communication delay increases, is needed [15].

The Internet architecture needs therefore to be greatly enhanced to allow the emergence of new services and applications. The aforementioned proposals try to address the major concerns about the current Internet architecture. Table 18.1 presents a summary of the main issues addressed by each proposal.

TABLE 18.1
Future Internet Proposals Comparison

Proposal	Clean Slate	Congestion	Mobility	Routing	Scalability	Security
4WARD	✓	✓	✓	✓	✓	✓
ANA	✓	✗	✓	✓	✓	✓
FARA	✗	✗	✓	✓	✗	✗
NIRA	✗	✗	✗	✓	✓	✗
PSIRP	✓	✓	✓	✓	✓	✓
XCP	✗	✓	✗	✗	✗	✗

18.2.2 ENERGY SUSTAINABLE INTERNET TECHNOLOGIES

This section reviews the main technologies and techniques that can be employed in order to reduce energy consumption on telecommunication networks, such as power management and network design, virtualization, pipeline forwarding, selectively connected end systems, or network elements ranking.

The constant growth of the Internet for several years has resulted in a significant increase in the amount of energy required to operate all the network devices, which may be working all day long. This huge energy consumption has become problematic, since the world environmental conditions are becoming more and more unpredictable because of the emission of GHGs to the atmosphere. This leads to the need to find good energy-saving solutions, not only to reduce environmental damage but also to reduce the associated energy costs [16,17].

Only recently has energy consumption become a priority problem to be solved in future Internet architectures, because of the rapid growth of energy, costs, consumers, broadband access, and other services offered by the ISPs. Energy efficiency is a problem that will affect both wired networks and service infrastructures. This is highly dependent on the arrival of new services, because of the traffic increase that may have originated from them [18]. Next, some of the prior work in the energy efficiency field for current and future Internet architectures will be discussed.

18.2.2.1 Power Management and Network Design

In legacy networks, energy consumption was not a major concern, not being important enough to be addressed in their design. The major concerns of those systems were mainly reliability, cost-effectiveness, robustness, service quality, and service availability. With the increase of data traffic and new applications, the Internet is consuming more and more energy. To prevent the increase in energy consumption, it is important to explore new solutions that will allow a better energy management. Hereafter, some energy-saving solutions will be discussed [17]:

- Energy-saving mode: The idea is to put equipment to sleep, since there is no need to waste energy when the equipment is not actually being used. This way, an interesting mechanism for energy saving is to put equipment to sleep when they are idle. This can be done at different levels: at the individual level, where switches, routers, or other devices are put to sleep; at the network level, combining sleep with routing changes and the use of bandwidth aggregation, so that when in low activity only the idle equipment are put to sleep; finally, at Internet level, this can be done by changing the network topology, allowing the adaptation of routes to different network loads.
- Adaptive link rate (ALR): In this approach, the link rate will be dynamically changed according to its utilization. This is done by exploiting the variable periods of idleness between consecutive burst of packets. This way, the equipment has the ability to dynamically reduce the link rate, because of lack of utilization, a technique that is being adopted by IEEE Energy Efficient Ethernet (EEE) [19].
- System redesign: The idea behind this concept is to design new network architectures and protocols, taking into account the energy consumption constraint. Embedding energy-saving mechanisms directly in new architectures has a tremendous impact in reducing energy consumption. The design of new architectures and protocols must satisfy capacity needs for different network users. One idea may pass by limiting the packet processing that needs more energy to only a group of routers and the creation of new data link and routing protocols that are able to work in on–off networks [17].
- Reliability and energy consumption: In Ref. [20], the relationship between reliability and energy consumption is explored. A trade-off model between power utilization and network performance is defined. Reliability and power saving are deeply addressed in this model, with the goal of developing robust and energy-efficient networks.

- Optical technology: Nowadays, optical technology is widely used in the backbone of ISP networks. Developments in this kind of technology will lead to all-optical networks, which will eliminate the need for optical converters, resulting in overall reduction of energy consumption [21].
- Advanced CMOS technology and superconductors: another approach is to develop smaller chips that consume less energy, using CMOS technology and superconductors, achieving gains on energy consumption of approximately 40% [22].

18.2.2.2 Virtualization

The traditional paradigm used by ISPs is to run a single application in one server, for simplicity, resulting in high resource waste and consequently a lot of energy waste. Using virtualization, it is possible to run multiple applications in a smaller number of machines, reducing the necessary hardware to execute those applications. The less hardware is used, the less energy will be required to operate that hardware. Such virtualization concepts are currently widely adopted on cloud computing architectures.

Virtualization may be used not only from the server point of view but also at other levels such as storage, network, platform, application, and resource. For instance, using server virtualization, the physical server will be separated into multiple virtual servers. This can be achieved using different approaches such as virtual machine (VM), paravirtualization, or operating system virtualization [18].

When using VM or full virtualization technology, multiple VMs will share the same physical machine, called the host machine. It is the host operating system or VM monitor that is responsible for allocating the necessary resources for running the VMs. Each VM runs its services on top of a guest operating system, which provides the necessary abstractions for file access and network support for their running applications. This way, a VM system may run different VMs with different operating systems, giving the users the flexibility to create, copy, save, read, modify, share, migrate, and even roll back the execution state of the VM [23]. For instance, the possibility of replicating the same VM image in different hosts in an easy way makes system administrators' lives a lot easier.

The paravirtualization technology is used to reduce the performance issues of the full virtualization, since it does not replicate entirely the original guest running environment. In this case, the guest operating system must be modified to be able to run in the paravirtualized environment, redirecting all virtualization-sensitive operations to the VM monitor [24]. There is a front-end driver that handles all the guests' i/o requests and delivers them to the back-end driver, which will interpret these requests and makes a correspondence with the desired physical device [25].

The operating system virtualization consists of a single operating system kernel running on a server. All guest environments can solely use this specific operating system. On the other hand, networking virtualization uses all available resources and functionalities, combining them into a virtual network or even subdividing them into virtual networks. This allows optimizing the resource utilization of network equipment in order to reduce their energy consumption.

The use of virtualization in future Internet architectures can play a big role in the energy-saving field. There is still the need to evaluate which type of virtualization will allow better energy management [18].

18.2.2.3 Pipeline Forwarding

The pipeline forwarding mechanism is a packet-scheduling technique that combines simplicity and effectiveness using a global common time reference, in order to perform network traffic shaping. It does not need a large amount of network resources and offers good performance. It is also capable of offering QoS and good scalability [26,27]. Pipeline forwarding is used in various architectures, which are designed to reduce the overall network energy consumption in the future Internet, for example, the Greener Internet proposed in Ref. [16].

Using this technique, switches will be synchronized through the utilization of a time period, Time Frame (TF), which can be assumed as a sort of virtual container for IP packets. The duration of the TF can be obtained by using external sources, for example, the Universal Time Coordinated (UTC) from GPS or Galileo positioning systems, or it can also be distributed throughout the network. To allow QoS, the transmission capacity can be partially or totally allocated to one or more flows during the resource allocation period [16]. The pipeline forwarding behavior is managed by two simple rules:

1. The packets that will be sent in TF t by some node n must be put in their output ports buffer in TF $t - 1$.
2. When a packet p is transmitted in TF t by a node n, it must be also transmitted by the node $n + 1$ in TF $t + d_p$, where d_p is the forwarding delay.

The forwarding delay is calculated during the resource allocation period, which involves scheduling techniques. The pipeline forwarding uses a predefined schedule, Synchronous Virtual Pipe (SVP), for forwarding a pre-allocated number of bytes during one or more TFs along a path of subsequent UTC-based switches [16]. There are two main implementations of the pipeline forwarding:

* Time-driven switching: Using this technique, all the packets belonging to the same TF will be switched to the same output port. Therefore, it will not be necessary to perform header processing, resulting in low complexity and possible optical implementation.
* Time-driven priority: This technique is suitable for optical backbones, arranging the traffic in large-capacity SVPs that are handled by high-speed switches. If more flexibility is necessary, the time-driven priority will combine pipeline forwarding with IP routing. This way, packets that enter in the same switch input port during the same TF can be sent to different output ports, according to the established rules in IP routing.

18.2.2.4 Selectively Connected End Systems

Selectively connected end systems can manage their own network connectivity in response to internal or external events. This way, it is possible for them to predict changes in connectivity and to react in accordance. For example, the end systems may predict the loss of connectivity just by knowing that they are moving to an area that has low layer-two connectivity. Thus, using selective connectivity will allow hosts to go to sleep, achieving a substantial power saving without sacrificing their place in the network.

In terms of power management, the end system may have three different states, namely, on, off, and sleep. Analogously to networking, end systems will be characterized by having connectivity, no connectivity, or operating in selectively connected mode. The end system can enter in the sleep state without losing its place in the network. Occasionally, an end system in the sleep state may be required to go back on in order to perform some specific tasks. The architectural concepts and components of this solution are as follows [28]:

* Assistants: An assistant is a generic mechanism, which helps the host while he is in sleep mode, by performing the routine operations that normally are assigned to end systems. For instance, the assistant will allow the host to keep his connectivity by responding to keep-alive messages on his behalf.
* Exposing selective connectivity: For reasons concerning energy management, it is important for end systems to know each other's state, and hence for the host to expose his level of connectivity throughout the different layers of his protocol stack and to inform possible peers with which he may want to communicate. For example, an active end system may be induced to enter in sleep mode when he wants to communicate with a sleeping end system.
* Evolving soft state: There is a need to evolve the soft state, since it is difficult to renew the state for sleeping end systems. There are two solutions for resolving this problem:
  * Proxyable state—Using this state, it will be the assistant who will be responsible for managing the soft state.

TABLE 18.2
Summary of the Different Types of Rankings

Ranking	Topology Aware	Traffic Aware
Degree centrality	✓	✗
Betweenness centrality	✓	✗
Closeness centrality	✓	✗
Eigenvector centrality	✓	✗
Load	✗	✓

- Limbo state—This state is in between the soft state and statelessness. Soft state assumes that a host is not available, but there is the need to know if the host is completely turned off or only sleeping. This way, when the renovation of the state expires, the host will enter in the limbo state, allowing only the necessary information used for distinguishing the two states to be exchanged among the participants.
- Host-based control: The end system has control of how the other ones in the network will react to his selective connectivity. Whenever a host moves to the selective connected mode, it is necessary to delegate his tasks to the other participants.

18.2.2.5 Ranking Network Elements

In order to efficiently choose which network elements to be turned off, it is important to rank each one according to its importance in the network. This can be done by looking to the network topology or to the traffic volume passing through the network element.

The most widely used topology-based rankings are degree centrality, betweenness centrality, closeness centrality, and eigenvector centrality. The degree centrality is defined as the number of links connected to each node. The betweenness centrality represents the number of shortest paths in which a node participates. The closeness centrality gives the average distance between a node and all the other ones, in which the more critical nodes are the ones with the lowest closeness centrality. Lastly, the eigenvector centrality corresponds to the influence of a node in the network by taking into account the importance level of its neighbors [29] (Table 18.2).

The traffic volume-based rankings only take into consideration the amount of traffic that is routed by the network elements. An example of the application of this principle is presented in Ref. [30].

18.2.3 ENERGY-SAVING MODELS

This section reviews the main energy-saving models that have been proposed, presenting their comparative analysis. Indeed, in order to achieve significant reductions in the energy consumption of networks, the possibility of making routing and traffic engineering decisions on the basis of the utilization and criticality of the network elements must be explored. This way, it is possible to achieve a reduction of the overall energy consumption of the network by dynamically turning off network nodes and links when their resources are not required. Hereafter, some solutions based on the aforementioned concepts will be discussed.

18.2.3.1 Dynamic Link Metric

The basic idea of the algorithm presented in Ref. [31] is to aggregate traffic to the most used links. The links with no traffic load will be turned off, allowing some energy savings. Also, a threshold to avoid traffic congestion in a link by restraining the allowed amount of traffic that may pass in it is defined. A link is considered congested whenever its traffic load exceeds the threshold, making it necessary to switch back on some other link to carry the remaining traffic. This will be achieved by dynamically changing the weight of the link, on the basis of the traffic load, the desired threshold, and a configuration coefficient k, transferring the traffic load to the most commonly used links [31]. Whenever the traffic load exceeds the threshold, the weight of the link will be raised in order to reduce its traffic load. When the traffic load is below the threshold, the weight of the link will be decreased in order to increase its traffic load. In this case, the weight of the link with the higher utilization will only be decreased. The changes made to the weight of the links must be communicated to all nodes in the network topology. Finally, to power off a link, it must be taken into account that the link has no traffic load and the network remains fully connective without the link.

This approach explores the redundancy in the core networks to encounter the minimum set of links that need to be powered on in order to successfully route all the traffic, allowing a reduction in energy consumption by powering off the unused links. The downside of this approach is the decrease of network performance, especially in high-peak traffic hours, owing to the increase of the packet delay.

18.2.3.2 Green Open Shortest Path First Protocol

In Ref. [32], a solution that uses the topological information advertised by routers using the Open Shortest Path First (OSPF) protocol is proposed. The focus of this study is on making the OSPF protocol more "green," that is, energy aware. The OSPF protocol specifies that each router computes its own Shortest Path Tree (SPT) by applying the Dijkstra algorithm. Hence, only the links that belong to at least one SPT will be used to route data in the network.

The algorithm proposed in Ref. [32], Energy-Aware Routing (EAR), defines two sets of routers, the exporters and the importers. The exporters are responsible for computing the shortest routing paths and the importers will then compute their own SPTs on the basis of the SPTs calculated by the exporters, selecting the routing paths to be used. This way, it will be possible to reduce the number of links used for routing traffic. The EAR algorithm is composed of three phases:

1. Exporter router (ER) selection: ERs will be responsible for computing their SPT by applying the Dijkstra algorithm. The neighbors, importer routers (IRs), of the selected ERs will use these SPTs to identify possible links that can be switched off. The selection of the ERs is based on the information contained in the LSA database of each router. With this information, the routers with the highest degree will be selected. Routers that are not neighbors of another ER are the only ones that can be selected.
2. Modified path tree (MPT) evaluation: In this phase, the links to be switched off, according to the MPTs computed by the IRs, will be determined. The IRs will use a modified version of the Dijkstra algorithm, in which the root node will be his associated ER, instead of the node itself. The SPT computed by the IR will be the same as the one computed by his ER. After this, the IR will insert itself as root in the computed routing tree, creating a new routing tree called MPT.
3. Routing path optimization: After the completion of phase 2, each IR will have a list of links to be switched off. Turning off these links will generate a new network topology, which must be propagated to all network nodes. Hence, the IRs with at least one link to be switch off must send LSA messages to the network. At the end of this procedure, each network router will have the current information about the network topology.

The EAR algorithm is a solution that addresses the problem of energy efficiency in today's IP networks without taking into account QoS constraints. The main advantage of EAR is the full compatibility with the OSPF protocol, allowing energy to be saved in low-traffic periods. The best criteria for the selection of the ERs and its respective number to avoid network congestion, especially in high-peak traffic hours, are still being researched. Hence, it is possible to extend this algorithm to take into account the QoS constraints.

18.2.3.3 Sleep Coordination in Wired Core Networks

In Ref. [33], Ho and Cheung proposed a distributed routing protocol, General Distributed Routing Protocol for Power Saving (GDRP-PS), in which the goal is to put routers into sleep mode without compromising the QoS and network connectivity. This protocol will offer a similar operation as other distributed routing protocols in high-traffic hours, and in low-traffic hours, it will put some routers into sleep mode to save energy, taking into account network connectivity and QoS [34].

This protocol uses two types of routers: power-saving routers (PSRs) and traditional routers. The traditional routers will use the OSPF protocol. These types of routers are always powered on, even if there are no packets to be processed. Instead, the PSRs will have two different states: working and sleeping.

The algorithm starts by randomly choosing one coordinator, which will record the information about available PSRs and will also be responsible for coordinating the operations of the PSRs. Because of the constant monitoring of the PSRs, the coordinator will never be put to sleep. Furthermore, after a predefined period, a new coordinator will be randomly chosen, giving the opportunity for all the PSRs to be coordinators (fairness).

To change from the working state to the sleeping state, the PSR must detect that the network is idle by measuring the maximum utilization of all the links that are connected to it, U_{max}. A network is considered to be idle if the U_{max} is below a determined threshold, T_1. If the network is idle, then the PSR will verify if the network connectivity can be maintained in its absence. If so, the PSR will recompute its routing table and will send a message to the coordinator to get permission for entering in the sleeping state, since more than one PSR in the sleep state is not allowed. This is necessary to guarantee that the remaining PSRs will not be overloaded. In case of a positive response from the coordinator, the PSR will broadcast the rebuilt routing table and will enter the sleeping state for a period of time. If not, it will remain in the working state.

After waking up from the sleeping period, the PSR will rejoin the network by using the existing routing protocol and the routing tables of all the network nodes will be rebuilt. When the coordinator becomes aware of the waking up of the PSR, it will verify if its own maximum link utilization is greater than the threshold; that is, the network loading is high (high-peak hours). If so, the coordinator will send a wakeup message to the PSR; otherwise, it will do nothing. The PSR is expecting a confirmation message from the coordinator after a certain period of time. If it receives the confirmation message, the PSR will remain in the working state; otherwise, it will go back to the sleeping state for another period of time.

In this protocol, the process of putting routers in sleep mode to achieve energy savings was defined. According to the results presented in Ref. [33], the GDRP-PS is able to achieve a reduction of approximately 18% in the total energy consumption of the network.

18.2.3.4 Switching Off Network Elements

In Ref. [30], the possibility of switching off not only network links but also network nodes is explored. The goal of the proposed algorithm is to find the minimum set of routers and links that must be powered on so that the total energy consumption of the network can be reduced. Hereafter, the proposed heuristics to solve the aforementioned energy consumption problem will be explained, taking into account the parameters described in Table 18.3.

To reduce the total network energy consumption, the solution discovers the routers and links that can be turned off without jeopardizing the network connectivity. A complete knowledge of the network topology and of the average amount of traffic that is exchanged between all node pairs is assumed, and some flow conservation constraints are enforced.

The proposed algorithm will iteratively try to switch off a network element (node or link). At each iteration, the network element will be disabled and all the shortest paths will be recomputed. After this step, whether the network retains its connectivity and whether the traffic demand can be satisfied will be verified.

TABLE 18.3

Parameters in the Problem Formulation

Parameter	Description
x_{ij}	Link ij is on or off
y_i	Node i is on or off
PL_{ij}	Power consumption of the link ij
PN_i	Power consumption of the node i
f_{ij}^{sd}	Traffic from source to destination that is routed through the link ij
f_{ij}	Total amount of traffic that is routed through the link ij
t_{sd}	Average amount of traffic going from source to destination
c_{ij}	Capacity of the link ij

18.2.3.5 Dijkstra-Based Power-Aware Routing Algorithm (DPRA)

The DPRA [30] is a heuristic algorithm that consists of the partitioning of the traffic demand from a source node to a destination node. Then, the path that consumes the minimum power for the specified traffic demand will be computed, taking into account the resources that are already allocated. This will be executed for all node pairs and until all the traffic demand is allocated.

Each link of the network will be associated with a cost equal to the increase of the power consumption of the destination node, which can be calculated taking into account the traffic of the link and the energy profile of the destination node. Afterward, the maximum resources in use by each node and consequently by each link will be calculated, excluding the nodes and links whose available resources are not enough for the allocation of more traffic. Finally, the Dijkstra algorithm will be executed taking into account the newly calculated costs and the disabled network elements.

18.2.3.6 Green-Game

Green-Game [29] proposes a model to solve a resource consolidation problem by taking into account both traffic load and network topology. Using this information, it will be possible to rank the contribution of each node in the packet delivery process. This can achieve a good trade-off between performance and energy savings, since the ranking combines traffic awareness and topology awareness. Taking this into consideration, the Green-Game will try to find the set of nodes that can safely be turned off on low load networks.

The ranking of each node will be obtained by computing the Shapley value [35]. The Shapley value will rank nodes with a higher value when their absence disconnects the network and when their presence is very important in the packet forwarding process. In combination with the traffic load, the Shapley value can efficiently distinguish the network nodes by their importance. Hence, the network nodes with the lowest Shapley value will be possibly turned off.

The high complexity in the computation of the Shapley value makes it unsuitable for being applied in real networks. This way, some optimizations to reduce the computational complexity of the Shapley value was proposed in the Green-Game, so it can become practical in real networks.

The work developed in the Green-Game provides an efficient way of choosing which network elements to be turned off. The higher ranked network elements will most likely be the most used ones. By using this measure, it will be possible to reduce the impact of the energy-saving mechanism in the network performance.

18.2.3.7 Comparative Summary

The reviewed energy-saving models provide mechanisms that can put network elements in a power-saving mode. Table 18.4 presents a summary of the main characteristics for each energy-saving model.

TABLE 18.4
Energy-Saving Models Summary

Energy-Saving Model	Link Control	Node Control	Offline Scheme
Dynamic link metric	✓	✗	✗
Green OSPF	✓	✗	✗
GDRP-PS	✗	✓	✗
Switching off network elements	✓	✓	✓
DPRA	✓	✓	✓
Green-Game	✓	✓	✗

18.3   AN INTEGRATED APPROACH TO ENERGY-EFFICIENT ROUTING

This section proposes a new architecture for energy-efficient routing for current and future telecommunication networks and its experimental evaluation. The proposed solution embeds energy awareness to the IP network architecture and to the PSIRP network architecture. This is done by controlling the working state of network elements and by exploring their idleness periods. This way, it will be possible to turn off the unused network elements. Traffic aggregation will also be used to give an opportunity for underused network elements to be turned off. Hereafter, the implemented modules for the proposed architecture will be explained in more detail.

18.3.1   NETWORK ENERGY OPTIMIZATION

18.3.1.1   Topology Manager

The topology manager has at all times an overview of the network conditions. It provides the necessary abstractions for controlling the simulation environment and for managing the network. With these abstractions, it is possible to check for partitions in the network when turning off a link or node, which is of great importance to the energy-saving algorithm.

18.3.1.2   Energy Optimization in the IP Network

Energy optimization in the IP network is based on the IP stack and on the Dijkstra algorithm to find the shortest path between a pair of nodes in the network. In this sense, this architecture is an extension of the current IP architecture with an energy-saving module. This new module will allow the unused network elements (nodes or links) to be turned off in order to achieve a reduction in the overall network energy consumption.

18.3.1.3   Energy Optimization in the PSIRP Network

With this future Internet architecture, we will not rely on the current IP stack. The implemented architecture uses the PSIRP design specifications. However, the implemented architecture will not follow thoroughly all of the PSIRP specifications, with the forwarding functionality being the only one implemented. This model will be extended to allow the integration of an energy-saving module.

The PSIRP module was implemented inside the network node with the energy-saving model as an extension. The components belonging to both architectures are detailed below and shown in Figure 18.1.

18.3.1.4   Solution Modules

To allow the possibility to reduce the network energy consumption, several modules were implemented on the network node. The implemented modules are grouped as common, IP, and PSIRP.

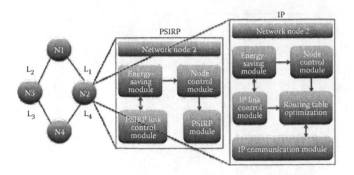

FIGURE 18.1   Module interaction in a network node with the IP stack or PSIRP.

18.3.1.4.1   Common Modules
- *Energy-saving module*: This module is responsible for applying the energy-saving algo-rithm to the network node. It makes decisions concerning the traffic aggregation to allow the possibility of redirecting traffic to the most used network elements. It is also respon-sible for the decisions regarding the working state (wake and sleep) of the network node. This way, it tries to achieve the best network configuration that will allow a reduction in the overall network energy consumption.
- *Node control module*: This module is responsible for controlling the working state of the network node. It allows the network node to be woken up and put to sleep according to the decisions made by the energy-saving algorithm. It also allows each of the interfaces belonging to the network node to be turned off individually.

18.3.1.4.2   IP Modules
- *IP link costs module*: This module is responsible for applying the OSPF routing protocol to gather/update all the network link costs and status to be used in the computation of the routing tables, since each network element will need to be aware of the changes that may occur in the network topology owing to the decisions made by the energy-saving algorithm.
- *IP routing table optimization*: This module is responsible for updating the routing tables of the network node according to the shortest path principle. This is done by computing the Dijkstra's algorithm over the weights of links that belong to the network. When updating the routing tables, the state of each network element, node or link, is also considered.
- *IP communications module*: This module uses the IP stack to provide communication capabilities to the network node. This way, the network node will have the ability to the send and receive data over its communication channels.

18.3.1.4.3   PSIRP Modules
- *PSIRP link costs module*: This module, instead of implementing the OSPF protocol, will gather/update the link costs by exchanging information with the topology manager. This way, every network node in the network has access to the new topology information.
- *PSIRP communication module*: This module hands over the communication capabilities to the network nodes. A simplified version of the PSIRP architecture was implemented. The rendezvous system and the forwarding mechanism were the only functionalities considered in the implemented architecture. In the following, the implemented functionalities are described.
  - *Rendezvous system*: The rendezvous system is responsible for managing the link identi-fiers used by each link that connects a pair of nodes. For each direction of the link, it will be given a different link identifier. It is also responsible for the creation of the *zFil-ter*, which represents the routing path between a pair of nodes. The *zFilter* contains the

link identifiers of each hop that belongs to the selected routing path. The *zFilter* is the result of performing an OR operation over the link identifiers. Finally, the rendezvous system will select the routing path between two nodes according to the shortest path.

- *Forwarding mechanism*: The forwarding mechanism represents the packet delivery process. When a network node wants to send a packet, it will check which of its interfaces belong to the given *zFilter*. After that, the packet will be sent through the selected interface. The matching of the link identifier against the *zFilter* has a low computational complexity, since it is based in performing an AND operation between the *zFilter* and link identifier. The forwarding process of the PSIRP is described below.

Algorithm – The PSIRP forwarding process

for all *LinkID's* do
  if *zFilter & LinkID == LinkID* **then**
    Forward the packet through the link
  end if
end for

18.3.2 ENERGY CONSUMPTION MODEL

The energy consumption model is used for calculating the overall network energy consumption. This will allow us to evaluate the energy savings when enabling the energy-saving module, under the set of assumptions assumed by the model. To accurately evaluate the energy savings, it is desirable that the energy consumption model can provide good estimates of the energy consumed by real network devices. The implemented energy consumption model for each network element was based on the work of Bianzino et al. [36] and can be summarized by Table 18.5. The C parameter represents the switching capacity of a network node, which is the double of the sum of the capacity of all its links (see Equation 18.1).

$$c(n) = 2 \sum_{(i,j) \in L} c_{ij} \tag{18.1}$$

Taking this into account, link *lu* and node *nu* utilization in a certain period of time can be calculated using Equations 18.2 and 18.3, respectively.

$$I(i,j) = (\alpha_{ij} - \beta_{ij}) \tag{18.2}$$

$$lu(i,j) = \frac{l_{ij}}{c_{ij} * T}$$

TABLE 18.5
Energy Consumption of the Network Elements

Network element	E_0 (W)	M (W)
Nodes	$0.85C^{2/3}$	$C^{2/3}$
(0–100) Mbps links	0.48	0.48
(100–600) Mbps links	0.90	1.00
(600–1000) Mbps links	1.70	2.00

$$l(n) = \sum_{(i,n) \in L} l_{in} + \sum_{(n,i) \in L} l_{ni}$$

$$c(n) = 2 \sum_{(i,j) \in L} c_{ij} \qquad (18.3)$$

$$nu(n) = \frac{l_n}{c_n * T}$$

The total network energy consumption can be defined as the amount of energy spent by all nodes and links that belong to the network topology and that are powered on. When a network element is powered on, it consumes a constant amount of energy, E_0, even when its utilization is zero. If its utilization is greater than zero, the energy consumption of the network element will be increased by a fraction (link/node utilization) of the difference between M and E_0. According to this model, Equation 18.4 represents the total network energy consumption:

$$E_T = \frac{1}{2} \sum_{(i,j) \in L} ((lu_{ij} + lu_{ji})E_{fij} + x_{ij}E_{0ij}) + \sum_{n \in N} (nu_n E_{fn} + x_n E_{0n}) \qquad (18.4)$$

18.3.3 Energy-Saving Algorithm

The proposed algorithm aims to reduce the overall network energy consumption by exploring the possibility of turning off the network elements (nodes or links) that are not being used. This has to be done taking into account the resulting impact in the performance of the network. To achieve the best trade-off between energy consumption and performance, the following questions were taken into consideration during the design of the energy-saving algorithm.

- How to aggregate traffic to most frequently used links?
- Which are the network elements that can be turned off? In which sequence?
- When to turn back on links, in order to reduce the impact in the network performance?

For the purpose of saving energy, the algorithm will turn off the network elements that are not being used. The network elements to be turned off must be carefully selected in order to reduce the inevitable impact in the network performance. Hereafter, how the algorithm tries to achieve the best trade-off between energy savings and network performance will be explained.

The algorithm is divided into two main functions: the traffic aggregation and the selection of the network elements to be turned off. First, the use of traffic aggregation will allow the possibility of turning off the underused links by transferring their traffic to other links that have higher utilization. Using this mechanism, it is possible to induce the underused links to an idle mode, allowing for a greater number of network elements to be turned off. Lastly, the selection algorithm will choose the network elements to be turned off and in which order. This will greatly affect energy savings and network performance. The detailed steps of the algorithm are enumerated below:

1. Check the utilization of every link: For each node in the network, the utilization of its entire links is analyzed. When the utilization of a link, lu, exceeds a threshold, its weight will be increased to reduce its utilization. The links with a utilization below the threshold will become candidates for a weight decrease (see Algorithm 1). Only the link with higher utilization among the candidates will have a decrease in weight, which may allow some

traffic to be aggregated into this link (see Equations 18.5 and 18.6). The modification of the link weight takes into account the remaining traffic, λ, which can be allocated to it. Lastly, the links that are not being used will be chosen as candidates for the turn-off procedure.

$$\lambda(i, j) = |1 - lu_{ij}| \tag{18.5}$$

$$\text{cost}'(i, j) = \begin{cases} \lambda_{ij} \times \text{cost}_{ij}, & lu_{ij} \leq \text{threshold} \\ \dfrac{\text{cost}_{ij}}{\lambda_{ij}}, & lu_{ij} > \text{threshold} \end{cases} \tag{18.6}$$

Algorithm 1 The traffic engineering algorithm

for $i = 1 \rightarrow N$ **do**
  for $j = 1 \rightarrow L$ **do**
    $u \leftarrow LinkUtilization\ (i, j)$
    if $u > threshold$ **then**
      $IncreaseWeight\ (i, j)$
    end if
    $j \leftarrow j + 1$
  end for
  Decrease weight of the most used link of the node below the threshold
  $i \leftarrow i + 1$
end for

Algorithm 2 Turning off network elements

$SortByRank\ (nodes)$
for $n = 1 \rightarrow N$ **do**
  if $CanGoToSleep\ (n)$ **then**
    $Sleep\ (n)$
  end if
  $n \leftarrow n + 1$
end for
$SortByRank\ (links)$
for $l = 1 \rightarrow L$ **do**
  if $CanBeTurnedOff\ (l)$ **then**
    $TurnOff\ (l)$
  end if
  $l \leftarrow l + 1$
end for

2. Ranking the network elements: In this step, a ranking is assigned to each network element that has been selected as candidate to be turned off. This ranking will reflect the importance of the network element in the network. This way, it will be possible to specify the sequence in which the network elements will be turned off, starting with the least important ones. In the ranking, the local centrality [37] measure will be used to classify

the importance of each network element to the network topology. This measure was chosen because of its low complexity and because it is more accurate than degree centrality. The local centrality of the network node v, $C_L(v)$, is then defined as

$$Q(u) = \sum_{w \in r(u)} N(w) \tag{18.7}$$

$$C_L(v) = \sum_{u \in r(v)} Q(u) \tag{18.8}$$

where $\Gamma(u)$ is the set of the nearest neighbors of node u and $N(w)$ is the number of the nearest and the next-nearest neighbors of node w. The ranking of network elements will take into consideration both the history of utilization and the local centrality of the network element. The ranking of links and nodes are described below:

a. Links: The ranking of links will be calculated using Equation 18.9, where H_{lu} is the history of utilization of a link. This computation will then be used to order the links by ranking. The links with the same ranking will be reordered by their local centrality C_L.

$$R_L(i, j) = (C_L(i) + C_L(j)) \times H_{lu}(i, j) \tag{18.9}$$

b. Nodes: The ranking of nodes will be calculated using Equation 18.10, where H_{nu} is the history of utilization of a node. As for the links, this computation will allow to order the network nodes according to their ranking. The nodes with the same ranking will be reordered by their local centrality C_L.

$$R_N(n) = C_L(n) \times H_{nu}(n) \tag{18.10}$$

3. Turn off the network elements: With the output of the previous step, each of the chosen network elements will be possibly turned off (see Algorithm 2). The links will be turned off if the network remains connective, that is, without causing partitions in the network. On the other hand, turning off nodes is not a trivial operation because it would cause packets to be lost, since the receivers of the packets would be unavailable. Because of this and of the existing technology, the nodes will only be put to sleep instead of fully switched off. Thus, when a node enters in sleep mode, it will wake up after a fixed period of time. A node can only enter in sleep mode if the following conditions are met:
a. No remaining traffic in any of its links.
b. The remaining nodes of the network can still communicate with each other.
c. All of its neighbors in sleep mode can rejoin the network in its absence.
   If the above conditions are all satisfied, then the node will go to sleep and all of its active links will be turned off. After the sleeping period, the node will verify if the network needs its presence and will enter in the pre-wake-up state, because of pending packets destined to it or the network performance has dropped too much. If its presence is not required, then the node will go to sleep again; otherwise, the node will be turned on and it will remain awake as long as the algorithm decides to put it to sleep.

Finally, the algorithm may be forced to turn back on some links to avoid network congestion in the case of high traffic demand. These links will be reconnected taking into account their significance in the network, starting with the most important ones.

18.3.4   Experimental Evaluation

Let us now discuss the experimental results of the aforementioned solution that integrates several energy-saving components, with the main goal of verifying the trade-off between energy savings and network performance. This evaluation will be carried out on both IP-based and PSIRP-based architectures. The system will be evaluated on two different topologies with both light and heavy traffic. The evaluation results will be obtained by network simulation, using NS3 simulator.

For the evaluation, two different network topologies were considered to be used in the experimental scenarios, which allow the adaptation of the energy-saving algorithm in different networks to be verified. To provide a more realistic scenario, two topologies from real networks, the Abilene and the COST-239 networks, were chosen. The chosen network topologies vary in the number of links. The weights of the links were randomly chosen between the values 1 and 10. Also, links with a capacity of 10 Mbps will be used. The main characteristics of the topologies are summarized in Table 18.6.

18.3.4.1   Testing Scenarios

Evaluation results of the solution will be presented in different network scenarios. The presented scenarios test the adaptation of the energy-saving algorithm to different loads of traffic. To perform this evaluation, the following metrics were defined:

- Throughput: Gives the packet delivery average rate.
- Delay: Gives the average time that a packet needs to go from the source to the destination.
- Link utilization: Gives the average link utilization of the network.
- Energy: Gives the overall energy consumption of the network.

The evaluation results were obtained by running the simulation 40 times for each experimental scenario, with the average and the standard deviation calculated. With these results, the trade-off between network performance and energy savings for each of the chosen experimental scenarios will be analyzed.

18.3.4.2   Light-Traffic Scenario

This scenario corresponds to the experimental evaluation of the solution with low traffic demand. In this situation, it will be possible to achieve good energy savings because there will be some unused network elements that will be turned off. Because of the low traffic demand, a major performance reduction of the network is not expected. Lastly, the traffic conditions of this scenario correspond to 100 kB of traffic size and 880 μs of Inter-Packet time interval. Table 18.7 presents experimental results obtained by injecting a small amount of traffic in both network topologies and by using the IP and PSIRP architectures. In this situation, the energy-saving algorithm is disabled, serving as basis for a comparison between the previously defined metrics in a low-traffic scenario. Table 18.8 presents the results with the energy-saving algorithm enabled.

TABLE 18.6

Network Topologies Used in Evaluation

Network	Nodes	Links	Degree	Reference
Abilene	11	14	2.55	[38]
COST-239	11	26	4.73	[39]

TABLE 18.7

The IP/PSIRP Evaluation without the Energy-Saving Algorithm in a Low-Traffic Scenario

IP	Abilene Network Topology	
Metric	**AVG ($\bar{x}$)**	**STD (σ)**
Throughput (Mbps)	8.536	0.166
Delay (ms)	2.530	0.192
Link utilization (%)	17.822	0.011
Energy (W)	12.318	0.198
IP	**COST 239 Network Topology**	
Metric	**AVG ($\bar{x}$)**	**STD (σ)**
Throughput (Mbps)	8.852	0.099
Delay (ms)	1.646	0.097
Link utilization (%)	6.872	0.003
Energy (W)	18.369	0.152
PSIRP	**Abilene Network Topology**	
Metric	**AVG ($\bar{x}$)**	**STD (σ)**
Throughput (Mbps)	8.830	0.165
Delay (ms)	2.340	0.189
Link utilization (%)	16.601	0.010
Energy (W)	12.267	0.127
PSIRP	**COST 239 Network Topology**	
Metric	**AVG ($\bar{x}$)**	**STD (σ)**
Throughput (Mbps)	9.375	0.075
Delay (ms)	1.407	0.068
Link utilization (%)	5.951	0.002
Energy (W)	18.349	0.130

Finally, Table 18.9 presents the comparative data between disabling and enabling the energy-saving algorithm. The results show that in a low-traffic scenario, energy savings above 25% at the cost of approximately 11% of throughput can be achieved. Also, in the worst cases, the energy-saving algorithm will increase average link utilization by ~9% and the delay by ~30%.

18.3.4.3 Heavy-Traffic Scenario

This section describes the evaluation results of the solution with high traffic demand. In this situation, the solution will be tested in a more demanding scenario, with a low reduction in the energy consumption being expected. Some network links may be turned off, but because of performance constraints, they will eventually be turned on again. Finally, the traffic conditions of this scenario correspond to 1 MB of traffic size and 400 μs of Inter-Packet time interval.

Table 18.10 presents the results obtained by injecting a large amount of traffic in both network topologies and by using IP and PSIRP architectures. In this situation, the energy-saving algorithm is disabled, serving as basis for a comparison between the previously defined metrics in a heavy-traffic scenario. On the other hand, Table 18.11 presents results with the energy-saving algorithm enabled.

TABLE 18.8

The IP/PSIRP Evaluation with the Energy-Saving Algorithm in a Low-Traffic Scenario

IP	Abilene Network Topology	
Metric	AVG ($\bar{x}$)	STD (σ)
Throughput (Mbps)	7.650	0.244
Delay (ms)	3.137	0.326
Link utilization (%)	18.909	0.013
Energy (W)	8.735	0.331

IP	COST 239 Network Topology	
Metric	AVG ($\bar{x}$)	STD (σ)
Throughput (Mbps)	7.793	0.223
Delay (ms)	2.132	0.237
Link utilization (%)	7.302	0.006
Energy (W)	8.493	0.430

PSIRP	Abilene Network Topology	
Metric	AVG ($\bar{x}$)	STD (σ)
Throughput (Mbps)	8.009	0.275
Delay (ms)	2.972	0.271
Link utilization (%)	18.028	0.010
Energy (W)	8.614	0.341

PSIRP	COST 239 Network Topology	
Metric	AVG ($\bar{x}$)	STD (σ)
Throughput (Mbps)	8.594	0.162
Delay (ms)	1.745	0.109
Link utilization (%)	6.280	0.003
Energy (W)	8.034	0.280

TABLE 18.9

Impact of the Energy-Saving Algorithm in a Low-Traffic Scenario

Abilene Network Topology		
Metric	IP	PSIRP
$\Delta_{\text{Throughput}}$ (%)	−10.386	−9.294
Δ_{Delay} (%)	24.013	27.013
$\Delta_{\text{LinkUtilization}}$ (%)	6.103	8.597
Δ_{Energy} (%)	−29.089	−29.783

COST 239 Network Topology		
Metric	IP	PSIRP
$\Delta_{\text{Throughput}}$ (%)	−11.97	−8.33
Δ_{Delay} (%)	29.53	24.06
$\Delta_{\text{LinkUtilization}}$ (%)	6.26	5.54
Δ_{Energy} (%)	−53.77	−56.22

TABLE 18.10
IP/PSIRP Evaluation without the Energy-Saving Algorithm in a High-Traffic Scenario

IP	Abilene Network Topology	
Metric	**AVG ($\bar{x}$)**	**STD (σ)**
Throughput (Mbps)	6.184	0.327
Delay (ms)	4.992	0.889
Link utilization (%)	39.427	0.006
Energy (W)	56.184	0.475
IP	**COST 239 Network Topology**	
Metric	**AVG ($\bar{x}$)**	**STD (σ)**
Throughput (Mbps)	8.616	0.076
Delay (ms)	1.755	0.056
Link utilization (%)	14.694	0.003
Energy (W)	83.777	0.140
PSIRP	**Abilene Network Topology**	
Metric	**AVG ($\bar{x}$)**	**STD (σ)**
Throughput (Mbps)	6.309	0.415
Delay (ms)	5.219	1.753
Link utilization (%)	37.080	0.007
Energy (W)	56.369	1.316
PSIRP	**COST 239 Network Topology**	
Metric	**AVG ($\bar{x}$)**	**STD (σ)**
Throughput (Mbps)	8.895	0.056
Delay (ms)	1.554	0.032
Link utilization (%)	13.271	0.002
Energy (W)	83.610	0.152

Table 18.12 presents the results of the comparison made between disabling and enabling the energy-saving algorithm. The results show that in a high-traffic scenario, energy savings of no more than 36% at the cost of approximately 11% of throughput cannot be achieved. Also, in the worst cases, the energy-saving algorithm will increase the average link utilization by ~5% and the delay by ~30%.

18.4 FAILURE RECOVERY MECHANISMS FOR CONVERGENT NETWORKS

In order to reduce network downtime and its associated impact and costs, effective reconfiguration mechanisms are also needed for immediate recovery in case of network failures. This section describes comparatively several different alternatives for implementing redundant mechanisms for failure recovery (e.g., due to natural disasters) and their implications on energy consumption by telecommunication networks.

We will employ energy-saving models described previously in order to qualitatively evaluate the failure recovery architectures in terms of energy consumption. We address energy-saving mechanisms that can be employed for reducing energy consumption while maintaining network connectivity. Optical and radio recovery strategies are described, together with their associated energy cost, applicable within the available state-of-the-art technologies. The recovery alternatives are

TABLE 18.11

IP/PSIRP Evaluation with the Energy-Saving Algorithm in a High-Traffic Scenario

Abilene Network Topology

Metric	AVG ($\bar{x}$)	STD (σ)
Throughput (Mbps)	5.647	0.201
Delay (ms)	5.588	0.629
Link utilization (%)	39.250	0.006
Energy (W)	52.379	0.855

COST 239 Network Topology

Metric	AVG ($\bar{x}$)	STD (σ)
Throughput (Mbps)	7.688	0.092
Delay (ms)	2.154	0.076
Link utilization (%)	14.973	0.002
Energy (W)	56.955	0.741

Abilene Network Topology

Metric	AVG ($\bar{x}$)	STD (σ)
Throughput (Mbps)	5.685	0.233
Delay (ms)	5.706	0.753
Link utilization (%)	37.829	0.007
Energy (W)	52.096	0.968

COST 239 Network Topology

Metric	AVG ($\bar{x}$)	STD (σ)
Throughput (Mbps)	7.949	0.097
Delay (ms)	2.008	0.072
Link utilization (%)	13.927	0.002
Energy (W)	54.849	0.698

TABLE 18.12

Impact of the Energy-Saving Algorithm in a High-Traffic Scenario

Abilene Network Topology

Metric	IP	PSIRP
$\Delta_{Throughput}$ (%)	−8.678	−9.892
Δ_{Delay} (%)	11.947	9.336
$\Delta_{LinkUtilization}$ (%)	−0.448	2.021
Δ_{Energy} (%)	−6.773	−7.580

COST 239 Network Topology

Metric	IP	PSIRP
$\Delta_{Throughput}$ (%)	−10.77	−10.63
Δ_{Delay} (%)	22.72	29.22
$\Delta_{LinkUtilization}$ (%)	1.90	4.94
Δ_{Energy} (%)	−32.02	−34.40

comparatively evaluated in terms of energy consumption, followed by the main conclusions from such evaluation.

18.4.1 CONVERGENT NETWORKS ARCHITECTURES

We consider hereafter both static architectures for FUTON (without active elements between RAs and CU) and dynamic architectures, whether the use of active elements allow modification of the network topology. Furthermore, we will consider the existence of redundant links (radio or optical) to achieve the targeted robustness and reliability and a significant increase of network availability [40]. Redundancy ensures that the network is available and running during scenarios when there is a failure at a certain point in the network. Though the fact remains that adding redundancy to the network forces a rise in the capital expenditure (CAPEX), OPEX reduction is achieved through this strategy, wherein network energy costs do not rise significantly. Failures especially on the optical fiber backbone (such as fiber cuts) can cause significant impairments all over a convergent heterogeneous network, affecting eventually large groups of antennas if redundant failure recovery mechanisms are not in place.

Figure 18.2 illustrates a network failure affecting users within the cell area under a remote antenna (RA_1) due to a failure in the fiber reaching out to it. But RAs are also prone to failures, ranging from equipment failure to storms, thunderstorms, or natural disasters. Hence, another scenario is depicted as well in Figure 18.2 where a failure occurs owing to RA_4 failure (or a fiber link and RA simultaneous failure, both probable in natural disasters).

These scenarios (and similar) affect the OPEX to a large extent. It is one of the most important aspects from the operator's point of view that recovery scenarios are present in such situations of link failures (e.g., owing to fiber cuts) or remote units malfunctioning/down (e.g., owing to RA failure). Focus can be placed for failures occurring on the optical part of the network between the (de) multiplexer (or splitter) element and the RAU, or failures occurring at the antenna sites.

18.4.2 OPTICAL RECOVERY MODELS

In order to identify faults, the RAUs contain software agents that continuously monitor key performance indicators and generate alarms when measuring abnormal values for such variables. The management module at the CU gathers the data from all RAUs on a periodic or event-driven basis and sends configuration messages to the appropriate network elements in order to compensate for such faults and to keep the network operational. Hereafter, we describe several failure recovery

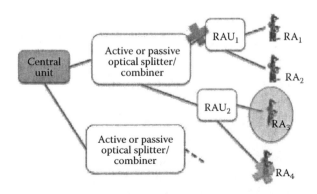

FIGURE 18.2 Architecture for radio network with an optical backhaul. An example of possible failures on the optical backhaul (darker cross) and on the radio part of the network (lighter cross) is shown.

mechanisms for failures on the optical backhaul and evaluate the impact on energy consumption on the adoption of such strategies.

18.4.2.1 Redundant Link, Common Wavelengths

Consider the scenario in which the CU has one wavelength dedicated to one particular RAU. Recovery for such scenario [4] was proposed by using a redundant optical link along with active elements (for merging at subcarrier level). One wavelength caters to a pair of RAUs downstream while another wavelength caters to a pair of RAUs upstream. The active elements (needed for subcarrier merging) are stationed in the active node (AN) and RAUs. Hence, a constant increase on energy consumption is introduced by the usage of an AN, independently of the occurrence of network failures.

Referring to Figure 18.3a, under normal conditions, RAU_1 receives downstream traffic via wavelength λ_0. Among λ_0 subcarriers, only half of these carry information intended for RAU_1, with these being filtered out using an electrical filter after an optical–electrical (OE) conversion. The procedure for RAU_2 is similar, but for the other subcarriers. In the upstream traffic, wavelengths λ_1' and λ_2' subcarriers from RAUs 1 and 2 are merged into a single wavelength λ_0', transporting all the subcarriers to the CU. This merging requires an active element, and so this is done at the AN through optical–electrical–optical (OEO) conversion.

In case of a failure in the link between the AN and a RAU (say RAU_1), the event is detected by the CU network manager because of an alarm generated by the RAU or because of a "RAU unreachable condition." Upon fault detection, the manager instructs the AN to stop acting as an active element. In addition, RAU_2, besides filtering half of the subcarriers (the ones of λ_2) for

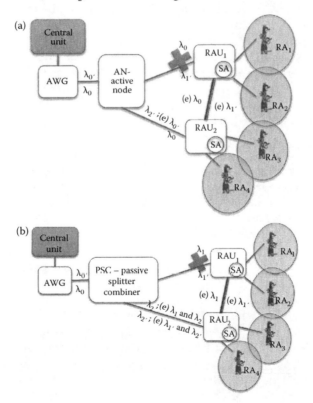

FIGURE 18.3 (a) Failure on an optical link, and usage of a redundant optical link. In case of failure (dark cross), the additional transmitted wavelengths up to the AN are marked with (e). (b) Failure on an optical link, and usage of a redundant optical link with redundant wavelengths.

itself, it is also configured to forward all the downstream wavelength λ_0 subcarriers to RAU_1 via a redundant link. Hence, the lasers on the optical link between RAU_1 and RAU_2 are activated, which will further consume additional energy. Furthermore, the lasers on the optical link between RAU_1 and AN can be deactivated, since the link is useless (e.g., owing to broken cable or laser damage).

For the upstream traffic, RAU_1 will use the redundant link to forward wavelength λ_1' to RAU_2. The active element of RAU_2 will merge these subcarriers along with its λ_2' to transmit wavelength λ_0', combining all the subcarriers. Hence, the increase on energy consumption by such recovery mechanism is mainly due to the usage of an active node.

Under heavy load, all links might be active, and the data are distributed to RAU_1 or to RAU_2 through more than one optical link so that the load is balanced. But usually the optical link is not the bottleneck, and such configuration would result in further energy consumption. Hence, according to Figure 18.2, usually only two out of the three links between AN, RAU_1, and RAU_2 will be active, and those two can be selected according to the approaches described in the previous sections. The same applies to the following recovery scenario.

18.4.2.2 Redundant Link through Additional Wavelength

In this scenario, recovery is achieved according to Ref. [4] by using a redundant optical link per RAU and transmission direction, along with passive elements (for combining different wavelengths). Each RAU is connected with the CU using two wavelengths, one per direction of downstream and upstream. The passive elements required for the wavelength combination are stationed in a passive splitter combiner (PSC) and the RAUs.

Under normal conditions, as shown in Figure 18.3b, RAUs 1 and 2 receive downstream traffic via wavelengths λ_1 and λ_2, respectively. The PSC splits λ_1 and λ_2 for forwarding to RAUs 1 and 2, respectively. In the upstream traffic, wavelengths λ_1' and λ_2' from RAU 1 and 2 are combined at the PSC, resulting to λ_0', which is forwarded to the CU. In case of a link failure between PSC and a RAU (say RAU_1), the PSC forwards the entire downstream wavelengths λ_1 and λ_2 to RAU_2. RAU_2 filters out λ_2, while forwarding both λ_1 and λ_2 to RAU_1 via the redundant link. Then RAU_1 will filter out λ_1. For the upstream traffic, RAU_1 will use the redundant link to forward wavelength λ_1' to RAU_2. The passive element at RAU_2 will merge this wavelength along with its upstream wavelength λ_2', thus transmitting all the upstream traffic to the CU.

For this solution, the active node AN is not used, being replaced by a passive PSC, which does not require energy supply. The increase on energy consumption, from the activation of the extra link, is compensated in turn by turning off the broken link. But the transmission of an additional wavelength in both directions between PSC and RAU_2 will result in an energy increase.

18.4.2.3 Asymmetric Redundant Link

This scenario corresponds to the usage of only one wavelength downstream and a wavelength per RAU upstream, which therefore is a combination of the previous two scenarios. Recovery is achieved by using a redundant optical link along with passive elements (for combining different wavelengths).

In case of a link failure between the PSC and a RAU (say RAU_1), RAU_2 besides filtering half of the subcarriers for itself will also be forwarding all the downstream wavelength λ_1 to RAU_1 via the redundant link, as for the downstream scenario in A, but without the need to employ an active node AN. For the upstream traffic, RAU_1 will use the redundant link to forward wavelength λ_1, to RAU_2. The passive element at RAU_2 will merge this wavelength along with its upstream wavelength λ_2', thus transmitting all the upstream traffic to the CU. This is the same upstream scenario as in B.

Overall, energy consumption will be lower than for scenario B, since in case of failure, the transmission of an additional wavelength between PSC and RAU_2 is only made in the upstream direction, so that transmission on the downstream direction will be more efficient.

18.4.3   RADIO RECOVERY MODELS

This section considers energy consumption variation from employing recovery models based on providing redundancy employing wireless links. First, we will look into a centralized version of radio recovery based on an adaptive power control strategy, as analyzed in Ref. [41] for compensating for variations in neighboring cells' activity. This scenario is especially useful for antenna equipment failures. A strategy is also presented based on cooperative antennas to compensate for link failures between a neighboring antenna and the associated RAU.

18.4.3.1   Transmission Power Adaptation

This recovery scenario covers cases such as performance degradation on site cells and antenna equipment failure [41]. The model, as described in Ref. [4], consists as well of having a CU network manager to receive alarms from RAUs, related to performance indicators or failures. Upon reception of such alarms, a collection of RAUs is selected and its transmission power is adjusted in order to compensate for the occurred changes. The transmission power may be lowered if the overall network does not require such coverage or raised if necessary to compensate for other antenna problems.

This scenario is illustrated in Figure 18.4a. A software agent (SA) at RAU_2 communicates through the control channel a cell performance variation to the CU manager. At the same time, the network manager, as a result of a failure on that antenna, does not receive RAU_1 control data. The network manager then sends configuration messages (using, for instance, a protocol such as SNMP) to the SAs in RAUs. These messages are used to adjust the transmission power of the neighboring cells' antennas in order to compensate for the neighboring cells' variations. Although such measure does not affect energy consumption on the optical part, the power required to further extend an antenna reach to cover the area spanned by an antenna on failure is very significant. On the one hand, if the

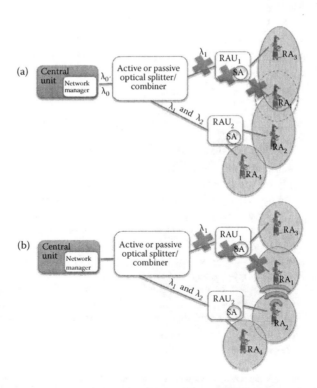

FIGURE 18.4 (a) Transmission power adaptation for compensating failures. (b) Cooperative antennas to compensate for failures.

radio equipment and optical network path to the failed antenna are turned off, energy consumption is reduced; on the other hand, the overall energy consumption significantly increases because of the need to cover the same region with a smaller number of antennas. Hence, energy consumption at both mobile terminal and the remote antenna providing protection increases. Furthermore, power level imposes limits on the RA coverage protection area.

18.4.3.2 Cooperative Antennas

Whenever there is a failure in the link between the CU to the RA, the concept of cooperative multiple-input and multiple-output antennas can be applied as a recovery mechanism for failures either at the RAU or on the optical link. Hence, the RA is assumed to be in a functioning state. Figure 18.4b illustrates the recovery scenario on the basis of cooperative antenna concepts. Following a failure in the link to the RA_1 antenna, the network manager configures the RAU of the neighboring antenna (RA_2) to activate its redundant equipment to communicate with the RA_1, which has a failed link.

Compared to the previous scenario, the neighboring RA communicates only with the RA on the path of a failed link. The RA with the failed link communicates with the CU and vice versa via the neighboring RA. Energy consumption on the optical part is reduced, since the broken path up to the RA can be turned off. However, such energy saving is achieved at the cost of the overall network performance, since one RA (RA_2) needs to transmit all users' data (for his coverage area as well as for the other RA_1). Additionally, energy consumption on the radio network increases, since more power is needed at RA_2 to reach RA_1, with a negligible effect on end users' devices.

18.4.4 COMPARATIVE RECOVERY EVALUATION

Let us now perform a comparative evaluation in terms of energy for the five recovery scenarios previously described (Table 18.13).

Radio recovery may be attractive in terms of low CAPEX owing to the fact that redundancy is guaranteed by extra equipment and not by fiber deployment. But though radio recovery is a possibility, it still depends on the distance between the RAs and the cost of extra radio equipment. The antenna transmitting at a higher output power will require higher energy consumption, which might not be a good choice. There are also legal limits of radiation related to health issues of the

TABLE 18.13

Comparative Energy Consumption for Failure Recovery Mechanisms

Redundant link, common wavelengths	Constant increase on energy consumption is introduced mainly because of the usage of an active node, independently of the occurrence of network failures.
Redundant link through additional wavelength	Increase on energy consumption from the activation of the extra link is compensated by turning off the broken link. But the transmission of an additional wavelength in both directions between PSC and RAU_2 will require an increase on energy.
Asymmetric redundant link	Energy consumption will be lower than for scenario B, because transmission on the downstream direction is more efficient.
Transmission power adaptation	Reducing energy consumption on the optical network. However, energy consumption significantly increases because of the need to cover the same geographic region with a smaller number of antennas. Energy consumption at users' mobile terminal also increases, representing a further cost burden to the end user (and lower QoS, since device battery will end faster). Power level limits the RA coverage protection area.
Cooperative antennas	Energy consumption on the optical part is reduced. However, such energy saving is achieved at the cost of the overall network performance. Additionally, energy consumption on the radio network increases, although having a smaller increase than for scenario IV.A. The effect on end users' mobile devices energy consumption is negligible.

human population closer to the antennas. Specific to the *transmission power adaptation* scenario, the mobile terminals of the end user would also need to transmit at a higher power to reach the neighboring RA, and this is achieved through faster battery consumption. The existing load on the neighboring RA has also to be considered. Compared to radio recovery, optical recovery is a choice with fewer hazards and energy consumption but, on the other hand, implies more CAPEX.

18.5 CONCLUSIONS

This chapter addressed the most significant problems of the current Internet architecture, giving special attention to the energy consumption issue. Throughout the chapter, the efforts that are being made by the research community to reduce the energy consumption of the Internet infrastructure are reviewed, as well as for one future Internet architecture model following a publish–subscribe paradigm. Also, a solution that enables energy awareness in both architectures by turning off unused network elements is proposed.

The implemented energy-saving solution makes traffic engineering decisions to aggregate traffic to most used links, which will allow the possibility of inducing underused links to an idle mode. The unused network elements (nodes or links) can be turned off if the network remains connected. Also, a ranking mechanism that classifies the importance of a network element is proposed. This mechanism is of extreme importance to achieve a good trade-off between energy savings and network performance, mainly because it will turn off in the first place the network elements that are less important to the packet delivery process.

The evaluation of the solution shows that significant energy savings can be achieved in a low-traffic scenario without too much impact on the network performance. In a heavy-traffic scenario, the proposed solution also manages to reduce the energy consumption but by a smaller percentage.

Indeed, energy efficiency is currently a very important issue, and it is expected to be of extreme importance for future Internet architectures as well. By integrating several strategies of the scientific state of the art, we show that it is possible to combine them in order to improve throughput in both current and future Internet architectures.

Future work will address the use of alternative optimization strategies in order to further improve the trade-off between energy efficiency and network throughput.

18.5.1 FUTURE RESEARCH AND POSSIBILITIES

Several limitations that were not yet addressed were found during the implementation and testing of the proposed solution:

- Energy consumption when waking up: It is assumed that the transition between the sleeping and the working states does not consume any energy, which is not true in a real-life situation. With this assumption, one cannot possibly know the impact on energy savings when waking up a node. Ideally, the energy-saving algorithm should aim to reduce at maximum the number of wakeups, in order to achieve good energy savings.
- Fixed sleeping period: The energy-saving algorithm assumes a fixed sleeping period for the network nodes, which may generate unnecessary wakeups on some of them. The sleeping period should take into consideration the number of transitions between the pre-wakeup and working states. Hence, the nodes that do not pass consecutively from the sleep state to the working state after the expiration of the sleeping period should see their sleeping expiration period increased; otherwise, it should decrease. Ideally, the sleeping expiration period should be dynamic, allowing further energy savings.
- Publish/subscribe data: The mechanism for publishing and subscribing data in the PSIRP architecture was not implemented. Only the forwarding procedure was implemented, which means that the publisher and the subscriber are well known. Because of this, it is not

possible to evaluate the overhead in the network that is caused by the publish and subscribe messages.

- LIT implementation: The utilization of bloom filters in the PSIRP architecture can cause numerous false positives when matching the LITs against the *zFilter*. This is very problematic in very dense networks where the probability in generating a false positive is very high. Therefore, one can investigate the usage of LIT to reduce the probability of false positives, which was not implemented in the solution.

Other future research possibilities include studying the applicability of the energy-saving algorithm in conjunction with a distance-vector routing protocol. As opposed to link-state routing protocols, every network node only has partial knowledge of the network topology.

Finally, we would like to implement and evaluate the energy-saving algorithm in real networks. In this work, the energy-saving algorithm was only evaluated by making use of network simulation, even though the energy-saving algorithm was designed taking into consideration a future implementation in real networks.

AUTHOR BIOGRAPHIES

Artur Miguel Arsénio is YDreamsRobotics CEO, an innovative company aimed at turning things into robots. He is also assistant professor in computer science at Instituto Superior Técnico (IST). Before, he headed Innovation at Nokia Siemens Networks Portugal. He also worked as senior solution architect, leading several international R&D teams for Siemens, and as chief engineer, he represented Siemens Networks on IPTV Standardization forums. Artur Arsénio received his doctoral degree in computer science from the Massachusetts Institute of Technology (MIT) in 2004. He received both his MSc and Engineering degrees in electrical and computer engineering from IST. He is the inventor of six international patent families and (co)authored more than 80 scientific publications. He is the recipient of several scientific and innovation awards. He is a co-founder and associate member at Beta-i Entrepreneurship Association, and he was a co-organizer of TEDxEdges, the first TEDx event ever in Portugal. He is co-founder and vice-chair of ACM SIGCOMM Portugal chapter, a Fulbright, and president of the MIT Alumni Association in Portugal.

Tiago Silva obtained a BS degree at the Computer Science and Engineering Department of Instituto Superior Tecnico/Technical University of Lisbon. He received his MS degree in computer networks by the same institution. His master thesis addressed research topics in the fields of energy efficient networks and future network architectures.

REFERENCES

1. Global e-Sustainability Initiative (GeSI). Smart 2020 report: Global ICT solution case studies. Technical report, http://www.smart2020.org/publications/, 2008. Last accessed February 2013.
2. D. Meyer, L. Zhang, and K. Fall. Report from the IAB workshop on routing and addressing. RFC 4984, September 2007.
3. A. Bianzino, C. Chaudet, D. Rossi, and J.-L. Rougier. A survey of green networking research. *IEEE Communications Surveys Tutorials*, 99:1–18, 2010.
4. B. Gangopadhyay, C. Santiago, and A. Arsenio. Comparative evaluation of failure recovery mechanisms for convergent networks. In *The 12th International Symposium on Wireless Personal Multimedia Communications*, Japan, 2009.
5. M. Okada, J. Kani, T. Watanabe, and N. Yoshimoto. An optimal investment strategy for tunable, pluggable, and tunable pluggable optical transceivers in static DWDM networks. *Optical Communication Networks*, 2009.
6. S. Pato, J. Pedro, J. Santos, A. Arsénio, P. Inácio, and P. Monteiro. On building a distributed antenna system with joint signal processing for next generation wireless access networks: The FUTON approach. In *The 7th Conference on Telecommunications*, Portugal, 2009.

7. G. Tselentis. *Towards the Future Internet*, pages 75–84. IOS Press, 2010.

8. A. Feldmann. Internet clean-slate design: what and why? *ACM Sigcomm Computer Communication Review*, 37(3), July 2007.

9. C. Jinzhou, W. Chunming, J. Ming, and Z. Dong. A review of future internet research programs and possible trends. *Institute of Electrical and Electronics Engineers*, 2010.

10. G. Tselentis. *Towards the Future Internet*, pages 91–101. IOS Press, 2009.

11. N. Niebert, S. Baucke, I. EI-Khayat, M. Johnsson, B. Ohlman, H. Abramowicz, K. Wuenstel, H. Woesner, J. Quittek, and L. Correia. The way 4ward to the creation of a future internet. *Institute of Electrical and Electronics Engineers*, 2008.

12. G. Bouabene, C. Jelger, C. Tschudin, S. Schmid, A. Keller, and M. May. The autonomic network architecture. *IEEE Journal on Selected Areas in Communications*, 28(1), January 2010.

13. D. Clark, R. Braden, A. Falken, and V. Pingali. Fara: Reorganizing the addressing architecture. In *ACM Sigcomm Workshops*, August 2003.

14. X. Yang, D. Clark, and A. Berger. Nira: A new inter-domain routing architecture. *IEEE/ACM Transactions on Networking*, 15(4), August 2007.

15. Y. Zhang and T. Henderson. An implementation and experimental study of the explicit control protocol (XCP). *Institute of Electrical and Electronics Engineers*, 2:1037–1048, 2005.

16. M. Baldi and Y. Ofek. Time for a "greener" internet. *Institute of Electrical and Electronics Engineers*, 2009.

17. H. Mellah and B. Sansò. Review of facts, data and proposals for a greener internet. In *ICST Broadnets*, 2009.

18. R. Bolla, R. Bruschi, F. Davoli et al. Energy efficiency in the future internet: A survey of existing approaches and trends in energy-aware fixed network infrastructures. *IEEE Communications Surveys*, 2010.

19. C. Gunaratne, K. Christensen, B. Nordman, and S. Suen. Reducing the energy consumption of ethernet with adaptive link rate (ALR). *IEEE Transactions on Computers*, 57(4), April 2008.

20. B. Sansò and H. Mellah. On reliability, performance and internet power consumption. *Institute of Electrical and Electronics Engineers*, 2009.

21. A. Gladisch, C. Lange, and R. Leppla. Power efficiency of optical versus electronic access networks. *Institute of Electrical and Electronics Engineers*, 2(143), September 2008.

22. G. Papadimitriou, C. Papazoglou, and A. Pomportsis. Optical switching: switch fabrics, techniques, and architectures. *IEEE Journal of Ligthwave Technology*, 21(2), February 2003.

23. Y. Li, W. Li, and C. Jiang. A survey of virtual machine system: Current technology and future trends. *Institute of Electrical and Electronics Engineers*, 2010.

24. M. Anisetti, V. Bellandi, A. Colombo, M. Cremonini, E. Damiani, F. Frati, J. Hounsou, and D. Rebeccani. Learning computer networking on open paravirtual laboratories. *IEEE Transactions on Education*, 50(4), November 2007.

25. B. Zhang, X. Wang, R. Lai, L. Yang, Y. Luo, X. Li, and Z. Wang. A survey on I/O virtualization and optimization. *Institute of Electrical and Electronics Engineers*, 2010.

26. M. Baldi and G. Marchetto. Pipeline forwarding of packets based on a low-accuracy network-distributed common time reference. *IEEE/ACM Transactions on Networking*, 17(9), December 2009.

27. M. Baldi, J. Martin, E. Masala, and A. Vesco. Quality-oriented video transmission with pipeline forwarding. *IEEE Transactions on Broadcasting*, 54(3), September 2008.

28. M. Allmany, K. Christensenz, B. Nordman, and V. Paxsony. Enabling an energy-efficient future internet through selectively connected end systems. In *ACM Sigcomm Hotnets*, November 2007.

29. A.P. Bianzino, C. Chaudet, D. Rossi, J. Rougier, and S. Moretti. The green-game: Striking a balance between QoS and energy saving. In *The 2011 23rd International Teletraffic Congress (ITC)*, pages 262–269, September 2011.

30. L. Chiaraviglio, M. Mellia, and F. Neri. Reducing power consumption in backbone networks. In *IEEE International Conference on Communications, 2009*. ICC '09. pages 1–6, June 2009.

31. S. Gao, J. Zhou, T. Aya, and N. Yamanaka. Reducing network power consumption using dynamic link metric method and power off links. June 2009.

32. A. Cianfrani, V. Eramo, M. Listanti, M. Marazza, and E. Vittorini. An energy saving routing algorithm for a green OSPF protocol. *Institute of Electrical and Electronics Engineers*, 2010.

33. K. Ho and C. Cheung. Green distributed routing protocol for sleep coordination in wired core networks. In *2010 6th International Conference on Networked Computing (INC)*, pages 1–6, May 2010.

34. R. Garroppo, S. Giordano, G. Nencioni, and M. Pagano. Energy aware routing based on energy characterization of devices: Solutions and analysis. *Institute of Electrical and Electronics Engineers*, June 2011.

35. S. Moretti and F. Patrone. Transversality of the Shapley value. *TOP*, 16:1–41, 2008.
36. A. Bianzino, C. Chaudet, F. Larroca, D. Rossi, and J. Rougier. Energy-aware routing: A reality check. In *The 2010 IEEE GLOBECOM Workshops*, pages 1422–1427, December 2010.
37. D. Chen, L. Lü, M. Shang, Y. Zhang, and T. Zhou. Identifying influential nodes in complex networks. *Physica A: Statistical Mechanics and Its Applications*, 391(4):1777–1787, February 2012.
38. M. Yu, M. Thottan, and L. Li. Latency equalization as a new network service primitive. *IEEE/ACM Transactions on Networking*, 20(1):125–138, February 2012.
39. P. Batchelor, B. Daino, P. Heinzmann, D.R. Hjelme, R. Inkret, H.A. Jager et al. Study on the implementation of optical transparent transport networks in the european environment—Results of the research project cost 239. *Photonic Network Communications*, 2:15–32, 2000.
40. S. Ballew. *Managing IP Networks with Cisco Routers*. O'Reilly, Cambridge, MA, 1997.
41. J. Tsai and H. Hsieh. Adaptive antenna power level control for wireless forward link data services, In *Wireless Communications and Networking Conference*, 2006.

Index

Page numbers followed by f and t indicate figures and tables, respectively.